Nanoparticles in Life Sciences and Biomedicine

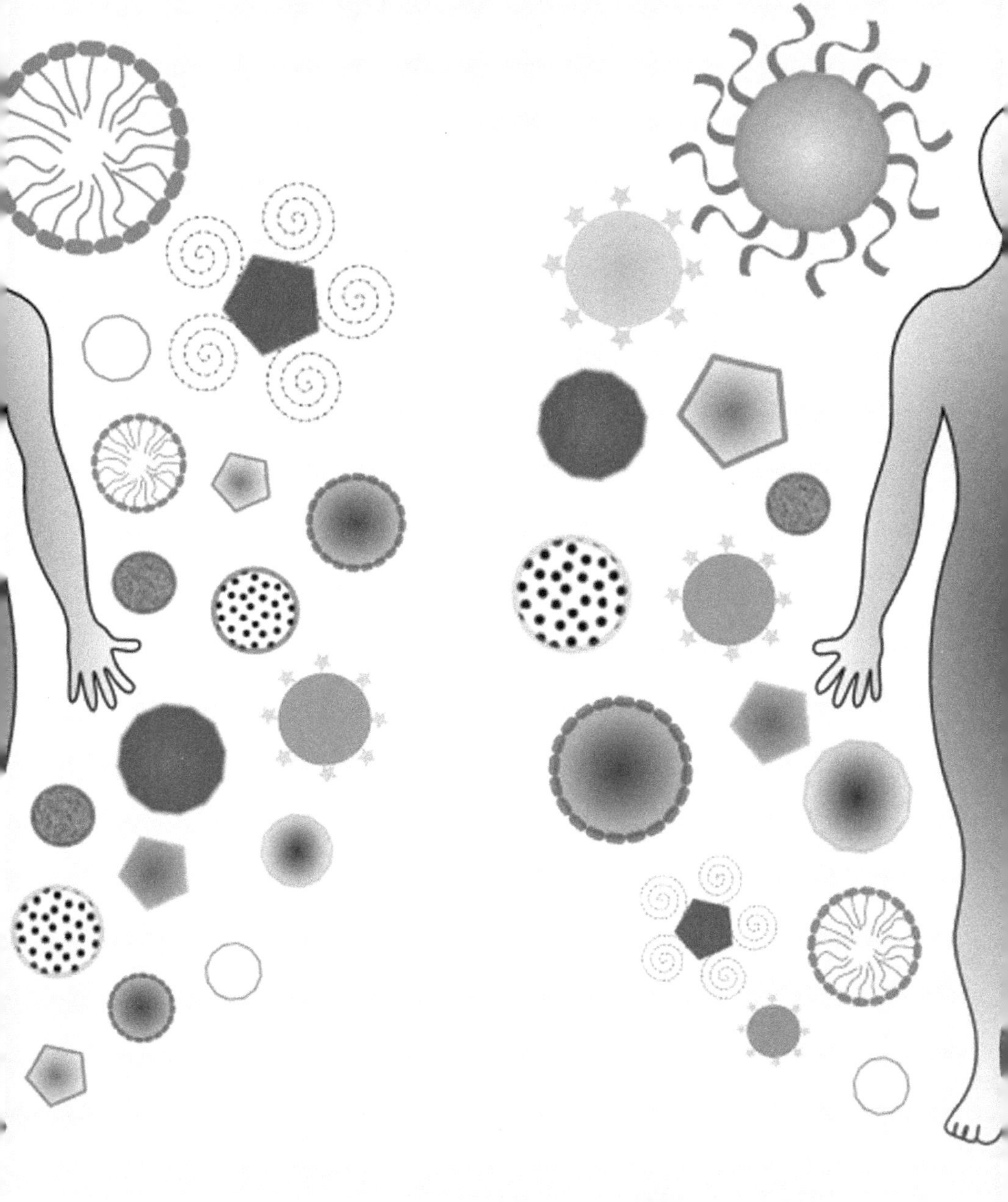

Nanoparticles in Life Sciences and Biomedicine

edited by

Ana Rute Neves
Salette Reis

Pan Stanford Publishing

Published by

Pan Stanford Publishing Pte. Ltd.
Penthouse Level, Suntec Tower 3
8 Temasek Boulevard
Singapore 038988

Email: editorial@panstanford.com
Web: www.panstanford.com

British Library Cataloguing-in-Publication Data
A catalogue record for this book is available from the British Library.

Nanoparticles in Life Sciences and Biomedicine

Cover image by José Plácido Lopes de Araújo

ISBN 978-981-4745-98-7 (Hardcover)
ISBN 978-1-351-20735-5 (eBook)

Ana Rute Neves

dedicates
this book to her
beloved sister,

Sílvia Pina Neves.

Contents

Preface

These days, the creation of new and more efficient therapies for improving human health greatly depends on drug delivery systems. Nanotechnology has emerged as a powerful strategy for the development of nanoparticles, such as nanoemulsions, liposomes, nanocrystals, and nanocomplexes, applied in the diagnosis, treatment, or theranostics of several diseases. Recent research and development in the nanotechnology field exploits several administration routes, like oral, colorectal, topical, pulmonary, and brain delivery, and deals with site-specific delivery strategies, permeability through biological barriers, internalization pathways, and potential nanocytotoxicity.

This book introduces several principles and knowledge in this field needed for the audience to understand science at the nanoscale. The book compiles and details in great depth current research and recent advances in drug delivery systems with several biomedical applications for a range of diseases and pathologies, from cancer to infectious diseases, passing through tuberculosis, rheumatoid arthritis, neurodegeneration, genetic conditions, and other important disorders, contributing toward improving the knowledge of researchers and possible future applications.

Ana Rute Neves and Salette Reis

University of Porto, Portugal

October 2017

Part I
Introduction

Chapter 1

Importance and Application of Nanotechnology for Improving Existing Therapy

Ana Rute Neves and Salette Reis
UCIBIO, REQUIMTE, Department of Chemical Sciences, Faculty of Pharmacy, University of Porto, Portugal
ananeves@ff.up.pt

Nowadays, the successful treatment of various illnesses and disorders greatly depends on drug delivery systems to create new and more efficient therapies. In this context, nanotechnology has emerged as a powerful strategy for the development of nanoparticles, with several biomedical applications for a range of diseases and infections, from diagnosis to therapy. Nanotechnology is defined as the engineering and manufacturing of materials, devices, and systems at an incredibly small scale from a few nanometers to a few hundred nanometers, without changing unique properties [1]. The prefix of nanotechnology derives from "nanos," the Greek word for "dwarf." The field of nanotechnology originated in 1959, when Richard P. Feynman (Nobel Prize in Physics, 1965) introduced the notion of nanoscale, with its famous expression "There's plenty of

Nanoparticles in Life Sciences and Biomedicine
Edited by Ana Rute Neves and Salette Reis

ISBN 978-981-4745-98-7 (Hardcover), 978-1-351-20735-5 (eBook)
www.panstanford.com

room at the bottom." In 1904, Paul Ehrlich (Nobel Prize in Physiology or Medicine, 1908) had already introduced the idea of the "magic bullet." According to what he envisaged, just like a bullet fired from a gun hits a specific target, there could be a way to specifically target body cells to treat diseases [2]. This concept of a magic bullet represents the beginning of a new era that still concerns the development and approval of several forms of drug-targeting delivery systems for the treatment of several diseases. In fact, the development of an effective nanodelivery system capable of protecting and transporting a specific bioactive compound to the desired site of action is one of the most challenging tasks of pharmaceutical research. This strategy would allow improving efficacy and reducing potential intrinsic side effects related to nonspecific action.

In this context, nanotechnology has gained the status of a "hot topic" and is considered nowadays a multidisciplinary field covering a host of knowledge areas, such as engineering, physics, chemistry, and biology, and presenting extensive applications in a range of fields, like microelectronics, construction, environment, sustainability, agriculture, and health care [3]. In particular, the great impact of nanotechnology on human health is receiving considerable acknowledgment due to its rapid growth and advances in drug delivery systems, biomaterials, biomedical devices, intelligent processes, and many other areas of biomedicine and applied life sciences. This phenomenon is revolutionizing pharmaceutical sciences, and many drugs are being reconsidered for the treatment of many diseases by loading them in nanodelivery systems.

Several considerations must be taken into account before developing nanosystems for biomedical applications. First of all, they must be made of biocompatible and biodegradable materials and provide sustained and controlled release of the bioactive agents as drug carriers [4]. Moreover, the ability to manipulate physical, chemical, and biological properties of these nanosystems is also an important feature to provide a rational design of nanoparticles for drug delivery, for diagnostic purposes with image contrast agents, and for enhancing the current therapy [5]. To achieve this, the active agent must be dissolved, entrapped, or encapsulated within the nanoparticles or adsorbed on their surfaces. Codelivery of two or more drugs can also be achieved for combination therapy when synergistic effects are well established. As a consequence, a nanodelivery system must improve a drug's bioavailability by enhancing its aqueous solubility, increase the residence time in circulation, and potentiate drug delivery targeted at a specific site of action. Hence, the nanosystem must improve the permeability of the drug through biological membranes, especially the intestinal, dermal, or blood–brain barriers, and provide controlled distribution of the drug, targeting only those cells that need to be treated and minimizing undesirable side effects on vital tissues [6, 7].

In the last decades, several nanobased drug delivery systems have been approved by the Food and Drug Administration (FDA) for the treatment of a wide spectrum of illnesses, like cancer, hypercholesterolemia, and infectious diseases. In fact, the number of patents and products in this field is increasing significantly and a large number of clinical trials are currently taking place, raising the hopes and interest in nanodelivery systems. In this respect, the market of nanotechnology and nanodelivery applications has been widely spread by the pharmaceutical industry. The majority of nanobased formulations already commercially available in the market include nanocrystals, liposomes, nanoemulsions, and nanocomplexes, being intended mainly for parenteral administration [8, 9].

This book aims to provide an overview of recent advances and achievements in nanobased drug delivery systems, providing a wide coverage of applications in the nanomedicine field and contributing toward improving the knowledge of researchers on designing and developing new generations of nanodelivery systems for possible future applications. For this purpose, this book will cover a wide range of topics on nanotechnology applied for the diagnosis, treatment, or theranostics of several pathologies and diseases, from cancer to infectious diseases, passing through tuberculosis, rheumatoid arthritis, and other inflammatory conditions. Gene therapy will also be considered as a promising therapeutic strategy to combat genetic or acquired diseases rather than just treating the symptoms. The ultimate goal will be the review of the most recent research and developments in the nanotechnology field, exploiting several administration routes, like oral, colorectal, topical, pulmonary, and brain delivery, and highlighting site-specific delivery strategies, permeability through biological barriers, internalization pathways, and potential cytotoxicity.

The novelty of this book will be the interdisciplinary way of introducing the topics, with the main focus on the therapy and diagnosis of different illnesses using nanotechnology. The book will cover an expanded scope in order to reach a wider audience interested in drug nanodelivery topics.

Acknowledgments

This work received financial support from the European Union (FEDER funds) and National Funds (*Fundação para a Ciência e a Tecnologia* and *Ministério da Educação e Ciência* [FCT/MEC]) under the partnership agreement PT2020 UID/MULTI/04378/2013 - POCI/01/0145/FEDER/007728. AR also thanks ICETA for her postdoctoral grant (FOOD_RL3_PHD_GABAI_02) under the project NORTE-01-0145-FEDER-000011.

References

1. Poole, C.P., et al. *Introduction to Nanotechnology*. John Wiley & Sons, Hoboken, New Jersey, 2003.
2. Silverstein, A. *Paul Ehrlich's Receptor Immunology*. Academic Press, New York, 2001.
3. Kelsall, R., et al. *Nanoscale Science and Technology*. Wiley, New York, 2005.
4. Wang, G., et al. Recent developments in nanoparticle based drug delivery and targeting systems with emphasis on protein based nanoparticles. *Expert Opin Drug Deliv*, 2008, **5**:499–515.
5. Rathod, K.B., et al. Glimpses of current advances of nanotechnology in therapeutics. *Int J Pharm Pharm Sci*, 2011, **3**:8–12.
6. Moghimi, S.M., et al. Nanomedicine: current status and future prospects. *FASEB J*, 2005, **19**:11–30.
7. Farokhzad, O.C., et al. Impact of nanotechnology on drug delivery. *ACS Nano*, 2009, **3**:16–20.
8. Hafner, A., et al. Nanotherapeutics in the EU: an overview on current state and future directions. *Int J Nanomed*, 2014, **9**:1005–1023.
9. Couvreur, P., et al. Nanotechnology: intelligent design to treat complex disease. *Pharm Res*, 2006, **23**:1417–1450.

Part II
Oral Drug Delivery Approaches

Chapter 2

Nanocarriers as a Strategy for Oral Bioavailability Improvement of Poorly Water-Soluble Drugs

Luíse L. Chaves,[a] Alexandre C. Vieira,[a] Domingos Ferreira,[b] Bruno Sarmento,[c,d,e] Salette Reis,[a] and Sofia A. Costa Lima[a]

[a]*UCIBIO, REQUIMTE, Department of Chemical Sciences, Faculty of Pharmacy, University of Porto, Portugal*
[b]*Laboratory of Pharmaceutical Technology, Department of Pharmaceutical Sciences, Faculty of Pharmacy, University of Porto, Portugal*
[c]*I3S - Institute of Research and Innovation in Health, University of Porto, Portugal*
[d]*INEB - Institute of Biomedical Engineering, University of Porto, Portugal*
[e]*CESPU - Institute for Advanced Research and Training in Health Sciences and Technologies and University Institute of Health Sciences, Gandra, Portugal*
luiselopes@gmail.com

The oral route is still considered the most-used way of administration due to its noninvasive nature and high patient compliance. However, the low solubility of drugs often results in poor bioavailability (BA) and, consequently, inconstant plasma concentrations. The use of nanoformulations is one of the most promising strategies to overcome this issue due to the significant enhancement in their solubility and intestinal permeation. The major purpose of this

Nanoparticles in Life Sciences and Biomedicine
Edited by Ana Rute Neves and Salette Reis

ISBN 978-981-4745-98-7 (Hardcover), 978-1-351-20735-5 (eBook)
www.panstanford.com

chapter is to outline the main nanosystems commonly used to improve solubility of poorly soluble drugs.

2.1 Introduction

The major portion of the drug delivery market is occupied by oral drug delivery systems [1]. The oral route is still considered the most convenient route for drug administration as it offers high patient compliance, allowing a flexible and controlled dosing schedule, especially important for chronic therapy [2]. It is associated with low costs for both industry and patient, when compared to other administration routes [3]. Apart from the highlighted attributes, the therapeutic efficacy of oral delivery systems is often concealed due to certain factors associated with physicochemical, anatomical, and biochemical, as well as physiological, constraints [4]. Despite this complexity, the fundamental events controlling oral drug absorption are the solubility/dissolution of the drug in the gastrointestinal (GI) environment and its permeability through the GI tract [4], which are the main causes of poor oral BA [5].

Nanocarriers have attracted increasing attention in recent years for oral chemotherapy, particularly for poorly soluble drugs [6]. Incorporation of drugs inside an appropriate nanocarrier seems to be a promising strategy since the biodistribution of the drug will be dictated by the properties of the nanocarriers and not by the drug molecule itself [7].

This chapter will address the most commonly used nanocarriers to improve the BA of poorly soluble drugs, focusing on the main challenges during their formulation, the mechanisms of internalization, and the characterization techniques for nanocarriers with poorly soluble drugs.

2.2 Oral Bioavailability

Oral BA is strictly related to pharmacokinetics of drugs and comprises two essential features: the rate of absorption, which means how fast the drug enters the systemic circulation, and the extent of absorption, which determines how much of the initial dosage reaches the circulation [6]. For both parameters, the solubility/dissolution rate is crucial since it determines how fast a drug reaches a maximum concentration in the luminal intestinal fluid, becoming available to be absorbed [8]. When a drug presents poor BA, the recommended dosage must be higher, as only a fraction of the drug enters the systemic circulation and reaches the site of action. On the other hand, a high drug dosage is often associated with side effects and increased costs [6]. Poorly water soluble drugs often exhibit slow dissolution in the GI tract, ultimately limiting their absorption [4].

Nowadays, poorly soluble compounds represent approximately 40% of the top oral marketed drugs. In addition, they stand for 90% of new chemical entities. This characteristic of poor water solubility seems to be a tendency in the drug discovery field due to the sources used for this purpose, such as combinatorial chemistry and high-throughput screening, in which the aim is to maximize drug-receptor interaction by stressing hydrophobic interactions [5].

A better understanding of the physicochemical and biopharmaceutical properties of drugs is not only helpful for developing new pharmaceutical products with already approved drugs but also important during the screening of new drug candidates, allowing the identification of potential absorption problems after oral administration [8]. In 1995, Amidon and coworkers established the Biopharmaceutical Classification System (BCS) based on the drugs' solubility and intestinal permeability [9]. According to the Food and Drug Administration (FDA), a drug substance is considered "highly permeable" when the extent of absorption in humans is determined to be 90% or more of an administered dose, while it is considered "highly soluble" when the highest dose strength is soluble in 250 mL or less of aqueous media over the pH range of 1–7.5 at 37°C [9, 10].

The BCS categorizes drug substances in four categories: high solubility/high permeability (class I), low solubility/high permeability (class II), high solubility/low permeability (class III), and low solubility/low permeability (class IV) [9]. Regarding formulation design, the BCS can provide an indication of the experimental work difficulty. For instance, pharmaceutical formulations containing drugs belonging to class I or III can be prepared with a simple strategy, while for class II and IV drugs, formulation designs are often required [8].

Recently, the application of nanotechnology to improve BA of poorly water soluble drugs through incorporation in nanocarriers is gaining attention, as drug molecule size is reduced to the nanometer range. By downsizing drug molecules, the thermodynamic and kinetic characteristics of the drug change [6], which leads to saturation solubility in the GI mucosa and a consequently greater gradient concentration for drug absorption [11]. The nanocarriers affect drug molecules' interfacial energy, resulting in colloidal instability and spontaneous aggregation into a more thermodynamically stable state [12].

2.3 Nanocarriers for Oral Delivery

A nanocarrier is generally defined as a nanometer-sized system consisting of at least two components in which one is the active compound [2].

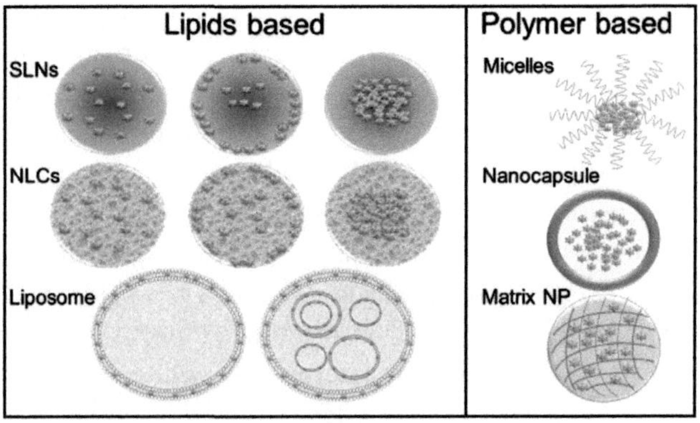

Figure 2.1 Different classes and types of nanocarriers.

The use of nanocarriers to encapsulate poorly soluble drugs offers additional advantages such as protection of the drug from GI degradation, improved circulation half-life, active and passive targeting of specific intestinal cells, reduced systemic side effects, codelivery of multiple drug combinations using a single nanocarrier, and the possibility to modulate drug release, which is often appropriate [7]. In addition, they may improve the oral BA by overcoming the first-pass effect and/or P-glycoprotein (Pgp) efflux and enhancing cellular uptake [3].

There are several nanocarriers (Fig. 2.1) already used for this purpose, namely lipid, phospholipid, and polymer based, which may be designed case by case in order to optimize their physicochemical features as well as in vivo efficacy [7]. In the case of lipophilic molecules only nanocarriers with enough high lipophilicity should be used.

The various types of nanoarchitectures mostly used to improve the BA of poorly soluble drugs upon oral administration and their features will be addressed in the subsequent sections [11].

2.3.1 Polymeric-Based Nanocarriers

2.3.1.1 Polymeric nanoparticles

Polymeric nanoparticles (NPs) are solid particles ranging from 10 to 1000 nm in size [13] and, compared to the lipid-based nanocarriers, have advantages such as higher stability in the GI tract, mainly in an acidic environment, and the possibility to achieve a more controlled drug release by modulating physicochemical characteristics of the drug [6].

Polymeric-based nanocarriers can be divided into two categories, namely nanocapsules, which are vesicular systems with "reservoir type" cores where the drug is confined, and nanospheres, which are matrix systems where the drug may be uniformly dispersed. In the case of nanocapsules, the inner phase may store the drug in a liquid (oily phase or water with a surfactant) or solid state as a molecular dispersion, depending on the preparation method and raw materials used, which is an advantage compared to the nanospheres, especially for poorly soluble drugs [13].

NPs may be composed of synthetic or natural polymers, and their choice affects directly the drug entrapment efficiency [14]. The synthetic polymers mostly used to formulate polymeric NPs are polyester derivatives such as polylactide (PLA), polylactide/polyglycolide copolymers such as poly(lactic-co-glycolic) acid (PLGA), polycaprolactones (PCLs), and polyacrylates (PCAs). PLGA is the most widely used polymer due to its biocompatibility and already described predictable biodegradability [6]. Natural polymers, also called biopolymers due to their biocompatibility, may also be used to obtain NPs. In general, biopolymers, such as alginate and albumin, exhibit lower immunogenicity than synthetic ones [14] and may also provide additional features, such as mucoadhesion and the possibility for chemical modification into a chitosan [6].

Table 2.1 Examples of polymeric nanocarriers as oral delivery systems with poorly soluble drugs, and their main characteristics

Drug	Therapeutic class	Excipient(s)	Diameter (nm)	EE/LC (%)	Ref.
Atazanavir*	Antiretroviral	Eudragit® RL 100	466	60/NA	[15]
Atorvastatin*	Antilipidemic	Zein	183	30/15	[16]
Celecoxib*	Anti-inflammatory	Ethyl cellulose	100–150	NA/10–50	[17]
Daidzein*	Hypertension, coronary heart disease, cerebral thrombosis, menopause syndrome	PLGA	309–323	81–83/ 1.3–1.8	[18]
Lopinavir*	Antiretroviral	PCL	195	94/NA	[19]
Resveratrol*	Chemopreventive, cardiopreventive, antioxidant, anti-inflammatory	PLGA	170	78/NA	[20]
Paclitaxel**	Antineoplasic	Polyanhydride	178–188	75–88/NA	[21]

*Belongs to class II
**Belongs to class IV

Regarding the method of fabrication, different methods can be reported, such as nanoprecipitation, emulsion/solvent evaporation, and emulsification, depending on the nature of the polymer as well as on the drug to be encapsulated [22]. One of the major disadvantages of polymeric NPs is the use of solvents, irrespective of the method of choice. The solvent choice is quite relevant for the successful encapsulation of poorly soluble drugs as their proportion directly influences particle size and drug entrapment efficiency [14].

When the objective is to encapsulate lipophilic drugs, the organic phase contains the polymer together with the drug. The emulsion-solvent evaporation is the most common method used to prepare polymeric NPs, though the encapsulation of water-insoluble drugs is yet quite challenging [23]. It is crucial to define and control the critical parameters of NP synthesis since they affect their physicochemical characteristics.

Polymeric NPs increase the rate and extent of oral absorption of class II (e.g., atazanavir, atorvastatin, celecoxib, daidzen, lopinavir, and resveratrol) and class IV (e.g., paclitaxel) drugs, resulting in augmentation in the values of drug plasma levels and thus significant enhancement in the rate and extent of BA of these drugs compared to those of the pure drug [17–23]. Table 2.1 illustrates some of the diverse polymeric-based nanocarriers that improve the BA of poorly soluble drugs.

2.3.1.2 Polymeric micelles

Polymeric micelles are nanosystems with an average hydrodynamic diameter less than 100 nm [24], composed of amphiphilic block copolymers, with self-assembly properties. They are characterized by a core-shell structure, in which the core is a dense grid formed by hydrophobic fragments, while the shell has hydrophilic characteristics [25]. One of the most commonly used hydrophilic fragments is poly(ethylene glycol) (PEG), although others have been explored, such as poly(*N*-vinyl pyrrolidone) (PVP) and PCA.

For the hydrophobic core, cationic or anionic polymers may be used, although biodegradable polymers, such as PCL, PLA, and poly(amino acids), are expected to be more promising for in vivo administration than the nondegradable ones [26]. They have the special feature of loading drugs with low solubility since their core serves as a reservoir due to the noncovalent hydrophobic drug/polymer interactions, thus conferring chemical stability on the drug inside the micelles [24]. Hydrophobic core characteristics are critical for determining the micelle loading capacity and may be assessed by evaluating the degree of polarity between the drug and the polymer hydrophobic fragments [24].

By physically entrapping poorly soluble drugs, polymeric micelles have the ability to increase their intrinsic water solubility and consequently drug BA at the targeted site of action. They generally have high thermodynamic stability and, compared to polymeric NPs, they do not need an organic solvent to be produced, which decreases the possibility of toxicity.

Besides polymer-drug miscibility, another important parameter that governs the entrapment efficiency is the hydrophilic-lipophilic balance of the block copolymers as well as the polymer/drug (P/D) ratio. As a general rule, block copolymers with longer hydrophobic fragments seem to promote higher lipophilic drug loading. Furthermore, the drug loading tends to increase with a decreasing P/D ratio. Optimization of this parameter for lipophilic drugs allows high drug loading with a minimum amount of polymer, as smaller amounts of polymer will be required for solubilizing higher amounts of the drug [24].

Despite the advantages of the use of micelles to deliver poorly soluble drugs, their stability, mainly in a physiological environment, it is still an issue. Once administered, the micelles must resist premature dissociation after dilution in the stomach and remain intact until they reach the intestinal environment. One of the parameters that dictate their stability is the critical micelle concentration (CMC) of the system. In general, lower CMC values denote more resistance to effects of dilution and therefore confer greater stability [24]. Another disadvantage of micellar systems is their biocompatibility and cytotoxicity. Clinical studies have shown that these systems may often lead to unexpected adverse side effects. Minimal options of copolymers that are safe are available and have been studied. On the other hand, there is still a lack of information about the effects that the amphiphilic copolymers produce after oral administration. Thus, it is important to get the know-how on the safety of micellar systems, by in vitro and in vivo studies [24].

Recently, class II (efavirenz) and class IV (oxcarbazepine, quercetin, and paclitaxel) drugs were successfully incorporated into polymeric micelles (Table 2.2). Reports indicate for paclitaxel micelles higher in vitro cytotoxicity when compared with commercial formulation and the effective targeting of M cells by micelles incorporating efavirenz with a higher anti-HIV activity as compared to that of the free drug. In vivo assays have been described with improved quercetin BA by polymeric micelles as compared to the BA of the free drug.

Table 2.2 Examples of micelles developed for oral delivery of poorly soluble drugs, and their main characteristics

Drug	Therapeutic class	Excipient(s)	Diameter (nm)	EE/LC (%)	Ref.
Efavirenz*	Antiretroviral	Carboxylated functional Pluronic F127	140	NA/>10	[27]
Oxcarbazepine**	Antiepileptic	Pluronic P84, F127, and F108	15–28	11–98/NA	[28]
Paclitaxel**	Antineoplasic	H40-PCL-*b*-PCA-*b*′-methoxy PEG/PEG-folate	450	62/10	[29]
Quercetin**	Anti-inflammatory, antioxidant, anticancer	Soluplus®	79	96/NA	[30]

*Belongs to class II
**Belongs to class IV

2.3.2 Lipid-Based Nanocarriers

Lipid-based nanocarriers seem to have the most suitable features to load poorly soluble drugs since they may offer high drug loading, present suitable long-term storage stability, and are feasible for scale-up compared to other colloidal carriers [31].

The variety of materials available, the biocompatibility of the lipids and other lipophilic substances like phospholipids, and their ability to improve oral BA have made lipid nanocarriers very attractive for oral delivery [1]. Lipid nanocarriers may enhance the solubility of lipophilic drugs through their interaction with the intestinal microenvironment and by modifying drug release, which prevents supersaturation of the drug and consequent precipitation in the intestinal lumen, leading to BA improvement [32].

Furthermore, lipid systems may stimulate intestinal lymphatic drug transport and the interaction with enterocyte-based transport processes, which can reduce undesired effects and may be of particular interest for dietary lipids and certain highly lipophilic compounds (e.g., lipid-soluble vitamins and drugs) [32].

On the other hand, lipid nanocarriers are not universal platforms to deliver poorly soluble drugs due to the fact that lipophilic drugs do not necessarily display high solubility in a lipid phase. Therefore, besides the solubilization of the drug in the lipid phase, the partition of the drug between aqueous and oil phases should be taken into consideration.

A number of nanocarriers based on biocompatible lipids and oils have been efficiently used to improve the BA of poorly soluble drugs. The most commonly used will be discussed in this section.

2.3.2.1 Liposomes

Liposomes are vesicular structures made of natural or synthetic amphiphilic phospholipids that spontaneously self-assemble into bilayers when in contact with an aqueous medium. Depending on the production conditions, they can form unilamellar or multilamellar vesicles, which makes them vary in size from 50 to >1000 nm. Inside the circular structure of the phospholipid, vesicles form an aqueous compartment. Thus, due to its amphipathic nature, a liposome is able to enclose hydrophilic drugs in the aqueous core and lipophilic drugs within the phospholipid membrane [14].

The most common production method of liposomes for drug delivery is based on the dehydration-rehydration technique, since it may provide a higher drug loading efficiency as compared to other methods [14]. Particularly for poorly soluble drugs, achieving high drug loading is quite challenging, as the aqueous core space is much higher in comparison to the lipid bilayer, which limits the accommodation of large amounts of drug. Thus, for drugs with very lipophilic characteristics, the utilization of multilamellar liposomes would be a better strategy, as their lipid bilayers' structural organization should entrap more lipophilic drugs, compared to unilamellar vesicles. The composition of the bilayer can be easily changed in order to achieve a suitable partition between the drug and the phospholipid layer, without disrupting the integrity of the liposomal system [33].

There are still challenges regarding their stability under physiological conditions, mainly in the GI tract [39]. Once in the gastric environment, the phospholipids are found to be highly susceptible to gastric acid, bile salts, and lipases, whose destructive effects commonly lead to losing of the liposomal integrity and leakage of the drug. In addition, even if the liposomes can reach the intestinal lumen, another challenge is the permeation of the vesicles across the GI epithelia due to the physicochemical characteristics of the epithelia and the presence of the mucus layer [40]. To overcome these limitations, many attempts have been made to ameliorate their stability by modifying constituents and to improve liposomal BA. Surface modification is one of the most-used strategies as this parameter may both alter stability and introduce new functionalities to improve cellular internalization [40]. Surface modification may be achieved by the attachment of surfactants such as bile salts or polysorbates (Tweens) or by coating with functional polymers such chitosan or PEG [40]. For instance, chitosan may increase the liposome residence time in the intestinal mucosa due to its mucoadhesive properties, leading to an enhanced BA [40].

Table 2.3 Examples of liposomes as oral delivery systems with poorly soluble drugs, and their main characteristics

Drug	Therapeutic class	Excipient(s)	Diameter (nm)	EE/LC (%)	Ref.
Daidzein*	Hypertension, coronary heart disease, cerebral thrombosis, menopause syndrome	Soybean phospholipid, sodium oleate, glycerol monostearate	45	92/NA	[34]
Resveratrol*	Chemopreventive, cardiopreventive, antioxidant, anti-inflammatory	Distearoyl phosphatidyl choline, cholesterol	493	92–100/NA	[35]
Cyclosporine A**	Immunosuppressant	Cholesterol, lecithin	63–72	~98/~9	[36]
Nimodipine**	Cardiovascular agent	Soybean phospholipid	317–379	73–85/NA	[37]
Sorafenib**	Antineoplasic	Lipoid e80 cholesterol	165	91/4	[38]

*Belongs to class II
**Belongs to class IV

Recently, polymersomes have been applied to improve oral BA of poorly soluble drugs, as they exhibit enhanced stability and tunable properties [41]. These polymeric nanocarriers are self-assembled vesicles similar in structure and function to liposomes. To improve the incorporation of low-water-soluble drugs an amphiphilic β-cyclodextrin-centered triarm star polymer (mPEG2k-PLA3k)3-CD is used, resulting in considerable drug loading capability.

In vivo assays reveal that class IV drugs (e.g., nimodipine, sorafenib, and cyclosporine A [CsA]) incorporated into liposomes exhibit higher drug plasma level and improved biological activity as compared to free drugs. A few reports indicate a higher pharmacokinetic profile of class II drugs (daidzein and resveratrol) loaded into liposomes than the plain drugs. Table 2.3 shows the potential of liposomes to enhance the BA of poorly soluble drugs for oral administration.

2.3.2.2 Solid lipid nanoparticles

Solid lipid nanoparticles (SLNs) are composed of lipids that are solid at room temperature and in an aqueous medium, the colloidal system must be stabilized by an emulsifier. The resultant nanosystem generally contains submicron particles between 50 and 1000 nm in size [31].

SLNs combine advantages of polymeric nanocarriers, such as the possibility to achieve controlled drug release and efficient encapsulation, with the biocompatibility of the material used, as in the case of lipid-based nanocarriers [33]. Additional advantages of SLNs include improved stability both in storage and in the GI tract, feasibility for scale-up, low cost, and production without the use of organic solvents [1].

A broad range of solid lipids with huge degrees of polarity are available to produce SLNs, ranging from nonpolar solid lipids, such as triglycerides and waxes, to glyceride mixtures, fatty acids, and emulsifying wax. Furthermore, many surfactants may also be used to stabilize the nanoformulation, such as phospholipids, bile salts, and nonionic surfactants (such as polysorbates and poloxamers) [32]. In addition, particularly for poorly soluble drugs, the possibility to screen different biocompatible lipid compositions according to their compatibility with the drug is an added value, as this kind of study may improve the physicochemical characteristics of the systems as well as drug entrapment.

Regarding the methods of production, SLNs are usually produced by the formation of an oil-in-water emulsion, followed by the solidification of the dispersed lipid phase. Other methods have been reported, such as high-shear homogenization, high-pressure homogenization, and ultrassonication. Still, the production of lipid nanocarriers with suitable diameters and narrow polydispersity indexes remains a challenge and a critical step during SLN design [3].

Depending on the method of production and on the constituents, the drug may be incorporated into the particles in different ways. For example, a high-pressure homogenization process usually produces particles with matrix-like behavior, in which the drug is dispersed molecularly in the lipid phase. On the other hand, in the case of a drug-enriched shell type, the drug is dispersed in the outer shell of the lipid phase, which happens when phase separation occurs during the cooling process. The drug may also stay in the inner core of the nanocarrier when the drug concentration is close to the saturation solubility in the lipid, and the drug starts to precipitate, while a lipid shell with a small amount of the drug is formed around the core [42].

The crystallinity of the solid lipids is one of the most important issues during SLN design, which is commonly associated with polymorphism and dynamic transition, leading to low drug incorporation, drug expulsion, and consequent drug release due to the lipid rearrangement and the tendency for particle aggregation [42].

In vitro studies describe improved cytotoxicity of class II (e.g., paclitaxel) drugs when loaded in SLNs as compared to the cytotoxicity of the free drugs and also enhanced drug cellular uptake, suggesting potential as an oral drug delivery system. Others report biocompatibility of SLN-containing class II drugs (risperidone) and enhanced in vitro permeability for the class IV drug curcumin loaded in SLNs. In vivo assays reveal oral BA improvement for class II drugs (e.g., efavirenz, raloxifene, miconazole, glibenclamide, and dapsone) when delivered by SLNs in comparison to the oral BA of plain drugs. Some class II and class IV compounds have been successfully incorporated into SLNs for enhanced oral BA and are detailed in Table 2.4.

Table 2.4 Examples of SLN as oral delivery systems with poorly soluble drugs, and their main characteristics

Drug	Therapeutic class	Excipient(s)	Diameter (nm)	EE/LC (%)	Ref.
Dapsone*	Antibiotic	Cetyl palmitate	308	68/17	[43]
Efavirenz*	Antiretroviral	Compritol 888 ATO Gelucire 44/14	160	86/39	[44]
Glibenclamide*	Antidiabetic	Precirol® ATO5	105–112	20–70/NA	[45]
Miconazole*	Antifungal	Precirol ATO5	22	90/NA	[46]
Raloxifene*	Prevention and treatment of osteoporosis	Glyceryl tribehenate	167	92/NA	[47]
Curcumin**	Antioxidant, anti-inflammatory, antibacterial, antifungal, chemopreventive	Compritol 888 ATO	270	80/2	[48]
Paclitaxel**	Chemotherapeutic agent	Stearic acid	251	71/NA	[49]

*Belongs to class II
**Belongs to class IV

2.3.2.3 Nanostructured lipid carriers

To overcome some drawbacks associated with SLNs, mainly regarding stability and loading capacity, a second generation of lipid-based nanocarriers was developed. Nanostructured lipid carriers (NLCs) are systems composed of a blend, in appropriate proportion, of solid and liquid lipids as a core matrix, in which the final physical state is still solid [50].

The presence of liquid and solid lipids allows the immobilization of higher amounts of drugs, compared to SLNs, thus increasing the loading capacity and preventing the premature expulsion of the drug, providing greater long-term stability. This happens due to the chemical differences between the two types of lipids, which lead to imperfections inside the matrix. The larger this imperfection, the higher is the entrapment efficiency, which can be modulated by varying the ratio of the solid lipid to the liquid lipid and their composition [50]. Another mechanism by which NLCs may improve drug loading is probably by the partition of the solubilized drug between solid lipids and liquid lipids. Many times, in SLNs the drug is soluble in molten lipids before preparation and reaches

Table 2.5 Examples of NLCs as oral delivery systems with poorly soluble drugs, and their main characteristics

Drug	Therapeutic class	Excipient(s)	Diameter (nm)	EE/LC (%)	Ref.
Budesonide*	Corticosteroid	Precirol ATO 5, miglyol 812	204	96/ NA	[51]
Resveratrol*	Chemopreventive, cardiopreventive, antioxidant, anti-inflammatory	Cetyl palmitate, miglyol 812	150–250	~70/ NA	[52]
Vinpocetine*	Vasodilators, treatment of chronic cerebral vascular ischemia, acute stroke, senile cerebral dysfunction, Alzheimer's disease	Compritol 888 ATO or monostearin, miglyol 812	136	95/2	[53]
Curcumin**	Antioxidant, anti-inflammatory, antibacterial, antifungal, chemopreventive	Precirol ATO 5, miglyol 812	280	95/ NA	[54]
Saquinavir**	Antiretroviral	Precirol ATO 5, miglyol 812	165–1090	99/1	[55]

*Belongs to class II
**Belongs to class IV

saturation, but during the cooling phase it may precipitate, resulting in immediate expulsion of the drug. In NLCs, this saturation may not occur as the solubilized drug may be partitioned within solid and liquid lipids, as they have different polarities, avoiding precipitation [50]. This feature is even more important when the drug used has very high lipophilicity, since lipids are able to more easily solubilize these types of drugs.

Unlike SLNs, NLCs may exhibit biphasic drug release, that is, an initial burst release followed by a second release phase that is more controlled. This occurs due to the presence of a considerable amount of drug in the outer oily layer of the nanocarriers, which is rapidly expulsed, while in the second release phase the drug is released at a constant rate from the lipid core. Modulation of the release profiles as a function of the lipid matrix composition is very advantageous. These singular features make, nowadays, NLCs the "smarter" lipid systems, having improved properties in contrast to other lipid-based formulations [50].

Table 2.5 presents some examples of these lipid nanocarriers for oral delivery of poorly soluble drugs from class II and IV. A few reports indicate in vitro potential of NLCs to improve oral BA of class IV (e.g., saquinavir) and class II drugs (e.g., resveratrol). In vivo studies reveal the contribution of NLCs in significantly improving the biological activity of class IV (e.g., curcumin) and class II (e.g., budenoside) drugs and in a few cases enhanced oral BA (e.g., vinpocetine).

2.4 Oral Absorption: Crossing the Intestinal Barrier

The understanding of how different nanocarriers interact with biological systems is important as it dictates the rate of absorption to reach systemic circulation and where the medicinal effects take place [56]. For the oral route, the absorption step involves several stages and is determined by the drug's physicochemical and formulation properties.

The human GI tract is a dynamic structure that, besides the role of absorption, acts as an efficient barrier against undesired bacteria and toxins. To overcome this barrier is still a challenge in the development of drug nanosystems for oral delivery [4]. Upon ingestion, the first obstacle for nanosystems is to retain their integrity in the complex gastric medium, which contains numerous compounds, such as bile salts, ions, lipids, cholesterol, and enzymes, and an extremely acidic pH environment [4]. Afterward, upon reaching the small intestine portion, where approximately 90% of the absorption occurs, nanocarriers can be absorbed or internalized.

In the intestinal lumen, nanocarriers find themselves at the mucus layer, which covers the epithelium and is constantly renewed. The mucus consists of mucin glycoproteins, enzymes, electrolytes, and water, and due to its cohesive and adhesive nature, the passage of nanocarriers with certain physicochemical properties may be a challenge [57]. To overcome this drawback, nanosystems with mucoadhesion properties need to be developed, which may increase the residence time at the absorption site [14].

Below the mucus layer, particles will find the intestinal epithelium, which is formed of a monolayer of a heterogeneous population of cells derived from undifferentiated cells called crypt. The cells from the crypt may differentiate into absorptive cells (enterocytes); mucus secreting cells (goblet); endocrine cells; Paneth cells, which secrete large amounts of protein-rich materials; and M cells, which compose a lymphoid region called Peyer's patches, which are specialized in antigen sampling [4]. Despite this heterogeneity, the enterocytes are the most predominant cells in the intestine. However, the mechanisms by which they absorb or internalize nanocarriers varies and will be addressed in the next section. In addition, M cells represent a target in the drug delivery framework due to their specialization in phagocytosis, which may be useful to transfer nanocarriers from the lumen to the basolateral membrane, where there are several populations of lymphocytes and mononuclear phagocytes, as macrophages, which serve as host cells for innumerous diseases [4].

2.4.1 Nanocarrier Absorption Mechanisms

Oral absorption gets initiated in the mouth and carries on in the stomach and the small intestinal, finally reaching the colon. This process occurs throughout the GI tract membranes by mechanisms of passive transport and carrier-mediated transport. The primary route of membrane permeation for many drugs and small size nanocarriers, in several cells of the intestinal membrane, is the passive diffusion. Drugs with low passive permeation are often transported through a transport carrier and ligand-targeted nanocarriers by an endocytosis-mediated mechanism.

A biological cell membrane is a fluidic hydrophobic barrier composed of a lipid bilayer, cholesterol, and membrane anchor proteins (e.g., transporters).

2.4.1.1 Passive transport

Passive diffusion occurs when substances (free drug or drug nanocarriers) diffuse across a cell membrane from a region of high concentration to one of low concentration. It can take place by two pathways: the paracellular pathway, in which substances diffuse through the intercellular space between the intestinal enterocytes [58], and the transcellular (or lipophilic) pathway, which requires substances to travel through the cell, passing through both the apical membrane and the basolateral membrane (Fig. 2.2).

The tight junctions between the intestinal enterocytes are negatively charged; thus positively charged substances cross through readily and negatively charged substances are repelled [59]. The paracellular route (Fig. 2.2, II) represents less than 0.01% of the total surface area of the intestinal membrane, and closer to the colon the junctional complex becomes tighter. The transcellular route (Fig. 2.2, I) is the major one responsible for the intestinal absorption of substances. It involves the movement of substances on the basis of a diffusion gradient and may include transcellular diffusion, active carrier-mediated transportation (Fig. 2.2, IV), and transcytosis (Fig. 2.2, V). The physicochemical properties of the substances control the rate of transport through the transcellular pathway [60].

The lipid bilayer portion of the cell membrane is fluidic, with no specific binding sites; thus passive transport is not saturable, not subject to inhibition, and nonsensitive to the molecule's stereospecific structure [61]. The lipid bilayer portion of a cell membrane is similar in most cells; thus passive diffusion occurs regardless of cell type but may be dependent on the membrane lipid composition.

Usually, the absorption of colloidal carriers (e.g., polymer-based nanocarriers) follows mechanistic pathways involving intracellular uptake by M cells of Peyer's patches (Fig. 2.2, VI) or/and intercellular/

paracellular uptake through the intestinal membrane [62]. By the intracellular pathway drug reaches systemic circulation through the lymphatic system. This pathway enables nanocarriers to avoid the hepatic first-pass effect, thus increasing plasma drug concentration. For lipid-based nanocarriers containing self-emulsifying agents, absorption can be driven by lipase-mediated chylomicron formation toward the lymphatic system [63]. The chylomicron formation also aids the absorption of water-insoluble molecules as it enhances the dissolution and assimilation of lipophilic molecules [64].

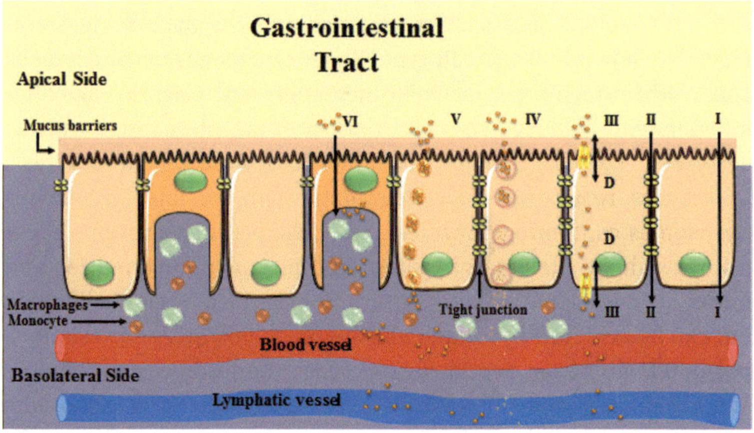

Figure 2.2 Schematic representation of the transport pathways across the intestinal barrier: (I) transcellular passive diffusion, (II) paracellular passive diffusion, (III) influx-/efflux-facilitated transport by membrane proteins, (IV) endocytosis with lysosome degradation, (V) transcytosis, and (VI) uptake by M cells.

Lipid-based nanocarriers can be transported through intracellular Peyer's patches uptake and by the chylomicron formation, evading first-pass effect and targeting the lymphatic system with the drugs; this has special relevance in lymphatic cancers and some infections' therapy.

2.4.1.2 Carrier-mediated transport

Molecules' transport across a membrane involving a protein is designated as carrier-mediated transport (Fig. 2.2, III). If the transport process requires energy it is named active transport (e.g., endocytosis and influx/efflux) and may involve transport against a concentration gradient. When the process is not driven by energy it is defined as facilitated transport (e.g., phagocytosis) and is based on the concentration gradient of a substrate.

Interaction of nanocarriers with the outer surfaces of the cells and cellular membranes leads to internalization into intracellular vesicles and initiates the endocytosis process. This invagination of plasma membrane surrounding the nanocarriers can have different sizes, compositions, and internal environments, according to the internalization pathway (e.g., endosomes, phagosomes, or macropinosomes). The endocytosis pathway is determined by the size, morphology, and surface chemistry of the nanocarriers and may also depend on the cell type. Phagocytosis; macropinocytosis; and clathrin-mediated, caveolin-mediated, and clathrin/caveolin-independent endocytosis are the internalization mechanisms described so far. Usually the internalization of nanocarriers in vesicles results in their degradation at the late endosomes (lysosomes containing enzymes and acidic pH) and prevents them from reaching their target sites (cytoplasm, mitochondria, and nucleus). The escape of nanocarriers from endosomes is challenging but can be achieved by using endosomolytic polymers, for example, polyamidoamine, or peptides to increase the intracellular delivery and avoid lysosomal degradation. Exocytosis of the vesicle contents is also a possible pathway for the excretion of nanocarriers from cells. Direct translocation of nanocarriers across the plasma membrane may occur independent of the metabolic activity of the cells. This pathway, named transduction, is mediated by peptides known as cell-penetrating peptides and is energy and receptor independent [65].

Carrier-mediated transport is selective and may involve transporters using the mechanism of active drug influx and efflux. More than 400 membrane transporters have been described. In carrier-mediated transport the physicochemical properties of the molecules and their potential affinity for various transport proteins play essential roles [66]. Nanocarriers coupled to affinity moieties that target GI surface markers (e.g., ICAM-1 and cobalamin receptor), involved in transport, may improve BA. Enterocytes express various transporters on the apical and basolateral membranes conducting the influx (solute carrier, SLC family) or efflux (ATP-binding cassette, ABC family) of endogenous molecules and xenobiotics. Contrary to the passive transport process, the carrier-mediated transport is saturable, subject to inhibition, and stereospecific and occurs in the specific cells expressing the transporter.

2.4.2 Factors Affecting Nanocarriers' Absorption

The rate and extent of oral drug absorption are influenced by physicochemical and biopharmaceutics factors. Physicochemical data are heterogeneous but indicate that ionization state, molecular weight, and lipophilicity affect small molecules' BA. Nanocarriers are designed to enhance drug solubility and protect the encapsulated therapeutic agents from the extreme conditions found in the GI tract. Particle size, surface charge, and chemical composition are essential parameters to consider in order to improve the BA of the oral drug.

The molecule's ionization determines its dissolution rate and passive diffusion across the GI tract. Therefore, the pH value is a crucial parameter in oral drug absorption and dissolution [67]. Physiological aspects influence the pH, as the luminal secretions alter the local environment. At the luminal side the pH is acidic and can be as low as 2.3. While in the duodenum, the secretion of pancreatic bicarbonate and bile neutralizes the pH and along the small intestine the pH progressively becomes alkaline. pH-sensitive nanocarriers can be designed taking advantage of this specific feature of the GI tract. According to their composition, drug release will occur under the intended pH and the body regions with physiological pH will not be affected [41].

The particle size matters in nanocarriers delivering drugs orally. Small particles (ca. 100 nm) have been reported to exhibit higher absorption as compared to larger (ca. 500 nm) particles [64]. However, larger nanocarriers show longer retention periods in Peyer's patches, facilitating transport to the lymphatic system.

Surface charge and hydrophobicity also affect oral BA and thus the nanocarrier's interaction with the GI tract. Nanocarriers with high hydrophobicity exhibit improved accumulation in Peyer's patches [64]. Contrary to size, nanocarrier charge is a crucial parameter for interaction with intestinal in vitro models [68]. M cells' uptake is highly affected by surface charge. Specific targeting of M cells, for example, the lectin ligand, also improves cellular uptake [69]. Positively charged nanocarriers present higher cellular toxicity compared to negatively charged ones and may even lead to reactive oxygen species production, and mitochondrial damage [70]. Most probably, these effects are related to the electrostatic interaction between the surface particles and cell membranes. In the intestinal membrane in vitro model, the receptor-mediated endocytic pathway plays a role in the internalization and transport of positively charged nanocarriers, while the lipid raft pathway mediates the uptake and transport of negatively charged nanocarriers [68].

The surface charge of nanocarriers alters upon absorption due to plasma protein adsorption. This phenomenon determines the biological nanocarriers' fate, as it can lead to phagocytosis, long-term blood circulation, or specific targeting. Protein adsorption on nanocarriers may reduce mean resident time and increase first-pass metabolism [71].

Oral drug delivery poses a significant challenge due to the short residence time of the formulations within the GI tract. Therefore to increase the effectiveness of therapy, delivery systems with ability to adhere to the oral mucosa or target specific parts of the GI tract would be advantageous. Although the mechanisms governing oral absorption remain unknown, various strategies have proved to be effective, for example, improvement of mucoadhesiveness, polymer coating, and ligand-mediated targeting of the epithelia. Polysaccharides and chitosan reinforce the mucoadhesion by interfering with the tight junctions, thus enhancing absorption of the drug. Coating lipid-based nanocarriers with chitosan leads to inversion of surface potential from negative to positive, preventing the pH-triggered aggregation of lipid nanocarriers in the stomach [31]. In fact, positively charged chitosan has a high affinity to bind with negatively charged mucin and mucosal surfaces [72]. Also, modification with tocopheryl PEG 1000 succinate and pluronics could enhance oral absorption by inhibition of intestine efflux pumps, such as Pgps [73].

Nanocarriers taken from the GI tract can reach the systemic circulation or be cleared by the immune system. To improve oral BA nanocarriers' recognition and elimination can be overcome by coating them with hydrophilic polymers, such as PEG [74]. The impact of PEGylation in inhibiting lipid digestion, mediated by gastric and pancreatic lipases, is dependent on the molecular weight and packing density of PEG molecules, as increasing the molecular weight and packing density of PEG headgroups results in improved steric hindrance of digestive enzymes, thus protecting the lipid core [75].

Ligand-mediated targeting of epithelia takes advantage of the specific receptors' expression on the cell surface. Active targeting of M cells, by receptor-mediated phagocytosis, can be achieved using lectins [76]. The potential of a transenterocytic phagocytic pathway has also been explored, targeting several receptors (e.g., vitamins, carbohydrate, and proteins) [69].

2.5 Nanocarriers' Characterization: Issues and Challenges

Proper characterization of nanocarriers is a critical step to control product quality, stability, and safety though it is not easy work due to the complexity and diversity of these formulations. Parameters must be studied according to the type of nanocarrier chosen as well as the nature of the materials used. Despite the diversity of the nanocarriers available, there are basic common characterization techniques for particle size and shape and the homogeneity of the system (polydispersity index). Accurate and sensitive techniques have already been used for these purposes, which are able to provide excellent qualitative and quantitative information [1].

The dynamic light scattering (DLS) is one of the most popular techniques used to estimate the size distribution of small particles in solution or suspension. The data obtained provide information about a population of nanocarriers in each sample solution, and not information about each particle. The main strengths of DLS are that the experiment duration is short, the technique is noninvasive, diluted samples can be measure in a wide range of concentrations, and even small amounts of higher-molecular-weight species can be detected. However, the high accuracy of the technique can, sometimes, lead to a wrong interpretation of the data as the presence of even a small percentage of aggregates may interfere in the scattering intensity, increasing the size of measured particle. In the case of nanocarriers with poorly soluble drugs, proper dilution sometimes is not enough to solubilize unentrapped drugs in suspension, and these particles are also measured, giving false results of bigger nanocarriers.

Besides the mentioned basic characterization parameters, there are some other critical parameters that must be studied more deeply and carefully.

Entrapment efficiency, together with drug loading and drug release kinetics, is not only important for therapeutic purposes, as it dictates the rate of drug that may reach the site of action, but also should be well determined, without having superestimation or false results, which are common for insoluble drugs [77]. Furthermore, the physical state of the drug in the nanocarriers, that is, in the amorphous state or in crystal form, may be important as it directly influences drug loading in nanocarriers as well as the release characteristics of the encapsulated drug [77].

The next sections will address in more detail these critical parameters and the possible issues that their determination may give rise to.

2.5.1 Polymorphism and Crystallinity

Drug stability, solubility, dissolution, efficacy, and, consequently, BA are parameters directly related to the physical state of the drug in nanocarriers. The drug may be encapsulated in a crystalline or amorphous state, which will depend on several parameters of the nanocarrier design. In addition, the crystalline form of the drug may change, leading to the formation of a polymorph of the drug.

Polymorphs are chemically identical, although the different arrangements of the molecules lead to different configurations of the crystal, which can vary from each other significantly in solubility and stability [78].

Several factors can alter the physical state of the drug, for instance, the use of an organic solvent as well as its type, the method of obtaining, and even some drying processes, such as lyophilization. Because most of the techniques applied to obtain nanocarriers with poorly soluble drugs use organic solvents, it is important to check whether polymorphs of the drug have been already reported and monitor this parameter during all the steps of the formulation. The assessment of the polymorphic form and the crystallinity of the drug may be performed using different techniques that may be correlated with each other, such as differential scanning calorimetry (DSC), X-ray diffraction (XRD), and microscopy [78].

2.5.2 Drug Entrapment

In general, encapsulation of a poorly soluble drug is influenced by several factors, such as the nature of the nanocarrier, matrix polymorphism, the method of preparation, surfactants, and the use of organic solvents to dissolve the drug [33]. As already mentioned, regardless of the nature of the nanocarrier matrix used, the miscibility between the drug and the carriers must be well determined, since the higher the affinity between both components, the higher seems to be the entrapment efficiency and the lower is the carrier/drug ratio necessary to achieve suitable formulation [33]. Furthermore, the entrapment of the drug inside nanocarriers is also dependent on the distribution and partition of the drug between the external phase, which is usually aqueous, and the matrix [33]. It has already been reported that poorly soluble drugs tend to precipitate in the external phase, when the method of preparation involves an oil-in-water emulsion, leading to low entrapment efficiency [77].

The entrapment efficiency may be determined by destroying the nanocarriers, with previous separation from the unentrapped drug (direct method), or by measuring only the drug present in the external phase (indirect method). In both cases, the obtained nanocarriers should be separated, despite the fact that the small size of these particles makes it difficult to separate the colloids and the dispersion medium and, for lipophilic drugs, the drug theoretically soluble is present as a precipitate, leading to an overestimation of the drug loading capacity.

Basically two methods are used to measure entrapment efficiency, namely centrifugation and ultrafiltration, depending on the nature of the material used. For instance, for polymeric-based nanocarriers, it is common to use the centrifugation method as with high speeds of centrifugation they form a pellet and the supernatant may be collected despite the force used to perform the technique. On the other hand, ultrafiltration is often successfully used for lipid-based nanocarrier as most of the lipids have low densities and are not able to form pellets at considerably high speeds of centrifugation. This technique utilizes membranes with different molecular weight cutoffs that retain the nanocarriers while the solubilized drug passes through the membrane [78]. None of the mentioned methods are completely suitable for estimating the entrapment efficiency of poorly soluble drugs, as the presence of precipitated drug is common, turning this characterization into one of the most difficult challenges during the design of the formulations.

2.5.3 Drug Release Profile

As already mentioned earlier, the solubility of the drug in an aqueous medium may be indicative of its in vivo performance, as the precipitation of the drug reduces the total amount of drug absorbed, decreasing its BA. Thus, the best way to mimic in vitro the behavior of the drug solubilization upon oral administration is through a dissolution test, also called release profile in the case of nanocarriers. The in vitro release profile, if properly performed, can inform the rate by which the drug becomes available to be absorbed in different conditions encountered in vivo such as different pH values, the presence of digestion proteins, and protein binding [79].

Some critical conditions must be established during the in vitro release assay in order to provide some predictive potential, such as the release media and sampling methods. The choice of the medium should simulate in vivo systems. Phosphate-buffered saline (PBS) is the most simple and common medium used for this purpose, despite significant underestimation of the intestinal fluid environment [79]. Thus, serum-containing buffers may be a reasonable choice of release medium for mimicking a physiological fluid, especially for poorly water soluble drugs, since they have a more complex composition that seems to affect solubilizing properties (e.g., fasted or fed state simulated intestinal fluids [FaSSIF and FeSSIF, respectively]) [79].

The first challenge to be solved during an in vitro drug release of nanocarriers containing poorly soluble drugs is to satisfy sink conditions, defined as the volume of medium at least 3-fold higher than the volume necessary to reach a supersaturated solution of the drug. It means that to obtain realistic results, a suitable volume of the release medium must be used, implying the use of a very low concentration of formulation in the case of poorly soluble drugs, compromising sometimes the accuracy of the quantification analysis. The use of simulated intestinal fluids is even more useful for achieving sink conditions compared to PBS due to the solubilizing effects of the proteins. Likewise, this difficulty may be alleviated by concentrating the samples prior to analysis or adding surfactants of cosolvents in the release medium to increase the drug solubility [80].

Besides the composition and the volume of the release media, it is important to select the more appropriate method, depending on the formulation to be studied. Centrifugation and dialysis are most widely used for this purpose, although both have critical limitations. The first method is based on the centrifugation of the samples at each time point, at high speeds, to separate the nanoformulations from the free drug. The main disadvantages of this method are that the pressure generated during the centrifugation can force the release of the drug and, for lipophilic drugs, precipitation of the drug in the pellet may occur. In addition, the compression of the pellet is often excessive, turning the resuspension difficult, or sometimes the separation is incomplete, leading to cumulative errors in the measurement of the released drug [80].

In the dialysis, the nanocarrier suspension is placed inside a dialysis bag with a specified molecular weight cutoff and the released drug molecules diffuse out of the bag, for further quantitative analysis. In this case, the major difficulty is to achieve real concentration of poorly soluble drugs in the receptor compartment despite premature precipitation of the drug inside the dialysis bag.

2.5.4 Permeability Assays

One of the last steps during formulation design, but not less important, is the evaluation of its permeability characteristics in order to compare the in vitro absorption profile with that of the pure drug. In the face of all the challenges associated with the conception of drug delivery systems containing poorly soluble drugs, there is a significant increase in the number of reports in this framework, applying techniques from the most basic models—including non-cellular-based (artificial membrane), cell-based, or tissue-based models (in vitro)—to more complex and time-consuming methods, such as in situ, in vivo intestinal perfusion, and in silico [4]. Nevertheless, the in vitro techniques are preferred as they are less laborious and less expensive as compared to in vivo animal studies and can represent an alternative to animal use mainly during initial stages of research.

Currently, there are a variety of human immortalized cell lines that are able to easily grow into confluent monolayer models, which can mimic with good correlation with in vivo results the intestinal epithelium for transport studies. The commonly used cells for this purpose are endothelial cells (e.g., Caco-2 cells). Caco-2 cells differentiate both structurally and functionally into cells resembling mature enterocytes and have also been shown to express several transport systems of different molecules [81]. The successful application of in vitro models to predict drug absorption across the intestinal mucosa depends on how closely they mimic the characteristics of the in vivo intestinal epithelium [4]. Thus, other cell lines began to be used in cocultures together with Caco-2 due to the need to develop models with greater predictability, which involve other mechanisms of absorption and have the ability to provide a more realistic mucus barrier [82]. In this context, several studies have been done applying cocultures of Caco-2 and mucus-secreting goblet cells (HT29) [83] or Caco-2 coculture with Raji B lymphocytes, which has been developed to mimic the M cells [84].

Despite the fact that good correlations have been encountered between in vivo/in vitro values of permeability with these models, there are still several difficulties associated with the prediction of absorption of poorly soluble drugs, especially for drugs belonging to BCS class IV. As already discussed, the absorption of these drugs is influenced to the extreme by the physiological environment, such as the presence of endogenous surfactants as well as the presence or absence of food, which are factors very difficult to mimic in vitro. In addition, the gastric emptying rate, the GI transit rate, and the GI pH cannot be incorporated in the result's interpretation [81].

Likewise, the limit of detection of analytical methods, together with the tendency for drug precipitation before permeation, is one more obstacle during in vitro permeability assays for poorly soluble drugs, leading to uncertain results. Besides, it is common to find high levels of toxicity in many cell lines, which turns the concentration of the drug in the sample even lower. One recent strategy to overcome the solubilization problem during in vitro permeability tests is the use of solubilizing agents (e.g., poloxamers and Tritons), which may partially eliminate the detection limit challenge, turning possible the estimation of the permeability [85].

2.6 Conclusions

Despite the vast and growing therapeutic applications of poorly soluble drugs it was demonstrated that there are still severe issues regarding the development of oral dosage forms. In this section, it could be seen that different systems, either polymeric or lipid based, have been studied as nanocarriers to improve oral BA of class II and class IV drugs. In this framework, the use of nanotechnology has been proven to be a promising and efficient strategy to improve solubility, dissolution kinetics, and oral BA of hydrophobic drugs. A few reports already describe in vivo assays with a significant enhancement of drug-loaded nanocarriers' oral BA as compared to that of the free drug. It was shown in this section that there are several different nanocarriers used for this purpose, with different features, which allows the development of tailor-made nanotherapeutics for different drugs. With rapid scientific and technological advances in nanosizing hydrophobic drugs, their potential can be vital for clinical applications. This whole field will thus require more attention in the future, in particular regarding in vitro–in vivo correlations, in order to elucidate in detail all effects involved and to provide an adequate basis for appropriate carrier selection.

Acknowledgments

Luíse L. Chaves thanks the CAPES Foundation, Ministry of Education of Brazil (0831-12-3), and Alexandre C. V thanks the CNPq (246514/2012-4). Sofia A. Costa Lima thanks Operação NORTE-01-0145-FEDER-000011 (Qualidade e Segurança Alimentar — uma abordagem (nano) tecnológica) for her investigator contract. The authors also thank National Funds from Fundação para a Ciência e a Tecnologia (FCT) and FEDER for financial support under Program PT2020 (project 007728-UID/QUI/04378/2013).

References

1. Das, S., et al. Recent advances in lipid nanoparticle formulations with solid matrix for oral drug delivery. *AAPS PharmSciTech*, 2011, **12**(1):62–76.

2. Plapied, L., et al. Fate of polymeric nanocarriers for oral drug delivery. *Curr Opin Colloid Interface Sci*, 2011, **16**(3):228–237.
3. Harde, H., et al. Solid lipid nanoparticles: an oral bioavailability enhancer vehicle. *Expert Opin Drug Deliv*, 2011, **8**(11):1407–1424.
4. Nunes, R., et al. Tissue-based in vitro and ex vivo models for intestinal permeability studies, in *Concepts and Models for Drug Permeability Studies*, Sarmento, B., ed. Woodhead Publishing, Cambridge, 2016, pp. 203–236.
5. Chaves, L., et al. Quality by design: discussing and assessing the solid dispersions risk. *Curr Drug Deliv*, 2014, **11**(2):253–269.
6. Pathak, K., et al. Oral bioavailability: issues and solutions via nanoformulations. *Clin Pharmacokinet*, 2015, **54**(4):325–357.
7. Narvekar, M., et al. Nanocarrier for poorly water-soluble anticancer drugs–barriers of translation and solutions. *AAPS PharmSciTech*, 2014, **15**(4):822–833.
8. Kawabata, Y., et al. Formulation design for poorly water-soluble drugs based on biopharmaceutics classification system: basic approaches and practical applications. *Int J Pharm*, 2011, **420**(1):1–10.
9. Amidon, G.L., et al. A theoretical basis for a biopharmaceutic drug classification: the correlation of in vitro drug product dissolution and in vivo bioavailability. *Pharm Res*, 1995, **12**(3):413–420.
10. FDA, Waiver of in vivo bioavailability and bioequivalence studies for immediate-release solid oral dosage forms based on a biopharmaceutics classification system: guidance for industry. *Food and Drug Administation*, Rockville, MD. http://www.fda.gov/downloads/drugs/guidancecomplianceregulatoryinformation/guidances/ucm070246, accessed on 12th June, 2015.
11. Desai, P.P., et al. Overcoming poor oral bioavailability using nanoparticle formulations: opportunities and limitations. *Drug Discovery Today*, 2012, **9**(2):e87–e95.
12. Merisko-Liversidge, E.M., et al. Drug nanoparticles: formulating poorly water-soluble compounds. *Toxicol Pathol*, 2008, **36**(1):43–48.
13. Akash, M.S.H., et al. Polymeric-based particulate systems for delivery of therapeutic proteins. *Pharm Dev Technol*, 2015, 1–12.
14. Zazo, H., et al. Current applications of nanoparticles in infectious diseases. *J Controlled Release*, 2016, **224**:86–102.
15. Singh, G., et al. Atazanavir-loaded Eudragit RL 100 nanoparticles to improve oral bioavailability: optimization and in vitro/in vivo appraisal. *Drug Deliv*, 2016, **23**(2):532–539.

16. Hashem, F.M., et al. Optimized zein nanospheres for improved oral bioavailability of atorvastatin. *Int J Nanomed*, 2015, **10**:4059–4069.

17. Morgen, M., et al. Polymeric nanoparticles for increased oral bioavailability and rapid absorption using celecoxib as a model of a low-solubility, high-permeability drug. *Pharm Res*, 2012, **29**(2):427–440.

18. Ma, Y., et al. The comparison of different daidzein-PLGA nanoparticles in increasing its oral bioavailability. *Int J Nanomed*, 2012, **7**:559–579.

19. Ravi, P.R., et al. Design, optimization and evaluation of poly-ε-caprolactone (PCL) based polymeric nanoparticles for oral delivery of lopinavir. *Drug Dev Ind Pharm*, 2015, **41**(1):131–140.

20. Singh, G., et al. Optimized PLGA nanoparticle platform for orally dosed trans-resveratrol with enhanced bioavailability potential. *Expert Opin Drug Deliv*, 2014, **11**(5):647–659.

21. Zabaleta, V., et al. Oral administration of paclitaxel with pegylated poly(anhydride) nanoparticles: permeability and pharmacokinetic study. *Eur J Pharm Biopharm*, 2012, **81**(3):514–523.

22. Moritz, M., et al. Recent developments in the application of polymeric nanoparticles as drug carriers. *Adv Clin Exp Med*, 2015, **24**(5):749–758.

23. Alshamsan, A. Nanoprecipitation is more efficient than emulsion solvent evaporation method to encapsulate cucurbitacin I in PLGA nanoparticles. *Saudi Pharm J*, 2014, **22**(3):219–222.

24. Lu, Y., et al. Polymeric micelles and alternative nanonized delivery vehicles for poorly soluble drugs. *Int J Pharm*, 2013, **453**(1):198–214.

25. Gothwal, A., et al. Polymeric micelles: recent advancements in the delivery of anticancer drugs. *Pharm Res*, 2016, **33**(1):18–39.

26. Elsabahy, M., et al. Design of polymeric nanoparticles for biomedical delivery applications. *Chem Soc Rev*, 2012, **41**(7):2545–2561.

27. Roy, U., et al. Preparation and characterization of anti-HIV nanodrug targeted to microfold cell of gut-associated lymphoid tissue. *Int J Nanomed*, 2015, **10**:5819–5835.

28. Singla, P., et al. A systematic physicochemical investigation on solubilization and in vitro release of poorly water soluble oxcarbazepine drug in pluronic micelles. *Colloids Surf, A*, 2016, **504**:479–488.

29. Tabatabaei Rezaei, S.J., et al. pH-responsive unimolecular micelles self-assembled from amphiphilic hyperbranched block copolymer for efficient intracellular release of poorly water-soluble anticancer drugs. *J Colloid Interface Sci*, 2014, **425**:27–35.

30. Dian, L., et al. Enhancing oral bioavailability of quercetin using novel soluplus polymeric micelles. *Nanoscale Res Lett*, 2014, **9**(1):1–11.
31. Luo, Y., et al. Solid lipid nanoparticles for oral drug delivery: chitosan coating improves stability, controlled delivery, mucoadhesion and cellular uptake. *Carbohydr Polym*, 2015, **122**:221–229.
32. Porter, C.J., et al. Lipids and lipid-based formulations: optimizing the oral delivery of lipophilic drugs. *Nat Rev Drug Discov*, 2007, **6**(3):231–248.
33. Narvekar, M., et al. Nanocarrier for poorly water-soluble anticancer drugs--barriers of translation and solutions. *AAPS PharmSciTech*, 2014, **15**(4):822–833.
34. Zhang, Z., et al. Daidzein-phospholipid complex loaded lipid nanocarriers improved oral absorption: in vitro characteristics and in vivo behavior in rats. *Nanoscale*, 2011, **3**(4):1780–1787.
35. Basavaraj, S., et al. Improved oral delivery of resveratrol using proliposomal formulation: investigation of various factors contributing to prolonged absorption of unmetabolized resveratrol. *Expert Opin Drug Deliv*, 2014, **11**(4):493–503.
36. Deng, J., et al. The studies of N-Octyl-N-Arginine-Chitosan coated liposome as an oral delivery system of Cyclosporine A. *J Pharm Pharmacol*, 2015, **67**(10):1363–1370.
37. Sun, C., et al. Liquid proliposomes of nimodipine drug delivery system: preparation, characterization, and pharmacokinetics. *AAPS PharmSciTech*, 2013, **14**(1):332–338.
38. Xiao, Y., et al. Sorafenib and gadolinium co-loaded liposomes for drug delivery and MRI-guided HCC treatment. *Colloids Surf B*, 2016, **141**:83–92.
39. Wu, W., et al. Oral delivery of liposomes. *Ther Deliv*, 2015, **6**(11):1239–1241.
40. Nguyen, T.X., et al. Recent advances in liposome surface modification for oral drug delivery. *Nanomedicine (Lond)*, 2016, **11**(9):1169–1185.
41. Hu, M., et al. Polymersomes via self-assembly of amphiphilic β-cyclodextrin-centered triarm star polymers for enhanced oral bioavailability of water-soluble chemotherapeutics. *Biomacromolecules*, 2016, **17**(3):1026–1039.
42. Müller, R.H., et al. 20 years of lipid nanoparticles (SLN & NLC): present state of development & industrial applications. *Curr Drug Discov Technol*, 2011, **8**(3):207–227.

43. Vieira, A.C., et al. Design and statistical modeling of mannose-decorated dapsone-containing nanoparticles as a strategy of targeting intestinal M-cells. *Int J Nanomed*, 2016, **11**:2601–2617.

44. Makwana, V., et al. Solid lipid nanoparticles (SLN) of Efavirenz as lymph targeting drug delivery system: elucidation of mechanism of uptake using chylomicron flow blocking approach. *Int J Pharm*, 2015, **495**(1):439–446.

45. Gonçalves, L.M.D., et al. Development of solid lipid nanoparticles as carriers for improving oral bioavailability of glibenclamide. *Eur J Pharm Biopharm*, 2016, **102**:41–50.

46. Aljaeid, B.M., et al. Miconazole-loaded solid lipid nanoparticles: formulation and evaluation of a novel formula with high bioavailability and antifungal activity. *Int J Nanomed*, 2016, **11**:441–447.

47. Ravi, P.R., et al. Lipid nanoparticles for oral delivery of raloxifene: optimization, stability, in vivo evaluation and uptake mechanism. *Eur J Pharm Biopharm*, 2014, **87**(1):114–124.

48. Righeschi, C., et al. Enhanced curcumin permeability by SLN formulation: the PAMPA approach. *LWT - Food Sci Technol*, 2016, **66**:475–483.

49. Baek, J.-S., et al. 2-Hydroxypropyl-β-cyclodextrin-modified SLN of paclitaxel for overcoming p-glycoprotein function in multidrug-resistant breast cancer cells. *J Pharm Pharmacol*, 2013, **65**(1):72–78.

50. Khan, S., et al. Nanostructured lipid carriers: an emerging platform for improving oral bioavailability of lipophilic drugs. *Int J Pharm Invest*, 2015, **5**(4):182.

51. Beloqui, A., et al. Budesonide-loaded nanostructured lipid carriers reduce inflammation in murine DSS-induced colitis. *Int J Pharm*, 2013, **454**(2):775–783.

52. Neves, A.R., et al. Novel resveratrol nanodelivery systems based on lipid nanoparticles to enhance its oral bioavailability. *Int J Nanomed*, 2013, **8**:177–187.

53. Zhuang, C.-Y., et al. Preparation and characterization of vinpocetine loaded nanostructured lipid carriers (NLC) for improved oral bioavailability. *Int J Pharm*, 2010, **394**(1–2):179–185.

54. Beloqui, A., et al. A comparative study of curcumin-loaded lipid-based nanocarriers in the treatment of inflammatory bowel disease. *Colloids Surf B*, 2016, **143**:327–335.

55. Beloqui, A., et al. Mechanism of transport of saquinavir-loaded nanostructured lipid carriers across the intestinal barrier. *J Controlled Release*, 2013, **166**(2):115–123.
56. Bannunah, A.M., et al. Mechanisms of nanoparticle internalization and transport across an intestinal epithelial cell model: effect of size and surface charge. *Mol Pharm*, 2014, **11**(12):4363–4373.
57. des Rieux, A., et al. Targeted nanoparticles with novel non-peptidic ligands for oral delivery. *Adv Drug Deliv Rev*, 2013, **65**(6):833–844.
58. Lennernas, H. Does fluid flow across the intestinal mucosa affect quantitative oral drug absorption? Is it time for a reevaluation? *Pharm Res*, 1995, **12**(11):1573–1582.
59. Karlsson, J., et al. Paracellular drug transport across intestinal epithelia: influence of charge and induced water flux. *Eur J Pharm Sci*, 1999, **9**(1):47–56.
60. Avdeef, A. Physicochemical profiling (solubility, permeability and charge state). *Curr Top Med Chem*, 2001, **1**(4):277–351.
61. Sugano, K., et al. Coexistence of passive and carrier-mediated processes in drug transport. *Nat Rev Drug Discov*, 2010, **9**(8):597–614.
62. Kreuter, J. Peroral administration of nanoparticles. *Adv Drug Deliv Rev*, 1991, **7**:71–86.
63. Sanjula, B. Effect of poloxamer 188 on lymphatic uptake of carvedilol-loaded solid lipid nanoparticles for bioavailability enhancement. *J Drug Target*, 2009, **17**:249–256.
64. Bargoni, A., et al. Solid lipid nanoparticles in lymph and plasma after duodenal administration to rats. *Pharm Res*, 1998, **15**:745–750.
65. Elsabahy, M., et al. Design of polymeric nanoparticles for biomedical delivery applications. *Chem Soc Rev*, 2012, **41**(7):2545–2561.
66. Varma, V.M., et al. Targeting intestinal transporters for optimizing oral drug absorption. *Curr Drug Metab*, 2010, **11**(9):730–742.
67. DeSesso, J.M., et al. Anatomical and physiological parameters affecting gastrointestinal absorption in humans and rats. *Food Chem Toxicol*, 2001, **39**(3):209–228.
68. Bannunah, A.M., et al. Mechanisms of nanoparticle internalization and transport across an intestinal epithelial cell model: effect of size and surface charge. *Mol Pharm*, 2014, **11**(12):4363–4373.
69. des Rieux, A., et al. Targeted nanoparticles with novel non-peptidic ligands for oral delivery. *Adv Drug Deliv Rev*, 2013, **65**(6):833–844.

70. Nangia, S., et al. Effect of nanoparticle charge and shape anisotropy on translocation through cell membranes. *Langmuir*, 2012, **28**(51):17666–17671.

71. Göppert, T.M., et al. Adsorption kinetics of plasma proteins on solid lipid nanoparticles for drug targeting. *Int J Pharm*, 2005, **302**:172–186.

72. Hombach, J., et al. Mucoadhesive drug delivery systems. *Handb Exp Pharmacol*, 2010, **197**:251–266.

73. Chen, D., et al. Comparative study of Pluronic F127-modified liposomes and chitosan-modified liposomes for mucus penetration and oral absorption of cyclosporine A in rats. *Int J Pharm*, 2013, **449**:1–9.

74. Knop, K., et al. Poly(ethylene glycol) in drug delivery: pros and cons as well as potential alternatives. *Angew Chem Int Ed Engl*, 2010, **49**:6288–6308.

75. Feeney, O.M., et al. 'Stealth' lipid-based formulations: poly(ethylene glycol)-mediated digestion inhibition improves oral bioavailability of a model poorly water soluble drug. *J Controlled Release*, 2014, **192**:219–227.

76. Zhang, X., et al. Ligand-mediated active targeting for enhanced oral absorption. *Drug Discov Today*, 2014, **19**(7):898–904.

77. Panyam, J., et al. Solid-state solubility influences encapsulation and release of hydrophobic drugs from PLGA/PLA nanoparticles. *J Pharm Sci*, 2004, **93**(7):1804–1814.

78. Kathe, N., et al. Physicochemical characterization techniques for solid lipid nanoparticles: principles and limitations. *Drug Dev Ind Pharm*, 2014, **40**(12):1565–1575.

79. Buckley, S.T., et al. Biopharmaceutical classification of poorly soluble drugs with respect to "enabling formulations. *Eur J Pharm Sci*, 2013, **50**:8–16.

80. Abouelmagd, S.A., et al. Release kinetics study of poorly water-soluble drugs from nanoparticles: are we doing it right? *Mol Pharm*, 2015, **12**:997–1003.

81. Antunes, F., et al. Models to predict intestinal absorption of therapeutic peptides and proteins. *Curr Drug Metab*, 2013, **14**(1):4–20.

82. Gamboa, J.M., et al. In vitro and in vivo models for the study of oral delivery of nanoparticles. *Adv Drug Deliv Rev*, 2013, **65**(6):800–810.

83. Béduneau, A., et al. A tunable Caco-2/HT29-MTX co-culture model mimicking variable permeabilities of the human intestine obtained by an original seeding procedure. *Eur J Pharm Biopharm*, 2014, **87**(2):290–298.

84. Lasa-Saracíbar, B., et al. In vitro intestinal co-culture cell model to evaluate intestinal absorption of edelfosine lipid nanoparticles. *Curr Top Med Chem*, 2014, **14**(9):1124–1132.

85. Fischer, S.M., et al. In-vitro permeability of poorly water soluble drugs in the phospholipid vesicle-based permeation assay: the influence of nonionic surfactants. *J Pharm Pharmacol*, 2011, **63**(8):1022–1030.

Chapter 3

Synthesis and Applications of Amphiphilic Chitosan Derivatives for Drug Delivery Applications

Daniella Silva,[a] Andreia Almeida,[b,c,d] Cláudia Azevedo,[b,c] Sérgio P. Campana-Filho,[a] and Bruno Sarmento[b,c,e]

[a]*São Carlos Institute of Chemistry, University of São Paulo, Brazil*

[b]*I3S - Institute of Research and Innovation in Health, University of Porto, Portugal*

[c]*INEB - Institute of Biomedical Engineering, University of Porto, Portugal*

[d]*Faculty of Engineering, University of Porto, Portugal*

[e]*IIFACTS - Institute for Research and Advanced Training in Health Sciences and Technologies, Gandra, Portugal*

bruno.sarmento@ineb.up.pt

Nowadays, it is well established that the increased interest in chitosan is mainly due to its biological activities, including biocompatibility, biodegradability, adhesiveness, nontoxicity, nonimmunogenicity, and antibacterial and antifungal bioactivity. Additionally, the ability of chitosan to adhere to mucosal surfaces and temporarily open the tight junctions between epithelial cells must be highlighted as a property of great interest to pharmaceutical applications. Moreover, the ease of carrying out chemical modifications on hydroxyl and

Nanoparticles in Life Sciences and Biomedicine
Edited by Ana Rute Neves and Salette Reis

ISBN 978-981-4745-98-7 (Hardcover), 978-1-351-20735-5 (eBook)
www.panstanford.com

amino groups present in chitosan chains allows one to insert new functional groups to improve or change a given physicochemical property and/or the polymer biological activities. Particularly, the addition of hydrophobic and hydrophilic groups allows the preparation of chitosan amphiphilic derivatives that are able to self-assemble and form stable micelles. Knowledge of the reaction routes can facilitate the preparation of chitosan derivatives with characteristics adequate for fulfilling specific needs for a given application, such as the encapsulation and controlled release of drugs. This section revises the literature on the regioselective synthesis of chitosan derivatives, focusing on the addition of hydrophobic groups to the chitosan chain through alkylation and acylation reactions, aiming to tailor the polymer solubility. As a viable and cost-effective strategy, amphiphilic derivatives with self-assembly properties has been widely used to prepare chitosan nanocarriers that have been investigated as vehicles for the delivery of drug, such as proteins and peptides, anti-inflammatory drugs, and anticancer drugs, as is shown here.

3.1 Introduction

Over the past years, many efforts have been made to develop effective drug delivery systems. Despite the well-established effectiveness of these systems, by the time a drug reaches the target, the drug may lose part of its therapeutic effectiveness and, moreover, side effects can result [1]. Additionally, the administration of hydrophobic drugs is a main challenge due to their low solubility in an aqueous environment [2]. Thus, the demand for a biomaterial able to encapsulate such drugs is a major focus of research currently.

Chitosan, a linear polysaccharide generally prepared from chitin, one of the most abundant polymers in nature [3], exhibits excellent properties, such as biocompatibility, biodegradability, nontoxicity, and mucoadhesiveness [4], which enable its applications in pharmacy, medicine, and tissue engineering. However, the solubility of chitosan, limited to moderately acidic aqueous media (pH 4–5), precludes several potential applications and chemical modifications must be carried out to improve the polymer's solubility as well as to change its hydrophobicity/hydrophilicity balance, enabling it to form micelles and nanoparticles (NPs). Recently, the formation of polymer-based micelles or NPs from chitosan derivatives has been a strong focus to allow the encapsulation and controlled delivery of drugs, not only because they exceed the drug solubility barrier, improve its therapeutic efficacy, and reduce side effects, but also due to increased circulation time as well as drug accumulation preferentially in the targeted tissue or organ [2].

The strategy of carrying out chemical modifications in chitosan with the aim to change or improve given physicochemical properties and biological activities is advantageous as it is possible to modify selectively the chitosan functional groups without any effects on its main structural backbone [5]. Moreover, a wide range of derivatives can be prepared with modified properties, which can be tailored according to the needs of specific applications. Indeed, chitosan derivatives can exhibit improved or novel properties and functionalities depending on the *locus* of as well as on the nature of the chemical modification to be explored.

On the basis of these considerations, this section focuses on chemical modifications of chitosan, highlighting those that allow the preparation of amphiphilic derivatives that are intended to encapsulate therapeutic agents for biomedical and biopharmaceutical applications.

3.2 Structure and Characterization of Chitosan

Chitosan, a linear polysaccharide that occurs in the cell wall of some fungi, is generally prepared from chitin, one of the most abundant natural polymers in the world [3]. Chitin is composed mainly of 2-acetamido-2-deoxy-D-glucopyranose (GlcNAc) units linked by β-(1,4) glycosidic bonds, while chitosan is a copolymer in which 2-amino-2-deoxy-D-glucopyranose (GlcN) units predominate (Fig. 3.1). Thus, these polymers possess different contents of GlcNAc and GlcN units, this structural difference resulting in distinct physicochemical properties. Chitin is insoluble in water as well as in most of common organic solvents, while chitosan is soluble in moderately acidic aqueous media. Indeed, the solubility of chitosan in such aqueous media is directly related to the content of GlcN units as the protonation of its amino groups generates positive charges and it confers water solubility to the polymer if the charge density is sufficiently high. Additionally, the content of GlcN units and the charge density as well strongly affect the conformation of chitosan in solution. The higher the GlcN content, the higher is the intrinsic viscosity of the polymer. The structural distinction between chitin and chitosan is expressed by the content of GlcNAc units, which is quantitatively given by the average degree of acetylation (DA), higher in chitin (DA > 80%) as compared to chitosan (DA < 60%). Thus, the conversion of chitin into chitosan is referred to as an *N*-deacetylation reaction since acetamido groups of GlcNAc units are converted into amino groups, increasing the content of GlcN units in the polymeric chains. The *N*-deacetylation of chitin can be achieved by enzymatic hydrolysis, but it is most frequently carried out by treating the polymer with concentrated aqueous alkaline solutions [6, 7]. As a consequence of their different structures and although sharing some structural similarity as both polymers, chitin and chitosan, possess primary (C3) and secondary (C6) hydroxyl groups, only chitosan possesses primary amino groups (C2). Thus, due the presence of primary amino groups pertaining to GlcN units, chitosan offers additional possibilities for carrying out chemical modifications as compared to chitin and other polysaccharides.

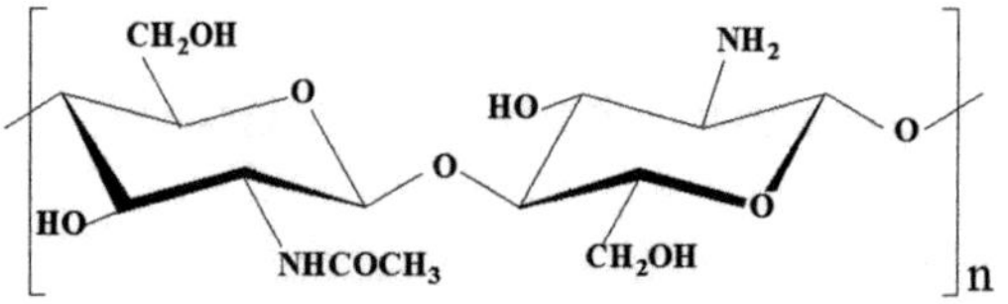

Figure 3.1 Chitosan structure, where *n* is the average degree of polymerization.

Chitosan is a polymer of great interest in pharmacy, medicine, dentistry, and tissue engineering due to its biological activities such as nontoxicity [8–11], antimicrobial activity [12–14], biocompatibility [15–17], biodegradability [18–20], mucoadhesiveness, and ability to open, temporarily, the tight junctions of the epithelium, thereby facilitating the permeation of drugs [4, 21–24]. These properties may double the therapeutic effects, making chitosan a polymer with pronounced potential for biomedical and biopharmaceutical applications, especially for drug delivery.

Although chitosan presents interesting properties, its solubility is limited to moderately acidic aqueous media, which limits some applications. However, the control of the structural characteristics of chitosan, mainly its average DA; the distribution of GlcNAc and GlcN units, which is expressed by the pattern of acetylation (PA); and also its average molecular weight and dispersity, can generate chitosans with excellent properties, including solubility in physiological media [25–29]. Moreover, chemical modifications are carried out to prepare derivatives of chitosan with improved properties, including solubility, aiming to expand its application possibilities. As a result, various derivatives of chitosan have been proposed.

Due to the presence of hydroxyl and amine groups, this polymer can be derivatized by reactions of alkylation [30–33], carboxymethylation [34–37], quaternization [30, 34–42], Schiff base [43, 44], and acylation [6, 7, 45–51]. Thus, the aim of this section is to revisit some methods to obtain chitosan derivatives with hydrophobic groups, the focus being on the preparation of amphiphilic derivatives of chitosan potentially applicable in drug delivery systems.

3.3 Chitosan Amphiphilic Derivatives

A large number of amphiphilic chitosan derivatives have been reported in the literature over the past years. Generally, the chemical derivatization of chitosan can be regioselective due to the presence of reactive sites C2-amine, C3-hydroxyl, and C6-hydroxyl. The properties of products of the regioselective reactions are strongly influenced by the distribution of substituent groups along the polymer chain and, therefore, it is possible to obtain *N*-substituted, *O*-substituted, or *N,O*-substituted chitosan derivatives. However, such substitutions may be controlled over the reaction conditions applied [31, 52], as is shown in this review.

3.3.1 Alkylation

Alkylation is one of the most frequent modifications of chitosan. Alkylation reactions occur through the grafting of alkyl chains in the structure of the chitosan. In other words, alkyl chitosan is obtained by the introduction of alkyl groups on the amine groups of chitosan by reductive amination of chitosan [53]. Generally, this substitution reaction is carried out in heterogeneous conditions and owing to the semicrystalline character of chitosan, amorphous regions will be more accessible than ordered ones, giving rise to substituted products displaying block distribution patterns.

The hydrophobic character of alkylated chitosan is dependent on the length of the chains, that is, the longer the alkyl chains, the higher is the polymer hydrophilicity [54]. Additionally, the higher the average degree of substitution (DS), the more hydrophobic will be the modified chitosan.

Desbrières et al. [53] developed a procedure to obtain alkylated derivatives of chitosan from its swollen structure and modify it after precipitation by neutralization, which improves the accessibility to the reactive sites. Generally, aldehydes and ketones are employed as alkylating agents and the Schiff bases resulting from the reaction with chitosan are converted to *N*-alkyl chitosan derivatives by reduction with sodium borohydride ($NaBH_4$) or sodium cyanoborohydride ($NaBH_3CN$) (Fig. 3.2), the reaction efficiency depending on the choice of the reducing agent [52]. Indeed, $NaBH_3CN$ is commonly used because it is more reactive and selective, but it is highly toxic [52].

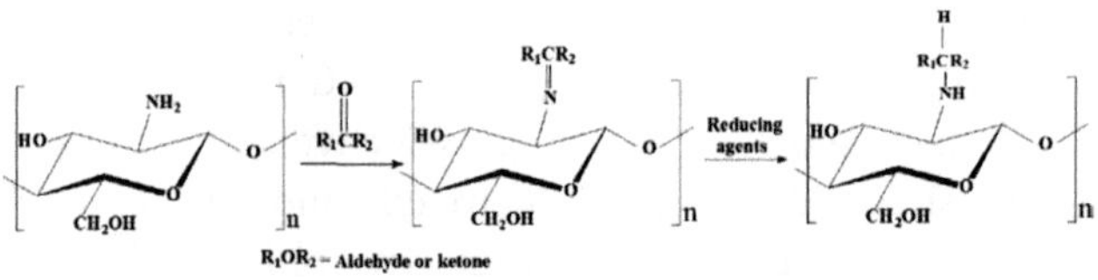

Figure 3.2 Reaction scheme to obtain *N*-alkyl derivatives of chitosan.

Chitosan modifications can dramatically alter its properties. Therefore, it is necessary to characterize the properties and medical safety of these new derivatives [55]. Indeed, the presence of hydrophobic interactions between the alkyl chains improves the properties of modified chitosans: the C12 is the minimum length of the alkyl chain for the hydrophobic behavior to be effective; the solutions of such chitosan derivatives are usually non-Newtonian, mainly when the length of alkyl chains increases; also the C12 alkyl chain length proved to be more efficient to improve film formation and mechanical properties; the efficiency of this modification increases with increasing DS [54].

Apart from their properties, alkyl chitosan derivatives have been widely used in drug delivery. In Section 3.5, examples are given of this modification.

3.3.2 Acylation

The addition of hydrophobic groups is also carried out by acylation reactions. Acylated chitosan was proposed as a drug carrier for drug delivery, essentially due to its hydrophobic association, which promotes self-assembly of NPs or micelles. The advantage of the acylation of chitosan over its alkylation is because acylation allows the introduction of new groups at sites C2-amine, C3-hydroxyl, and C6-hydroxyl [52]. In the case of *O*-acylation, the presence of the ester bond in the structure of the resulting chitosan derivative permits its degradation by the action of lipase enzymes, making it a biodegradable polymer [52].

The acylation of chitosan makes possible its solubilization in organic solvents due to the introduction of hydrophobic groups in the polymer. However, the solubility of acylated chitosan in water depends on the acyl chain length and on the DS [5, 52]. This means that short acyl chains and a relatively low DS result in water-soluble chitosan derivatives but increasing the DS will dramatically decrease the water solubility.

3.3.2.1 N-acylation reaction

The *N*-acylation of chitosan may be achieved by activation of carboxylic acids through reaction with carbodiimides as well as by using reactive carboxylic acid derivatives, such as anhydrides and acyl chlorides.

Carbodiimide reactions are carried out by activation of carboxylic acids in the presence of 1-ethyl-3-(3-dimethylaminopropyl) carbodiimide (EDC). This reaction gives rise to the intermediate *O*-acylisourea, which is capable of undergoing nucleophilic attack of the primary amine groups, such as those present in GlcN units of chitosan [52]. The primary amine connects to the carboxyl group through an amide bond and releases the by-product urea derivative, isourea. The reaction scheme is shown in Fig. 3.3.

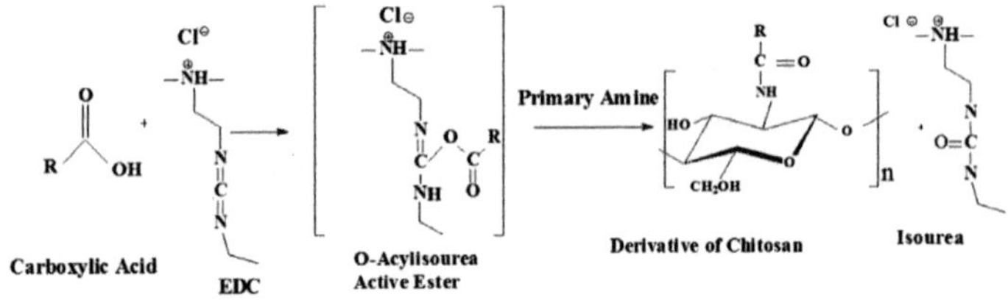

Figure 3.3 Acylation modification by carbodiimide reaction.

In carbodiimide reactions *N*-hydroxisuccimide (NHS) or its water-soluble sulfo-NHS has been commonly used with the EDC. Adding this, the efficiency of the reaction is improved through the formation of an amine-reactive intermediate that is more stable than the *O*-acylisourea. The reaction scheme addition of NHS is shown in Fig. 3.4.

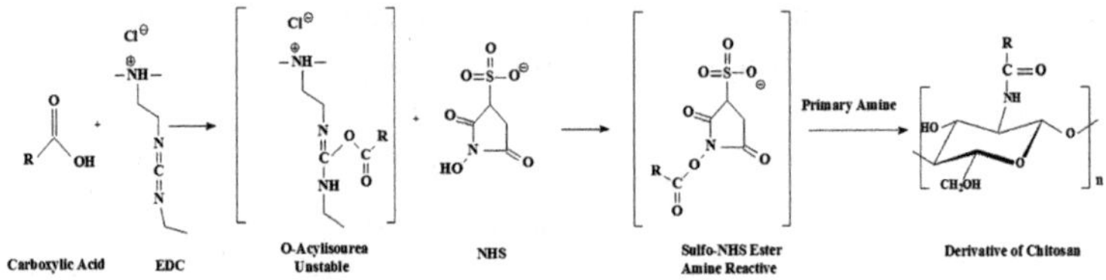

Figure 3.4 Acylation synthesis from the use of EDC and NHS.

Lee et al. [56] carried out the chitosan modification with EDC, and it is possible to verify that with an increase in the amount of deoxycholic acid and EDC there is an increase of the DS. However, the DS ranges from 2.8 to 5.1 per 100 anhydroglucosamine units of chitosan, indicating that the limited solubility of deoxycholic acid in the reaction medium affected the achievement of derivatives with a higher DS. The same was observed in the reaction with linoleic acid, resulting in a DS of only 1.8% [57].

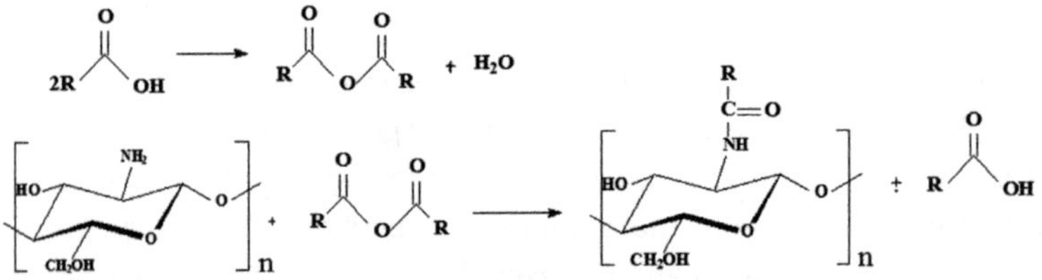

Figure 3.5 *N*-acyl chitosan reaction by anhydrides.

Other methodology for obtaining acylated chitosan is by reaction with anhydrides, which results in selective *N*-substitution. This reaction can be carried out in homogeneous [58] or heterogeneous [59] conditions. The reaction scheme is shown in Fig. 3.5.

3.3.2.2 O-Acylation reaction

The reactions presented above aim to produce the *N*-acylation of chitosan. However, *O*-acylation reactions may also occur, as many of these reactions are not fully selective for amine groups. Thus, the preparation of derivatives with well-defined structures can be achieved through the control of reaction conditions, allowing the preparation of advanced functional materials from chitosan.

Since amine groups are stronger nucleophiles as compared to primary and secondary hydroxyl groups, the preparation of *N*-phthaloyl chitosan derivative is a promising advance as besides being a regioselective reaction, it is easily carried out under mild conditions. The phthaloyl group is commonly used as a protector of amine groups and after protection *O*-substitution reactions are carried out. Finally, the phthaloyl group is removed by reaction with hydrazine, regenerating the amine groups [60]. Figure 3.6 shows the phthaloyl protecting group and reaction of *O*-acylation.

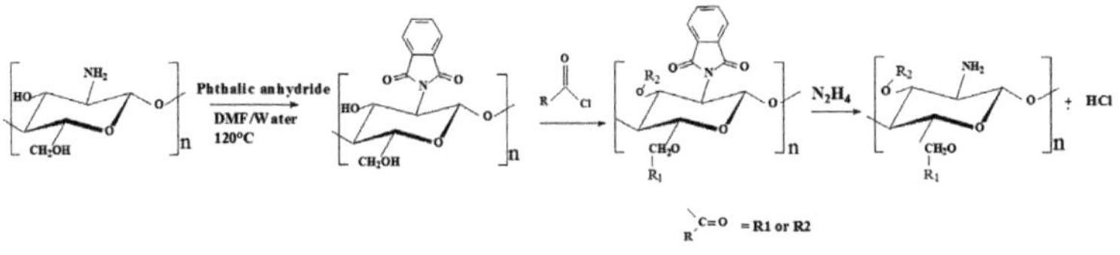

Figure 3.6 Reaction of protection through the phthaloyl group.

In spite of the reactions with phthaloyl being effective, this approach needs several steps, including the protection and deprotection of the amine groups, favoring the occurrence of side reactions as well as a low reaction yield. Therefore, a new method was developed using methanesulfonic acid (CH_3SO_3H) to protect the amine groups, allowing the occurrence of selective *O*-acylation by reaction with acyl chlorides [52]. As a strong acid, methanesulfonic acid is able to protonate the amine groups through the formation of salts, resulting in a substitution preferentially in the hydroxyl groups of chitosan, as shown in Fig. 3.7.

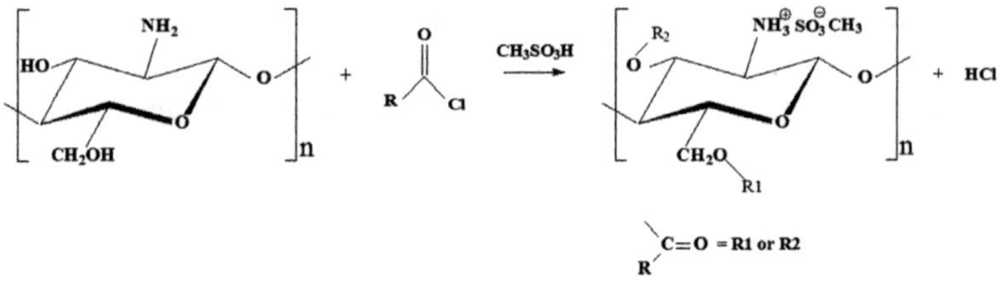

Figure 3.7 Reaction of *O*-substitution through the protection of amine groups with CH_3SO_3H.

3.3.2.3 Other chemical modifications

Besides the modifications previously referred to, there are many other modifications that can be made in the chitosan backbone. Among them, hydroxyalkylation and carboxyalkylation are common modifications used to prepare chitosan derivatives to be used in drug delivery systems because they improve chitosan water solubility.

Hydroxyalkyl chitosan is prepared by reacting chitosan with epoxides such ethylene oxide or with glycidol. Depending on the epoxide and reaction conditions, mainly temperature and pH, the reaction may occur at the amine or hydroxyl groups to yield *N*-hydroxyalkyl, *O*-hydroxyalkyl chitosan, or a mixture of both [5, 52]. The ratio of *O*-/*N*-substitution depends on the choice of catalyst (NaOH or HCl) and the reaction conditions [61].

On the other hand, carboxyalkyl chitosan is obtained by the introduction of acidic groups in the chitosan backbone. It is possible to produce *N*- and *O*-carboxyalkyl chitosan using different reaction conditions [52]. Through the introduction of carboxyl groups into the amine groups of chitosan it is possible to obtain a chitosan derivative with pH-dependent behavior due to charge densities in the molecular chain [5].

3.4 Chitosan-Based Nanocarriers as Drug Delivery Systems

Chemical modifications of chitosan are the most efficient and well-known method to create self-assembled systems. These changes lead to the formation of amphiphilic chitosan derivatives by the addition of hydrophobic groups, such as alkyl and acyl groups, to the chitosan backbone, as we noted previously.

Owing to the introduction of hydrophobic substituents in the chitosan chain, the resulting amphiphilic chitosan derivative is capable of forming micelles [62–64] or NPs [65–67] by self-assembly in an aqueous medium. These target drug delivery systems are developed to prevent drug degradation, as well as their loss, to avoid the harmful side effects and to increase the BA of the drug at the required site of interest.

3.4.1 Polymer-Based Micelles

Molecular chains of amphiphilic chitosan derivatives, when in contact with aqueous solvents, tend to rearrange through micelle formation or micelle-like aggregates via intra- or intermolecular association between hydrophobic moieties [52].

Micellar systems consist in molecules able to self-organize as a consequence of specific interactions between the constituents themselves. Polymer-based micelles show potential due to the possibility of solubilization in an aqueous environment. Also, polymeric micelles are able to reduce the side effects of encapsulated drugs, being protected from environmental stimuli such as enzymes and gastric pH [68–70]. Likewise, these systems have high loading capacity, improved stability in the bloodstream, therapeutic potential, and longevity [69].

For being an amphiphilic system, the hydrophobic moiety is protected from the aqueous medium, and therefore the hydrophilic part is in contact therewith (see Fig. 3.8). The fact that they are organized into core-shell-type structures makes them the choice for drug delivery applications. The stability of these processes is because of the decrease of Gibbs free energy, which makes these systems thermodynamically more stable; this process is called self-assembly [49, 71]. The hydrophobic core provides space for the encapsulation of hydrophobic drugs, such as paclitaxel or doxorubicin (DOX), and the hydrophilic shell allows the inclusion of hydrophilic drugs and also provides stability, protecting the drugs from the biological environment [69].

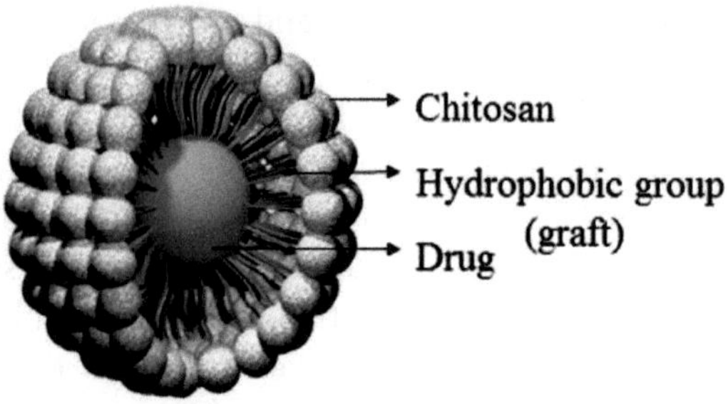

Figure 3.8 Schematic representation of drug encapsulation by self-assembled nanosystems by chemically modified chitosan.

It is already described that polymer-based micelles have high encapsulation efficiency despite the reduced size of the micelles (5–100 nm) [72]. Due to their uniformity, the time of drug circulation in the blood is higher [70, 73], providing better permeability to anticancer drugs by improving their delivery from the deep blood vessels to tumors [74]. Additionally, these carriers are typically accumulated at extensive vasculature sites, such as solid tumors and sites of inflammation, due to enhanced permeability and retention (EPR) effect [70, 73].

There are two concepts that have to be explained about micelles: the first is the critical micelle concentration (CMC), which is the minimum concentration required for a given polymer to form micelles by self-assembly. The second is the critical aggregation concentration (CAC), which is the minimum concentration required for a given polymer to aggregate. Typically, the CMC values are higher than the CAC values [52]. The CMC of the polymer-based micelles depends on the hydrophobic character of the molecule [72] and the molecular weight of the blocks [68]. There are several methods for calculating the CMC, such as by conductivity measurements, interfacial tension, and fluorescence spectroscopy [69]. However, in the case of polymer-based micelles, the CMC is very low, fluorescence spectroscopy being the best choice because of its high sensitivity [69].

3.4.2 Polymer-Based Nanoparticles

As micelles, NPs, especially those based on amphiphilic chitosan derivatives, have been extensively used in the drug delivery field.

These NPs can be defined as spherical biodegradable and nanosized colloidal systems [75]. Typically they consist of an amphiphilic diblock of copolymers (an hydrophobic block and an hydrophilic one) that can self-assemble into NPs in an aqueous solution [76]. Similarly, the hydrophobic block is used as a space for carrying poorly water soluble drugs, promoting the solubilization of hydrophobic drug and avoiding degradation, while the hydrophilic block allows recognition by the immune system, prolonging drug circulation time in the body as well as preventing drug interactions with the biological environment [66]. When the drugs are encapsulated or physically retained within a polymeric matrix they are termed "nanospheres" and when they are trapped within a cavity surrounded by a polymeric membrane, they are termed "nanocapsules" [77].

Polymer-based NPs are claimed to have many advantages too, such as reduced toxicity, higher uptake by cancer cells, more homogeneous size distribution, higher loading capacity for poorly water soluble drugs, good stability, and sustained release of drugs by diffusion through the polymer matrix or by erosion and degradation of particles [76]. Also, after the administration of the NPs, they preferentially accumulate in solid tumors due to the EPR effect, which is due to leaky tumor vessels and the lack of effective lymphatic drainage system [78, 79].

The generation of NPs includes natural or synthetic biocompatible and biodegradable polymers. However, this section will focus only on NPs obtained from amphiphilic chitosan derivatives, some examples being described in the following section.

3.5 Applications of Chitosan Amphiphilic Derivatives in Drug Delivery

Currently, new chemical formulations for drugs are being developed, but the low solubility in aqueous media and unacceptable levels of toxicity for drugs or excipients in the formulation have been a limitation. The first formulations developed had in their composition solubilizing agents [80]. However, they exhibited secondary effects. Thereby, the advance of studies in this area led to the development of new types of formulations, such as systems based in NPs and micelles, with the aim of avoiding the use of such solubilizing agents [81].

Polymer-based micelles and NPs produced from chitosan derivatives have been considered as promising carriers for bioactive molecules. Accordingly, some applications are presented below for these amphiphilic derivatives of chitosan in drug delivery systems.

3.5.1 Anti-Inflammatory Drugs

The main purpose of the specific application of anti-inflammatory drugs is to achieve the desired pharmacologic response at a selected site without undesirable interactions. Therefore, amphiphilic derivatives of chitosan are widely used.

Meloxicam (MX) was loaded into polymer-based micelles from amphiphilic chitosan derivatives, *N*-naphthyl-*N*-*O*-succinyl chitosan (NSCS), *N*-octyl-*N*-*O*-succinyl chitosan (OSCS), and *N*-benzyl-*N*-*O*-succinyl (BSCS) chitosan, obtained by reduction with sodium borohydride ($NaBH_4$) [82]. Among the hydrophobic groups, the *N*-octyl substituent resulted in the chitosan derivative presenting the highest loading efficiency, great stability, and the lowest MX release in simulated gastric fluid medium as compared to the other chitosan derivatives [82]. Besides that, all polymer-based micelles produced from the above-mentioned amphiphilic derivatives of chitosan were spherical in shape and had low toxicity against Caco-2 cells. Additionally, the results indicated that all polymer-based micelles can incorporate hydrophobic drugs, especially MX, with high efficiency and good stability, particularly OSCS-based micelles [82].

As another example, *N*-palmitoyl chitosan, prepared by reacting chitosan with palmitoyl anhydride and presenting the DS in the range of 1.2%–14.2%, formed stable micelles in an aqueous solution. Such a micellar system was loaded with ibuprofen, and the results indicated that the loading capacity was approximately 10% [63]. In addition, the results of the CAC indicated that these micelles can remain stable in diluted conditions, the CAC values decreasing with increasing DS. The results showed that the drug release strongly depended on the pH and temperature [63]. In other words, at low pH and a high temperature the drug release was faster. Nevertheless, the derivative with a lower DS showed a particle size above 200 nm [63]. This method of preparation of water-soluble *N*-palmitoyl chitosan used to encapsulate ibuprofen was successful and contribute to the development of a new drug delivery system.

In a study carried out by Wu et al. [83], a new chitosan amphiphilic derivative was prepared, namely deoxycholic acid–phosphorylcholine chitosan (DCA-PCCS), which was used to encapsulate quercetin as a hydrophobic model drug. To obtain this derivative, a combination of the Atherton–Todd reaction for coupling phosphorylcholine and a carbodiimide coupling reaction for linking deoxycholic acid to the amino groups of chitosan was necessary. The resultant DCA-PCCS micelles had spherical shapes, an average diameter of 281.7 ± 4.2 nm, and a zeta potential of around 5.67 ± 0.55 mV. The CMC of these micelles was 1.77×10^{-2} mg/mL, presenting low cytotoxicity and good hemocompatibility [83]. Nevertheless, the drug loading and release experiments for quercetin showed the micelles to be a potential nanocarrier for the delivery and release of hydrophobic drugs, avoiding unfavorable biological responses.

Marques et al. [64] used a nonsteroidal drug as an antitumor agent to combat breast cancer through micellar nanocarriers composed of two different amphiphilic derivatives of chitosan, namely CS-DOCA and CS-Leu, formed through the inclusion of deoxycholic acid and leucine, respectively, in the chitosan backbone. They used ibuprofen to test the anticancer activity in breast cancer cells, wherein the first amphiphilic derivative (CS-DOCA) is more substituted than CS-Leu. The results showed spherical micelles with a narrow size range between 171 and 173 nm, CMC of 0.101 mg/mL for CS-DOCA and 0.065 mg/mL for CS-Leu, high efficiency of encapsulation of ibuprofen for both micellar systems, and the internalization of micelles by tumor cells, confirmed by confocal microscopy [64]. This study achieved a reduction of viability in breast cancer cells by around 13% for a concentration of copolymer of 200 μg/mL for CS-DOCA [64], while the other amphiphilic chitosan derivative was not so effective (Table 3.1). However, such results show these systems as promising options to treat breast cancer. Additionally, the versatility of the systems can be achieved for other types of carriers for drug delivery.

Table 3.1 Overview of amphiphilic chitosan derivatives used to encapsulate anti-inflammatory drugs

Chitosan derivative	Modification	System	Active agent	Ref.
NSCS, OSCS, BSCS	Alkylation	Micelles	Meloxicam	[82]
N-palmitoyl chitosan	Acylation	Micelles	Ibuprofen	[63]
DCA-PCCS	Acylation	Micelles	Quercetin	[83]
CS-DOCA, CS-Leu	Acylation	Micelles	Ibuprofen	[64]

3.5.2 Anti-Cancer Drugs

Amphiphilic chitosan derivatives have already proven to be excellent systems for drug delivery when loaded with anticancer agents such as paclitaxel, camptothecin, and DOX (Table 3.2).

The derivative *N*-succinyl-*N*-octyl chitosan (SOC) was synthesized by *N*-substitution reaction in the presence of sodium borohydride. The three derivatives obtained had DS values ranging between 28.6% and 52.5%, and CMC values were 3.1×10^{-5}, 2.4×10^{-5}, and 5.9×10^{-6} g/mL [84]. The CMC decreases with an increasing DS of the octyl content. The encapsulation of DOX was achieved by dialysis and was affected by the amount of octyl groups, increasing when DS increased [84]. Additionally, the average size of these micelles was dependent on the amount of the octyl chains, ranging between 100 and 200 nm. These results showed a sustained release pattern, also dependent on the amount of the octyl chains and drug loading content: the higher the drug content, the slower the drug release [84]. Additionally, these polymer-based micelles exhibit more toxicity than the free drug, which may make them useful as a carrier of DOX in anticancer treatment.

Zhang et al. [85] developed *N*-octyl-*N*-trimethyl chitosan derivatives (OTMCS) by introducing the octyl group in the chitosan backbone followed by *N*-methylation, resulting in a derivative possessing an average DS of 8% and 54% for octyl and trimethyl substituents, respectively [85]. This new derivative, OTMCS, can form micelles by self-assembly and is capable of encapsulate hydrophobic drugs such as 10-hydroxycamptothecin, enhancing their solubility in aqueous media. The results showed an average particle size ranging between 24 and 278 nm and drug loading content between 4.1% and 32%, depending of the DS [85]. Additionally, the results showed a sustained-release behavior of the drug and good stability of the system. The conclusion is that these micelles could be promising carriers for hydrophobic drugs, improving their solubility, stability, and release.

In another example from the same authors, *N*-octyl-*O*-sulfate chitosan (NOCS) was prepared via reductive amination. NOCS micelles have been used to study the release of paclitaxel. The loading of paclitaxel and formation of micelles occur via dialysis using ethanol and water as the solvent [86]. The solvents used to dissolve both, the polymer and the drug, and the feed weight ratio of paclitaxel to the polymer significantly affected the entrapment efficiency, ranging between 37.6% and 59.1% [86]. However, the highest entrapment efficiency was observed when the feed weight ratio of paclitaxel to the polymer was 1:1.5–1:1.7 [86]. The paclitaxel-loaded micelles were around 250 nm in size and spherical in shape, without aggregation, and with high paclitaxel loading, of around 25% [86], which still makes NOCS a promising nanocarrier for hydrophobic drugs. Therefore, the authors used these same micelles for in vivo studies [87]. The biodistribution study revealed that drug retention was higher in the liver, followed by kidney and lung 15 min. after administration. However, after 8 h, the concentration of paclitaxel became higher in the spleen, liver, lung, and kidney [87]. Still, the brain and spinal cord were exposed to low concentrations of paclitaxel. The antitumor efficacy in different animal tumor models of paclitaxel-loaded micelles seems to be more efficient in inhibition of tumor growth at certain concentrations than when Taxol® is used. However, at a concentration of 10 mg/kg both have similar behaviors. Regarding toxicity, only Taxol® shows toxic effects [87]. These results indicate that the use of NOCS loaded with paclitaxel is more beneficial than the conventional system containing Cremophor EL, overcoming the limitations and providing the necessary therapeutic efficacy.

Galactosylated *O*-carboxymethyl chitosan-graft-stearic acid (Gal-OCMCS-g-SA) was synthesized by coupling the carboxyl group of stearic acid (SA) with the amine group of *O*-carboxymethyl chitosan (OCMCS) in the presence of EDC and NHS [66]. This derivative was synthesized for liver-targeting delivery of DOX and has a DS of stearic groups ranging from 4.6% to 12.6% and galactose DS was 13.1%. Also, the particle size and polydispersity index decreased with an increasing DS, through a nonlinear relation [66]. These results indicated that the NPs with spherical shapes had a drug loading content of around 13% and an entrapment efficiency of around 77%, depending on the DS. The in vitro release studies showed that the DOX-loaded Gal-OCMCS-g-SA NPs had a pH-dependent release of DOX, which may be beneficial to the accumulation of the drug in tumor tissues [66]. Additionally, the hemolysis test showed good safety of these micelles in blood-contacting applications, which indicates the material as a potential candidate for the treatment of cancer, especially for liver targeting.

Balan et al. [88] used different amounts of hydrophobic groups of palmitoyl chloride (PC) to prepare, through a nucleophilic acyl substitution reaction, a novel amphiphilic derivative where free amine groups of chitosan are able to react with the carboxyl groups of acyl halides to form an amide bond. *N*-palmitoyl chitosan and magnetite were used to load DOX into magnetic nanocapsules, using a double-emulsion method. Magnetic nanocapsules exhibited suitable magnetic saturation, superparamagnetic behavior, and a good ability to incorporate a chemotherapeutic agent [88], properties which can be exploited in many different areas of biomedicine. These magnetic nanocapsules showed a DS directly related with molar ratio. At 1/0.5, 1/1, and 1/2 chitosan/PC molar ratio, the DS (%) was 7.69 ± 0.42, 12.35 ± 1.60, and 25.57 ± 2.33, respectively. Consequently, due to the fact that parts of amine groups have been substituted, the zeta potential decreases because there are fewer amines [88]. These systems showed a narrow size distribution of 215 ± 23 nm, encapsulation efficiency of 73%, and drug loading of 1.54%. Also, in vitro drug release exhibited biphasic drug release, with an initial burst effect in the first 6 h, followed by a constant release for up to 6 days [88]. Cytotoxicity studies revealed that the empty magnetic nanocapsules do not have cytotoxicity at the highest concentration tested, which makes them good for used as a drug delivery system.

Oleoyl chloride has been used to synthesize oleoyl-chitosan (OCH), and this chitosan derivative was applied to load NPs with DOX. In this research, the DS of the OCH derivative was 11% and the diameters of empty and loaded with DOX NPs were 255.3 nm and 315.2 nm, respectively [67]. The results showed that these NPs had encapsulation efficiency around 53% and the drug release was fast and complete at pH 3.8, whereas at pH 7.4 there was a sustained release after an initial burst release. Additionally, the toxicity of OCH NPs was investigated by MTT assay and the results showed low toxicity of the material at all concentrations and the hemolysis test showed a nontoxic level. Finally, this system showed inhibitory effects on the growth of four cancer cell lines (A549, Bel-7402, HeLa, and SGC-7901) at more concentration levels than that of DOX [67].

Gu et al. [62] produced chitosan-graft-poly(ε-caprolactone) (CS-*g*-PCL) through acylation modification to encapsulate 5-fluorouracil (5-Fu) as an anticancer system. The polymer-based micelles prepared had a particle size in the 61.4–108.6 nm range depending on the DS, with spherical shapes, after introduction of the drug and the encapsulation efficiency was above 90% for a drug loading content of 14.8% [62]. The in vitro drug release of 5-Fu-loaded micelles occurred in a sustained manner and with a controlled release behavior. Additionally, the cytotoxicity test made by the MTT assay showed no cytotoxicity against A549 and MCF-7 cells for empty micelles and the loaded micelles as well. However, the 5-Fu-loaded micelles present lower toxicity when compared with the free drug [62].

All of the examples presented above appear to be quite promising systems for loading and delivery of hydrophobic drugs, particularly anticancer drugs. Still much research needs to be done until we can use each of these systems in the treatment of human cancer.

Table 3.2 Overview of amphiphilic chitosan derivatives used to encapsulate anticancer drugs

Chitosan derivative	Modification	System	Active agent	Ref.
SOC	Alkylation	Micelles	Doxorubicin	[84]
OTMCS	Alkylation	Micelles	10-Hydroxy-camptothecin	[85]
NOCS	Alkylation	Micelles	Paclitaxel	[86]
Gal-OCMCS-g-SA	Acylation	Nanoparticles	Doxorubicin	[66]
N-palmitoyl chitosan	Acylation	Nanocapsules	Doxorubicin	[88]
OCH	Acylation	Nanoparticles	Doxorubicin	[67]
CS-*g*-PCL	Acylation	Micelles	5-fluorouracil	[62]

3.5.3 Proteins and Peptides

Amphiphilic NPs have many potential applications for the administration of therapeutic molecules, mainly for hydrophobic drugs but also for anionic drugs and proteins (Table 3.3).

Kim et al. [89] prepared NPs by the conjugation of glycol chitosan with hydrophobic 5β-cholanic acid to be used as Arg-Gly-Asp (RGD) peptide delivery carriers. These hydrophobically modified glycol chitosan (HGC) NPs had a low CAC (0.047 mg/mL), and the average diameters of the empty and loaded NPs were around 200 nm and 335 nm, respectively. The peptides used in this experiment were loaded into the hydrophobic core of the NPs by the solvent evaporation method. This system had a good loading efficiency (85.0 ± 7.0%) of the RGD peptides into NPs [89]. The in vivo biodistribution of the free RGD and the RGD NPs showed that the latter remains in a higher concentration in the tumor compared with the free peptide. Also, antitumor efficacy of these NPs showed to be significantly improved in models of mice bearing tumors [89].

Carboxymethyl chitosan (CMCS) was prepared by reacting CMCS with oleoyl chloride, through acylation modification, to form oleoyl-carboxymethy-chitosan (OCMCS). This system was produced by incorporation and incubation methods to encapsulate several extracellular products (ECPs), including proteases, haemolysin, enterotoxins, and acetylcholinesterase [90]. The OCMCS NPs observed by transmission electron microscopy (TEM) presented spherical shapes and diameters ranging between 238 and 482 nm, depending on the increase of the molecular weight. Increasing the chitosan molecular weight in OCMCS NPs led to an increased hydrodynamic mean diameter and zeta potential. Additionally, in the incubation method, the particle size of the OCMCS NPs increases further. The biodistribution of OCMCS NPs in carps were detected in liver, spleen, heart, and intestine after 48 h of administration, which indicates penetration into intestinal epithelial cells because of the mucoadhesive properties of this system [90]. The loading efficiency by the incorporation method was between 51.9% and 62.5%, increasing according to the increase of the molecular weight. However, the loading efficiency was around 43.2%–49.7% by the incubation method. The results showed an initial burst release of around 50% of the proteins encapsulated in the first 4–8 h, followed by a sustained release [90].

The micelles already described above [83], based on DCA-PCCS, were similarly produced again to form nanocomplexes of bovine serum albumin (BSA) as the model protein [91]. The results showed a DS of the hydrophobic moiety of 3% and encapsulation efficiency of 80.5% for a loading content of 15.6% of the BSA. The hydrodynamic diameter was 213.0 ± 3.8 nm and the zeta potential 13.4 ± 0.39 mV for nanocomplexes. The empty NPs of DCA-PCCS showed a larger size and lower zeta potential [91]. The NPs showed excellent cytocompatibility (above 80% for all concentrations), good hemocompatibility, and a rapid release behavior at the first 12 h, followed by a slow release at 72 h [91].

A model protein that has received much attention is insulin, because it was found that insulin-loaded chitosan NPs enhanced nasal absorption of this protein, having a good loading capacity and fast release profile [92].

N-octyl-*N*-arginine chitosan (OACS) was synthesized first by the formation of *N*-octyl chitosan and then by the introduction of hydrophilic arginine groups to prepare micelles through self-assembly of positively charged polymers and negatively charged insulin [93]. The particle size and zeta potential of this system was around 257–327 nm and +4.61 to +6.31 mV, respectively, with an octyl DS in the range of 11.2%–24.7%. The encapsulation efficiency ranges from 39.4% to 75.0% for drug loading values ranging between 8.1% and 13.4% for spherical micelles [93]. Studies performed to evaluate the enzymatic stability of insulin in the gastrointestinal (GI) tract showed a protection varying from 85% to 51% for 0.5 h and 4 h, respectively. On the other hand, free insulin showed less protection, varying from 50% to 6% for 0.5 h and 4 h, respectively [93]. The protection that is given to the insulin through encapsulation in micelles is evident. The transport experiments executed in Caco-2 cells showed that these micelles could pass through the Caco-2 monolayer more efficiently than the free insulin by virtue of permeation-enhancing properties of the arginine guanidino group [93]. Also, the cytotoxicity test showed no apparent cytotoxic effect at all concentrations.

Table 3.3 Overview of amphiphilic chitosan derivatives used to encapsulate peptides and proteins

Chitosan derivative	Modification	System	Active agent	Ref.
HGC	Acylation	Nanoparticles	RGD peptide	[89]
OCMCS	Acylation	Nanoparticles	ECPs	[90]
DCA-PCCS	Acylation	Nanocomplexes	BSA	[91]
OACS	Alkylation/ Acylation	Micelles	Insulin	[93]

3.6 Concluding Remarks

The ability to deliver effectively a drug to a desired location is directly associated with nanotechnology that enables the creation of drug delivery vehicles using polymeric materials.

Chitosan, a natural polymer with excellent properties, is proposed here as the ideal candidate for the hydrophilic part of polymeric micelles or NPs. Due to the presence of reactive groups (NH_2 and OH) in its backbone, chitosan can be easily modified to improve its properties or to open the use of chitosan in new applications.

On the other hand, polymer-based micelles or NPs can encapsulate hydrophobic, cationic, and anionic therapeutic molecules and can significantly increase their solubility in an aqueous environment. They can also be drug carriers to the desired region while maintaining stability thereof over the routing and minimize the side effects of drugs.

A wide range of applications is presented using amphiphilic chitosan derivatives, such as anti-inflammatory drugs, anticancer drugs, and peptides or proteins. Therefore, the ability of amphiphilic derivatives of chitosan to self-assemble has a great interest in the pharmaceutical field and it is expected that in a few years these nanosystems will become commercial products that can treat various diseases.

Acknowledgments

This work was financed by the European Regional Development Fund (ERDF) through the Programa Operacional Factores de

Competitividade - COMPETE 2020 - Operacional Programme for Competitiveness and Internationalisation (POCI), Portugal 2020, and by Portuguese funds through Fundação para a Ciência e a Tecnologia (FCT)/Ministério da Ciência e Tecnologia e Inovação in the framework of the project "Institute for Research and Innovation in Health Sciences" (POCI-01-0145-FEDER-007274). The work was also funded by VESPU (NanoGum-CESPU-2014). The authors also thank the Brazilian agency Coordenação de Aperfeiçoamento de Pessoal de Nível Superior (CAPES) for financial support through the scholarship in the program Ciência sem Fronteiras (CSF), process number 99999.006830/2015-03.

References

1. Khorsand, B., et al. Intracellular drug delivery nanocarriers of glutathione-responsive degradable block copolymers having pendant disulfide linkages. *Biomacromolecules*, 2013, **14**(6):2103–2111.
2. Xun, W., et al. Self-assembled micelles of novel graft amphiphilic copolymers for drug controlled release. *Colloids Surf B*, 2011, **85**(1):86–91.
3. Honarkar, H., et al. Applications of biopolymers I: chitosan. *Monatsh Chem*, 2009, **140**(12):1403–1420.
4. Sosnik, A., et al. Mucoadhesive polymers in the design of nano-drug delivery systems for administration by non-parenteral routes: a review. *Prog Polym Sci*, 2014, **39**(12):2030–2075.
5. Mourya, V., et al. Chitosan-modifications and applications: opportunities galore. *React Funct Polym*, 2008, **68**(6):1013–1051.
6. Varum, K.M., et al. Determination of the degree of N-acetylation and the distribution of N-acetyl groups in partially N-deacetylated chitins (chitosans) by high-field n.m.r. spectroscopy. *Carbohydr Res*, 1991, **211**:17–23.
7. Tsigos, I., et al. Chitin deacetylases: new, versatile tools in biotechnology. *Trends Biotechnol*, 2000, **18**:305–312.
8. Hernández-valdepeña, M.A., et al. Suppression of the tert -butylhydroquinone toxicity by its grafting onto chitosan and further cross-linking to agavin toward a novel antioxidant and prebiotic material. *Food Chem*, 2016, **199**:485–491.
9. Berezin, A.S., et al. Chitosan-isoniazid conjugates: synthesis, evaluation of tuberculostatic activity, biodegradability and toxicity. *Carbohydr Polym*, 2015, **127**:309–315.

10. Yu, L., et al. A sensitive and low toxicity electrochemical sensor for 2, 4-dichlorophenol based on the nanocomposite of carbon dots, hexadecyltrimethyl ammonium bromide and chitosan. *Sens Actuators, B*, 2015, **224**:241–247.

11. Shukla, S., et al. In vitro toxicity assessment of chitosan oligosaccharide coated iron oxide nanoparticles. *Toxicol Rep*, 2015, **2**:27–39.

12. Hafsa, J., et al. Antioxidant and antimicrobial proprieties of chitin and chitosan extracted from Parapenaeus Longirostris shrimp shell waste. *Ann Pharm Fr*, 2016, **74**(1):27–33.

13. Madureira, A.R., et al. Production of antimicrobial chitosan nanoparticles against food pathogens. *J Food Eng*, 2015, **167**:210–216.

14. Kaya, M., et al. Porous and nanofiber α-chitosan obtained from blue crab (Callinectes sapidus) tested for antimicrobial and antioxidant activities. *LWT - Food Sci Technol*, 2016, **65**:1109–1117.

15. De Souza, R., et al. Biocompatibility of injectable chitosan-phospholipid implant systems. *Biomaterials*, 2009, **30**:3818–3824.

16. Roy, P., et al. Chitosan–nanohydroxyapatite composites: mechanical, thermal and bio-compatibility studies. *Int J Biol Macromol*, 2015, **73**:170–181.

17. Sarhan, W.A., et al. High concentration honey chitosan electrospun nanofibers: biocompatibility and antibacterial effects. *Carbohydr Polym*, 2015, **122**:135–143.

18. Muzzarelli, R., et al. Biological activity of chitosan: ultrastructural study. *Biomaterials*, 1988, **9**:247–252.

19. Bagheri-Khoulenjani, S., et al. An investigation on the short-term biodegradability of chitosan with various molecular weights and degrees of deacetylation. *Carbohydr Polym*, 2009, **78**:773–778.

20. Chang, S.-H., et al. Plasma surface modification effects on biodegradability and protein adsorption properties of chitosan films. *Appl Surf Sci*, 2013, **282**:735–740.

21. Liu, M., et al. Efficient mucus permeation and tight junction opening by dissociable "mucus-inert" agent coated trimethyl chitosan nanoparticles for oral insulin delivery. *J Controlled Release*, 2016, **222**:67–77.

22. Hsu, L.-W., et al. Effects of pH on molecular mechanisms of chitosan-integrin interactions and resulting tight-junction disruptions. *Biomaterials*, 2013, **34**:784–793.

23. Uchegbu, I.F., et al. Chitosan amphiphiles provide new drug delivery opportunities. *Polym Int*, 2014, **63**:1145–1153.

24. Sonaje, K., et al. Effects of chitosan-nanoparticle-mediated tight junction opening on the oral absorption of endotoxins. *Biomaterials*, 2011, **32**:8712–8721.

25. Younes, I., et al. Cytotoxicity of chitosans with different acetylation degrees and molecular weights on bladder carcinoma cells. *Int J Biol Macromol*, 2016, **84**:200–207.

26. Younes, I., et al. Influence of acetylation degree and molecular weight of homogeneous chitosans on antibacterial and antifungal activities. *Int J Food Microbiol*, 2014, **185**:57–63.

27. Hanuza, J., et al. Determination of N-acetylation degree in chitosan using Raman spectroscopy. *Spectrochim Acta A Mol Biomol Spectrosc*, 2015, **134**:114–120.

28. Mellegård, H., et al. Antibacterial activity of chemically defined chitosans: influence of molecular weight, degree of acetylation and test organism. *Int J Food Microbiol*, 2011, **148**:48–54.

29. Novoa-Carballal, R., et al. Chitosan hydrophobic domains are favoured at low degree of acetylation and molecular weight. *Polymer (UK)*, 2013, **54**:2081–2087.

30. Kurita, Y., et al. Reductive N-alkylation of chitosan with acetone and levulinic acid in aqueous media. *Int J Biol Macromol*, 2010, **47**:184–189.

31. Kurita, Y., et al. N-Alkylations of chitosan promoted with sodium hydrogen carbonate under aqueous conditions. *Int J Biol Macromol*, 2012, **50**:741–746.

32. Benediktsdóttir, B.E., et al. N-alkylation of highly quaternized chitosan derivatives affects the paracellular permeation enhancement in bronchial epithelia in vitro. *Eur J Pharm Biopharm*, 2014, **86**:55–63.

33. dos Santos Alves, K., et al. Chitosan derivatives with thickening properties obtained by reductive alkylation. *Mater Sci Eng, C*, 2009, **29**:641–646.

34. Bidgoli, H., et al. Effect of carboxymethylation conditions on the water-binding capacity of chitosan-based superabsorbents. *Carbohydr Res*, 2010, **345**:2683–2689.

35. Kong, C.-S., et al. Carboxymethylations of chitosan and chitin inhibit MMP expression and ROS scavenging in human fibrosarcoma cells. *Process Biochem*, 2010, **45**:179–186.

36. Chen, W., et al. Synthesis and antioxidant properties of chitosan and carboxymethyl chitosan-stabilized selenium nanoparticles. *Carbohydr Polym*, 2015, **132**:574–581.

37. Patrulea, V., et al. Optimized synthesis of O-carboxymethyl-N,N,N-trimethyl chitosan. *Carbohydr Polym*, 2015, **122**:46–52.

38. Huang, J., et al. Effect of quaternization degree on physiochemical and biological activities of chitosan from squid pens. *Int J Biol Macromol*, 2014, **70**:545–550.

39. Zhou, Y., et al. Potential of quaternization-functionalized chitosan fiber for wound dressing. *Int J Biol Macromol*, 2013, **52**:327–332.

40. Stepnova, E.A., et al. New approach to the quaternization of chitosan and its amphiphilic derivatives. *Eur Polym J*, 2007, **43**:2414–2421.

41. dos Santos, D.M., et al. Response surface methodology applied to the study of the microwave-assisted synthesis of quaternized chitosan. *Carbohydr Polym*, 2016, **138**:317–326.

42. Senra, T.D., et al. Extensive *N*-methylation of chitosan: evaluating the effects of the reaction conditions by using response surface methodology. *Polym Int*, 2015, **64**:1617–1626.

43. Baran, T., et al. Cu(II) and Pd(II) complexes of water soluble O-carboxymethyl chitosan Schiff bases: synthesis, characterization. *Int J Biol Macromol*, 2015, **79**:542–554.

44. Yuan, B., et al. Schiff base – chitosan grafted l-monoguluronic acid as a novel solid-phase adsorbent for removal of congo red. *Int J Biol Macromol*, 2016, **82**:355–360.

45. Delezuk, J.A.M., et al. Ultrasound-assisted deacetylation of beta-chitin: influence of processing parameters. *Polym Int*, 2011, **60**:903–909.

46. Badawy, M.E.I., et al. Synthesis and fungicidal activity of new N,O-acyl chitosan derivatives. *Biomacromolecules*, 2004, **5**(2):589–595.

47. Badawy, M.E.I., et al. Fungicidal and insecticidal activity of O-acyl chitosan derivatives. *Polym Bull*, 2005, **54**:279–289.

48. Pavinatto, A., et al. Interaction of O-acylated chitosans with biomembrane models: probing the effects from hydrophobic interactions and hydrogen bonding. *Colloids Surf B*, 2014, **114**:53–59.

49. Le Tien, C., et al. N-acylated chitosan: hydrophobic matrices for controlled drug release. *J Controlled Release*, 2003, **93**:1–13.

50. Hirano, S., et al. Novel N-saturated-fatty-acyl derivatives of chitosan soluble in water and in aqueous acid and alkaline solutions. *Carbohydr Polym*, 2002, **48**:203–207.

51. Hirano, S., et al. Selective N-acylation of chitosan. *Carbohydr Res*, 1976. **47**:315–320.

52. Aranaz, I., et al. Chitosan amphiphilic derivatives: chemistry and applications. *Curr Org Chem*, 2010, **14**:308–330.

53. Desbrières, J., et al. Hydrophobic derivatives of chitosan: characterization and rheological behaviour. *Int J Biol Macromol*, 1996, **19**:21–28.

54. Rinaudo, M., et al. Specific interactions in modified chitosan systems. *Biomacromolecules*, 2005, **6**(5):2396–2407.

55. Larsson, M., et al. Biomedical applications and colloidal properties of amphiphilically modified chitosan hybrids. *Prog Polym Sci*, 2013, **38**(9):1307–1328.

56. Lee, K.Y., et al. Physicochemical characteristics of self-aggregates of hydrophobically modified chitosans. *Langmuir*, 1998, **14**:2329–2332.

57. Liu, C.-G.G., et al. Linolenic acid-modified chitosan for formation of self-assembled nanoparticles. *J Agric Food Chem*, 2005, **53**:437–441.

58. Hu, Y., et al. Self-aggregation and antibacterial activity of N-acylated chitosan. *Polymer*, 2007, **48**:3098–3106.

59. Jiang, G.-B., et al. Preparation of polymeric micelles based on chitosan bearing a small amount of highly hydrophobic groups. *Carbohydr Polym*, 2006, **66**:514–520.

60. Kurita, K. Controlled functionalization of the polysaccharide chitin. *Prog Polym Sci (Oxford)*, 2001, **26**:1921–1971.

61. Gruber, J.V., et al. Synthesis of N-[(3'-hydroxy-2', 3'-dicarboxy)-ethyl] chitosan: a new, water-soluble chitosan derivative. *Macromolecules*, 1995, **28**(26):8865–8867.

62. Gu, C., et al. Preparation of polysaccharide derivates chitosan-graft-poly (ε-caprolactone) amphiphilic copolymer micelles for 5-fluorouracil drug delivery. *Colloids Surf B*, 2014, **116**:745–750.

63. Jiang, G.-B., et al. Novel polymer micelles prepared from chitosan grafted hydrophobic palmitoyl groups for drug delivery. *Mol Pharm*, 2006, **3**(2):152–160.

64. Marques, J.G., et al. Synthesis and characterization of micelles as carriers of non-steroidal anti-inflammatory drugs (NSAID) for application in breast cancer therapy. *Colloids Surf B*, 2014, **113**:375–383.

65. Min, K.H., et al. Hydrophobically modified glycol chitosan nanoparticles-encapsulated camptothecin enhance the drug stability and tumor targeting in cancer therapy. *J Controlled Release*, 2008, **127**(3):208–218.

66. Guo, H., et al. Self-assembled nanoparticles based on galactosylated O-carboxymethyl chitosan-graft-stearic acid conjugates for delivery of doxorubicin. *Int J Pharm*, 2013, **458**:31–38.

67. Zhang, J., et al. Self-assembled nanoparticles based on hydrophobically modified chitosan as carriers for doxorubicin. *Nanomedicine*, 2007, **3**:258–265.

68. Biswas, S., et al. Recent advances in polymeric micelles for anti-cancer drug delivery. *Eur J Pharm Sci*, 2016, **83**:184–202.

69. Ahmad, Z., et al. Polymeric micelles as drug delivery vehicles. *RSC Adv*, 2014, **4**(33):17028–17038.

70. Kwon, G.S., et al. Amphiphilic block copolymer micelles for nanoscale drug delivery. *Drug Dev Res*, 2006, **67**(1):15–22.

71. Ishihara, M., et al. Biological, chemical, and physical compatibility of chitosan and biopharmaceuticals, in *Chitosan-Based Systems for Biopharmaceuticals: Delivery, Targeting and Polymer Therapeutics*, Sarmento, B., das Neves, J., eds. John Wiley & Sons, Chichester, 2012, pp. 93–106.

72. Torchilin, V.P. Micellar nanocarriers: pharmaceutical perspectives. *Pharm Res*, 2007, **24**(1):1–16.

73. Kazunori, K., et al. Block copolymer micelles as vehicles for drug delivery. *J Controlled Release*, 1993, **24**(1):119–132.

74. Kwon, G.S., et al. Block copolymer micelles as long-circulating drug vehicles. *Adv Drug Deliv Rev*, 1995, **16**(2):295–309.

75. Pérez-Herrero, E., et al. Advanced targeted therapies in cancer: drug nanocarriers, the future of chemotherapy. *Eur J Pharm Biopharm*, 2015, **93**:52–79.

76. Hu, C.-M.J., et al. Nanoparticle-assisted combination therapies for effective cancer treatment. *Ther Deliv*, 2010, **1**(2):323–334.

77. Hillaireau, H., et al. Nanocarriers' entry into the cell: relevance to drug delivery. *Cell Mol Life Sci*, 2009, **66**(17):2873–2896.

78. Park, J.H., et al. Self-assembled nanoparticles based on glycol chitosan bearing hydrophobic moieties as carriers for doxorubicin: in vivo biodistribution and anti-tumor activity. *Biomaterials*, 2006, **27**(1):119–126.

79. Duncan, R. Polymer conjugates for tumour targeting and intracytoplasmic delivery. The EPR effect as a common gateway? *Pharm Sci Technol Today*, 1999, **2**(11):441–449.

80. Sui, W., et al. Self-assembly of an amphiphilic derivative of chitosan and micellar solubilization of puerarin. *Colloids Surf B*, 2006, **48**:13–16.

81. Martin, L., et al. The release of model macromolecules may be controlled by the hydrophobicity of palmitoyl glycol chitosan hydrogels. *J Controlled Release*, 2002, **80**:87–100.

82. Woraphatphadung, T., et al. pH-Responsive polymeric micelles based on amphiphilic chitosan derivatives: effect of hydrophobic cores on oral meloxicam delivery. *Int J Pharm*, 2016, **497**:150–160.

83. Wu, M., et al. In vitro drug release and biological evaluation of biomimetic polymeric micelles self-assembled from amphiphilic deoxycholic acid–phosphorylcholine–chitosan conjugate. *Mater Sci Eng, C*, 2014, **45**:162–169.

84. Xiangyang, X., et al. Preparation and characterization of N-succinyl-N′-octyl chitosan micelles as doxorubicin carriers for effective anti-tumor activity. *Colloids Surf B*, 2007, **55**:222–228.

85. Zhang, C., et al. Polymeric micelle systems of hydroxycamptothecin based on amphiphilic N-alkyl-N-trimethyl chitosan derivatives. *Colloids Surf B*, 2007, **55**:192–199.

86. Zhang, C., et al. Self-assembly and characterization of paclitaxel-loaded N-octyl-O-sulfate chitosan micellar system. *Colloids Surf B*, 2004, **39**:69–75.

87. Zhang, C., et al. Pharmacokinetics, biodistribution, efficacy and safety of N-octyl-O-sulfate chitosan micelles loaded with paclitaxel. *Biomaterials*, 2008, **29**:1233–1241.

88. Balan, V., et al. Doxorubicin-loaded magnetic nanocapsules based on N-palmitoyl chitosan and magnetite: synthesis and characterization. *Chem Eng J*, 2015, **279**:188–197.

89. Kim, J.-H., et al. Self-assembled glycol chitosan nanoparticles for the sustained and prolonged delivery of antiangiogenic small peptide drugs in cancer therapy. *Biomaterials*, 2008, **29**(12):1920–1930.

90. Liu, Y., et al. Preparation and evaluation of oleoyl-carboxymethy-chitosan (OCMCS) nanoparticles as oral protein carriers. *J Mater Sci: Mater Med*, 2012, **23**(2):375–384.

91. Wu, M., et al. Self-assemblied nanocomplexes based on biomimetic amphiphilic chitosan derivatives for protein delivery. *Carbohydr Polym*, 2015, **121**:115–121.

92. Andrade, F., et al. Solid state formulations composed by amphiphilic polymers for delivery of proteins: characterization and stability. *Int J Pharm*, 2015, **486**(1):195–206.

93. Zhang, Z.-H., et al. N-octyl-N-arginine chitosan micelles as an oral delivery system of insulin. *J Biomed Nanotechnol*, 2013, **9**(4):601–609.

Chapter 4

Oral Administration of Lipid-Based Delivery Systems to Combat Infectious Diseases

Rita M. Pinto,[a,*] Daniela Lopes,[a,*] Cláudia Nunes,[a] Bruno Sarmento,[b,c] and Salette Reis[a]

[a]*UCIBIO, REQUIMTE, Department of Chemical Sciences, Faculty of Pharmacy, University of Porto, Portugal*

[b]*I3S - Institute of Research and Innovation in Health, University of Porto, Portugal*

[c]*CESPU - Institute for Advanced Research and Training in Health Sciences and Technologies and University Institute of Health Sciences, Gandra, Portugal*

shreis@ff.up.pt

According to the World Health Organization (WHO), infectious diseases are still one of the leading causes of death worldwide, englobing direct and indirect associations (e.g., cancers caused by infectious diseases). Furthermore, the increase of resistance with loss of drug efficacy and the lack of new antimicrobial drugs have become a worldwide challenge to provide adequate treatment against infectious diseases. The drawbacks associated with this

*Rita M. Pinto and Daniela Lopes contributed equally to the chapter.

Nanoparticles in Life Sciences and Biomedicine
Edited by Ana Rute Neves and Salette Reis

ISBN 978-981-4745-98-7 (Hardcover), 978-1-351-20735-5 (eBook)
www.panstanford.com

problem lead daily to high costs to society and to health care systems. Therefore, microparticles and NPs emerged as a promising drug delivery strategy to improve the current treatment against infectious diseases. Among all, lipid particles have been highlighted for their BA and their ability to incorporate both lipophilic and hydrophilic drugs, protecting them against in vivo degradation. Further, they can be produced on an industrial scale and can be orally administered, which play a key role in the increase of therapeutic compliance. However, different characteristics of the gastrointestinal tract must be taken into account in the design of delivery systems that aim for oral absorption and stability. This section aims to review these characteristics as well as the design of lipid micro- and nanoparticles to be applied in the treatment of infectious diseases.

4.1 Introduction

At the beginning of the twentieth century, infectious diseases were the leading cause of death at a worldwide level [1]. However, with the development of antimicrobial drugs and specially with the production of penicillin in the 1940s, the fight against microbial agents emerged [2]. Consequently, this phenomenon led to a significant decrease in morbidity and mortality resultant from infectious diseases [2]. Nevertheless, in 1999, the World Health Organization (WHO) reported that infectious diseases were the cause of 32% of deaths at a global level, most of the cases being from developing countries [3, 4]. The same organization also reported that infectious diseases were responsible for killing over 14 million people worldwide [5, 6]. Nowadays, the developed world is also being severely affected by infectious diseases, as a consequence of antimicrobial resistance (AMR) [7]. In the United States, AMR is estimated to be responsible for US$20 billion in extra costs to the health care system and US$35 billion in costs to society per year [7, 8]. Besides, this phenomenon leads to more than 8 million additional days in hospitals [7]. AMR is also a concern in the European Union, since it is estimated that this phenomenon causes 25,000 deaths and costs more than US$1.5 billion per year [9]. These costs include both health care expenses and loss in productivity [9].

AMR is a natural evolutionary process in microorganisms, since AMR genes existed long before the therapeutic application of antimicrobial drugs [7, 10]. For instance, in bacteria the occurrence of mutations in an infection cycle is highly probable due to the large number of bacteria, the rapid generation time, and the high intrinsic rate of mutation [11]. However, AMR is not exclusive to bacteria, since other microorganisms, such as parasites, viruses, and fungi, can acquire AMR [7]. In the last decades, this phenomenon has been rapidly evolving and spreading due to the wide use of antimicrobial agents in humans and animals [7]. In fact, the ability of antimicrobial drugs to induce genetic mutations in microbial pathogens is already proven [12]. This association is more significant when sublethal concentrations are used [12]. Consequently, there is no assurance that new antimicrobial drugs can respond to the increasing rates of resistance in a timely manner [2, 7, 13]. Thus, AMR is becoming a worldwide threat to efficient treatment of infectious diseases [14].

There are also other drawbacks of the current antimicrobial therapy. Oral administration is usually the primordial route due to its commodity and security [15]. However, it is also associated with poor bioavailability (BA) due to poor aqueous solubility and/or poor permeability of drugs [15]. The acidic pH of the stomach can also reduce the BA of drugs through their degradation [15]. Further, hepatic first-pass metabolism contributes to reducing the concentration of drugs in the systemic circulation, since drug metabolism occurs during this process [15, 16]. Consequently, infection conditions usually require high doses of antimicrobial drugs in order to achieve an adequate concentration in serum, which commonly lead to side effects [17]. However, even applying an aggressive antimicrobial treatment, the complete eradication of the infection can be difficult to achieve due to drugs' degradation or low efficacy [17]. The therapeutic effect of most antimicrobial agents is also compromised by intrinsic factors [18]. These factors include poor cellular penetration, limited intracellular retention, inefficient subcellular distribution, and decreased intracellular activity [18].

Novel and effective strategies are needed to overcome these limitations. One field that has been evolving in the last years is nanotechnology. Nanoparticles (NPs) protect drugs from the acidic conditions of the stomach, improve their half-life, and maintain a sustained drug release at the target site, leading to higher concentrations of antimicrobial drugs near the microorganism, even with lower doses [19]. Hence, these delivery systems can minimize drugs' side effects and improve their therapeutic efficacy, decreasing the potential development of microbial resistance [19]. NPs are also capable of loading multiple antimicrobial agents within their structures [8]. As a consequence, AMR mechanisms are unlikely to develop, since it requires multiple and simultaneous gene mutations in the microbial DNA [8]. Combining these advantages with the possibility of vectorization to a specific target, NPs have been developed to be applied in several diseases, including infectious diseases. This section aims to summarize several approaches to combat infectious diseases by resorting to lipid-based delivery systems for oral administration. Further, studies to assess pharmacokinetic properties and therapeutic efficacy will also be discussed.

4.2 Oral Administration

Oral administration is the most convenient and safest route for drug administration [15]. This route is cost effective, has fewer complications, and is less intrusive than intravenous administration [15, 20]. Further, the possibility of self-administration and its feature as a patient-friendly route increase the therapeutic compliance [21]. However, the selection of the administration route should be dependent on the purpose and on the target site [22]. For instance, to target muscles or skin, a topical route is more advantageous [22]. On the other hand, to target the liver or the spleen, the intravenous route is preferential, but with a higher risk of local and vascular infection and patient discomfort [15, 20, 22].

Although oral administration is the most common route, it implies contact with an extremely acidic medium, with digestive enzymes, and with biliary salts [23, 24]. These drawbacks are associated with low BA of drugs in vivo, because of their degradation, poor solubility, and low ability to cross biological barriers [21]. One strategy to overcome these limitations is to use NPs to protect the drug from the harsh conditions of the environment and to enhance its permeability [21]. Still, it is necessary to take into account all the conditions and barriers of the GI tract (Fig. 4.1) in order to surpass its restraints and, if possible, to take advantage of its characteristics to design the NP [21].

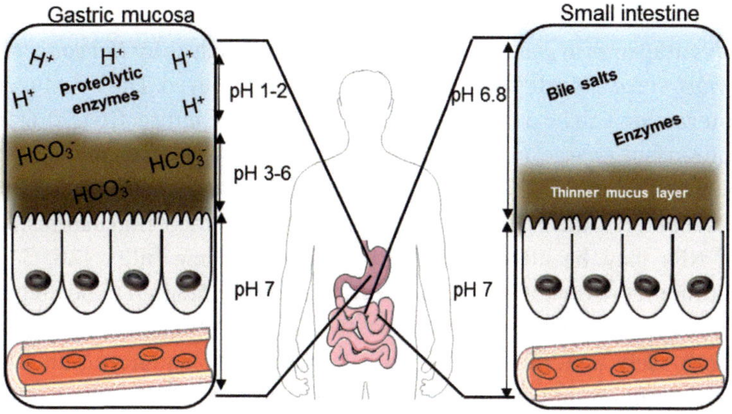

Figure 4.1 Schematic representation of the main differences between the gastric mucosa and the small intestine barrier. The pH of the lumen and the thickness of the mucus layer are the major differences.

The mucus layer is the first barrier for oral administration, and it is composed of water, salts, lipids, phospholipids, and different macromolecules, among which mucins are highlighted [25, 26]. It is divided in two different layers: the outer layer, which is loosely attached and has larger pores, and the inner layer, which is more packed [25]. Although the pore of the mucin fibers is around 100 nm, they are also able to pack, creating larger pores [25]. This barrier works simultaneously as a physical barrier to prevent the penetration of foreign substances and microorganisms and as a permeable gel that allows the diffusion of several molecules, including nutrients [25]. The mucus layer protects epithelial cells, which have restricted permeability [24]. Still, NPs are able to cross epithelial cells by transcellular or paracellular pathways [21].

In general, three different strategies have been used to enhance NP absorption through the mucus layer: (i) mucoadhesiveness, with attraction to a mucosa membrane due to several interactions established (e.g., electrostatic and H-bonds), (ii) mucolytic NPs, which have the disadvantage of favoring infections by disrupting the mucus barrier, and (iii) muco-penetrating NPs, by changing the size or their surface characteristics [19, 21, 27]. NPs are easily entrapped within the protective outer mucus layer by steric difficulties or mucoadhesion features [24, 28]. As a consequence of high mucoadhesiveness, NP penetration and mobility within the mucosa may be affected [29]. Nevertheless, this entrapment can be very advantageous in gastric or enteric infections, such as by *Helicobacter pylori,* since a local release of the drug can be achieved. The NP-mucus interaction can be a consequence of electrostatic interactions, due to the surface charge, or of hydrophobic interactions [30]. However, it is also important to be aware that given the fast turnover of the gastric mucosa and the natural clearance of the GI tract, mucoadhesion of NPs may be affected, decreasing their permeability [28, 31]. Although the turnover time of the GI mucus is estimated to be higher than in other physiological sites, it is still very fast, especially in the loosely adherent layer [32]. NPs with higher penetration and establishment of interactions with deeper layers of the mucus may be a promising alternative [29, 30]. For that purpose, particles with a hydrophilic and uncharged surface may prevent the establishment of interactions with the mucus [27]. One promising strategy is the use of poly(ethylene glycol) (PEG), which is a hydrophilic compound and has an additional advantage of mucoadhesiveness and stability [24]. However, to treat bacterial infections, this strategy has been avoided since PEG prevents the fusion and the contact with bacterial membranes [33].

Mucin molecules are organized into a porous structure, which should be taken into account when the purpose is local penetration [27]. In general, the drug delivery system should be small enough to be able to overcome the steric obstruction caused by the mucus [27]. Furthermore, smaller particles are promptly distributed throughout the human body due to their ability to cross membrane or mucosa pores and to circulate in small vessels like capillaries [22]. Small size NPs are also highlighted by their therapeutic efficacy [22]. Using muco-penetrating NPs, it is possible to release the drug near the epithelial cells or even improve their intracellular delivery or uptake [24].

One important advantage of the GI tract is its large surface of absorption, which in total reaches around 300–400 m^2 [21]. The thickness varies along the GI tract, with the thinnest firmly adherent layer in the small intestine [21, 23]. Thus, this area is primary in the absorption of NPs [21]. One of the strategies used by microorganisms to attach to and colonize the mucosa of the GI tract is the alteration of its bioequilibrium [25]. More specifically, they are able to disturb the equilibrium between the turnover rate and the secretion of mucus, which determines its thickness [34]. Learning from microorganisms' abilities, intelligent delivery systems can be designed, for instance, using ligands to epithelial receptors (e.g., bacterial lipopolysaccharide [LPS] or flagellin), which lead to the inhibition of the synthesis of mucus glycoproteins [25]. It is also possible to use lectins, vitamins, sugars, or adhesins to target the GI tract [24].

The GI tract has also several enzymes, such as lipases, that are able to degrade lipids [30]. Furthermore, the pH of the GI tract varies from 1.0 in the stomach lumen to 7.0 in the colon [30]. Even in the stomach, a pH gradient exists, given the low pH of the lumen and the physiologic pH near epithelial cells [35]. On the other hand, in the intestine, the pH is higher (around 6.8) [30]. These characteristics should be taken into consideration when designing particles for oral delivery, due to their instability under harsh conditions. Moreover, the intestinal barrier has a high immunologic ability, leading to a high loss of NPs by the action of the immune cells [30]. Nevertheless, their nanoparticulate state promotes the uptake by M cells of Peyer's patches [36]. Consequently, lipid NPs enable the system to bypass the effect of first-pass metabolism, through lymphatic absorption [36, 37].

It is inevitable to mention that a growing concern is the toxicity of NPs delivered through the oral route. The topical contact between NPs and epithelial cells can lead to a local toxicity, and the specific properties of the NP surface can interact with the GI tract, changing the cellular uptake of the NPs [21]. Usually, in vitro studies are performed; however, they are insufficient for a translation to a toxicity in the human body [24]. Studies of the effect of NPs on genes or on the immune system after oral administration are still lacking [24]. Lipid NPs are classified as biodegradable, since the human body is able to degrade these NPs [38]. However, even biodegradable NPs should be studied, since the products of their degradation can have toxic effects [21, 24]. The cytotoxicity of different lipid NPs was evaluated, and they all showed 90% cell viability, using Caco-2 cells [39]. In vivo studies of orally administered lipid NPs composed of glycerol monostearate, tristearin, and Pluronic-F68 as a surfactant were also performed in rats, and they did not cause any damage to the intestinal epithelium [40]. Lipids are usually safe, and the use of surfactants and cosurfactants is more worrisome [21, 41]. Nevertheless, lipid NPs are still identified as well tolerated in living systems [41].

4.3 Lipid-Based Delivery Systems

4.3.1 Lipid-Based Nanoparticles

In the last decades, many colloidal drug carriers have been studied to improve drug solubility by oral administration [42]. For the majority of the NPs, a low-cost large-scale production method is currently not available [42]. Nevertheless, lipid-based NPs present the best features of other colloidal systems and have emerged as a strategy to overcome their limitations [42]. These NPs are usually composed of a matrix of physiological or physiologically related lipids characterized by their versatility, biocompatibility, and biodegradability [36, 38, 42]. Lipids are natural materials that can be degraded by natural processes, such as enzymatic activity [38]. Due to these processes, complex lipids are easily and completely degraded in the human body [38]. Therefore, the excipients that compose the matrix of lipid-based NPs are *generally recognized as safe* (GRAS) for oral and topical administration [41].

Lipid-based NPs exhibit other outstanding advantages, including improved kinetic stability, drug solubility, and controlled drug release [36, 38, 43, 44]. According to the production process, lipid-based NPs can be easily scaled up and their production can avoid the use of organic solvents [36]. Besides, lipid-based NP production is cost effective, which increases the researchers' interest in these nanocarriers [36, 41].

Different types of lipid-based NPs have been engineered, such as liposomes, solid lipid nanoparticles (SLNs), nanostructured lipid carriers (NLCs), lipid–drug conjugate (LDC) NPs, and micelles. In this section, the most commonly studied, which are represented in Fig. 4.2, will be discussed.

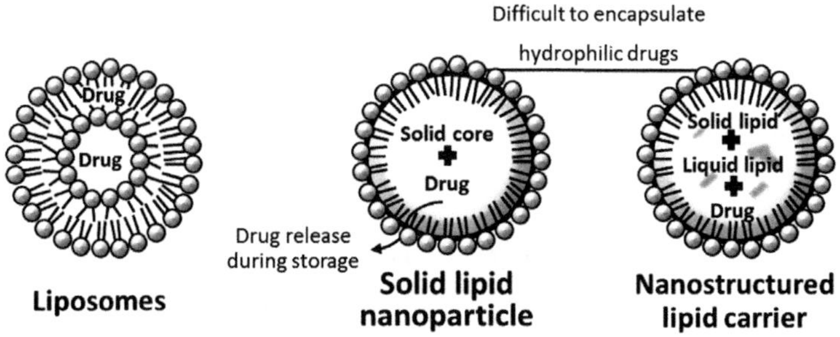

Figure 4.2 The three most common lipid-based NPs: (left) liposomes, (center) solid lipid nanoparticle, and (right) nanostructured lipid carrier.

Liposomes can incorporate both lipophilic and hydrophilic drugs, since they are composed of amphipathic lipids, which self-assemble into spherical vesicles with an aqueous core [45]. Liposomes have been the most studied drug delivery system to deliver antimicrobial drugs [2]. Their major advantages are their low toxicity, their versatility, and the possibility of coencapsulation [45]. Given their similarity to cell membranes, they are able to fuse with microbes [2, 46]. In fact, it has already been shown that the development of AMR is significantly reduced when the contact with the microorganism is by fusion [47].

SLNs were introduced in 1991 as an alternative to colloidal drug carriers, such as liposomes, emulsions, and polymeric microparticles and NPs [48]. SLNs are nanospheres with a mean particle size from 50 to 1000 nm, consisting of a matrix composed of lipids in a solid state at both room and body temperatures [36, 41, 49]. Fatty acids, waxes, monoglycerides, diglycerides, and triglycerides are widely used to construct the rigid core of SLNs [44]. To stabilize the solid matrix, various nontoxic surfactants can be added during SLN preparation, such as poloxamers, polysorbates, and polyvinyl alcohol [44, 50, 51].

To overcome the drawbacks of SLNs, namely the limited drug loading capacity and an early drug release during storage due to a structural reorganization, NLCs were studied [38, 42, 44]. The second generation of lipid-based NPs are composed of a lipid matrix with both solid and liquid lipids [44]. NLC matrices are less ordered and have imperfections due to the presence of the liquid lipid, providing more spaces to incorporate drugs when compared to SLNs

[41, 44]. Consequently, the drug loading capacity is enhanced, while drug release during storage is minimized [42, 44]. Besides, these nanocarriers are less susceptible to gelation during both preparation and storage, having a lower water content when compared to SLNs [38, 44].

4.3.2 Preventing Infectious Diseases by Oral Vaccines

Vaccines have been essential to humans and to the general public health, especially in terms of the global eradication of several infections, such as small pox in 1979 [52]. A more recent discovery was the possibility of using the oral route to administer vaccines against intestinal infections like cholera, rotavirus, and salmonella [53]. The advantages of oral vaccines are unquestionable and include higher compliance, reduction of severe side effects, and the possibility of self-administration, which is important in regions where access to medical care and instrumentation is lacking [53, 54]. Furthermore, regulatory issues related with oral administration are less strict than for injectable vaccines [53]. In fact, sterilization is not required for oral vaccines, facilitating their manufacture [54]. It is also less risky for medical personnel, since the probability of getting a transmittable disease is significantly reduced [53]. Another important advantage is the possibility to induce an immune response at the mucosa, since most of the pathogens (around 90%) infect humans by penetrating mucosal tissues, such as the lung and the GI tract [53, 54]. Although injectable vaccines are more effective in achieving a systemic immune response, they lack in terms of inducing a specific and local mucosa response to work as an immunological barrier to the invasion of pathogens [54]. Nevertheless, the oral route has several drawbacks associated with the high rate of failure in clinical trials, leading to the low number of licensed oral vaccines [53]. One limitation, and probably the most important one, is the instability that results from the features of the GI tract already mentioned, such as the low pH, the poor permeability, and the presence of several enzymes [23, 24]. This is particularly important since the efficacy of peptide or protein antigens is affected by proteolytic degradation [55]. NPs can be potent antigen delivery systems, by protecting and delivering the antigens only near the immune cells [56]. Moreover, due to the ability of NPs to interact with and activate immune pathways, they can work as an adjuvant and enhance the immune response to the loaded antigen [56]. The use of delivery systems has also been associated with the possibility of single doses being the focus of interest of institutions like WHO, due to the opportunity of having simplified immunization schedules [52]. Besides, the incorporation of different compounds in the same NP, with a controlled release of each one, may also help the simplification of the immunization schedules [57].

Despite controversy, in general, smaller particles in the nanometer range are shown as the most promising ones for the immune stimulation, due to their ability to cross intestinal barriers [52]. However, as mentioned in Section 4.2, Peyer's patches allow a way of penetration through lymphatic absorption [36]. In the GI tract, the ileum has a predominance of these patches, making it a preferential region for the absorption of oral vaccines [53]. This pathway of absorption allows the use of bigger NPs with high efficacy [52]. If this pathway is chosen, it is important to design the NP according to that purpose. For instance, the functionalization with PEG is known to increase the stability of liposomes in vivo [58]. However, Minato et al. showed by administering PEG-modified and unmodified liposomes to mice that, for higher doses, the systemic immunity was higher in unmodified liposomes [58]. This may be a consequence of the protecting effect promoted by PEG, which suppresses the uptake by macrophages and probably the uptake by Peyer's patches [58]. Other properties, such as the sustained release and the physicochemical properties of the surface, are also important for a strong immune response [52]. For instance, surface hydrophobicity is essential for a good interaction between antigen presenting cells and NPs [59]. Cationic surfaces are also recognized as a higher potentiator of the antigen presenting cells, due to electrostatic interactions established with negative membranes [56]. However, they are still associated with higher toxicity in vivo [53]. Another key factor recently suggested by Liu et al. was the location of the antigen within the NP [60]. They used cationic lipid-poly(lactide-co-glycolide) acid (PLGA) hybrid NPs to test different locations of the antigen: (i) adsorbed, (ii) encapsulated, and (iii) both adsorbed and encapsulated [60]. They reported, through in vivo studies, higher immunologic efficiency when the antigen was both adsorbed and encapsulated, providing simultaneously an initial antigen exposure and a controlled release and a long-term exposure of the antigen encapsulated [60].

Liposomes have been indicated as a potent adjuvant for immunization against several bacteria and viruses [55]. The use of multilamellar liposomes instead of unilamellar may be a better option due to their higher stability against detergents and their higher loading capacity of antigens [58]. Due to their composition, which includes an aqueous core and a hydrophobic bilayer, liposomes can load different types of antigens, including peptides and DNA [61]. For instance, Wang et al. proved that liposomes loading mycobacterium DNA were able to induce antigen-specific mucosal and humoral immune response against tuberculosis after oral administration to mice [62]. Furthermore, since they are usually endocytosed by antigen presenting cells, they are able to promote an effective adjuvant induction of the immune system, even when carrying weaker antigens [53, 61]. The similarity of these lipid nanocarriers to cell membranes allows their use in the encapsulation of toxins that interact with cell membranes in vivo [63]. Thus, this strategy has been used to develop toxoid vaccines [63]. It is also possible to use functionalization to improve the targeting and the uptake by immune cells [53]. Wang et al. used mannosylated liposomes for oral immunization against hepatitis B virus [64]. Due to the presence of mannose, liposomes were internalized by mononuclear cells via mannose receptors [64].

The use of other components, such as fucose and Ulex europaeus agglutinin-1 (UEA-1), has already shown enhanced uptake by M cells [53]. Moreover, the conjugation with specific antibodies also enhances the uptake in the GI tract [53].

Archaeosomes, which are liposomes composed of lipids from archaebacterial membranes, have also been used due to their adjuvant effect and their good tolerance [65]. Given their structure, archaeosomes are very stable, resisting oxidation, enzymes, bile salts, and even temperature and low pH [61]. Further, they are less permeable and their usefulness as oral vaccine carrier has been proved [61, 65]. Other types of liposomes, such as lipoplex, lipopolyplex, and niosome, have also been used for vaccine formulations [66]. A comprehensive overview of the application of these liposomal structures is described elsewhere [66, 67].

Nanoemulsions have also been used due to their safety and immunogenicity, already proved in a phase I study [61]. These emulsions vary in size, from dozens to hundreds of nanometers, and can exist in an oil-in-water structure or the inverse, a water-in-oil structure [56]. A combination of strategies is also possible. For instance, Liau et al. used liposomes in a double emulsion for oral vaccines [68]. Due to their compartmental structure, with liposomes incorporated inside the water vacuoles of the double emulsion, they were able to achieve a prolonged release of the fluorescein isothiocyanate (FITC)-labeled ovalbumin-encapsulated nanoparticles [68]. Another strategy was designed by Hollmann et al. by creating liposomes coated with S-layer proteins from different bacteria [69]. The coating enhanced the liposomes' stability against bile salts, pH, and thermal shock [69].

Although the study of oral vaccine delivery systems has been exponentially increasing, there is still a long way to go for clinical application in humans [53]. Lipid NPs containing immunostimulatory compounds are still pointed out as one of the most advanced nanocarriers [52].

4.3.3 Treating Infectious Diseases

As previously mentioned, oral delivery is the most cost-effective routes and simultaneously the most comfortable to the patient [15]. However, due to the severe conditions of the GI tract, several lipid-based delivery systems have been widely studied to improve antimicrobial drugs' pharmacokinetics and to treat a wide range of infections. A summary of the advantages and drawbacks of oral administration and the usefulness of lipid-based NPs is presented in Fig. 4.3. Infectious diseases are disorders promoted by the invasion of pathogenic agents, including bacteria, viruses, fungi, and parasites. Herein we will give a few examples of how lipid-based NPs are advantageous for treating infectious diseases.

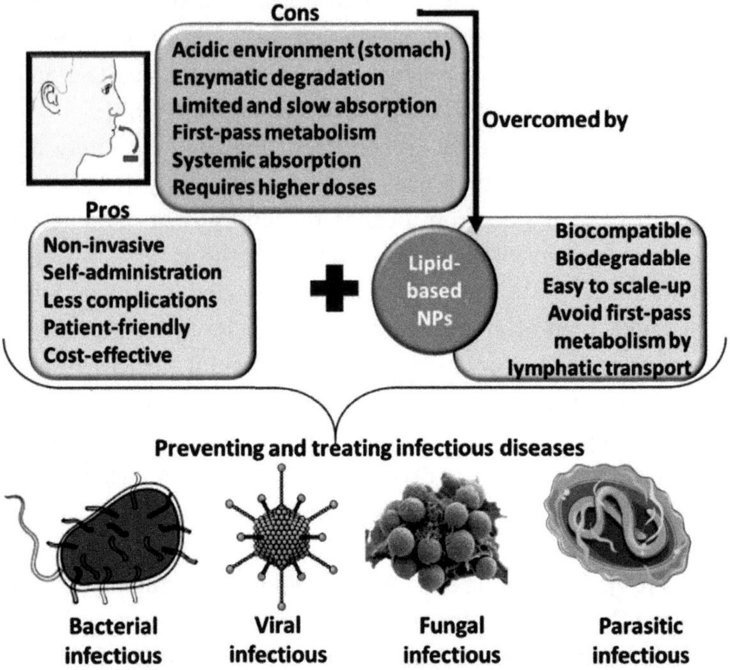

Figure 4.3 Schematic summary of the pros and cons of the oral administration and application of lipid-based NPs to overcome their limitations.

4.3.3.1 Bacterial infections

Given the limitation of the clinical use of antibiotics, namely their low BA [15, 16], researchers expect to find new and more efficient therapeutic approaches against bacterial infections. Lipid NPs and liposomes have been used to increase BA and aqueous solubility of many antibiotics.

Fluoroquinolones are widely used in clinics due to their safety and antibacterial efficacy [70, 71]. Norfloxacin is a synthetic third-generation fluoroquinolone with bactericidal activity against most of the gram-negative pathogens, being also active against some gram-positive bacteria [71, 72]. Nonetheless, this antimicrobial agent is associated with a poor aqueous solubility and low permeability [73]. Therefore, it has a low BA when administered by the oral route [71, 73]. To enhance its oral BA and pharmacological activity, Dong et al.

synthetized SLNs composed of SA as the solid lipid and polyvinyl acetate (PVA) as the emulsifier [73]. Plasma concentrations were higher for SLNs when compared with those of the free drug at all time points [73]. Antibacterial activity studies concluded that the SLNs remained effective against *Staphylococcus aureus* for a longer period of time when compared to that of the free drug [73]. Furthermore, cytotoxicity studies revealed that SLNs were biocompatible and presented low toxicity [73]. In conclusion, SLNs effectively enhanced BA and antibacterial activity of the drug, which can lead to lower doses and, consequently, decreased side effects and toxicity [73].

Besides antibiotics, many investigators are focusing their research on natural compounds with antimicrobial activity against a broad range of bacteria, such as free fatty acids [74]. For instance, linolenic acid has been highlighted due to its huge potential against all strains of *H. pylori*, including the ones resistant to metronidazole, which is one of the antibiotics usually administered in the standard treatment [75]. Therefore, Thamphiwatana et al. developed liposomal formulations to encapsulate linolenic acid in an innovative therapeutic approach [75]. To synthetize the liposomes, linolenic acid, L-α-phosphatidylcholine (EggPC), and cholesterol were mixed in a ratio of 3:6:1 [75]. The authors observed by fluorescence imaging that the liposomes were able to fuse with *H. pylori* [75]. To support this result, in vivo antibacterial studies were performed using C57BL/6 mouse as a model [75]. The animals were infected with the bacteria and further treated with the liposomal formulations, the standard triple therapy and PBS or bare liposome as control [75]. The results from this experiment showed that mice treated with the formulations had a reduced *H. pylori* burden of around 2.5 orders of magnitude compared to the negative controls [75]. On the other hand, triple therapy antibiotics only reduce it by around 1.4 orders of magnitude [75]. Moreover, a significant portion of liposomal formulations were able to accumulate within the mucus layer for at least 24 h, without compromising the gastric mucosal integrity [75]. Toxicological effects were evaluated in uninfected mice and demonstrated that the developed formulation was safe due to no apparent increase in gastric epithelial apoptosis [75]. Taking into account the previous results, it can be concluded that liposomes encapsulating linolenic acid have a huge potential to treat infections caused by *H. pylori* [75].

4.3.3.2 Viral infections

Viral infections are identified as one of the primary causes of death around the world [76]. Among the most worrisome viruses, the ones that are able to cause persistent infections that lead to cancer (e.g., hepatitis B virus and hepatitis C virus) or to immunodeficiency (e.g., HIV) are highlighted [76]. The main limitations of the current therapies are associated with the emergence of new viruses and the resistance mechanisms of the common viruses [77]. Since viruses are able to use the machinery of the host cell, the possibility of finding virus-specific proteins or enzymes that could be inhibited is extremely low [77]. Furthermore, when a virus-specific compound is found, it is usually particular for that virus, hindering the design of a broad-spectrum antivirus [77]. Another main drawback is the lack of antivirals able to combat latent infections [77].

Several NPs have been developed to improve the current antiviral treatment, including nanoemulsions, SLNs, and liposomes [78]. The advantages are well recognized. For instance, SLNs loading lapinavir, a poorly absorbed antiretroviral drug, proved their benefit by improving the drug's oral BA [79, 80]. These results were proven through in vivo studies, analyzing blood and lymph samples of Wistar rats [79, 80]. The higher oral BA was linked to intestinal lymphatic uptake, which avoids both the P-glycoprotein (Pgp) efflux pump and metabolism at the liver by cytochrome P450 [79]. The same benefit was highlighted by Makwana et al., when encapsulating efavirenz in SLNs composed of Gelucire 44/14, Compritol 888 ATO, Lipoid S75, and Poloxamer 188 [81]. In fact, SLNs were able to induce the secretion of chylomicron by enterocytes and, consequently, they increased the uptake through intestinal lymphatics [81]. This pathway of absorption is a key aspect when thinking of coadministration with other drugs that interact by inhibition or acceleration of metabolism at the liver [81].

NLCs have also been investigated as options to load antiviral drugs. For instance, Beloqui et al. loaded saquinavir in NLCs composed of Precirol ATO5, Miglyol 812, Tween 80, and Poloxamer 188 [82]. By changing their proportion, they reported sizes that varied from 165 nm to 1090 nm [82]. In vitro permeability studies showed that the amount of surfactant and the size were critical to the mechanism of transcytosis and to overcome the Pgp efflux pump [82]. A year later, they evaluated the advantages of coating with dextran-protamine [83]. Using Caco-2 cell monolayers and Caco-2/HT29-methotrexate (MTX) cell monolayers as models for enterocytes and mucus, respectively, they showed that dextran coating improved the permeability of the NPs [83]. In the mucus model, the improved effect was observed only when the surface was almost neutrally charged [83].

Liposomes have also been extensively used, and they have already proved their usefulness in lowering drug cytotoxicity as well as in improving the plasma half-life of antiviral drugs [78]. For instance, Zhong et al. synthetized liposomes composed of soya lecithin and cholesterol to incorporate a codrug composed of lamivudine and ursolic acid, which are used against the hepatitis B virus [84]. According to what was observed in the in vitro release, pharmacokinetic studies in rats showed that the time of circulation of the drug was enhanced when administered as a liposomal suspension [84]. Furthermore, the absorption was significantly improved, with a relative BA of 11-fold higher than the codrug suspension [84]. It is also possible to improve the stability of the nanosystem in the GI tract using tetraether lipids [85]. That strategy was applied by Uhl et al. to encapsulate Myrcludex B [85]. This peptide has activity against hepatitis B virus, and it is currently under evaluation in clinical trials [85]. However, its use is limited to subcutaneous application due to its degradation on oral administration [85]. In vivo studies revealed an improvement in the oral uptake of this peptide when incorporated in the liposomes [85].

The advantages of NPs in the treatment of viral infections are recognized, especially concerning the improvement of BA and the possibility to reduce the number of doses [77]. Consequently, the treatment can become a more cost-effective solution [77]. However, there are still ethical and scientific limitations that must be overcome to translate this technology into the clinic [77].

4.3.3.3 Fungal infections

In the last few decades, fungal infections have been an increasing concern, special due to nosocomial infections, which are a crucial cause of morbidity and mortality in hospitalized patients [86, 87]. One of the predominant nosocomial fungal pathogens is the *Candida* spp., among other fungi [87].

Due to the necessity to find new and more promising therapies against opportunistic fungi, a few studies reporting the use of both lipid NPs and liposomes were recently published. For instance, a recent work performed by Aljaeid et al. was focused on the development of SLNs as a drug delivery system to encapsulate miconazole, a broad-spectrum antifungal drug with poor aqueous solubility [88, 89]. The final formulation was composed of Precirol ATO5, Cremophor RH40, lecithin, and dicetylphosphate [90]. It showed an initial in vitro release of 45% of the drug (in the first 2 h), followed by a sustained release [90]. Nonetheless, an in vivo pharmacokinetic study was also performed, using albino male rabbits [90]. This experiment confirmed the results obtained in vitro, revealing a 2.5-fold increase of the drug's BA when compared to the capsule formulation currently existing in the market [90]. Furthermore, antifungal activity against *Candida albicans* was evaluated, showing that miconazole capsules exhibited enhanced antifungal activity compared to that of the powder alone [90]. However, SLNs showed the maximum inhibition zone compared to the previous two formulations [90]. Hence, the authors concluded that SLNs loading miconazole for oral administration enhanced its BA and were more efficient in the treatment of *C. albicans* infections [90].

Due to the relevance of miconizole as an antifungal agent, the encapsulation of this drug was evaluated in a study performed by Mendes and his coworkers [91]. Miconazole was encapsulated in NLCs to target the oral mucosa [91]. The optimal formulation was composed of Gelucire 43/01 as the solid lipid and Miglyol 812 as the lipid liquid, being posteriorly tested against *C. albicans* [91]. According to the results, the NLCs required lower concentrations of the drug to inhibit the total agent growth when compared to the concentrations of the drug needed in the free form [91]. The placebo formulations were also evaluated, and it was observed that the NPs per se have antifungal activity [91]. In a further experiment, the NLCs were incorporated in a hydrogel composed of glycerin and benzalkonium chloride to improve the local delivery to the oral mucosa [91]. The hydrogel showed a similar antimicrobial effect against *C. albicans* using a 17-fold lower drug dose compared to the oral gel available in the market [91]. Therefore, a lower drug concentration is needed in each administration and the dosage frequency decreases, minimizing undesirable side effects [91].

Besides miconazole, other active substances were also encapsulated in lipid NPs to improve their therapeutic efficiency against fungal infections, such as oral candidiasis. For instance, Garg et al. formulated both SLNs and NLCs to load eugenol, a phenolic compound of clove oil [92]. In this work, SA was used as a solid lipid and caprylic triglyceride as the liquid lipid [92]. The authors verified that the particle size decreased from 332 nm to 87.7 nm when the liquid lipid was added to the formulation [92]. The antifungal activity of the formulations was further evaluated both in vitro and in vivo [92]. In vitro studies revealed that the eugenol-loaded NPs were able to inhibit *C. albicans* cellular growth after 24 h of incubation [92]. On the other hand, SLNs alone were not efficient [92]. To confirm these results, in vivo studies were performed in immunosuppressed rats orally infected with the fungal agent [92]. In in vivo studies, the authors observed that the animals treated with eugenol-loaded SLNs presented significant reduction of the log CFU (colony-forming units) value, concluding that SLNs improve the therapeutic efficacy of eugenol against oral candidiasis [92].

Amphotericin B is also a relevant antifungal agent. Consequently, several liposomal-based drug delivery systems have been used to encapsulate the drug. However, oral delivery of liposomes presents some difficulties due to the hostile GI environment. Still, the drug delivery systems must be stable along the digestive tract. With this purpose, Skiba-Lahiani et al. developed oral amphotericin B–loaded liposomal formulations composed by vegetal ceramides [93]. The structural integrity of the vesicles was evaluated in the GI tract environment using an "artificial stomach-duodenum" model [93]. It was observed that the mean size increased after 30 min. of incubation in the digestive medium [93]. Nonetheless, the presence of ceramides in the formulations seems to limit these size variations when compared to formulations without ceramides [93]. Besides, these formulations were able to resist the hostile simulated GI environment composed of bile salts and digestive enzymes for a longer period of time [93]. In conclusion, liposomes with vegetal ceramides in their composition showed enhanced stability in the GI environment, which can be an advantage for oral administration of amphotericin B [93].

4.3.3.4 Parasitic infections

Parasitic infections are easily transmitted through contaminated food and are associated with high morbidity and mortality, especially in developing countries [94]. Despite the efforts, the fight against parasitic diseases is still unsatisfactory. The development of an effective vaccine is hindered by the fact that the majority of parasites do not induce a significant immune response [95]. Hence, the combat against parasitic infections is reduced to antiparasitic drugs [95]. Nevertheless, the drawbacks of the current treatment plans are several, including the toxicity and ineffectiveness of antiparasitic drugs, the costs of the treatment, and the resistance developed by parasites [96]. In general, antiparasitic drugs have low oral BA due to their degradation under the conditions of the GI tract combined with their low permeability [97]. Furthermore, some parasites are intracellular, hampering the exposure to drugs [96].

The possibility of using metallic-oxide-based NPs has been extensively studied due to their inhibitory effect against several infectious diseases [96]. NPs as drug delivery systems are also promising solutions for these types of diseases. The possibility of encapsulating synergic antimicrobials in the same delivery system, the high load capability with protection of the loaded drug, and the selective targeting may reduce the duration of the treatment, decrease toxic effects, and increase their efficacy [95]. For instance, Omwoyo et al. developed SLNs composed of SA to load dihydroartemisinin against malaria [98]. This drug is insoluble and unstable in acidic pH, which leads to a poor BA and a consequent insufficient efficacy [98]. The authors reported that the small size, of around 260 nm, allowed their uptake in the GI tract in a passive manner [98]. The protection of the drug against the severe conditions of the GI tract led to in vitro and in vivo efficacy [98]. When compared to the plain drug, the efficacy increased by 24% [98]. Therefore, this system may allow the reduction of the dose and its frequency [98]. Another artemisinin derivative, viz. arteether, was also encapsulated in SLNs composed of glycerol monostearate and three different surfactants (soya lecithin, Tween 80, and Pluronic F68) in different ratios [99]. Given its hydrophobicity, arteether is usually administered as an oily intramuscular injection [99]. In vivo studies showed a better pharmacokinetic profile after oral administration when the drug was incorporated into SLNs [99]. Thus, the use of this NP may allow a different and more patient-friendly route of administration [99]. Arteether was also incorporated in NLCs to allow their oral administration [100]. In vivo studies showed an almost complete eradication of the *Plasmodium berghei* from blood samples of mice [100]. Further, a longer survival period was also achieved with these NPs [100]. Interestingly, this drug had already been incorporated in liposomes, with higher BA (relative BA around 98% for the NPs versus 32% for the oral suspension) [101].

Lipid-based NPs can also reduce the cytotoxicity of antiparasitic drugs. As an example, de Souza et al. encapsulated praziquantel, a commonly used drug against schistosomiasis, in SLNs composed of SA and Poloxamer 188 [102]. Besides the higher efficacy of the drug delivery system compared to that of the free drug, the authors also reported a lower toxicity in HepG2 cells, which are human hepatoma cells [102]. Since prazinquantel is associated with hepatotoxicity, these in vitro studies showed that the drug-loaded NP is a promising solution for decreasing praziquantel toxicity in vivo [102]. Moreover, the delivery system has a lower intestinal permeation than the plain drug, indicating that SLNs act as a reservoir, allowing a higher dose in mesenteric veins of the intestine, where the parasites are located [102].

Liposomal formulations targeting macrophages have also been studied to treat intracellular infections, such as leishmaniosis [97]. The use of sugars, viz. mannose and fucose, allows the targeting of the mannose-fucose receptors of macrophages [97]. This strategy has also been used for SLNs, with a high antileishmanial activity [97]. An overview of the advantages of SLNs/NLCs versus liposomes in the treatment of parasitic diseases is well described by Date et al. [95].

4.4 Evaluating Lipid-Based Nanoparticles

4.4.1 Studies to Assess Pharmacokinetic Properties

Oral administration requires that NPs be able to overcome the harsh conditions of the GI tract and to cross its barriers to reach the target. These properties have to be fully studied to evaluate the pharmacokinetic properties of the designed NP (Fig. 4.4).

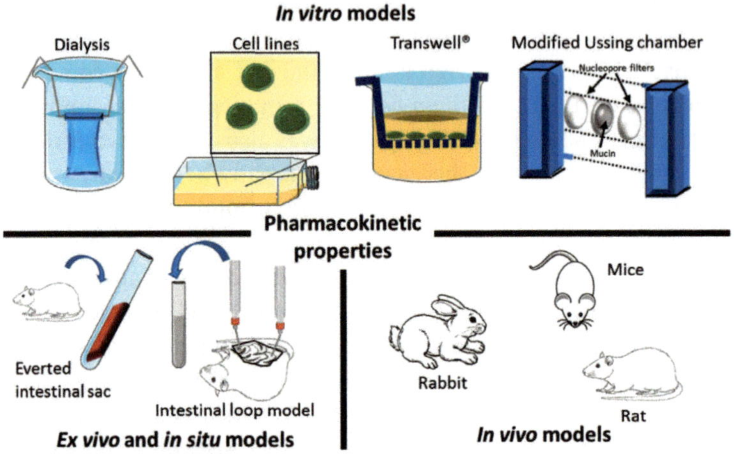

Figure 4.4 Schematic representation of the most common assays performed to evaluate the pharmacokinetic properties of NPs after oral administration. These assays include in vitro, ex vivo, in situ, and in vivo models.

In vitro techniques are the simplest ones to perform and are usually the first step after NP optimization. One of the most common experiments to evaluate the pharmacokinetics of drug delivery systems involves the observation of drug release in vitro. This evaluation is usually performed using the dialysis method, which is widely applied to lipid NPs. In this method, drug-loaded lipid NPs are added to a dialysis bag and dialyzed against a receiver solution [51]. For instance, Xie and coworkers applied this method, where ofloxacin-loaded SLNs were suspended in 0.1 M hydrochloric acid in a dialysis bag and further dialyzed against a higher volume of 0.1 M hydrochloric acid at 37°C under magnetic stirring [51]. At different time points, samples were taken from the receiver solution to determine the amount of drug diffused through the dialysis bag [51]. However, to have a more comprehensive and reliable medium, simulated gastric and intestinal fluids can be used [30]. Nevertheless, these assays are extremely simplistic, since the GI tract has enzymes and ionic strengths that affect the chemical and colloidal stability of the delivery system [30]. The complexity can be increased using cell models, namely Caco-2 cells and cocultures with Caco-2 cells and HT29 cell lines [27]. By combining both cell lines, a more complete model is obtained, with both absorptive and goblet cells [27]. Using a Transwell setup, the transport of the NPs across the layer of cells can be assessed [30]. To study the possibility of uptake by M cells directly to the mucosal immune system, a coculture of Caco-2 cell lines and lymphocytes can be used [30].

The interaction with the mucus is also a key aspect of the pharmacokinetic, since it is the first barrier of the GI tract. It is also possible to use mucus models to evaluate the interaction established between the delivery system and the mucus. These models can be as simplistic as mucin solutions or as complex as those more similar to in vivo, by resorting to mucus provided by horses, pigs, or humans [103]. A comprehensive overview regarding mucus models is provided elsewhere [103]. Several techniques can be used in this type of studies, from zeta potential, size measurements, and rheological measurements to more complex permeability studies [28]. Besides the most common Transwell device, an Ussing chamber with a donor and an acceptor compartment separated by a vertical layer of mucus is also an alternative to mucus-NP interactions studies [28]. A very recent review was published with a summary of in vitro models of the intestinal barrier and the mechanisms of transport of lipid NPs [104]. More recent and advanced techniques are multiple particle tracking and the gut-on-a-chip setup. The first one uses video microscopy to track NPs, providing useful information regarding their qualitative movement in different environments [28]. The second novelty is a model able to mimic the properties of the intestine and its microbial flora, with fluid flow and mechanical movements [30].

To improve in vitro models and to overcome their limitations, especially regarding the lack of tissue interactions, ex vivo and in situ techniques have been performed [27]. Ex vivo techniques include the isolation of intestinal tissues, which may ultimately damage their properties [27]. On the other hand, in in situ models the connection of the excision with the animal is kept during the experiment and only after a variable period of time the animal is sacrificed [27, 30].

In vitro studies are useful for a preliminary study; however, in vivo models must be used to assess NPs absorption, biodistribution, tissue accumulation, and the ability of the animal to excrete the NPs [24]. The animals used for these experiments can be various, such as rats [73], mice [51], and rabbits [90]. In the work elaborated by Aljaeid and Hosny, albino male rabbits were randomly divided in two groups [90]. One group worked as a control, where the rabbits were treated with the commercially available formulation in the market [90]. The other group was treated with the optimal solid lipid formulations developed by the authors [90]. At specific time points of the experiment, blood samples were collected and analyzed using high-performance liquid chromatography [90]. Furthermore, lymph samples of animal models can also be evaluated to look for an uptake by the intestinal lymphatic pathway [79].

Despite their unquestionable usefulness, animal models have their limitations that should be taken into account when translating the NP into clinical practice. An interesting and comprehensive overview by Kararli compares the GI tract of humans with the animal models commonly used in the preclinical studies [105]. For instance, the most common in vivo models, viz. rats and mice, have less mucin than humans [106]. On the other hand, models such as rabbits have a different quantity of Peyer's patches and rodents have less acidic stomachs [106].

4.4.2 Studies to Assess Therapeutic Efficacy

In this section, several examples demonstrate that NPs are able to enhance the therapeutic efficacy of antimicrobial agents. In fact, in general, drug-loaded NPs are more effective than the plain drugs [22]. This effect may happen by the improvement of the drug's BA, by improvement in its local concentration through controlled and targeted release, or even by a change in the route of administration to a more patient-friendly one [22]. The assessment of the therapeutic efficacy is then a crucial and key aspect to validate the usefulness of the delivery system. Indeed, it would be pointless to have a higher BA if it would not increase the drug's therapeutic efficacy.

In vitro assays are valuable tools to assess the therapeutic efficacy of antimicrobial NPs. For bacteria, in vitro antimicrobial activity studies are based on the determination of the minimum inhibitory concentration (MIC) and of the minimum bactericidal concentration (MBC) [107]. The MIC can be defined as the lowest concentration of a antimicrobial agent required to inhibit visible growth of an inoculum of the bacterium in a broth dilution test [108]. On other hand, MBC is the minimal concentration of a antimicrobial agent able to kill ≥99.9% of the initial inoculum within 24 h, and it is determined from broth dilution MIC tests by subculturing onto antibiotic-free agar media [108]. For instance, in the work of Dong et al. serial dilutions of NP suspensions were mixed with *S. aureus* to a final bacterial concentration of 5.5×10^5 CFU/mL [73]. The mixture was posteriorly incubated at 37°C for 12 h and 24 h, with shaking [73]. Usually, the MIC is determined by a visual comparison of the medium turbidity of the cultures and the control tubes [108]. Alternatively, the MIC value can be determined by spectroscopy [110]. MBC is measured by subculturing the broths previously used for MIC measurement onto fresh agar plates at 37°C [73]. Therefore, the microbial colonies are counted when they are observed by the naked eye [73]. Besides the determination of MIC and MBC, studies of NP-pathogen interactions can be performed, such as agglutination assays, epifluorescence studies, TEM technique, and quantification of bacterial adhesion using luminescence [19]. These studies are reported in more detail elsewhere [19].

In the case of fungal infections, the determination of the MIC can also be applied. For instance, Mendes et al. evaluated the antifungal activity of miconazole against *C. albicans* by comparing the MIC of miconazole encapsulated in NLCs to its free form [91]. The same type of assay is also performed with parasites, such as for antimalarial efficacy [98] and against *Schistosoma mansoni* [102].

For viruses, the feasibility of in vitro studies depends on the virus. The recommendation is that in vitro studies be performed if there are currently available cell cultures in which the virus can survive and complete its life cycle [111]. For those viruses where in vitro studies can be performed, these studies can be a good complement before the in vivo studies. As an example, Ron-Doitch et al. infected HaCaT cells with herpes simplex virus 1 (HSV-1) and then evaluated cell viability [112]. In such assays, the EC_{50} (half maximal effective concentration) is used instead of the "inhibitory" concentration once the NPs' effect is determined by their ability to protect an infected cell [112].

Nonetheless, similar to what has been described for pharmacokinetic assays, in vivo efficacy against microbial agents should also be assessed since BA strongly affects NP efficacy. Before the evaluation of the therapeutic efficacy of the delivery system, it is important to establish the infective dose of the pathogen, which is the dose necessary to establish the infection without inducing severe or lethal infections in the animals [113]. On the other hand, the lethal dose is responsible for deaths in 100% of the cases [113]. To perform this experiment, animals are divided into groups and infected with different concentrations of the pathogen. As an example, Wang et al. tested different concentrations of *Escherichia coli* [113]. The concentration that induces the deaths of all animals was considered the lethal dose [113]. When administering lower concentrations, some mice survived and bacterial counts were determined in several organs: heart, liver, lung, spleen, and kidneys [113]. Below the lethal concentration the bacteria were undetectable in all organs [113]. Therefore, this concentration was considered the infective dose [113]. A posterior mortality protection study was performed by the authors using 70 mice divided into 7 groups [113]. All the animals were injected intraperitoneally with the lethal infection inoculum size of bacteria and further treated with a single dose of norfloxacin-loaded SLNs [113]. Deaths were recorded over a 72 h period [113]. As in bacterial infections, the in vivo therapeutic efficacy of NPs against other pathogenic agents (viz. viruses, fungi, and parasites) can be evaluated by infecting animals with the pathogen of interest. After the infection, the survival rate, the manifestation of symptoms, or the rate of the pathogenic agent can be evaluated.

4.5 Conclusions and Future Perspectives

Among the scientific community, there is no doubt regarding the value of the oral route. Nevertheless, its limitations have also been highlighted, especially concerning the lower efficacy of drugs due to their lower BA after oral administration. Lipid NPs have been extensively studied to overcome several drawbacks of the current treatment of diverse diseases.

Their ability to protect the drug from the harsh conditions of the GI tract is a key advantage of drug delivery systems. This may be particularly interesting for the encapsulation of antimicrobial peptides. This emerging field includes naturally occurring molecules, produced by organisms such as plants and insects [114]. Their activity against several bacteria have been proved [114]. Nevertheless, being peptides, their stability in the GI tract may be compromised, which can be overcome by NPs. However, NPs have to be designed in conformity with the purpose, mainly concerning their stability under acidic conditions and in the presence of degradative enzymes and their difficulty in crossing the mucus barrier. Several strategies have been made in order to overcome these limitations, such as mucus-penetrating NPs (reviewed in "Developments of Mucus Penetrating Nanoparticles" [27]) and mucoadhesive NPs [106]. However, despite the efforts, there are only a few NPs in the market [22]. One of the drawbacks that hinder the use in the clinic is the translation from the assays performed in the academic field to the administration to humans [22]. For instance, the most common in vivo models have several differences when compared to the anatomy and physiologic of the GI tract in humans [106]. All these factors, combined with different transit times and different commensal colonizations, hamper the translation to the clinic [106]. Furthermore, other limitations rise, such as the difficulty in scaling up and the cost benefit, which are most of the time far from desirable [22]. Another drawback is the lack of toxicity studies [21]. Despite the well-recognized safety of both the oral route and the lipid-based delivery systems, the low cytotoxicity should not be interpreted as nontoxicity in vivo, since other factors, such as genotoxicity and immunogenic toxicity, must be taken into account [24]. These limitations lead to a huge discrepancy between the thousands of papers published reporting NPs against infectious diseases and the only 276 clinical trials till date with such NPs [22]. The future encompasses an interdisciplinary collaboration [22]. To have an ideal NP, it must be effective and nontoxic. Thus, besides the collaboration of health care researchers, such as physicists and pharmacists, other fields (e.g., biologists, virologists, and toxicologists) must be involved [22, 77].

At the end, and despite the extra cost of NPs in comparison with the current therapies, all the advantages of delivery systems may save financial resources. In fact, the efficacy is highly enhanced when NPs are used, lowering the frequency of dosage and the incidence of side effects. NPs may have an additional interest in cases where two or more drugs are used in the first-line treatment, facilitating the fight against viruses like HIV and bacteria such as *H. pylori* [22].

Acknowledgments

Daniela Lopes and Cláudia Nunes thank Fundação para a Ciência e a Tecnologia (FCT) for funding through the PhD scholarship and investigator grant, namely PD/BD/105957/2014 and IF/00293/2015, respectively. This work was supported by FCT through the FCT PhD programs and by Programa Operacional do Capital Humano (POCH), specifically by the BiotechHealth Programme (Doctoral Programme on Cellular and Molecular Biotechnology Applied to Health Sciences), reference PD/00016/2012. Additionally, this work received financial support from the European Union (FEDER funds POCI/01/0145/FEDER/007728) and National Funds (FCT/MEC, Fundação para a Ciência e Tecnologia and Ministério da Educação e Ciência) under the partnership agreement PT2020 UID/MULTI/04378/2013.

References

1. Cohen, M.L. Changing patterns of infectious disease. *Nature*, 2000, **406**(6797):762–767.
2. Huh, A.J., et al. "Nanoantibiotics": a new paradigm for treating infectious diseases using nanomaterials in the antibiotics resistant era. *J Controlled Release*, 2011, **156**(2):128–145.
3. The world health report 1999: making a difference. 1999, World Health Organization.
4. Fonkwo, P.N. Pricing infectious disease: the economic and health implications of infectious diseases. *EMBO Rep*, 2008, **9**(Suppl 1):S13–S17.
5. Taylor, L.H., et al. Risk factors for human disease emergence. *Philos Trans R Soc Lond B Biol Sci*, 2001, **356**(1411):983–989.
6. The world health report 2000, health systems: improving performance. 2000, World Health Organization.
7. Antimicrobial resistance global report on surveillance: 2014 summary. 2014, World Health Organization.
8. Pelgrift, R.Y., et al. Nanotechnology as a therapeutic tool to combat microbial resistance. *Adv Drug Deliv Rev*, 2013, **65**(13–14):1803–1815.
9. Antibiotic resistance fact sheet. 2015, World Health Organization. [cited 2016; Available from: http://www.who.int/mediacentre/factsheets/antibiotic-resistance/en/.
10. McBain, A.J., et al. Biocide tolerance and the harbingers of doom. *Int Biodeter Biodegrad*, 2001, **47**(2):55–61.

11. Walsh, C. Molecular mechanisms that confer antibacterial drug resistance. *Nature*, 2000, **406**(6797):775–781.
12. Sköld, O. *Antibiotics and Antibiotic Resistance*. John Wiley & Sons, Hoboken, New Jersey, 2011.
13. Hart, C.A. Antibiotic resistance: an increasing problem? *BMJ*, 1998, **316**(7140):1255–1256.
14. Mathers, C., et al. The global burden of disease: 2004 update. 2008, World Health Organization.
15. Harde, H., et al. Solid lipid nanoparticles: an oral bioavailability enhancer vehicle. *Expert Opin Drug Deliv*, 2011, **8**(11):1407–1424.
16. Pond, S.M., et al. First-pass elimination. Basic concepts and clinical consequences. *Clin Pharmacokinet*, 1984, **9**(1):1–25.
17. Beaulac, C., et al. Eradication of mucoid Pseudomonas aeruginosa with fluid liposome-encapsulated tobramycin in an animal model of chronic pulmonary infection. *Antimicrob Agents Chemother*, 1996, **40**(3):665–669.
18. Xie, S., et al. Biodegradable nanoparticles for intracellular delivery of antimicrobial agents. *J Controlled Release*, 2014, **187**:101–117.
19. Lopes, D., et al. Targeting strategies for the treatment of Helicobacter pylori infections, in *Nano Based Drug Delivery*, Naik, J. ed. IAPC Publishing, Croatia, 2015, pp. 339–366.
20. MacGregor, R.R., et al. Oral administration of antibiotics: a rational alternative to the parenteral route. *Clin Infect Dis*, 1997, **24**(3):457–467.
21. Araújo, F., et al. Safety and toxicity concerns of orally delivered nanoparticles as drug carriers. *Expert Opin Drug Metab Toxicol*, 2014, **11**(3):1–13.
22. Zazo, H., et al. Current applications of nanoparticles in infectious diseases. *J Controlled Release*, 2016, **224**:86–102.
23. Bakhru, S.H., et al. Oral delivery of proteins by biodegradable nanoparticles. *Adv Drug Deliv Rev*, 2013, **65**(6):811–821.
24. Ojer, P., et al. Toxicity evaluation of nanocarriers for the oral delivery of macromolecular drugs. *Eur J Pharm Biopharm*, 2015, **97**(Pt A):206–217.
25. Gamazo, C., et al. Mimicking microbial strategies for the design of mucus-permeating nanoparticles for oral immunization. *Eur J Pharm Biopharm*, 2015, **96**:454–463.

26. Boltin, D., et al. Pharmacological and alimentary alteration of the gastric barrier. *Best Pract Res Clin Gastroenterol*, 2014, **28**(6):981–994.

27. Liu, M., et al. Developments of mucus penetrating nanoparticles. *Asian J Pharm Sci*, 2015, **10**(4):275–282.

28. Griessinger, J., et al. Methods to determine the interactions of micro- and nanoparticles with mucus. *Eur J Pharm Biopharm*, 2015, **96**:464–476.

29. Arora, S., et al. Amoxicillin loaded chitosan-alginate polyelectrolyte complex nanoparticles as mucopenetrating delivery system for h. pylori. *Sci Pharm*, 2011, **79**(3):673–694.

30. Gamboa, J.M., et al. In vitro and in vivo models for the study of oral delivery of nanoparticles. *Adv Drug Deliv Rev*, 2013, **65**(6):800–810.

31. Lopes, D., et al. Eradication of Helicobacter pylori: past, present and future. *J Controlled Release*, 2014, **189**:169–86.

32. Boegh, M., et al. Mucus as a barrier to drug delivery - understanding and mimicking the barrier properties. *Basic Clin Pharmacol Toxicol*, 2015, **116**:178–186.

33. Thamphiwatana, S., et al. Nanoparticle-stabilized liposomes for pH-responsive gastric drug delivery. *Langmuir*, 2013, **29**(39):12228–12233.

34. Cone, R.A. Barrier properties of mucus. *Adv Drug Deliv Rev*, 2009, **61**(2):75–85.

35. Allen, A., et al. Gastroduodenal mucus bicarbonate barrier: protection against acid and pepsin. *Am J Physiol Cell Physiol*, 2005, **288**(1):C1–C19.

36. Battaglia, L., et al. Lipid nanoparticles: state of the art, new preparation methods and challenges in drug delivery. *Expert Opin Drug Deliv*, 2012, **9**(5):497–508.

37. Fricker, G., et al. Phospholipids and lipid-based formulations in oral drug delivery. *Pharm Res*, 2010, **27**(8):1469–1486.

38. Pardeike, J., et al. Lipid nanoparticles (SLN, NLC) in cosmetic and pharmaceutical dermal products. *Int J Pharm*, 2009, **366**(1–2):170–184.

39. Silva, A.C., et al. Preparation, characterization and biocompatibility studies on risperidone-loaded solid lipid nanoparticles (SLN): high pressure homogenization versus ultrasound. *Colloids Surf B*, 2011, **86**(1):158–165.

40. Rawat, M.K., et al. In vivo and cytotoxicity evaluation of repaglinide-loaded binary solid lipid nanoparticles after oral administration to rats. *J Pharm Sci*, 2011, **100**(6):2406–2417.
41. Severino, P., et al. Current state-of-art and new trends on lipid nanoparticles (SLN and NLC) for oral drug delivery. *J Drug Deliv*, 2012, **2012**:750891.
42. Das, S., et al. Recent advances in lipid nanoparticle formulations with solid matrix for oral drug delivery. *AAPS PharmSciTech*, 2011, **12**(1):62–76.
43. Anton, N., et al. Design and production of nanoparticles formulated from nano-emulsion templates - a review. *J Controlled Release*, 2008, **128**(3):185–199.
44. Ali Khan, A., et al. Advanced drug delivery to the lymphatic system: lipid-based nanoformulations. *Int J Nanomed*, 2013, **8**:2733–2744.
45. Drulis-Kawa, Z., et al. Liposomes as delivery systems for antibiotics. *Int J Pharm*, 2010, **387**(1–2):187–198.
46. Blecher, K., et al. The growing role of nanotechnology in combating infectious disease. *Virulence*, 2011, **2**(5):395–401.
47. Obonyo, M., et al. Antibacterial activities of liposomal linolenic acids against antibiotic-resistant Helicobacter pylori. *Mol Pharm*, 2012, **9**(9):2677–2685.
48. Muller, R.H., et al. Solid lipid nanoparticles (SLN) for controlled drug delivery - a review of the state of the art. *Eur J Pharm Biopharm*, 2000, **50**(1):161–177.
49. Wilczewska, A.Z., et al. Nanoparticles as drug delivery systems. *Pharmacol Rep*, 2012, **64**(5):1020–1037.
50. Abed, N., et al. Nanocarriers for antibiotics: a promising solution to treat intracellular bacterial infections. *Int J Antimicrob Agents*, 2014, **43**(6):485–496.
51. Xie, S., et al. Preparation and evaluation of ofloxacin-loaded palmitic acid solid lipid nanoparticles. *Int J Nanomed*, 2011, **6**:547–555.
52. Correia-Pinto, J.F., et al. Vaccine delivery carriers: insights and future perspectives. *Int J Pharm*, 2013, **440**(1):27–38.
53. Davitt, C.J., et al. Delivery strategies to enhance oral vaccination against enteric infections. *Adv Drug Deliv Rev*, 2015, **91**:52–69.
54. Marasini, N., et al. Oral delivery of nanoparticle-based vaccines. *Expert Rev Vaccines*, 2014, **13**(11):1361–1376.

55. Almeida, A.J., et al. Solid lipid nanoparticles as a drug delivery system for peptides and proteins. *Adv Drug Deliv Rev*, 2007, **59**(6):478–490.
56. Zhao, L., et al. Nanoparticle vaccines. *Vaccine*, 2014, **32**(3):327–337.
57. das Neves, J., et al. Nanotechnology-based systems for the treatment and prevention of HIV/AIDS. *Adv Drug Deliv Rev*, 2010, **62**(4–5):458–477.
58. Minato, S., et al. Application of polyethyleneglycol (PEG)-modified liposomes for oral vaccine: effect of lipid dose on systemic and mucosal immunity. *J Controlled Release*, 2003, **89**(2):189–197.
59. Yue, H., et al. Polymeric micro/nanoparticles: particle design and potential vaccine delivery applications. *Vaccine*, 2015, **33**(44):5927–5936.
60. Liu, L., et al. Immune responses to vaccines delivered by encapsulation into and/or adsorption onto cationic lipid-PLGA hybrid nanoparticles. *J Controlled Release*, 2016, **225**:230–239.
61. Newsted, D., et al. Advances and challenges in mucosal adjuvant technology. *Vaccine*, 2015, **33**(21):2399–2405.
62. Wang, D., et al. Liposomal oral DNA vaccine (mycobacterium DNA) elicits immune response. *Vaccine*, 2010, **28**(18):3134–3142.
63. Fang, R.H., et al. Engineered nanoparticles mimicking cell membranes for toxin neutralization. *Adv Drug Deliv Rev*, 2015, **90**:69–80.
64. Wang, T., et al. Mannosylated and lipid A-incorporating cationic liposomes constituting microneedle arrays as an effective oral mucosal HBV vaccine applicable in the controlled temperature chain. *Colloids Surf B*, 2015, **126**:520–530.
65. Li, Z., et al. Archaeosomes with encapsulated antigens for oral vaccine delivery. *Vaccine*, 2011, **29**(32):5260–5266.
66. Watson, D.S., et al. Design considerations for liposomal vaccines: influence of formulation parameters on antibody and cell-mediated immune responses to liposome associated antigens. *Vaccine*, 2012, **30**(13):2256–2272.
67. Wang, S., et al. Intranasal and oral vaccination with protein-based antigens: advantages, challenges and formulation strategies. *Protein Cell*, 2015, **6**(7):480–503.
68. Liau, J.J., et al. A lipid based multi-compartmental system: liposomes-in-double emulsion for oral vaccine delivery. *Eur J Pharm Biopharm*, 2015, **97**(Pt A):15–21.

69. Hollmann, A., et al. Characterization of liposomes coated with S-layer proteins from lactobacilli. *Biochim Biophys Acta*, 2007, **1768**(3):393–400.

70. Appelbaum, P.C., et al. The fluoroquinolone antibacterials: past, present and future perspectives. *Int J Antimicrob Agents*, 2000, **16**(1):5–15.

71. Breda, S.A., et al. Solubility behavior and biopharmaceutical classification of novel high-solubility ciprofloxacin and norfloxacin pharmaceutical derivatives. *Int J Pharm*, 2009, **371**(1–2):106–113.

72. Holmes, B., et al. Norfloxacin. A review of its antibacterial activity, pharmacokinetic properties and therapeutic use. *Drugs*, 1985, **30**(6):482–513.

73. Dong, Z., et al. Preparation and in vitro, in vivo evaluations of norfloxacin-loaded solid lipid nanopartices for oral delivery. *Drug Deliv*, 2011, **18**(6):441–450.

74. Desbois, A.P., et al. Antibacterial free fatty acids: activities, mechanisms of action and biotechnological potential. *Appl Microbiol Biotechnol*, 2010, **85**(6):1629–1642.

75.Thamphiwatana, S., et al. In vivo treatment of Helicobacter pylori infection with liposomal linolenic acid reduces colonization and ameliorates inflammation. *Proc Natl Acad Sci USA*, 2014, **111**(49):17600–17605.

76. Galdiero, S., et al. Silver nanoparticles as potential antiviral agents. *Molecules*, 2011, **16**(10):8894–8918.

77. Lembo, D., et al. Nanoparticulate delivery systems for antiviral drugs. *Antivir Chem Chemother*, 2010, **21**(2):53–70.

78. Shao, J., et al. Nanodrug formulations to enhance HIV drug exposure in lymphoid tissues and cells: clinical significance and potential impact on treatment and eradication of HIV/AIDS. *Nanomedicine*, 2016, **11**(5):545–564.

79. Negi, J.S., et al. Development of solid lipid nanoparticles (SLNs) of lopinavir using hot self nano-emulsification (SNE) technique. *Eur J Pharm Sci*, 2013, **48**(1–2):231–239.

80. Aji Alex, M.R., et al. Lopinavir loaded solid lipid nanoparticles (SLN) for intestinal lymphatic targeting. *Eur J Pharm Sci*, 2011, **42**(1–2):11–18.

81. Makwana, V., et al. Solid lipid nanoparticles (SLN) of Efavirenz as lymph targeting drug delivery system: elucidation of mechanism of uptake using chylomicron flow blocking approach. *Int J Pharm*, 2015, **495**(1):439–446.

82. Beloqui, A., et al. Mechanism of transport of saquinavir-loaded nanostructured lipid carriers across the intestinal barrier. *J Controlled Release*, 2013, **166**(2):115–123.

83. Beloqui, A., et al. Dextran-protamine coated nanostructured lipid carriers as mucus-penetrating nanoparticles for lipophilic drugs. *Int J Pharm*, 2014, **468**(1–2):105–111.

84. Zhong, Y., et al. Preparation and evaluation of liposome-encapsulated codrug LMX. *Int J Pharm*, 2012, **438**(1–2):240–248.

85. Uhl, P., et al. A liposomal formulation for the oral application of the investigational hepatitis B drug Myrcludex B. *Eur J Pharm Biopharm*, 2016, **103**:159–166.

86. Jarvis, W.R. Epidemiology of nosocomial fungal infections, with emphasis on Candida species. *Clin Infect Dis*, 1995, **20**(6):1526–1530.

87. Perlroth, J., et al. Nosocomial fungal infections: epidemiology, diagnosis, and treatment. *Med Mycol*, 2007, **45**(4):321–346.

88. Jain, S., et al. Design and development of solid lipid nanoparticles for topical delivery of an anti-fungal agent. *Drug Deliv*, 2010, **17**(6):443–451.

89. Pedersen, M., et al. Formation and antimycotic effect of cyclodextrin inclusion complexes of econazole and miconazole. *Int J Pharm*, 1993, **90**(3):247–254.

90. Aljaeid, B.M., et al. Miconazole-loaded solid lipid nanoparticles: formulation and evaluation of a novel formula with high bioavailability and antifungal activity. *Int J Nanomed*, 2016, **11**:441–447.

91. Mendes, A.I., et al. Miconazole-loaded nanostructured lipid carriers (NLC) for local delivery to the oral mucosa: improving antifungal activity. *Colloids Surf B*, 2013, **111**:755–763.

92. Garg, A., et al. Enhancement in antifungal activity of eugenol in immunosuppressed rats through lipid nanocarriers. *Colloids Surf B*, 2011, **87**(2):280–288.

93. Skiba-Lahiani, M., et al. Development and characterization of oral liposomes of vegetal ceramide based amphotericin B having enhanced dry solubility and solubility. *Mater Sci Eng C Mater Biol Appl*, 2015, **48**:145–149.

94. Torgerson, P.R., et al. World Health Organization estimates of the global and regional disease burden of 11 foodborne parasitic diseases, 2010: a data synthesis. *PLoS Med*, 2015, **12**(12):e1001920.

95. Date, A.A., et al. Parasitic diseases: liposomes and polymeric nanoparticles versus lipid nanoparticles. *Adv Drug Deliv Rev*, 2007, **59**(6):505–521.

96. Yah, C.S., et al. Nanoparticles as potential new generation broad spectrum antimicrobial agents. *Daru*, 2015, **23**:43.

97. Pham, T.T., et al. Strategies for the design of orally bioavailable antileishmanial treatments. *Int J Pharm*, 2013, **454**(1):539–552.

98. Omwoyo, W.N., et al. Development, characterization and antimalarial efficacy of dihydroartemisinin loaded solid lipid nanoparticles. *Nanomedicine*, 2016, **12**(3):801–809.

99. Dwivedi, P., et al. Pharmacokinetics study of arteether loaded solid lipid nanoparticles: an improved oral bioavailability in rats. *Int J Pharm*, 2014, **466**(1–2):321–327.

100. Ali, Z., et al. Development and characterization of arteether-loaded nanostructured lipid carriers for the treatment of malaria. *Artif Cells Nanomed Biotechnol*, 2014, **44**(2):545–549.

101. Bayomi, M.A., et al. In vivo evaluation of arteether liposomes. *Int J Pharm*, 1998, **175**:1–7.

102. de Souza, A.L., et al. In vitro evaluation of permeation, toxicity and effect of praziquantel-loaded solid lipid nanoparticles against Schistosoma mansoni as a strategy to improve efficacy of the schistosomiasis treatment. *Int J Pharm*, 2014, **463**(1):31–37.

103. Groo, A.C., et al. Mucus models to evaluate nanomedicines for diffusion. *Drug Discov Today*, 2014, **19**(8):1097–1108.

104. Beloqui, A., et al. Mechanisms of transport of polymeric and lipidic nanoparticles across the intestinal barrier. *Adv Drug Deliv Rev*, 2016, 106(Pt B):242–255.

105. Kararli, T.T. Comparison of the gastrointestinal anatomy, physiology, and biochemistry of humans and commonly used laboratory animals. *Biopharm Drug Dispos*, 1995, **16**:351–380.

106. Ensign, L.M., et al. Oral drug delivery with polymeric nanoparticles: the gastrointestinal mucus barriers. *Adv Drug Deliv Rev*, 2012, **64**(6):557–570.

107. Andrews, J.M. Determination of minimum inhibitory concentrations. *J Antimicrob Chemother*, 2001, **48**(Suppl 1):5–16.

108. French, G.L. Bactericidal agents in the treatment of MRSA infections-the potential role of daptomycin. *J Antimicrob Chemother*, 2006, **58**(6):1107–1117.

109. Hazzah, H.A., et al. Gelucire-based nanoparticles for curcumin targeting to oral mucosa: preparation, characterization, and antimicrobial activity assessment. *J Pharm Sci*, 2015, **104**(11):3913–3924.

110. Severino, P., et al. Sodium alginate-cross-linked polymyxin B sulphate-loaded solid lipid nanoparticles: antibiotic resistance tests and HaCat and NIH/3T3 cell viability studies. *Colloids Surf B*, 2015, **129**:191–197.

111. Guidance for industry; Antiviral product development - conducting and submiting virology studies to the agency. Department of Health and Human Services (Food and Drug Administration), 2006.

112. Ron-Doitch, S., et al. Reduced cytotoxicity and enhanced bioactivity of cationic antimicrobial peptides liposomes in cell cultures and 3D epidermis model against HSV. *J Controlled Release*, 2016, **229**:163–171.

113. Wang, Y., et al. Preparation and stability study of norfloxacin-loaded solid lipid nanoparticle suspensions. *Colloids Surf B*, 2012, **98**:105–111.

114. Steckbeck, J.D., et al. Antimicrobial peptides: new drugs for bad bugs? *Expert Opin Biol Ther*, 2014, **14**(1):11–14.

Chapter 5

Oral Administration of Nanoparticles and Gut Microbiota–Mediated Effects

Ana Raquel Madureira and Manuela Pintado
CBQF - Center for Biotechnology and Fine Chemistry - Associated Laboratory, Superior School of Biotechnology, Portuguese Catholic University, Porto, Portugal
rmadureira@porto.ucp.pt

Nowadays, there is the knowledge that compounds orally delivered by diet or medication influence the gut microbiota growth and metabolism, and, in turn, any changes in gut microbiota will influence the human body physiologic processes. When developing a new oral delivery system that is supposed to reach the intestine, the interaction of these systems with the gut bacteria has to be studied. Gut bacteria will metabolize these systems and can influence the bioavailability of the compounds that will be delivered. Hence, we are assisting in an increase in the number of studies on this subject, using in vitro fermentation models or even in vivo models with animals. This section will focus on the studies that have been done so far in what concerns the modulation of gut microbiome by oral delivery nanosystems.

Nanoparticles in Life Sciences and Biomedicine
Edited by Ana Rute Neves and Salette Reis

ISBN 978-981-4745-98-7 (Hardcover), 978-1-351-20735-5 (eBook)
www.panstanford.com

5.1 The Gastrointestinal Tract

The human gastrointestinal (GI) tract is divided into four specific regions: mouth, stomach, small intestine, and colon (Fig. 5.1). The first step of digestion occurs in the mouth, where a food bolus is generated and is carried through the esophagus to the stomach. In the stomach, the low pH is the predominant feature, which results from the high concentration of hydrochloric acid (HCl). Here, protein and lipid digestion takes place in the presence of gastric enzymes. Mucosal surfaces are protected from the acid by the existence of mucus. The small intestine is divided into three parts: duodenum, which is the short section that receives the digestive secretions from the pancreas and liver; the jejunum; and the ileum. Here, the two most important processes of digestion occur: the breakdown of macromolecules and the absorption of water and nutrients. Also in the duodenum, the chime is neutralized with sodium bicarbonate ($NaHCO_3$). This neutralization favors the enzyme's activity, especially pancreatin. Bile is produced in the liver and helps in the digestion of lipids by emulsifying dietary fats into small droplets, which in turn are ideal for the pancreatic lipase activity. The absorption of water and nutrients occurs in the inner lining of the small intestine, which has a big absorbent surface covered with villi. This event occurs by several mechanisms to prevent the build-up of digestion products in the lumen of the small intestine, which could inhibit the activity of the enzymes. All the compounds that are not absorbed go to the large intestine, where the absorption of water and electrolytes by the colon takes place, as well as the fermentation of polysaccharides and proteins by colonic microbiota, reabsorption of bile salts, and elimination of feces [1, 2]. Microbiota inhabits the intestine, where bacteria, fungi, viruses, and archaea form a complex ecosystem and live in a close relationship with the host. In the intestine, mucus is also present close to the lumen, without direct contact to the epithelium, forming a barrier that prevents colonization by other microbes and the diffusion of gastric acid to the epithelial surface. The mucus viscosity is influenced by pH, and therefore, a higher pH value facilitates the infiltration of microorganisms, which can lead to inflammation in the GI tract [3].

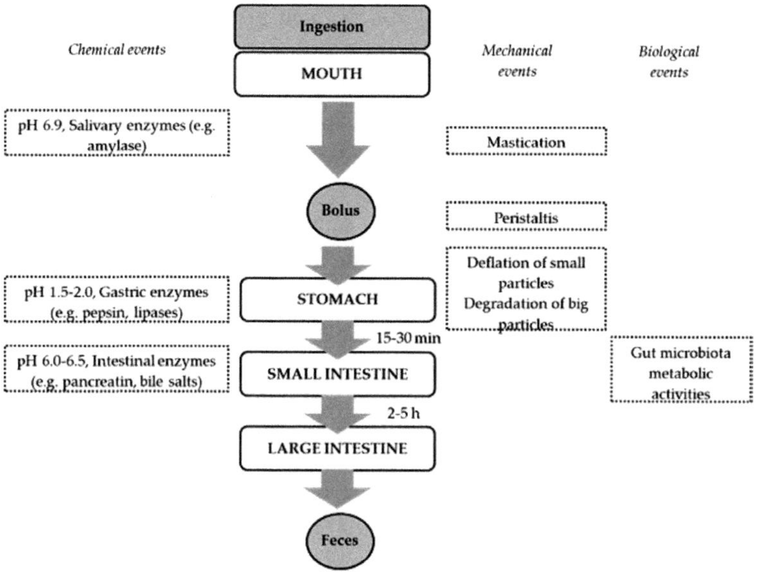

Figure 5.1 Schematic representation of the main steps of digestion, including the major chemical, mechanical, and biological processes.

5.2 Gut Microbiota Composition and Functions

The human GI tract is populated by ca. 100 trillion bacteria, representing up to a thousand different genera and species. The bacterial numbers increase from the duodenum (10^2 bacteria/gram) to the colon (10^{12} bacteria/gram), and the total weight of the bacteria can reach ca. 1 kg in an average adult, consisting of up to 5000 species [5]. Figure 5.2 represents the distribution of the main gut bacterial groups in the human GI tract. These communities are variable in composition and numbers along the digestive tract, and each individual harbors a unique specie collection adapted along life and affected by the lifestyle and nutritional diet.

In mice and human GI tracts, the majority of gut bacteria belong to the phyla Bacteroidetes and Firmicutes, although there are host-bacteria-specific differences at the genus level. Bacteria belonging to Proteobacteria, Actinobacteria, Verrucomicrobia, Cyanobacteria, and Deferribacteres groups and Fusobacteria are also present in humans. Nevertheless, the microbiota can also include potentially virulent species—pathobiont—which can cause disease when intestinal homeostasis is perturbed. The molecular mechanisms by which pathobionts cause disease remain poorly understood. In the normal gut, the majority of bacteria are commensals that only habit the lumen, and up to 50% are nonculturable. Yet, their quantification can be done using novel and modern techniques based on monoculture-based sequencing technologies [5].

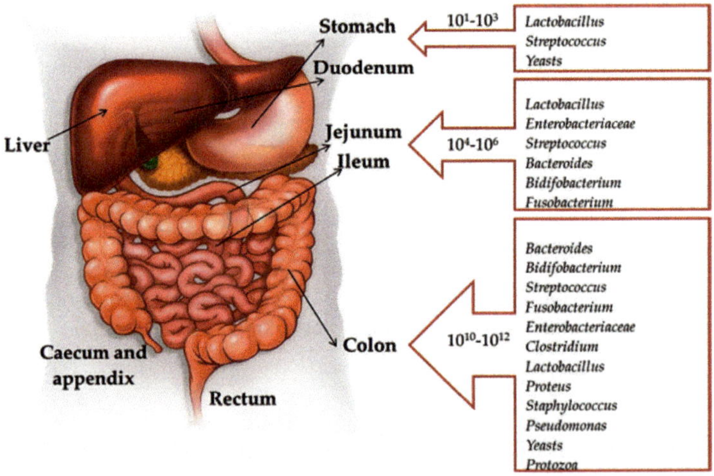

Figure 5.2 Major bacterial gut groups present in each human GI tract organ. The average numbers of bacterial groups are presented in colony-forming units (CFUs)/mL.

The small intestine microbial population plays important roles in human health and diseases. The majority of the human microbiome is harmless or beneficial to the host and acts as a protector against pathogens, provides nutrients and energy, and fosters development [6]. The GI tract functions are tremendously increased in the caecum and colon owing to these bacteria. The bacteria from *Lactobacillus* and *Bifidobacterium* species belong to Firmicutes and Actinobacteria divisions, respectively, and are pointed out as the beneficial gut bacteria associated with the control of GI tract functions, such as regulation of intestinal transit and inhibition of growth of potential pathogenic bacteria *Salmonella* spp. and *E. coli*, for example, by the production of inhibitory organic acids. Besides the functional roles in normal digestion, they also play immunological functions, such as conjugation of bile acids; prevention of pathogenic bacteria growth; production of butyrate, which regulates colonic enterocyte health; production of vitamins B12 and K; detoxification (or toxification) of certain ingested drugs or plant toxins; and immune system maturation [7–9].

5.2.1 Please Do Not Disturb Gut Microbiota!

The gut suffers uncontrollable changes daily, such us, the decrease of gut oxygen content over time, environment pollution, and antibiotics that we are exposed to. All these disturbances may lead to a new unhealthy state and positive and negative feedback that makes it difficult to correct [10]. Also, it can lead to microbial imbalance (dysbiosis) and may promote susceptibility to diseases. Nevertheless, diet is one of the most important controllable changes that remarkably influence gut microbiota populations. The daily promotion of specific substrates can select microbes that, on the other hand, favor the growth and metabolism of one bacteria group over others. For example, if we change the diet of mice from a low-fat, plant polysaccharide-rich diet to a high-fat, high-sugar "Western" diet, it will lead to changes in the metabolic activity of microbiota in just 1 day [11, 12]. Yet, in humans these changes take more time [13].

A disordered and impaired microbiota community is associated with conditions such as obesity; inflammatory bowel disease; intestinal infections; cancerous lesions of the intestine, liver, and pancreas, autoimmune diseases; and many behavioral and psychiatric issues [10, 14–20].

Owing to a considerable growth of orally delivered medicines and functional foods that are incorporated within nanoparticles (NPs), studies on digestion of these products and the impact of GI tract conditions on the stability, the loaded compound's release, and bioavailability (BA) are important. The interactions with the gut microbiota will also have an important and decisive role in the metabolism of these NPs. These bacteria can metabolize these nanosystems and interfere with their adsorption by the intestinal epithelium, or on the other hand, these NPs can change the microbiota positively or negatively, creating other physiological states. Thus, there is a need for targeted toxicological investigations of the influence of ingested compounds, such as NPs, on the gut microbiota.

5.2.2 Oral Delivery of Nanoparticles and Interactions with Gut Microbiota

The occurrence of interactions between NPs and gut microbiota is still uncertain from the existing literature, and it is important to understand whether these are detrimental, positive, or inconsequential. The first important concerns are related to the toxicological origin of NPs, if they are absorbed and if they accumulate in body organs. However, the particles and the compounds that are released before being absorbed may also have toxic effects, by inducing changes in the normal microbiota. Additionally, there is the possibility of microbiota interference or promotion of NP absorption, such as the case of gram-negative bacteria that induce adherence of NPs to LPSs and enhance their delivery. Finally, lumenal NPs may affect gut microbial metabolism, potentially influencing nutrient absorption or xenobiotic metabolism [21].

Most of the studies available use animals as model systems and assess gut microbial alterations, such as growth rate and feed conversion. Nevertheless, some new studies also use feces from human donors [22–24].

In addition to the effects of NPs in microbiota, it is also important to study the NPs' path along the entire GI tract, from mouth to colon, to understand the chemical changes that these NPs suffer during all digestion. After oral ingestion, there are mechanical forces and prominent pH changes along the GI tract that need to be considered. The physical (contractions, peristaltic movements, temperature, mucus viscosity, interfacial interactions, etc.) and chemical (pH, enzymes, mucus composition, etc.) parameters may affect the NPs' size and surface properties. In the stomach, high energetic contractions have been measured, but effects on NP agglomeration and aggregation are still unknown [25]. The changes of pH along the GI tract occur mainly in the fasted state, but pH is usually buffered to a range of 2–6 in the presence of food. Low pH can increase dissolution of particles, and enzymes in the digestive fluids can induce denudation of particles. As an example, AgNP agglomerates in synthetic gastric fluid by partial dissolution and consequently releases of Ag+ [26]. After dissolution of Ag particles in the acidic environment, the formation of AgCl may occur due to the combination of Ag^+ with Cl^- released from the stomach environment. Smaller particles tend to agglomerate to a higher degree [27]. Another example is the SLNs, which are affected considerably by the intestinal juice (pancreatin and bile salts), and dissolution followed by aggregation was observed in vitro by our research group [28].

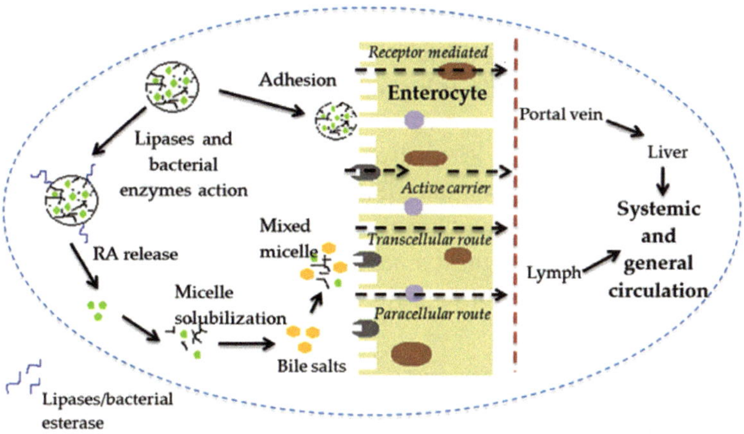

Figure 5.3 Hypothetical faith of solid lipid nanoparticles (SLNs) when reaching the small intestine, using as an example those loaded with rosmarinic acid (RA).

In Fig. 5.3, an example of the fate of NPs (e.g., SLNs) when reaching the intestinal phase is shown. As can be seen, SLNs reach the intestine and are immediately attacked by the lipases and bacterial enzymes that induce NP destabilization, since these enzymes have high activity against lipids, which are the main components of SLNs. Then, the solubilization by bile salts occurs and the loaded compound is absorbed by a transcellular route. SLNs that are smaller and not disrupted tend to adhere to the enterocytes and are absorbed. The ones that are not absorbed and the remaining materials and metabolites that result from the dissolution of the SLNs will proceed to the colon and will be metabolized by the colon bacteria. Therefore, the fate of NPs is strongly dependent on their composition and size, and, consequently, on the dissolution process by the intestinal enzymes. In addition, the presence of mucus on the epithelium also influences the contact of NPs with bacteria in the lumen. The epithelial cells are covered by a mucus layer that consists of a firmly adherent layer and a loosely adherent layer that can reach a total thickness of up to 1000 μm, producing a strong barrier that prevents penetration of both bacteria and NPs into cells [29]. This restriction is made by bonding of mucus fibers through ionic and hydrophobic interactions and by size filtering [30]. The residence time of food is shorter (3–5 h) in the stomach, while in the large intestine it is longer (20–30 h). Therefore, the food residence time, in conjugation with the thickness of the mucus layer, explains why the absorption of the NPs is higher in the large intestine than in the stomach [31].

In this section most of the discussed studies are with silver NPs. This happens because Ag, silica (SiO_2), TiO_2, and ZnO NPs are added as ingredients to foods and health care products. TiO_2, ZnO, and SiO_2 are produced in the highest amounts, while AgNPs are used in the highest number of products [32].

5.3 Studies of the Effects of Orally Delivered Nanoparticles

In Table 5.1 are resumed the principal in vivo and in vitro approaches used to study NPs' effects on gut bacteria growth and metabolism. In the case of in vitro studies, these can include the use of gut microbiota isolates in monocultures or cells infected with gut microbiota. Gut bacteria can be isolated from animal or human feces or intestinal tissues. Their species and strains can be confirmed with microbiological or molecular biology techniques by performing culture media growth or by DNA extraction followed by polymerase chain reaction (PCR) real-time sequencing. Another method is the determination of optical density by a spectophotometric apparatus or by a microplate reader using 650 nm wavelength for the determination of the minimal inhibitory concentration (MIC). To choose the best method we have to take into account the optimal growth conditions of each bacteria genus, since most of the gut bacteria are restrictive fastidious anaerobes. This is very important when using the total feces extracts, since all the procedures should be made in specific anaerobic atmosphere environment (5% H_2, 10% CO_2, and 85% N_2). Afterward, microorganisms are grown in specific media and stored at −80°C until their use. To perform the studies with NPs, dried NPs can be mixed with the growth media and in some cases include simple incubation at the optimal temperature, media, and time of bacteria growth.

Table 5.1 In vitro and in vivo approaches available to study the effects of nanoparticles in gut microbiota

Study format	Analyzed samples	Techniques
In vitro	Animal and human intestinal bacteria isolates Animal and human fecal material Cell culture lines infected with gut microbiota	Classic microbiology DNA extraction and PCR real-time sequencing
In vivo	Animal feces and intestinal/cecum tissues (mice, rats, pigs, avian, and fish species)	Acute and chronic oral administrations (direct feeding and gavage) DNA extraction and PCR real-time sequencing

On the other hand, in vivo studies include rats and mice that are mostly studied, but other studies are also available in fish, birds, or pigs. As an example, zebrafish (*Danio rerio*) microbiota has become an important model species for the study of microbial communities in vertebrate intestines [30]. All these animals are usually fed acute or chronically with the NP formulation in study. These feedings can be made by gavage using liquid-state NPs, or the NPs can be mixed with the daily diet. At sampling times, feces can be collected and analyzed or some studies use the cecum tissues of the animals after necropsy. Further, the bacteria DNA is extracted using standardized kits and sequenced by PCR real time.

Most of the studies found use metal NPs, such as silver and copper NPs, as can be seen in Table 5.2, but there are also studies available with carbon nanotubes (CNTs) and SLNs, as performed by our research group. In general, most studies are related with the determination of the antimicrobial activity of NPs on intestinal pathogens and the effects on all gut microbes.

Table 5.2 Studies available on the effects of NPs on gut microbiota

Model	Nanoparticles	Microbiota effects/interactions	Target microbiota	Refs.
In vitro (monocultures)	Carbon nanotubes	Antimicrobial effects; indirect immunomodulator	*Lactobacillus acidophilus*, *Escherichia coli*, *Staphylococcus aureus*, *Enterococcus faecalis*, *Bacillus subtilis*, *Ochrobactrum* sp.	[33, 34]
	Graphene oxide NPs	Growth promoter	*Bifidobacterium adolescentis*, *Escherichia coli*, *Staphylococcus aureus*	[36]
In vitro (human fecal material)	Silver NPs	Antimicrobial effects	Bacterial isolates, *Bacteroides*, *Roseburia* spp.	[37]
		Growth promoter	*Escherichia coli*, *Raoutella* spp.	
	Zinc, cerium, and titanium oxide NPs	Phenotypic changes	General groups	[38]
	SLNs	Antimicrobial effects (empty SLN)	General groups	[39]
		Growth promoter (loaded with herbal extracts)	*Bifidobacterium* spp., *Lactobacillus* spp.	
In vivo (mice and rats)	Copper-loaded chitosan NPs	Antimicrobial effects	Coliforms	[40]
		Growth promoter	*Bifidobacterium* spp., *Lactobacillus* spp.	
	Silver NPs	No effect detected	NA	[41, 42]
		Antimicrobial effects	Firmicutes, *Lactobacillus* spp.	[43]

Model	Nanoparticles	Microbiota effects/interactions	Target microbiota	Refs.
(*Continued*) In vivo (mice and rats)	Silver NPs	Population number changes	*Bacteroides*, Firmicutes	[44]
	SLN (acute)	Dose- and lipid-related changes	Firmicutes, Bacteroidetes	[45]
	SLN (chronic)	Growth promoter (loaded with herbal extracts)	Firmicutes, *Lactobacillus* spp.	[46]
In vivo (fish)	Copper-loaded NPs and silver NPs	Antimicrobial effects	*Cobacterium somerae*	[47]
	Shewanella putrefaciens Pdp11	Growth promoter	Lactic acid bacteria	[48]
		Changes in microbiota groups	Proteobacteria, Cyanobacteria, Vibrio	
In vivo (avian species)	Copper-loaded chitosan NPs	Antimicrobial effects	*Escherichia coli*	[49]
		Growth promoter	*Bifidobacterium*, *Lactobacillus* spp.	
	Silver NPs	Growth promoter	Lactic acid bacteria (except *Lactobacillus*)	[50]
In vivo (pigs)	Silver NPs	Antimicrobial effects	Firmicutes, *Escherichia coli*	[23]

5.3.1 In vitro Studies

The major part of the studies predicts that NPs' effects on gut microbiota are dependent on the NPs' dose and physical properties. For example, CNTs, single walled and multiwalled, pristine and functionalized, and short and long, were tested at doses ranging from 20 to 100 μg/l in isolated gut bacteria such as *Lactobacillus acidophilus*, *E. coli*, *S. aureus*, and *Enterococcus faecalis*. The antibacterial activity observed showed to be dose dependent, and the physicochemical properties of CNTs, such as rigidity, diameter, and length, were also associated with antibacterial activity. The shape of bacteria also had a role in their resistance, since rod-shaped bacteria were more resistant than spherical ones [33]. Later, another study also confirmed the antibacterial activity of single-walled carbon nanotubes (SWCNTs) [34], but at higher concentrations than the previous study. In this study, CNTs were found to have activity against *S. aureus*, *Bacillus subtilis*, *E. coli*, and *Ochrobactrum* sp. for 24 h at a concentration range of 10–50 mg/l. This effect was also dose and CNT length dependent. The mechanism of action was associated with modifications of the bacterial membrane. Changes in the bacterial membrane associated with antimicrobial activity were also observed by our research group when studying the antimicrobial activity of chitosan NPs loaded with phenolic compounds against intestinal pathogenic bacteria, such as *E. coli*, *Bacillus cereus*, *S. typhimurium*, and *Yersinia enterocolitica* [35].

Also, it was observed that the exposed bacteria may develop different adaptative mechanisms and that the presence of CNTs may cause a selective pressure on some species of the gut microbiota. Recently, but with a different approach, cell culture lines such as the human enterocyte-like Caco-2 were exposed to SWCNTs, multiwalled carbon nanotubes (MWCNT), graphene oxide (GO), AgNPs, TiO_2 NPs, and SiO_2 NPs. In this study, SWCNTs have broad-spectrum antibacterial effects. Low-dose-SWCNT treatment attenuated the human enterocyte-like Caco-2 cells from damage by *E. coli* and *S. aureus* infection by suppressing NLRP3 inflammasome activation. Some were also found to act as prebiotic, that is, to promote the growth and proliferation of gut probiotic bacteria. For example, GO anaerobic membrane scaffolds were found to promote the proliferation of anaerobic bacterium species, such as *Bifidobacterium adolescentis*, in the intestinal tract. In turn, these probiotic bacteria enhanced its antagonistic activity against pathogens *E. coli* and *S. aureus*. The author concluded that oral exposure of AgNPs can increase production of pro-inflammatory cytokine in the small intestine and colon and change the composition and diversity of gut microbiota [36].

5.3.1.1 Human feces volunteer donors

The effect of poly(*N*-vinyl pyrrolidone)-capped 10 nm AgNPs in mixtures of 33 different bacterial isolates from a healthy human donor was studied. A negative impact on gut microbiota was observed by an increase in *E. coli* numbers, and a decrease in *Bacteroides* and other regular gut bacteria was observed at concentrations of 100 mg/l. In addition, this study showed a dependence on the NP size (10–110 nm) [37].

Microbiota isolates from fecal material of human volunteer donors were cultured in a colon reactor with zinc oxide (ZnO) and cerium oxide (CeO_2) NPs at 0.01 μg/l and of 3 mg/l titanium oxide (TiO_2) NPs. This study showed that NPs affected phenotypes, including short-chain fatty acid (SCFA) production, hydrophobicity, sugar content of the extracellular polymeric substance, and electrophoretic mobility, which indicates that changes in the community's stability were affected by the NPs. Three distinct partition phases of the microbial community phenotype were observed, initial conditions, a transition period, and a homeostatic phase, all three significantly different from the control traits. The most active type of NPs were TiO_2, definitely because of its lack of dissociation and greater stability [38].

Our research group performed a study with SLNs produced with witepsol and carnauba waxes (WSLNs and CSLNs) using fecal material from three human volunteer donors for 24 h in batch reactors [39]. These SLNs had particle sizes of 310–950 nm for WSLNs and of 300–1000 nm for CSLNs. The empty SLNs had a negative impact on the main microbiota groups, associated with the fact that bacteria do not use the fatty acids present in the composition of waxes as a metabolic substrate source for survival. On the other hand, loaded SLNs with herbal extracts, which are rich in phenolic compounds and sugars, induced significant growths in the main bacterial groups, higher than even the free herbal extracts (that contain sugars). These results showed that SLNs slowly released the phenolic compounds at lower concentrations than free extracts and that were not negative to microbiota bacteria and also allowed a microbiota adaptation period to the compounds. Also, the size of SLNs comparable with other NPs is high, which results in a lower surface area of interaction with bacterial cells [39].

5.3.2 Animal Microbiota Studies

Chitosan NPs loaded with copper were fed to Sprague–Dawley® rats, and it was found that populations of *Lactobacillus* and *Bifidobacterium* increased while coliform numbers decreased [40]. Later, the same type of NPs were tested in avian broiler chickens and the same effects were observed [49]. Here the suppression of coliforms was a growth promoter of *Lactobacillus* and *Bifidobacterium*, and not a direct prebiotic effect.

The same type of NPs and also AgNPs were tested in zebrafish microbiota. In one such study it was shown that the microbial community of a vertebrate intestine can be disrupted by dietary exposure to NPs. Besides the richness in bacterial species some species were not present in microbiota exposed to NPs, such as *C. somerae* (OTUs Z4 and Z6), a common member of the microbial community of fish intestine [47].

Several studies with AgNPs in rats and mice showed no changes in microbiota. In Wistar rats, for example, using very small AgNPs of 14 nm at doses of 4.5 and 9.0 mg/kg/day does not provoke changes in the ratio of *Bacteroides* to Firmicutes [41]. The lack of obvious changes in the microbiota composition has also been reported after exposure of mice to 20 and 110 nm AgNPs at concentrations of 10 mg/kg/day [42].

Nevertheless, some studies showed changes in animals' microbiota population dependent on NPs' physical characteristics and dose. Changes were detected in ileal-mucosal microbial populations after AgNPs' oral gavage to rats of different sizes. In addition, the induction of intestinal gene expression and an apparent shift in the gut microbiota toward greater proportions of gram-negative bacteria were detected [43]. The smaller NPs (10 nm) induced a decrease in populations of Firmicutes phyla, along with a decrease in the *Lactobacillus* genus.

In bigger animals, such as weaning pigs, besides causing an increase in weight, the feeding of these animals with 110 nm AgNPs caused a decrease in Firmicutes at the highest concentration of 36 mg/kg. In addition, 60–100 nm AgNPs also reduced coliforms in the gut microbiota [23].

Using another type of NP, our research group, following the in vitro studies with SLNs (witepsol and carnauba) loaded with rosmarinic acid (RA), previously mentioned, performed acute and chronic feeding of Wistar rats. The SLNs were given by gavage to rats in a first acute administration for 14 days and then in another study for 6 weeks. In the acute study, two dosages were given, 1 mg/kg and 10 mg/kg, and it was possible to observe that the highest dosage induced decreases in *Bacteroides* and *Bifidobacterium* groups, where the most innocuous of all groups were the SLNs produced with witepsol and in lower dose. Nevertheless, chronic feeding with the highest concentration showed that in 6 weeks the gut microbiota adapted to SLN presence and no significant changes were observed between SLNs produced with witepsol and that with carnauba. Actually, SLNs loaded with herbal extracts exhibited a prebiotic effect, promoting an increase in probiotic bacteria groups.

The studies discussed until now have used free NPs, but there are also studies that evaluated the effect of NPs when incorporated in foods. Hence, some in vivo studies in mice evaluated the effect of the ingestion of silver NPs mixed in food for 28 days at doses relevant for human diet [44]. In the case of AgNPs, *Bacteroides* decreased their numbers while Firmicutes increased, and these changes were dose dependent. Japanese quail also received AgNPs in water at doses of 25 mg/kg and 10 nm in size and greater proportions of Firmicutes phyla, and an increase in lactic acid bacteria along with a decrease in the *Lactobacillus* genus were observed [50]. These studies show that AgNPs affect rat, mice, and avian microbiota and that these effects can be influenced by dose and size. In general, smaller NPs can have a higher antimicrobial effect. Also, the incorporation of NPs in foods also induces the same effects as those produced by free NPs.

With the intention of changing the microbiota, gilthead seabream (*Sparus aurata L.*) fish specimens were fed with encapsulated probiotic bacteria. The commercial diet was enriched with *Shewanella putrefaciens* Pdp11 before being encapsulated in calcium alginate beads. Changes in Proteobacteria after 30 days of treatment, with the absence of predominant bands related to Cyanobacteria and Vibrio genera, were found. The presence of bands of lactic acid bacteria, especially *Lactococcus* and *Lactobacillus* strains, was detected but not in the control group (not receiving probiotic treatment) [48].

The oral administration of NPs to animals has also shown the development of other symptoms and effects that are correlated with the changes in gut microbiota. For example, the oral exposure of rats to ZnO NPs induced not only liver damage but also behavioral changes in the treated animals. Since behavioral changes due to alterations of the gut microbiota have been reported, it might be hypothesized that ZnO caused the behavioral effects by affecting the gut microbiota [51–53]. Also, the development of colitis associated with alterations in gut microbiota has been reported in mice exposed to ambient particulate matter (which contains a substantial portion of carbon-based NPs) [54].

5.4 Conclusions and Future Perspectives

In general, with the published studies it can be concluded that the microbiota population is affected by the oral administration of NPs. In vitro, when using antibacterial NPs, a dose and length dependence of the effect is observed, where the morphology of bacteria also has influence, since the bacterial membrane is affected. In some cases, the increment of certain bacteria groups (*Lactobacillus* and *Bifidobacterium* spp.) leads to a decrease of others (pathogenic bacteria), exerting here a prebiotic effect. In vivo, in some studies the opposite happens, where the presence of NPs promotes the increase of the called pathogenic bacteria and the decrease of probiotic groups. A dependence on the dose and on the loaded compound is also detected. Small NPs have also shown to promote more changes in the microbiota populations jointly with higher doses, certainly because of the higher surface contact with bacterial membranes.

Also, the development of adaptative mechanisms was observed and the presence of certain NPs may cause selective pressure on some species of the gut microbiota, with selective survival of species more able to adapt to their presence. This situation resembles that observed for oral antibiotics, which cause GI symptoms due to an imbalance produced in microbiota. In some instances, the outcome may be severe colitis, for example, when there is the suppression of commensal bacteria antagonizing the pathogen bacterium *Clostridium difficile*.

Overall, NPs cause nonlethal changes to the microbial community phenotype. Nevertheless, there is a gap in the knowledge of the subsequent influence on the overall health. For example, Fondevila et al. observed that weight increase with AgNPs can be considered positive effect in agricultural animals [23]. In fact, the authors propose AgNPs as an alternative to antibiotics, but this is certainly not true for humans, especially in the light of the obesity epidemics in the Western world.

It should be emphasized that inferences about the clinical outcomes should be considered highly speculative at this stage. Furthermore, as shown in Table 5.2, the same NPs may cause different effects, for example, AgNPs cause a dominance of gram-negative strains in one study and of Firmicutes in another one.

Furthermore, there are differences observed in adult rats, mice, humans, quails, and weaning pigs due to differences in composition and stability of the gut microbiota. Different results can be explained by the fact that the samples used for microbiome analysis can be from the luminal content or from the gut tissue. Hence, the site of specimen collection influences the results, as the number and composition of gut microbiota change depending on the location in the intestinal tract. In addition, the choice of method for the analysis of the microbiota generates method-specific results and may lead to additional bias [55].

These findings suggest biomedical application of nanomaterials for gut microbiota-related therapeutic strategies to fight against human diseases. In addition, the interference of gut microbiota may act as a potential new mechanism by which NP exposure induces gut inflammation. On the basis of these findings, ingested NPs appear unlikely to have acute or severe toxic effects at typical levels of exposure; however, more subtle or chronic effects bear further investigation. This is particularly true with respect to intestinal permeability or oxidative stress and host-gut microbial balance, which have not been adequately explored. Also, it would be informative to assess the effects of administered NPs on gut microbes with a DNA-based technique like pyro sequencing, which eliminates culture bias and allows more sensitive detection of rare members of the microbiome [5, 41]. Future studies should be directed toward evaluation of dose-dependent effects, assessments of the autochthonous and allochthonous microbiota, duration of disruption after exposure, and the implications of microbial disruption for host health. Studies of the physiological consequences of NP-induced change in the gut microbiota are difficult to assess. Fecal transplantation from exposed to unexposed rats might be a way to address this effect in a manner similar to what has already been done for the evaluation of antibiotic treatment.

With increasing recognition of the importance of adequate materials characterization and adequate metadata, future investigations of these and other areas may be more easily applied to risk assessment and human health.

Acknowledgments

The authors acknowledge Fundação para a Ciência e a Tecnologia (FCT) for funding research work through the project NANODAIRY (PTDC/AGR-ALI/117808/2010) and through National Funds from FCT through the project PEst-OE/EQB/LA0016/2013, PTDC/AGR-TEC/2227/2012. Ana Raquel Madureira acknowledges FCT for the postdoctoral scholarship SFRH/BPD/71391/2010. There is no conflict of interests.

References

1. Guerra, A., et al. Relevance and challenges in modeling human gastric and small intestinal digestion. *Trends Biotechnol*, 2012, **30**(11):591–600.
2. Pedersen, P.B., et al. Characterization of fasted human gastric fluid for relevant rheological parameters and gastric lipase activities. *Eur J Pharm Biopharm*, 2013, **85**(3):958–965.
3. Engstrand, L., et al. Helicobacter pylori and the gastric microbiota. *Best Pract Res Clin Gastroenterol*, 2013, **27**(1):39–45.
4. Madureira, A.R., et al. Current state on the development of nanoparticles for use against bacterial gastrointestinal pathogens. Focus on chitosan

nanoparticles loaded with phenolic compounds. *Carbohydr Polym*, 2015, **130**:429–439.

5. Zoetendal, E.G., et al. Molecular ecological analysis of the gastrointestinal microbiota: a review. *J Nutr*, 2004, **134**(2):465–472.
6. Candela, M., et al. Interaction of probiotic Lactobacillus and Bifidobacterium strains with human intestinal epithelial cells: adhesion properties, competition against enteropathogens and modulation of IL-8 production. *Int J Food Microbiol*, 2008, **125**(3):286–292.
7. Fröhlich, E.E., et al. Cytotoxicity of nanoparticles contained in food on intestinal cells and the gut microbiota. *Int J Mol Sci*, 2016, **17**(4):509.
8. Collado, M.C., et al. Gut microbiota: a source of novel tools to reduce the risk of human disease? *Pediatr Res*, 2014, **77**(1–2):182–188.
9. Palm, N.W., et al. Immune–microbiota interactions in health and disease. *Clin Immunol*, 2015, **159**(2):122–127.
10. Lozupone, C.A., et al. Diversity, stability and resilience of the human gut microbiota. *Nature*, 2012, **489**(7415):220–230.
11. Turnbaugh, P.J., et al. The effect of diet on the human gut microbiome: a metagenomic analysis in humanized gnotobiotic mice. *Sci Transl Med*, 2009, **1**(6):6ra14–6ra14.
12. Wang, J., et al. Dietary history contributes to enterotype-like clustering and functional metagenomic content in the intestinal microbiome of wild mice. *Proc Natl Acad Sci USA*, 2014, **111**(26):E2703–E2710.
13. Wu, G.D., et al. Linking long-term dietary patterns with gut microbial enterotypes. *Science*, 2011, **334**(6052):105–108.
14. Stappenbeck, T.S., et al. Developmental regulation of intestinal angiogenesis by indigenous microbes via Paneth cells. *Proc Natl Acad Sci USA*, 2002, **99**(24):15451–15455.
15. Mazmanian, S.K., et al. An immunomodulatory molecule of symbiotic bacteria directs maturation of the host immune system. *Cell*, 2005, **122**(1):107–118.
16. Flint, H.J., et al. Obesity and colorectal cancer risk: impact of the gut microbiota and weight-loss diets. *Open Obes J*, 2010, **2**:50–62.
17. Calcinaro, F., et al. Oral probiotic administration induces interleukin-10 production and prevents spontaneous autoimmune diabetes in the non-obese diabetic mouse. *Diabetologia*, 2005, **48**(8):1565–1575.
18. Kamada, N., et al. Role of the gut microbiota in immunity and inflammatory disease. *Nat Rev Immunol*, 2013, **13**(5):321–335.

19. Diamond, B., et al. It takes guts to grow a brain. *Bioessays*, 2011, **33**(8):588–591.
20. Samuel, B.S., et al. Effects of the gut microbiota on host adiposity are modulated by the short-chain fatty-acid binding G protein-coupled receptor, Gpr41. *Proc Natl Acad Sci USA*, 2008, **105**(43):16767–16772.
21. Cattani, V.B., et al. Lipid-core nanocapsules restrained the indomethacin ethyl ester hydrolysis in the gastrointestinal lumen and wall acting as mucoadhesive reservoirs. *Eur J Pharm Sci*, 2010, **39**(1):116–124.
22. Ahmadi, J. Application of different levels of silver nanoparticles in food on the performance and some blood parameters of broiler chickens. *World Appl Sci J*, 2009, **7**(S1):24–27.
23. Fondevila, M., et al. Silver nanoparticles as a potential antimicrobial additive for weaned pigs. *Anim Feed Sci Technol*, 2009, **150**(3):259–269.
24. Ahmadi, F., et al. The impact of silver nano particles on growth performance, lymphoid organs and oxidative stress indicators in broiler chicks. *Glob Vet*, 2010, **5**(6):366–370.
25. Bellmann, S., et al. Mammalian gastrointestinal tract parameters modulating the integrity, surface properties, and absorption of food-relevant nanomaterials. *Wiley Interdiscip Rev Nanomed Nanobiotechnol*, 2015, **7**(5):609–622.
26. Axson, J.L., et al. Rapid kinetics of size and pH-dependent dissolution and aggregation of silver nanoparticles in simulated gastric fluid. *J Phys Chem C*, 2015, **119**(35):20632–20641.
27. Mwilu, S.K., et al. Changes in silver nanoparticles exposed to human synthetic stomach fluid: effects of particle size and surface chemistry. *Sci Total Environ*, 2013, **447**:90–98.
28. Madureira, A.R., et al. Insights into the protective role of solid lipid nanoparticles on rosmarinic acid bioactivity during exposure to simulated gastrointestinal conditions. *Colloids Surf B*, 2016, **139**:277–284.
29. Fröhlich, E., et al. Models for oral uptake of nanoparticles in consumer products. *Toxicology*, 2012, **291**(1):10–17.
30. Fröhlich, E., et al. Mucus as barrier for drug delivery by nanoparticles. *J Nanosci Nanotechnol*, 2014, **14**(1):126–136.
31. Jani, P.U., et al. Titanium dioxide (rutile) particle uptake from the rat GI tract and translocation to systemic organs after oral administration. *Int J Pharm*, 1994, **105**(2):157–168.

32. Vance, M.E., et al. Nanotechnology in the real world: redeveloping the nanomaterial consumer products inventory. *Beilstein J Nanotechnol*, 2015, **6**(1):1769–1780.

33. Chen, H., et al. Broad-spectrum antibacterial activity of carbon nanotubes to human gut bacteria. *Small*, 2013, **9**(16):2735–2746.

34. Zhu, B., et al. Modification of fatty acids in membranes of bacteria: implication for an adaptive mechanism to the toxicity of carbon nanotubes. *Environ Sci Technol*, 2014, **48**(7):4086–4095.

35. Madureira, A.R., et al. Production of antimicrobial chitosan nanoparticles against food pathogens. *J Food Eng*, 2015, **167**:210–216.

36. Chen, H., et al. Nanoparticle effects on gastrointestinal microbiome. *Nanomed Nanotechnol Biol Med*, 2016, **2**(12):457.

37. Das, P., et al. Nanosilver-mediated change in human intestinal microbiota. *J Nanomed Nanotechnol*, 2014, **5**(5):1.

38. Taylor, A.A., et al. Metal oxide nanoparticles induce minimal phenotypic changes in a model colon gut microbiota. *Environ Eng Sci*, 2015, **32**(7):602–612.

39. Madureira, A.R., et al. Fermentation of bioactive solid lipid nanoparticles by human gut microflora. *Food Funct*, 2016, **7**(1):516–529.

40. Han, X.Y., et al. Changes in composition a metabolism of caecal microbiota in rats fed diets supplemented with copper-loaded chitosan nanoparticles. *J Anim Physiol Anim Nutr*, 2010, **94**(5):e138–e144.

41. Hadrup, N., et al. Subacute oral toxicity investigation of nanoparticulate and ionic silver in rats. *Arch Toxicol*, 2012, **86**(4):543–551.

42. Wilding, L.A., et al. Repeated dose (28-day) administration of silver nanoparticles of varied size and coating does not significantly alter the indigenous murine gut microbiome. *Nanotoxicology*, 2015, 1–8.

43. Williams, K., et al. Effects of subchronic exposure of silver nanoparticles on intestinal microbiota and gut-associated immune responses in the ileum of Sprague-Dawley rats. *Nanotoxicology*, 2015, **9**(3):279–289.

44. van den Brule, S., et al. Dietary silver nanoparticles can disturb the gut microbiota in mice. *Part Fibre Toxicol*, 2016, **13**(1):38.

45. Madureira, A.R., et al. Safety profile of solid lipid nanoparticles loaded with rosmarinic acid for oral use: in vitro and animal approaches. *Int J Nanomed*, 2016, **11**:3621–3640.

46. Madureira, A.R., Nunes, S., Campos, D., Marques, C., Alcala, C., Sarmento, B., Gomes, A.M.P., Reis, F., Pintado, M. Oral chronic delivery

of polyphenolic solid lipid nanoparticles: toxicity studies and gut microbiota modulation. Unpublished.

47. Merrifield, D.L., et al. Ingestion of metal-nanoparticle contaminated food disrupts endogenous microbiota in zebrafish (Danio rerio). *Environ Pollut*, 2013, **174**:157–163.

48. Cordero, H., et al. Modulation of immunity and gut microbiota after dietary administration of alginate encapsulated Shewanella putrefaciens Pdp11 to gilthead seabream (Sparus aurata L.). *Fish Shellfish Immunol*, 2015, **45**(2):608–618.

49. Wang, C., et al. Effects of copper-loaded chitosan nanoparticles on growth and immunity in broilers. *Poultr Sci*, 2011, **90**(10):2223–2228.

50. Sawosz, E., et al. Influence of hydrocolloidal silver nanoparticles on gastrointestinal microflora and morphology of enterocytes of quails. *Arch Anim Nutr*, 2007, **61**(6):444–451.

51. Wang, T., et al. Lactobacillus fermentum NS9 restores the antibiotic induced physiological and psychological abnormalities in rats. *Benef Microbes*, 2015, **6**(5):707–717.

52. Stilling, R., et al. Microbial genes, brain & behaviour–epigenetic regulation of the gut–brain axis. *Genes Brain Behav*, 2014, **13**(1):69–86.

53. Hsiao, E.Y., et al. Microbiota modulate behavioral and physiological abnormalities associated with neurodevelopmental disorders. *Cell*, 2013, **155**(7):1451–1463.

54. Kish, L., et al. Environmental particulate matter induces murine intestinal inflammatory responses and alters the gut microbiome. *PLoS One*, 2013, **8**(4):e62220.

55. Brooks, J.P., et al. The truth about metagenomics: quantifying and counteracting bias in 16S rRNA studies. *BMC Microbiol*, 2015, **15**(1):66.

Chapter 6

Oral Nanotechnological Approaches for Colon-Specific Drug Delivery

Rute Nunes,[a,b] **Bruno Sarmento,**[a,b,c] **Salette Reis,**[d] **and Pedro Fonte**[d,e]

[a]*I3S - Institute of Research and Innovation in Health, University of Porto, Portugal*
[b]*INEB - Institute of Biomedical Engineering, University of Porto, Portugal*
[c]*CESPU - Institute for Advanced Research and Training in Health Sciences and Technologies and University Institute of Health Sciences, Gandra, Portugal*
[d]*UCIBIO, REQUIMTE, Department of Chemical Sciences, Faculty of Pharmacy, University of Porto, Portugal*
[e]*CBIOS - Research Center for Biosciences & Health Technologies, Lusófona University, Lisboa, Portugal*
pedro.fonte@ulusofona.pt

Colon-specific drug delivery is an interesting approach to treating local conditions of the colon or to achieving systemic absorption of drugs. Despite several efforts, the delivery of drugs into the colon by the oral route has proved to be an arduous task. Most of conventional approaches take advantage of physiological features of the GI tract to reach the colon. However, the presence of disease and/or intra-/intervariability of subjects leads to modest and some inconsistency in the efficacy of these systems. The potential of nanotechnology-

Nanoparticles in Life Sciences and Biomedicine
Edited by Ana Rute Neves and Salette Reis

ISBN 978-981-4745-98-7 (Hardcover), 978-1-351-20735-5 (eBook)
www.panstanford.com

based systems for drug delivery is now widely recognized and may be a valuable tool to circumvent the gaps left by conventional oral delivery systems. The development of nanosystems for the targeted delivery of drugs to treat colon diseases, such as inflammatory bowel disease (IBD), has attracted considerable attention over the years. Later, the development of nanosized systems for colon-specific drug delivery extended to other areas, such as colorectal cancer, infectious and endocrine diseases, and mucosal vaccines. In this section, the main anatomophysiological challenges faced by orally administered delivery systems before reaching the colon and the advantages and limitations of colon-specific drug delivery are discussed. The different nanotechnological approaches and materials used to achieve colon-specific drug delivery are also described. Finally, the latest achievements of nanotechnology-based systems for colon-specific drug delivery to treat specific health conditions are debriefed.

6.1 Introduction

The oral route is still the most convenient one for drug delivery, with good patient compliance, especially in cases of treatment of chronic diseases and in repeated administrations [1, 2]. It is a convenient pathway, allowing self-administration and dispensing with the sterilization of formulations, which reduces the production costs [3]. Colon-specific drug delivery has been explored in the last few years not only to improve the local treatment of diseases affecting the colon, minimizing systemic side effects, but also to enhance the systemic bioavailability (BA) of drugs [4]. Absorption in the colon is favored by a longer retention time of colonic contents compared to that in the upper gastrointestinal (GI) tract and the ability of colonic mucosa to facilitate the absorption of drugs, making it a good site for drug delivery [5]. It may be argued that the rectal route is a faster pathway to reach the colon, but the inconvenience of rectal administrations, the intestinal transit, and the small amount of rectal fluids needed to release the drug from formulations, make the oral route still the preferable.

Besides the advantages, oral drug delivery presents some limitations that need to be addressed. The challenge is to deliver drugs into the colon without significant degradation and absorption in the upper GI tract, allowing higher drug concentrations to reach the colon [4, 6]. Indeed, the harsh environment of the upper GI tract, namely the acidic pH of stomach and the intense enzymatic activity, leads to the possible degradation of drugs, hampering their BA. Therefore, more sensible drugs, such as therapeutic proteins and peptides, may be preferentially delivered to the colon [7].

As a result of the developments in the nanotechnology field in the last few decades, nanoparticles (NPs) have been proposed to improve the oral BA of drugs [8]. NPs are promising oral drug delivery systems widely used in nanomedicine, mainly due to their ability to deliver the drug in a sustained or controlled way, protecting the loaded drug from the harsh environment of the stomach [9, 10]. Nanocarriers may be also tailored using specific surface ligands, to ensure the drug delivery to specific tissues or cells. These delivery systems are usually made of natural and synthetic polymers, or other materials, such as lipids (e.g., liposomes, SLNs, and NLCs) [10–13]. However, the latter are not frequently used for colon-specific drug delivery due to some instability in the GI tract. Several polymers have been used to design nanocarrier systems, and the combination of two or more in a nanocarrier for colon-specific drug delivery has been demonstrated to be useful to overcome the different challenges of the GI tract. These systems may be orally administered in suspension form or be incorporated in solid dosage forms after NP formulation dehydration performed for example by lyophilization [14]. This technique has the potential to enhance the long-term stability of NPs and the stability of the loaded drug, being particularly important in the case of therapeutic proteins and peptides [15–19].

In this section, the different parameters involved in colon-specific drug delivery by NPs are reviewed. Thus, the colon anatomophysiological features and their influence on the advantages and limitations of colon delivery will be discussed. More importantly, the different nanotechnological approaches and their application in specific health problems are fully debriefed.

6.2 Colon Anatomophysiological Features

A drug intended to be delivered specifically to the colon should not be absorbed or suffer degradation during the passage through the upper GI tract, that is, the stomach and the small intestine. For an effective specific drug delivery, the drug should also be released in the colon and absorbed without significant degradation. To meet these requirements and design a carrier to deliver drugs specifically to the colon, it is crucial to fully know the anatomophysiological features of the GI tract, particularly the colon, since they influence the drug delivery and BA [20]. The GI tract is constituted of the stomach, the small intestine, and the large intestine (colon).

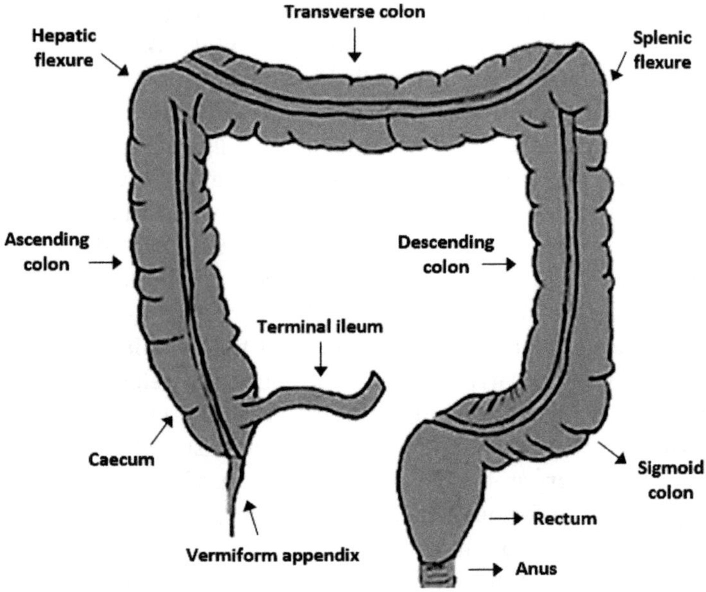

Figure 6.1 Scheme of the large intestine.

The colon is a cylindrical tube reaching a length of about 1.5 m and is subdivided into four parts: ascending, transverse, descending, and sigmoid colon (Fig. 6.1) [22]. It is responsible for the absorption of water and electrolytes present in colonic contents, mostly in the ascending segment, whereas the transverse and descending sections are responsible for the storage of indigestible food residues [23–25]. This organ is extremely vascularized and has an intrinsic nervous system, the enteric nervous system, responsible for peristaltic movements and secretion. Its architecture is based on glands formed by mucinous and columnar cells, and the colorectal mucosa is subdivided into the epithelium; the lamina propria; and the muscularis mucosae, which divides the deeper submucosa from the mucosa [26]. The colon does not have villi, but the crescentic folds increases its surface to about 1300 cm^2, necessary to absorb approximately 90% of the water present in the colon [23]. The drugs are passively absorbed in the colon by the transcellular or paracellular route. By the transcellular route, drugs pass the intestinal epithelium through cells, which is likely to occur for lipophilic drugs, whereas hydrophilic drugs are more likely to take the paracellular route by passing through tight junctions between cells.

Along with the anatomical features, the physiological factors of colon, which are different from other parts of the GI tract, influence colon-specific drug delivery [20]. These factors are the intestinal transit time and colonic residence time, colonic fluid volume and viscosity, colonic pH, and colonic microbiome.

The residence time of formulations in the colon is directly related with the intestinal transit time. The colonic residence time should be enough to allow an adequate drug release, absorption, and BA. Thus, the variability of the transit of formulations and food through the colon presents challenges in the design of drug delivery systems. The transit time in small intestine is about 4 h, whereas the transit time in colon ranges from 6 to 70 h [27, 28]. The slower transit time in the colon allows drugs to come in contact with the mucosa for a longer time than in the small intestine, compensating for the lower surface area of absorption. The transit time of delivery systems is mainly dependent on their features, presence of food in the GI tract, and the administration time. Larger dosage forms, such as capsules, have been described to have lower transit times than smaller dosage forms, such as small particles, and the colonic transit is also slower during sleep [29]. In addition, some pathological conditions of the GI tract influence the intestinal transit time and consequently the colonic residence time. For instance, patients with ulcerative colitis have shorter colonic time (24 h) compared to healthy subjects (52 h on average) [30]. In patients with irritable bowel syndrome the orocecal transit time is also delayed [31].

The high capacity of the colon to absorb water leads to low colonic fluid volume, ranging from 1 to 44 mL (on average about 13 mL), which hampers the drug dissolution and consequently its absorption and BA [32]. The increase of colonic fluid viscosity along the colon may also lead to these problems [33]. The absorption of drugs also depends on the secretion of the intestinal fluid, affecting the viscosity of the mucous-gel layer. The pH differences along the GI tract have been explored to target drug delivery at the colon. After the strong acidic environment of the stomach, the pH increases from pH 6 to pH 7.4 from the duodenum to the ileum [34], decreasing again to below pH 6 in the cecum and rising along the colon to pH 6.7 in the rectum [35]. These standard pH levels may present some variability between subjects, regarding the microbial metabolism, food and water intake, and disease state [36]. The colonic pH influences the solubility of drugs in the colonic fluid and influences also the drug release from pH-sensitive formulations.

The intestinal microbiome plays an important role in digestion and in intestinal health [37]. The colon has more than 400 different species of anaerobic and aerobic microorganisms, containing metabolic enzymes [38], which can metabolize drugs, turning them inactive or even leading to toxic metabolites [39]. However, some colon-specific drug delivery systems depend on enzymes in microflora to release drugs in the colon, since they may degrade matrices and coatings of formulations. For instance, polysaccharides such as pectins, guar gum, chitosan, and others are used in colon-specific delivery systems because they are resistant to gastric and small intestine enzymes but are metabolized in the colon by anaerobic microorganisms [40]. The microbiome is variable across subjects, depending mainly on genetic and environmental factors, but the dominant microflora seems to be consistent in the colon [41]. However, some factors, such as lifestyle, diet, diseases, and drug intake, may lead to dysbiosis [42].

6.3 Advantages and Limitations of Colon-Specific Drug Delivery

Colon-specific delivery of drugs has been used mainly due to its advantages, since it allows both local (e.g., colorectal cancer) and systemic drug delivery, with lower dosing and just a few side effects. It is particularly indicated for drugs intended to treat colonic diseases (e.g., amylin and oxyprenolol) and for drugs with suboptimal absorption in the upper GI tract (e.g., isosorbide dinitrate). The advantages of the colon as a target for drug delivery comes from its long transit time, neutral pH, and low proteolytic enzyme activity [43]. The latter and the avoidance of chemical and enzymatic degradation in the upper GI tract turn the colon into a good target for delivery of therapeutic peptides and proteins [44].

Besides the advantages, colon-specific drug delivery has some limitations, mainly caused by the aforementioned anatomophysiological features of the colon. The obvious challenge is the location of the colon, in the distal part of the GI tract, because of which the oral administered delivery system has to pass through the upper GI tract to reach the colon. For an intuitive understanding of both the advantages and limitations of colon-specific drug delivery, they are briefly summarized in Table 6.1.

Table 6.1 Advantages and limitations of colon-specific drug delivery

Advantages
• Poorly absorbed drugs may have improved bioavailability in the colon. • There is longer retention time compared to that in the upper GI tract and the colon is highly responsive to absorption enhancers, which improves the uptake of poorly absorbed drugs. • Local treatment of colonic diseases requires smaller drug amounts and reduced dosage frequency, compared to systemic administration, decreasing the cost of expensive drug regimens. • There is low incidence of drug side effects and interactions. • There is less proteolytic activity compared to that in the upper GI tract, facilitating the bioactivity of peptides, proteins, and vaccines. • This type of drug delivery avoids gastric irritation caused by some drugs, such as nonsteroidal anti-inflammatory drugs. • This type of drug delivery minimizes the effect of first-pass metabolism. • There is improvement of patients' compliance to treatment.
Limitations
• The bioavailability of the drug may be decreased by its binding to dietary residues and intestinal secretions. • There is low bioavailability of poorly soluble drugs due to the low colonic fluid volume and its high viscosity. • There is a possibility of incomplete drug release. • The colonic microbiome may lead to drug metabolization and inactivation or create toxic degradation products. • The physiological features involved in drug release, solubilization, and absorption may be changed in diseased states.

6.4 Nanocarriers as Tools for Colon-Specific Drug Delivery

Generally, the drug candidates for colon-specific delivery are those presenting poor absorption in the small intestine; those suffering degradation in the upper GI tract; and those intended to have local therapeutic effect, such as in the case of inflammatory bowel disease (IBD), colon cancer, irritable bowel syndrome, colonic dysmotility, parasitic diseases, diverticulitis, and diarrhea [8, 45–47]. Colon-specific delivery is also useful for enhancing the systemic absorption of macromolecules due to the longer transit time and lower P-glycoprotein (Pgp) expression and CYP3A4 and proteolytic activity compared to those of the small intestine [8]. However, as mentioned before, the colon as the distal part of the GI tract presents significant challenges. NP-based drug delivery systems are useful tools to overcome such problems, avoiding or minimizing the release of drug in the stomach and small intestine and targeting the drug release at the colon.

The selection of a suitable nanocarrier to deliver drugs to the colon depends mainly on the physicochemical properties of the drug and on the therapeutic target. Thus, the chemical nature, stability, and partition coefficient of drugs as well as the use of absorption enhancers influence the carrier selection. The nanocarrier has to be properly designed to accommodate and deliver the drug, since the size, charge, and surface features of NPs influence the delivery of drugs to the colon. Generally, NPs are produced using polymers both from synthetic and natural sources and designed to have a burst or a prolonged/sustained release of the drug after reaching the colon. In the design of NPs, the physiopathology of the disease and affected parts of the colon in the case of local delivery or the physiological features of the healthy colon when the formulation is used for systemic therapy must be considered. In this regard, NPs may be engineered with specific ligands on the surface for site-specific drug targeting [48].

In this section the most common strategies for colon-specific drug delivery, namely by means of pH-sensitive polymer NPs, microbial-triggered-release NPs, and time-controlled-drug-release NPs, are discussed.

6.4.1 pH-Sensitive Polymer Nanoparticles

Drugs loaded into NPs may be specifically delivered to the colon using the pH differences along the GI tract. The pH-dependent delivery approach is based in the solubility of polymers at different pH ranges. Thus, polymers insoluble at low pH are used, and they become increasingly soluble with pH increase. Since polymers are insoluble at low pH, they can protect drugs from the gastric enzymes and acidic environment of the stomach and small intestine to some extent, being frequently used as enteric coatings in oral dosage forms. They can also prevent nausea and gastric irritation caused by drugs prematurely released in the stomach. However, when designing a nanocarrier for colon-specific drug delivery, the potential partial solubilization of polymers during passage through the small intestine, which may lead to drug release and premature inactivation, should be avoided.

The most commonly used polymers for colon-specific delivery are methacrylic acid copolymers (Eudragit®), dissolving at pH of 6 and 7, and the Eudragit® S100, Eudragit® L100, and Eudragit® FS 30D [49, 50]. However, some colonic diseases, such as Crohn's disease and ulcerative colitis, may change the colonic pH to about 5.3 and 2.7–5.5, respectively [45, 51]. For this purpose, Eudragit® L100-55, dissolving at pH 5.5, and Eudragit® E100, dissolving at pH below 5, may be also used [49, 50]. Overall, to avoid drug release in the upper GI tract and targeted drug delivery at the colon different polymers and ratios are used.

pH-sensitive Eudragit® S100 NPs loaded with celecoxib, curcumin, and both were orally administered to rats for treatment of IBD [52]. Using a trinitrobenzenesulfonic acid (TNBS)-induced rat colitis model, celecoxib-curcumin NPs demonstrated an improvement in disease treatment compared to celecoxib-curcumin suspension and NPs loading each drug separately. Additionally, the polymer NPs loading the drugs showed significantly lower inflammatory cell activity compared to drugs administered in suspension form. In another study, NPs were produced combining a pH-sensitive polymer, Eudragit® FS 30D, and a water-swellable polymer, Eudragit® RS100, and this combination prolonged the budesonide (BDS) release at colonic pH for treatment of IBD [53]. The NPs combining both polymers and Eudragit® FS 30D NPs were fluorescent labelled to evaluate their biodistribution and administered orally to mice after inducing colitis using dextran sulfate sodium (DSS).

The NPs combining both polymers revealed significantly higher fluorescence intensity than that of Eudragit® FS 30D NPs, and the fluorophore presence was detected for at least 10 h.

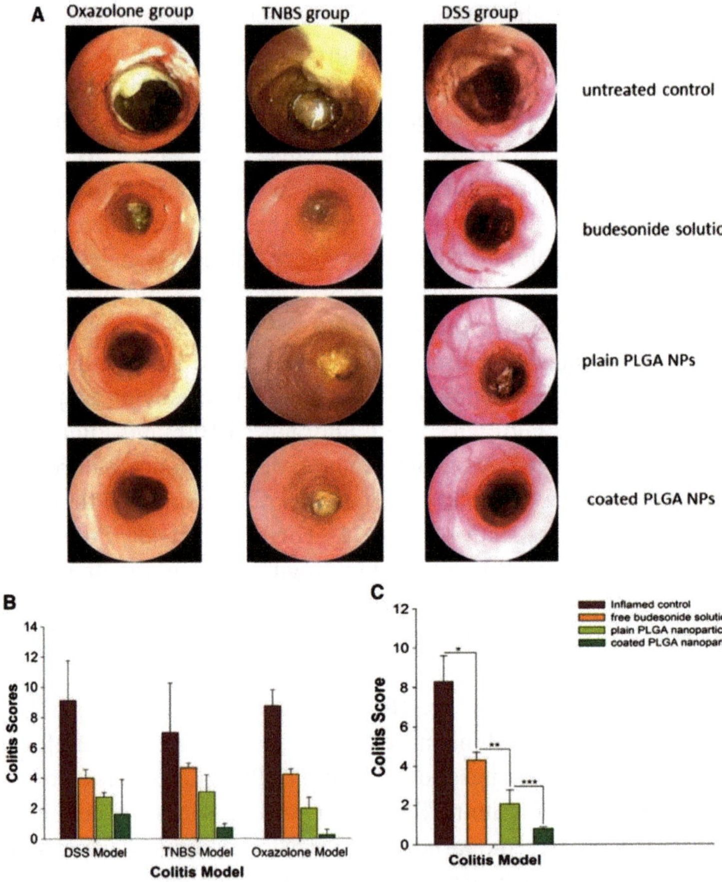

Figure 6.2 Efficacy of Eudragit® S100–coated budesonide-loaded PLGA nanoparticles, budesonide-loaded PLGA nanoparticles, and budesonide solution, after oral administration to different colitis-induced murine models. (A) High-resolution miniendoscopy of the intestine, (B) colitis scores upon treatment, and (C) combined colitis scores of the different colitis-induced models. Coated nanoparticles showed a significantly higher efficacy ($p < 0.001$) when compared to uncoated nanoparticles. NPs stands for nanoparticles. Reprinted from Ref. [8], Copyright (2016), with permission from Elsevier.

Additionally, the BDS-loaded Eudragit® FS 30D/Eudragit® RSPO NPs significantly decreased the IBD symptoms, compared to BDS-loaded Eudragit® FS 30D NPs and to controls with no BDS loaded.

Eudragit® polymers may be also used together with other polymer types as coating materials for colon-specific drug delivery. Previously, BDS-loaded poly(lactic-co-glycolic) acid (PLGA) NPs were coated with Eudragit® S100 to be used for IBD treatment [54]. In vitro release studies demonstrated that the presence of the polymer coating was important for decreasing the release rate of the drug in a pH 1.2 buffer, which increases the NPs' stability in the stomach. The in vivo efficacy of NPs was evaluated after oral administration to oxazolone-, TNBS-, and DSS-induced colitis murine models. The Eudragit® S100–coated PLGA NPs showed better efficacy in decreasing IBD symptoms compared to uncoated NPs and the BDS solution (Fig. 6.2). Additionally, fluorescence imaging analysis revealed that the Eudragit® S100–coated NPs were better retained in the colon compared to uncoated NPs. In another study, a similar nanocarrier-loading curcumin showed higher in vivo efficacy compared to a curcumin suspension [55].

6.4.2 Microbial-Triggered Drug Release Nanoparticles

The colon bears more than 400 different bacterial species, with a usual microbial load of 10^{10}–10^{12} CFU/mL (where CFU is colony-forming unit) [56, 57]. The bacteria are mainly anaerobic, producing different enzymes such as arabinosidase, glucuronidase, galactosidase, xylosidase, azoreductase, nitroreductase, urea hydroxylase, and deaminase, able to metabolize substrates such as the proteins and carbohydrates not digested in the upper GI tract. Some materials that are substrates of such enzymes may therefore be used for colon-specific drug delivery. Polymers based on azo-containing monomers, such as azo-polyurethanes, and on natural carbohydrates, such as alginic acid, pectin, and guar gum, have been used to produce NPs for colon-specific drug delivery [21, 58]. These polymers are degraded in the colon by bacteria or enzymes, releasing the loaded drug. However, differences in the microbiota between healthy and diseased colon have to be considered when designing NPs with microbial enzyme substrates. For instance, changes in bacterial diversity occur in IBD, with an increase in

Enterobacteriaceae sp., such as *E. coli*, and a reduction in bacteria producing anti-inflammatory substances, such as *Faecalibacterium* sp. [59].

Polysaccharides are easily digested in the colon by bacterial enzymes, so they may be used for colon-specific delivery [60]. Some examples are amylose, arabinogalactan, chitosan, chondroitin sulfate, cyclodextrin, dextran, guar gum, pectin, sodium alginate, and xanthan gum [61–63]. Most of them are used in food and in drug formulations as excipients, being overall considered for human administration. Previously, Eudragit® S100 and azopolyurethane were combined to produce NPs and evaluate the influence of pH- and enzyme-sensitive materials in colon-specific delivery [64]. Eudragit® S100/azopolyurethane NPs and Eudragit® S100 NPs were fluorescent labelled and orally administered to rats. After 8 h, the Eudragit® S100/azopolyurethane NPs delivered about 17% and 42% of total fluorescence to the cecum and colon, respectively. The Eudragit® S100 NPs delivered less than 10% of the fluorophore to the cecum and colon, demonstrating that the presence of colonic enzymes such as azoreductase may help to degrade polymers and enhance the drug delivery to the colon.

6.4.3 Time-Dependent Drug Release Nanoparticles

The use of time-dependent drug release NPs for colon drug delivery relies on GI tract transit time as the determinant factor. These systems are designed to pass through the harsh environment of the stomach following a lag phase of predefined duration and delivery of the drug ate a specific site of the GI tract [65]. NPs are composed of different materials, mostly based on polymers with gradual water swellability. The most commonly employed are cellulose derivatives (Fig. 6.3) and polymethacrylate-based copolymers (Eudragit®) (Fig. 6.4), although other polymers can be also used [6, 66]. The drug delivery occurs by different mechanisms (swelling, osmosis, etc.) or a combination of mechanisms and is not influenced by the pH, ionic strength, or presence of enzymes. The nature of the employed polymer, namely the molecular weight, viscosity grade, hydrophilicity, and swellability degree, will determine the duration of the lag phase [67].

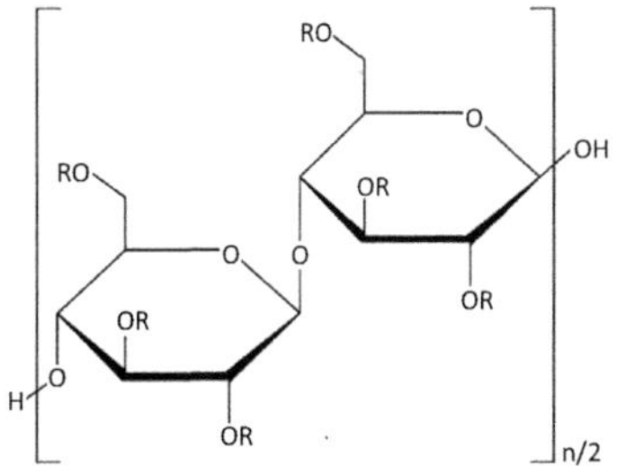

Cellulose derivative	R groups
Hydroxypropyl methylcellulose	H, CH_3, $H_3CH(OH)CH_2$
Hydroxypropyl cellulose	H, $[CH2CH(CH3)O]_mH$*
Hydroxyethyl cellulose	H, $[—CH2CH2O—]_mH$*

*m is a common integral number of cellulose derivatives

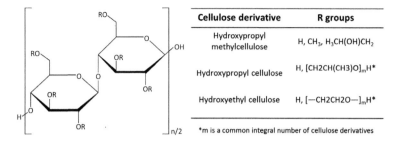

Figure 6.3 Chemical structure of the cellulose derivatives hydroxypropylmethyl cellulose, hydroxypropyl cellulose, and hydroxyethyl cellulose, frequently used as time-dependent polymers [68].

Although the development of the first conventional time-dependent systems (Time Clock® and Pulsicap®) occurred about two decades ago, no products have reached the market until now. Despite the innovation, the performance of these systems showed to be impaired by the subjects' inter- and intravariability, due to physiological (e.g., presence of food and feces) and/or pathological conditions [4]. Perhaps for this reason, the time-dependent drug release approach is not very explored in nanosystems for colon drug delivery. The only significant report of use of this approach in nanocarriers for colon-specific drug delivery was given by Naeem et al. [53]. In this study a dual-stimuli polymeric nanosystem with a pH- and time-dependent (pH/time NPs) release profile was designed. Eudragit® FS 30D (poly(methyl acrylate-co-methyl methacrylate-co-methacrylic acid, 7:3:1) was used as a pH-dependent polymer and Eudragit® RS100 (ammonio methacrylate copolymer, type B) as a time-dependent controlled release polymer (Fig. 6.4).

The in vitro release profile of time NPs was suboptimal compared with pH/time NPs and pH NPs. They suffered a fast and premature release of the loaded drug at pH 1.2 and 6.5. However, the presence of the time-dependent polymer in pH/time NPs conferred advantage to this system over the single pH NPs. This was due to the sudden and almost complete release of the drug from pH NPs at pH 7.4, while pH/time NPs provided a sustained release over 24 h, avoiding the initial burst release. The results of in vivo studies in mice with DSS-induced colitis were in accordance with this pattern. Overall, and similar to what happens with conventional stimuli-responsive systems, it seems that the use of a single stimulus may not be enough to improve the site specificity of drug delivery into the colon. The integration of systems with two or more stimuli in the same nanocarrier seems to be necessary to circumvent these problems.

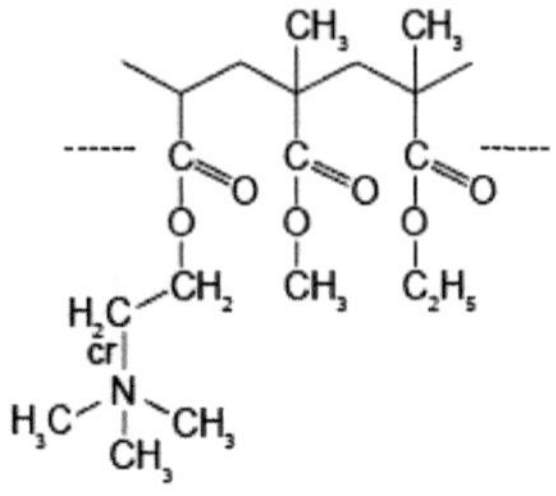

Polymer	Characteristics
Eudragit® RS 100	▪ Insoluble in water ▪ Low permeability ▪ pH independent swelling
Eudragit® RL 100	▪ Insoluble in water ▪ High permeability ▪ pH independent swelling

Figure 6.4 Chemical structures of Eudragit® RS 100 and Eudragit® RL 100 [69]. These polymethacrylate-based copolymers are composed of monomers of ethyl acrylate and methyl methacrylate and low content of methacrylic acid ester with quaternary ammonium groups. The molar ratio of ethyl acrylate, methyl methacrylate, and trimethylammonioethyl methacrylate is approximately 1:2:0.2 in Eudragit® RL 100 and approximately 1:2:0.1 in Eudragit® RS 100.

6.5 Applications of Nanoparticles for Colon-Specific Drug Delivery

6.5.1 Inflammatory Bowel Disease

In the last two decades, several attempts have been made to improve the delivery of drugs to the colon for the management of IBD. Briefly, IBD is the generic name for a group of chronic GI diseases, among which the most prevailing are Crohn's disease and ulcerative colitis [70]. The common feature is the chronic inflammation of the GI tract, characterized by periods of relapse and remission. Currently, there is no cure for IBD and current therapies are intended to control the disease progress and treat symptoms. They include anti-inflammatory and immunosuppressive agents [71]. Despite relevant progress in colon-specific drug delivery with the development of prodrugs or the use of dosage forms such as delayed or controlled release systems, these conventional strategies are far from ideal

[72]. One main challenge in the treatment of IBD relates with the fact that the biological environment and physiological processes are altered (Table 6.2), leading to unspecific and unpredictable performance of these systems since their design is usually based on normal physiological parameters [73]. Consequently, they may have some problems, such as systemic toxicity, poor tissue retention and distribution, insufficient drug accumulation within affected tissue, delivery of drug at undesirable sites of the GI tract, and incomplete drug release. These drawbacks lead to adverse effects, lower drug concentrations at the inflammation site, inconsistent efficacy, and patient intervariability [74].

Nanotechnology-based systems hold the promise to circumvent these problems, attracting great attention in the last few years for the delivery of drugs to the colon for the treatment of IBD [74]. The general advantages of these systems are widely known. They are able to (i) protect the drug against the harsh biological environment, (ii) overcome biological barriers, (iii) provide a sustained/controlled release, (iv) target specific sites or cells for the delivery, (v) improve the biodistribution and pharmacokinetics of drugs, (vi) and improve the toxicity profile of the loaded drug, among others [45]. Some of these advantages assume particular importance for the treatment of IBD, namely the NPs' ability to accumulate in the inflamed and disrupted tissue [76]. The specific accumulation of NPs in inflamed tissue was first documented by Lamprecht et al. [77]. When fluorescent polystyrene particles of different sizes (0.1, 1, and 10 μm) were used, higher amounts of particles were found in the inflamed intestinal tissue when compared with healthy intestinal tissue of rats. Although this trend was observed for all tested sizes (0.1, 1, and 10 μm), the results were clearly better for particles of size 0.1 μm and 1 μm. A higher amount of 0.1 μm particles was also found in inflamed tissue compared to the healthy tissue (15% versus 5%). This effect was explained by the particles' adhesion to the intestinal mucus, since its removal led to a decrease in fluorescence. More recent studies showed that negatively charged NPs (as the case of polystyrene particles) present a strong mucoadhesive behavior [78, 79]. Further, PLGA NPs loaded with the anti-inflammatory drug rolipram, ranging in size from 300 nm to 500 nm and having zeta potentials of –3 mV and –20 mV, were developed [80]. The NP formulations showed efficacy similar to that of the drug in solution in

Table 6.2 Physiological and microbial changes in the GI tract in IBD

GI regions	Transit time		Luminal pH		Microbiota		Mucus layer	
	Healthy	IBD	Healthy	IBD	Healthy	IBD	Healthy	IBD
Duodenum			6	7.4	Clostridiales			
Jejunum	2–6 h	↑30%	7	7	*Streptococcus* *Bacteroides* Actinomycinae	↑*Escherichia coli* ↓Clostridiales	0–37 μm	↑*
Ileum			7.4	7.4	*Lactobacillus* *Corynebacteria*			
Caecum Colon and rectum	52 h (6–70 h)	24 h	6–6.7	2.3–5.5	Firmicutes *Bacteroides* Proteobacteria Actinobacteria	↑Bacteroidetes ↓*Bifidobacteria* ↓*E. coli* ↑*Eubacteria* ↑*Peptostreptococcus*	110–150 μm	↑*

*Particularly in ulcerated areas
Source: [51, 75]

reducing the inflammation after daily oral administration for 5 days to Wistar rats induced with colitis, but with a better toxicity profile. Moreover, rolipram-loaded NPs were able to keep the inflammation at lower levels even 5 days after administration, contrasting with a strong relapse for the group administered with the drug solution. It was also observed that reducing the size of NPs led to an enhanced and selective delivery of molecules to affected areas, which is attributed to the epithelial enhanced permeability and retention (EPR) effect and the uptake by local inflammatory cells [76, 81].

Lipid-based nanocarriers have also showed potential usefulness in IBD treatment. NLCs (200 nm) encapsulating the anti-inflammatory drug BDS (BDS-NLCs) proved to be more effective than the drug solution in reducing the inflammation levels in vitro (activated macrophages) and in vivo [82]. In the latter, only BDS-NLCs were effective in decreasing the intestinal inflammation in a mice model of DSS-induced colitis after oral administration on nonconsecutive days (day 1, 3, and 5), whereas the drug solution showed no anti-inflammatory activity. In addition to the increased activity in vivo, BDS-NLCs provided higher sustainable drug levels in affected tissues than in the healthy ones. Guada et al. developed SLNs using Precirol® ATO 5 as the lipid matrix and different surfactants, encapsulating the immunosuppressive drug cyclosporine A (CsA) [83]. The performance of CsA-SLNs, with sizes ranging between 100 and ~200 nm and negative zeta potentials ranging between –30 mV and –15 mV, depending on the surfactant used, were compared in vivo with the marketed oral formulation of CsA, Sandimmune Neoral®. Besides favorable properties previously tested in vitro (physical and chemical stability, noncytotoxicity, and immunosuppressive activity) [84], neither the CsA-SLNs nor Sandimmune Neoral® presented relevant anti-inflammatory activity in DSS-induced colitis mice after 7 days of oral daily administration [83].

Despite promising results in providing efficient and selective targeting of inflamed tissues, NPs still present some drawbacks that may impair full efficacy in the treatment of IBD. These are related with the burst release characteristic of such systems as well as degradation in the upper GI tract, before reaching the target sites. The inclusion of stimuli-responsive polymers in nanotechnology-based systems is a promising strategy to circumvent problems associated with the passage across the upper GI tract [85]. The idea of having a system joining the advantages of nanosized materials and the properties of the conventional stimuli-dependent systems all in one product is tempting. This can be achieved by using stimuli-dependent materials in the production of NPs or incorporating the NPs into a secondary carrier with a stimuli-dependent profile. Makhlof et al. produced BDS-loaded NPs using PLGA and the pH-sensitive polymer Eudragit® S100 and compared their efficacy with BDS-Eudragit® S100 microparticles [85]. PLGA/Eudragit® S100 NPs displayed a comparable pH-dependent drug release profile compared to microparticles. However, NPs displayed a better therapeutic effect when orally administered to rats after TNBS colitis induction. PLGA/Eudragit® S100 NPs also exhibited enhanced ability to target inflamed tissues and lower systemic absorption compared with microparticles. Ali et al. used the same drug model and encapsulated it in PLGA NPs and PLGA NPs coated with Eudragit® S100 [54]. The latter avoided the initial burst observed for PLGA NPs in the acidic pH of stomach (pH 1.2), releasing the drug only at pH 7.4. In vivo results clearly showed benefits of the pH-sensitive coated PLGA NPs over the uncoated ones. Still, the latter showed to be superior over the orally administered drug in solution.

Another group developed the same type of pH-sensitive NPs (100 nm, −40 mV), based on PLGA and Eudragit® S100 but encapsulating curcumin [55]. This system was able to improve the poor BA of curcumin, increasing the drug's permeability across Caco-2 monolayers. Curcumin-PLGA/Eudragit® S100 NPs also exhibited enhanced anti-TNF-α secretion activity in lipopolysaccharide (LPS)-activated macrophages compared to the free drug. Further, the anti-inflammatory activity was confirmed in vivo, by the suppression of myeloperoxidase and TNF-α in a mouse with DSS-induced colitis. Curcumin-loaded NPs successfully accumulated in colonic tissues. However, they failed to demonstrate selective targeting of inflamed tissues.

Lately, the combination of two (dual) or more (multi) responsive systems in nanotechnology-based systems is another approach being exploited for IBD treatment [86]. These systems respond to a combination of two or more stimuli and are intended to avoid premature release of drugs in the stomach and small intestine, ensuring a sustained drug release throughout the colon and simultaneously selective accumulation in the target tissues. Naeem et al. developed enzyme-/pH-sensitive NPs using a polymeric mixture of azopolyurethane as the enzyme-dependent polymer and Eudragit® S100 as the pH-sensitive polymer (ES-Azo NPs) [64]. ES-Azo NPs remained intact in the stomach pH and avoided the burst release at pH 7.4, seen for single pH-sensitive NPs (ES NPs), providing a sustained release. The burst release of ES NPs at pH 7.4 favored premature drug release in the ileum, hindering its reaching more distal segments of the intestine. In their turn, enzyme-sensitive NPs exhibited higher accumulation (5.5.-fold) than ES NPs in colonic tissues of rats after oral administration. Further, the authors used this system to encapsulate BDS and the therapeutic efficacy was evaluated in a TNBS-induced colitis model. A better therapeutic effect was observed for ES-azopolyurethane NPs when compared with ES NPs and BDS free drug.

Another strategy proposed by the same group was pH- and time-dependent (pH/time) polymeric NPs comprising Eudragit® FS 30D (poly(methyl acrylate-co-methyl methacrylate-comethacrylic acid, 7:3:1) as a pH-dependent polymer and Eudragit® RS100 (ammonio methacrylate copolymer, type B) as a time-dependent controlled release polymer [53]. pH-dependent and time-dependent NPs were also produced and compared with the pH/time NPs. The formulations displayed sizes around 250 nm, and different release patterns of BDS were obtained. Time-dependent NPs showed an independent pH drug release, suffering a premature, but sustained, drug release at pH 1.2 (stomach pH) and 6.5 (small intestine pH). On the contrary, both pH-dependent and pH-/time-dependent NPs avoided the initial burst drug release in acidic conditions. However, once at pH 7.4 (ileum and colon) pH-dependent NPs suffered a quick and almost complete release of the drug, while pH/time NPs showed a sustained drug release over 24 h. This behavior may explain the better in vivo distribution of pH-/time-dependent NPs in the GI tract of mice with DSS-induced colitis. The burst release observed for pH-dependent NPs at pH 7.4 may lead to systemic absorption at the ileum level with consequent unwanted side effects and lower drug concentrations at the affected tissues, reducing the therapeutic efficacy.

The inclusion of NPs into stimuli-sensitive platforms as a second carrier is an emergent approach to circumvent these problems. In this case, a two-phase release system is proposed, and the second carrier can comprise microparticles, hydrogels, or capsules. Some examples that can be found in the literature are presented in Table 6.3. A particular and innovative colon-targeting platform is the NP-in-microparticle oral delivery system, called NiMOS [87]. This system was specially designed for oral administration of plasmids and siRNA and is composed of Type B gelatin NPs entrapped in poly(ε-caprolactone) microspheres (Fig. 6.5). The NiMOS system proved the ability to effectively deliver siRNA-TNFα [88], dual siRNA-TNFα and siRNA-cyclin [89], and plasmid DNA encoding IL-10 [90] to the inflammation site in the colon, silencing the target gene and reducing the inflammation in colitis-induced models [91, 92].

Table 6.3 Examples of nanoparticles included in a second system used as a carrier for colon-specific delivery

Nanocarrier system	Drug	Second carrier	Stimuli	Model	Ref.
PLGA (~250 nm)	Tacrolimus	pH-sensitive microspheres based on Eudragit® P4135F (30–60 μM)	pH	TNBS-induced colitis rat model	[93]
PLA (~370 nm)	Tripeptide Lys-Pro-Val	Alginate-chitosan (7:3 w/w) hydrogel	Colonic enzymes	DSS-induced colitis mice model	[94]
PLA (~450 nm)	CD98 siRNA/ poly(ethylene imine)	Alginate-chitosan (7:3 w/w) hydrogel	Colonic enzymes	DSS-induced colitis mice model	[95]
PEG-urocanic acid-modified chitosan (~200 nm)	Single-chain CD98 antibody	Alginate-chitosan hydrogel	Colonic enzymes	DSS-induced colitis mice model	[96]

PLGA, poly(lactic-co-glycolic) acid; PLA, poly(lactic acid); PEG, poly(ethylene glycol).

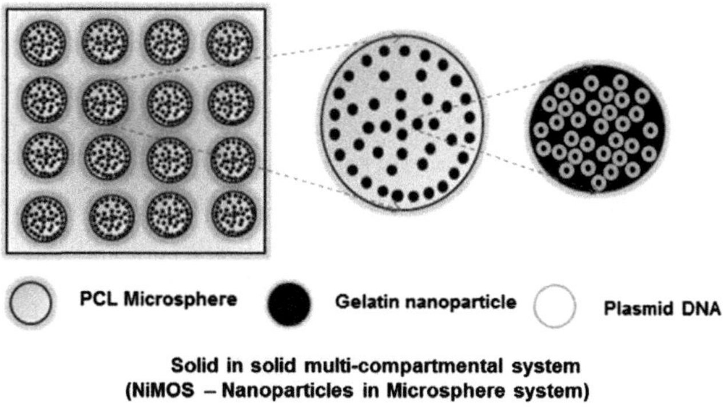

Figure 6.5 Schematic representation of a NiMOS system. Reprinted from Ref. [87], Copyright (2013), with permission from Elsevier.

6.5.2 Colorectal Cancer

Colorectal cancer is one of the most prevalent, being responsible for high mortality rates [97]. Current therapy usually involves the parenteral administration of chemotherapeutic agents. However, the systemic administration of anticancer drugs for colorectal cancer may lead to cytotoxic effects on noncancerous cells and suboptimal drug concentration at cancer cells, producing severe side effects and low therapeutic efficacy [98]. The obvious alternative to parenteral route is the local delivery of the drug into cancerous tissues, and for colorectal cancer, this can be achieved by rectal (suppositories, gels, or enemas) or oral administration. O*ral* administration is by *far* the most *preferable route* for drug delivery. However, as mentioned earlier, to be administered by the oral route, the drug must be formulated in a dosage form capable of preventing early drug release in the upper GI tract, providing stability during the journey throughout the GI tract, and promoting the drug release at the action site and ensuring drug uptake into cancerous cells. Admittedly, this is not an easy task to achieve with conventional dosage forms.

Ma et al. produced indomethacin-loaded Eudragit® S100 NPs (116 nm) and incorporated them in alginate pellets [99]. The objective was the creation of a colon-specific delivery system responsive either to enzymatic degradation by colonic microflora (alginate) or to pH stimuli. Eudragit® S100 pH-responsive NPs released indomethacin at the simulated colonic fluid and at simulated gastric fluid levels. However, its incorporation into alginate pellets resulted in an improve drug release in the simulated colonic fluid (~60%). The authors also studied the interactions established between Eudragit® RD30 NPs and the surrounding matrix when loaded into alginate microcapsules (15–80 μm). The formation of complexes between the positively charged NPs and the negatively charged alginate chains was observed, culminating in the aggregation of the former, even after the addition of a stabilizer, before the NPs' inclusion in alginate. This behavior impaired the release of NPs from microcapsules in simulated colonic fluid. However, it should be noted that the release medium did not include colonic bacteria or enzymes, which may give misleading results. The same research group developed unloaded Eudragit® RS PO NPs incorporated in Eudragit® S100–coated chitosan–hypromellose microcapsules (Fig. 6.6) [100]. To investigate their potential as colon-specific drug delivery systems, NPs were labelled with Cy5 dye, and the in vivo biodistribution and retention after oral administration in mice were studied. Eudragit® S100–coated microcapsules improved the distribution of NPs, enabling their permanence in the colonic region for 24 h after administration. Contrarily, uncoated Eudragit® RS PO NPs exhibited labile passage through the GI tract, being prematurely excreted in feces. However, the developed NPs-in-microcapsules system failed to provide colon specificity since a great extension of NP release occurred at the small intestine level. The use of mice to perform this type of in vivo studies is also debatable, since the pH values of their GI tracts may be very different from those of humans. In this case, the rat model seems to be a better choice [101].

Singh et al. encapsulated the anticancer drug 5-fluorouracil (5-Fu) into NPs comprising guar gum and xanthan gum and coadministered with probiotics [102]. 5-Fu-loaded NPs exhibited preferential release in the presence of cecal content, whereas the coadministration of probiotics promoted the replacement of the intestinal microbiota, which is important for the drug release from NPs and is damaged by 5-Fu toxicity action. Previously, the same anticancer drug was loaded in pH-sensitive NPs produced with synthetic methacrylic acid-ethyl hexyl acrylate copolymer [103]. The developed 5-Fu-loaded NPs (150 nm) exhibited a good pH-dependent release pattern, showing an increased drug release with the pH increase, followed by a sustained release over time at colonic pH. More importantly, 5-Fu-loaded NPs promoted higher cytotoxicity to the cancer colorectal cell line HCT-116 than the free drug. This result may be explained by the higher drug uptake observed for 5-Fu-loaded NPs in HCT-116 cells. The efficiency of NPs in causing apoptosis was further confirmed by immunoblot analysis.

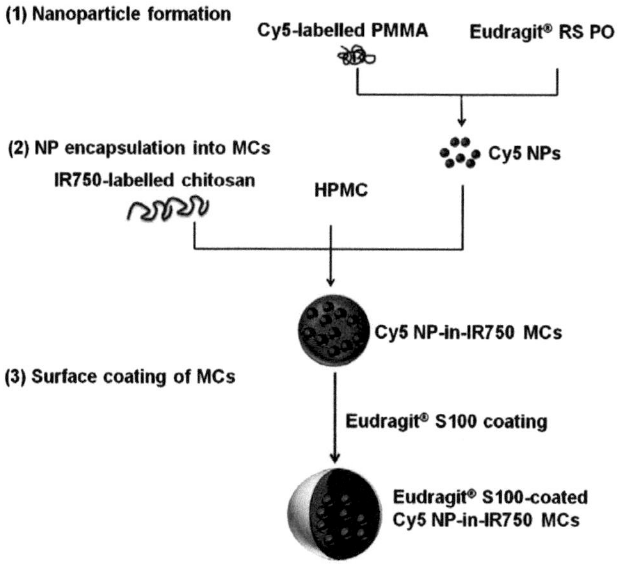

Figure 6.6 Schematic representation of nanoparticles-in-microcapsule system production. HPMC, hydroxypropyl methylcellulose; PMMA, poly(methyl methacrylate); NP, nanoparticles; MCs, microcapsules. Reprinted from Ref. [100], Copyright (2015), with permission from Elsevier.

In a similar study, the anticancer drug methotrexate was encapsulated in guar gum NPs and the surface modified with folic acid (MTX-FA-GGNP) [104]. Folate receptors are overexpressed in several human cancers, including colorectal cancer, but scarce in healthy tissues [105]. The guar gum NPs showed a pH-dependent release profile in simulated colonic fluid added with rat cecal contents, providing protection against early degradation at the stomach and small intestine levels. The cytotoxicity effect in cancer cells was higher for MTX-FA-GGNP than for methotrexate-loaded NPs unmodified with folic acid and free drug, evidencing the role of folic acid in the drug uptake in cancer cells. The biodistribution studies using fluorescein isothiocyanate (FITC) confirmed the presence of MTX-FA-GGNP in the colon, although some uptake occurred also in the stomach and the small intestine.

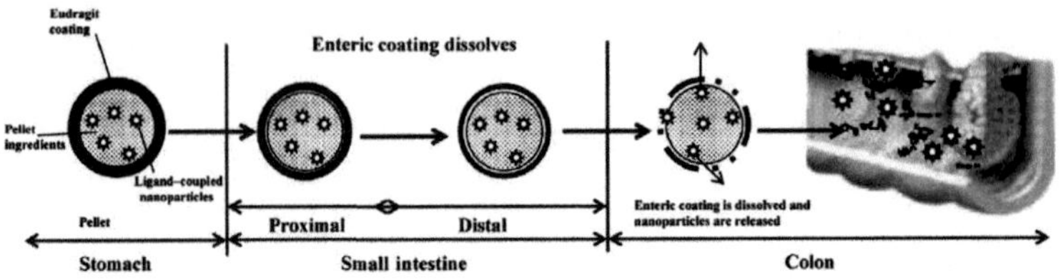

Figure 6.7 Mechanistic scheme of colon-specific targeting of Eudragit®-coated pellets encapsulating hyaluronic acid–coupled oxaliplatin-loaded chitosan nanoparticles. Reprinted from Ref. [106], Copyright (2010), with permission from Elsevier.

Jain et al. [106] used hyaluronic acid at the oxaliplatin-loaded chitosan NP surface to target drug delivery at colorectal cancer cells, since hyaluronic acid receptors are present in most cancer cells [107]. The NPs were obtained by ionotropic gelation and covalently conjugated the hyaluronic acid molecules to the free amino group of chitosan at the NP surface. Then, NPs were embedded into pellets by the extrusion-spheronization technique using microcrystalline cellulose (Avicel®). Later, the pellets containing the NPs were coated with the pH-sensitive polymer Eudragit® S100 (Fig. 6.7). In vitro and in vivo evaluations were performed to compare the hyaluronic acid–coupled oxaliplatin-loaded chitosan NPs (152 nm) with uncoupled oxaliplatin-loaded chitosan NPs (136 nm) [108]. The in vivo biodistribution studies showed both systems were able to avoid drug release during the passage across the upper GI tract after oral administration to mice. In the colon, and especially at the tumor level, they were able to provide sustained levels of drug for up to 24 h, whereas no signs of the drug were detected 8 h after administration of the drug solution. At the tumor level, higher concentrations of the drug were achieved for hyaluronic acid–coupled oxaliplatin-loaded chitosan NPs over the uncoupled NPs, demonstrating the targeting potential of the hyaluronic acid coupled system. Overall, the system consisting oxaliplatin-loaded chitosan NPs, either coupled or uncoupled with hyaluronic acid molecules, in pellets coated with

Eudragit® S100 provided specific and sustained drug delivery to the colon over time, minimizing undesired side effects. The presence of hyaluronic acid molecules at the NP surface improved the drug targeting at cancer cells, protecting the healthy tissues from the cytotoxicity action of the drug [106].

6.5.3 Vaccines

Oral vaccines may be an interesting approach for mucosal immunization against virus infections at the colorectal level. Despite great achievements, immunization by the parenteral route fails to elicit protective mucosal immunization. The immunization at the mucosal sites may be also more effective if administered at the entry sites of pathogens [109, 110]. In this case, intracolorectal administration is suitable although clinically impractical. The oral route is the most convenient, but the degradation or activity loss occurring during passage through the stomach is problematic when using the new generation of vaccines comprising proteins, synthetic peptides, and plasmid DNA [111]. Nanotechnology-based systems may be the holy grail to solve this issue, but just a few works have been undertaken so far.

Zhu et al. [112] encapsulated a peptide vaccine into PLGA NPs (300–500 nm) and the surface was coated with the pH-sensitive Eudragit® FS 30D polymer, resulting in microparticles (10–50 μm). While the uncoated PLGA NPs failed to bypass the stomach or fairly reach the small intestine, resulting in insignificant colonic responses, the NP-releasing microparticle system not only reached the colon but also provided higher mucosal uptake through the colon after oral administration in mice. Moreover, the coated NPs loading the vaccine were able to protect BALB/c mice from colorectal challenge with the *vaccinia* virus after oral delivery. Interestingly, this system provided protection not only in the colorectal but also in the vaginal mucosa.

6.5.4 Intestinal Infections

The colonic delivery of drugs encapsulated into nanocarriers may be useful to locally treat intestinal infections, including those caused by parasites and bacteria. Recently, Cerchiara et al. produced chitosan-based NPs by ionic gelation and nano-spray-drying, to be used for colonic delivery of the glycopeptide antibiotic vancomycin, for the treatment of intestinal infections [113]. NPs produced by nano-spray-drying revealed a better release profile for colonic delivery of vancomycin compared to those produced by ionic gelation. Despite using the same components and chitosan/tripolyphosphate molar ratios for both techniques, the nano-spray-drying technique allowed a better drug yield and association efficiency, as well as a more suitable release profile for colonic delivery over the ionic gelation technique.

6.5.5 Systemic Absorption

The relatively constant environment of the colon, in comparison with the upper GI tract, makes it a suitable locale for the systemic absorption of several drugs, including anti-inflammatory, antihypertensive, antianginal, antiarthritic, and antiasthmatic drugs [5, 21]. Also, the delivery of peptides and proteins in the colon to achieve systemic absorption has been tried for several decades [44, 114, 115]. Despite several efforts, only moderate success has been achieved so far. However, the onset of nanotechnologies has awakened the interest in the field and some attempts have been made to deliver these molecules, especially insulin, into the colon [116, 117].

6.6 Conclusion and Future Perspectives

Oral colon-specific drug delivery attracted significant attention during the last two decades. Several works have been carried out, and some degree of success was achieved with conventional colon-specific systems. However, the advent of nanotechnologies brought new insights and opportunities to this research field. They not only allowed the exploration of the potential of delivering drugs specifically to the colon in a controlled and sustained manner but also increased the chances of the systems to maintain their stability during the passage throughout the harsh environment of the upper GI tract. Different approaches have been developed to treat specific health problems, and the success of colon-specific delivery systems is influenced by the physicochemical properties of drugs, nanocarrier type, and GI tract physiological factors. Besides important developments, no relevant nanomedicine-based product for colon-specific drug delivery has reached the market so far.

We believe the progress in this field will focus more on the development of nanotechnological approaches for active targeting of specific cells in diseased/healthy tissues by NPs and the design of nanosystems capable of improving the permeation of drugs through the colon mucosa (e.g., mucus-penetrating NPs and mucus-adhesive NPs). Also, the increasing use of therapeutic proteins and peptides and gene therapies requires the development of more sophisticated oral delivery systems.

A suitable system for efficient and safe colonic drug delivery will probably result from combined approaches. However, there is still a long way to go concerning the use of innovative approaches in clinical practice. The inter- and intravariability of colonic physiological conditions of subjects in healthy and diseased conditions needs to be considered when designing nanocarriers for colon-specific delivery. Furthermore, the choice of adequate preclinical in vivo models and more clinical studies in humans are required to attest the efficacy of delivery systems. Another important factor, often neglected in the research works reported so far, is the study of formulations' toxicity; acute and chronic toxicity studies have to be performed to confirm the safety of delivery systems. The inclusion of nanocarriers in conventional oral platforms such as tablets, capsules, pellets, or granules for easier administration needs also to be regarded.

Acknowledgments

The authors would like to thank Fundação para a Ciência e a Tecnologia (FCT), Portugal (PTDC/SAL-FCT/104492/2008), and (SFRH/BD/78127/2011) for financial support. This work was also financed by ERDF through the Programa Operacional Factores de Competitividade – COMPETE and by Portuguese funds through FCT in the framework of the project Pest-C/SAU/LA0002/2013 and cofinanced by North Portugal Regional Operational (ON.2 – O Novo Norte) in the framework of Project SAESCTN-PIIC&DT/2011 under the National Strategic Reference Framework (NSRF).

References

1. Fonte, P., et al. Oral insulin delivery: how far are we? *J Diabetes Sci Technol*, 2013, **7**(2):520–531.
2. Castro, P.M., et al. Oral films as breakthrough tools for oral delivery of proteins/peptides. *J Controlled Release*, 2015, **211**:63–73.
3. Lee, V.H.L., et al. Oral drug delivery, in *Drug Delivery and Targeting: For Pharmacists and Pharmaceutical Scientists*, Hillery, A.M., Lloyd, A.W., Swarbrick, J., eds. Taylor & Francis, London, 2001, pp. 145–183.
4. Van den Mooter, G. Colon drug delivery. *Expert Opin Drug Deliv*, 2006, **3**(1):111–125.
5. Philip, A.K., et al. Colon targeted drug delivery systems: a review on primary and novel approaches. *Oman Med J*, 2010, **25**(2):79–87.
6. Rubinstein, A. Colonic drug delivery. *Drug Discov Today Technol*, 2005, **2**(1):33–37.
7. Sinha, V., et al. Oral colon-specific drug delivery of protein and peptide drugs. *Crit Rev Ther Drug Carrier Syst*, 2007, **24**(1):63–92.
8. Date, A.A., et al. Nanoparticles for oral delivery: design, evaluation and state-of-the-art. *J Controlled Release*, 2016, **240**:504–526.
9. Hans, M.L., et al. Biodegradable nanoparticles for drug delivery and targeting. *Curr Opin Solid State Mater Sci*, 2002, **6**(4):319–327.
10. Soppimath, K.S., et al. Biodegradable polymeric nanoparticles as drug delivery devices. *J Controlled Release*, 2001, **70**(1–2):1–20.
11. Almeida, A., et al. Solid lipid nanoparticles as a drug delivery system for peptides and proteins. *Adv Drug Deliv Rev*, 2007, **59**(6):478–490.
12. García-Fuentes, M., et al. Design of lipid nanoparticles for the oral delivery of hydrophilic macromolecules. *Colloids Surf B*, 2003, **27**(2–3):159–168.
13. Fonte, P., et al. Polymer-based nanoparticles for oral insulin delivery: revisited approaches. *Biotechnol Adv*, 2015, **33**(6 Pt 3):1342–1354.
14. Fonte, P., et al. Facts and evidences on the lyophilization of polymeric nanoparticles for drug delivery. *J Controlled Release*, 2016, **225**:75–86.
15. Soares, S., et al. Effect of freeze-drying, cryoprotectants and storage conditions on the stability of secondary structure of insulin-loaded solid lipid nanoparticles. *Int J Pharm*, 2013, **456**(2):370–381.
16. Fonte, P., et al. Effect of cryoprotectants on the porosity and stability of insulin-loaded PLGA nanoparticles after freeze-drying. *Biomatter*, 2012, **2**(4):329–339.

17. Fonte, P., et al. Stability study perspective of the effect of freeze-drying using cryoprotectants on the structure of insulin loaded into PLGA nanoparticles. *Biomacromolecules*, 2014, **15**(10):3753–3765.

18. Fonte, P., et al. Co-encapsulation of lyoprotectants improves the stability of protein-loaded PLGA nanoparticles upon lyophilization. *Int J Pharm*, 2015, **496**(2):850–862.

19. Fonte, P., et al. Annealing as a tool for the optimization of lyophilization and ensuring of the stability of protein-loaded PLGA nanoparticles. *Int J Pharm*, 2016, **503**(1–2):163–173.

20. Amidon, S., et al. Colon-targeted oral drug delivery systems: design trends and approaches. *AAPS PharmSciTech*, 2015, **16**(4):731–741.

21. Zhang, L., et al. Polysaccharide-based micro/nanocarriers for oral colon-targeted drug delivery. *J Drug Target*, 2016, **24**(7):579–589.

22. Ellis, H. Anatomy of the caecum, appendix and colon. *Surgery (Oxford)*, 2011, **29**(1):1–4.

23. Sandle, G. Salt and water absorption in the human colon: a modern appraisal. *Gut*, 1998, **43**(2):294–299.

24. Lamont, J.T. The large intestine: physiology, pathophysiology and disease. *Gastroenterology*, 1992, **102**(5):1820.

25. Edwards, C. Rectal drug delivery physiology of the colorectal barrier. *Adv Drug Deliv Rev*, 1997, **28**(2):173–190.

26. Ponz de Leon, M., et al. Pathology of colorectal cancer. *Dig Liver Dis*, 2001, **33**(4):372–388.

27. Coupe, A.J., et al. Variation in gastrointestinal transit of pharmaceutical dosage forms in healthy subjects. *Pharm Res*, 1991, **8**(3):360–364.

28. Rao, K.A., et al. Objective evaluation of small bowel and colonic transit time using pH telemetry in athletes with gastrointestinal symptoms. *Br J Sports Med*, 2004, **38**(4):482–487.

29. Stubbs, J.B., et al. A noninvasive scintigraphic assessment of the colonic transit of nondigestible solids in man. *J Nucl Med*, 1991, **32**(7):1375–1381.

30. Hebden, J.M., et al. Limited exposure of the healthy distal colon to orally-dosed formulation is further exaggerated in active left-sided ulcerative colitis. *Aliment Pharmacol Ther*, 2000, **14**(2):155–161.

31. Rana, S.V., et al. Small intestinal bacterial overgrowth and orocecal transit time in patients of inflammatory bowel disease. *Dig Dis Sci*, 2013, **58**(9):2594–2598.

32. Schiller, C., et al. Intestinal fluid volumes and transit of dosage forms as assessed by magnetic resonance imaging. *Aliment Pharmacol Ther*, 2005, **22**(10):971–979.

33. Shameem, M., et al. Oral solid controlled release dosage forms: role of GI-mechanical destructive forces and colonic release in drug absorption under fasted and fed conditions in humans. *Pharm Res*, 1995, **12**(7):1049–1054.

34. Bratten, J., et al. New directions in the assessment of gastric function: clinical applications of physiologic measurements. *Dig Dis*, 2006, **24**(3–4):252–259.

35. Sasaki, Y., et al. Improved localizing method of radiopill in measurement of entire gastrointestinal pH profiles: colonic luminal pH in normal subjects and patients with Crohn's disease. *Am J Gastroenterol*, 1997, **92**(1):114–118.

36. Ibekwe, V.C., et al. Interplay between intestinal pH, transit time and feed status on the in vivo performance of pH responsive ileo-colonic release systems. *Pharm Res*, 2008, **25**(8):1828–1835.

37. Macfarlane, G.T., et al. Fermentation in the human large intestine: its physiologic consequences and the potential contribution of prebiotics. *J Clin Gastroenterol*, 2011, **45** Suppl:S120– S127.

38. Rowland, I.R. Factors affecting metabolic activity of the intestinal microflora. *Drug Metab Rev*, 1988, **19**(3–4):243–261.

39. Kang, M.J., et al. Role of metabolism by human intestinal microflora in geniposide-induced toxicity in HepG2 cells. *Arch Pharm Res*, 2012, **35**(4):733–738.

40. Sinha, V.R., et al. Polysaccharides in colon-specific drug delivery. *Int J Pharm*, 2001, **224**(1–2):19–38.

41. Sartor, R.B. Genetics and environmental interactions shape the intestinal microbiome to promote inflammatory bowel disease versus mucosal homeostasis. *Gastroenterology*, 2010, **139**(6):1816–1819.

42. Albenberg, L.G., et al. Diet and the intestinal microbiome: associations, functions, and implications for health and disease. *Gastroenterology*, 2014, **146**(6):1564–1572.

43. Yang, L., et al. Colon-specific drug delivery: new approaches and in vitro/in vivo evaluation. *Int J Pharm*, 2002, **235**(1–2):1–15.

44. Haupt, S., et al. The colon as a possible target for orally administered peptide and protein drugs. *Crit Rev Ther Drug Carrier Syst*, 2002, **19**(6):499–551.

45. Collnot, E.M., et al. Nano- and microparticulate drug carriers for targeting of the inflamed intestinal mucosa. *J Controlled Release* 2012, **161**(2):235–246.

46. Wolk, O., et al. New targeting strategies in drug therapy of inflammatory bowel disease: mechanistic approaches and opportunities. *Expert Opin Drug Deliv*, 2013, **10**(9):1275–1286.

47. Patel, M., et al. Therapeutic opportunities in colon-specific drug-delivery systems. *Crit Rev Ther Drug Carrier Syst*, 2007, **24**(2):147–202.

48. Yun, Y., et al. Nanoparticles for oral delivery: targeted nanoparticles with peptidic ligands for oral protein delivery. *Adv Drug Deliv Rev*, 2013, **65**(6):822–832.

49. Yoshida, T., et al. pH- and ion-sensitive polymers for drug delivery. *Expert Opin Drug Deliv*, 2013, **10**(11):1497–1513.

50. Thakral, S., et al. Eudragit: a technology evaluation. *Expert Opin Drug Deliv*, 2013, **10**(1):131–149.

51. Hua, S., et al. Advances in oral nano-delivery systems for colon targeted drug delivery in inflammatory bowel disease: selective targeting to diseased versus healthy tissue. *Nanomedicine*, 2015, **11**(5):1117–1132.

52. Gugulothu, D., et al. pH-sensitive nanoparticles of curcumin-celecoxib combination: evaluating drug synergy in ulcerative colitis model. *J Pharm Sci*, 2014, **103**(2):687–696.

53. Naeem, M., et al. Colon-targeted delivery of budesonide using dual pH- and time-dependent polymeric nanoparticles for colitis therapy. *Drug Des Devel Ther*, 2015, **9**:3789–3799.

54. Ali, H., et al. Budesonide loaded nanoparticles with pH-sensitive coating for improved mucosal targeting in mouse models of inflammatory bowel diseases. *J Controlled Release*, 2014, **183**:167–177.

55. Beloqui, A., et al. pH-sensitive nanoparticles for colonic delivery of curcumin in inflammatory bowel disease. *Int J Pharm*, 2014, **473**(1–2):203–212.

56. Esseku, F., et al. Bacteria and pH-sensitive polysaccharide-polymer films for colon targeted delivery. *Crit Rev Ther Drug Carrier Syst*, 2011, **28**(5):395–445.

57. Sinha, V.R., et al. Microbially triggered drug delivery to the colon. *Eur J Pharm Sci*, 2003, **18**(1):3–18.

58. Van den Mooter, G., et al. Use of azo polymers for colon-specific drug delivery. *J Pharm Sci*, 1997, **86**(12):1321–1327.

59. Schippa, S., et al. Dysbiotic events in gut microbiota: impact on human health. *Nutrients*, 2014, **6**(12):5786–5805.

60. Dev, R.K., et al. Novel microbially triggered colon specific delivery system of 5-Fluorouracil: statistical optimization, in vitro, in vivo, cytotoxic and stability assessment. *Int J Pharm*, 2011, **411**(1–2):142–151.

61. Oliveira, G.F., et al. Chitosan–pectin multiparticulate systems associated with enteric polymers for colonic drug delivery. *Carbohydr Polym*, 2010, **82**(3):1004–1009.

62. Al-Hilal, T.A., et al. Oral drug delivery systems using chemical conjugates or physical complexes. *Adv Drug Deliv Rev*, 2013, **65**(6):845–864.

63. Rajpurohit, H., et al. Polymers for colon targeted drug delivery. *Indian J Pharm Sci*, 2010, **72**(6):689–696.

64. Naeem, M., et al. Enzyme/pH dual sensitive polymeric nanoparticles for targeted drug delivery to the inflamed colon. *Colloids Surf B*, 2014, **123**:271–278.

65. Amidon, S., et al. Colon-targeted oral drug delivery systems: design trends and approaches. *AAPS PharmSciTech*, 2015, **16**(4):731–741.

66. Friend, D.R. New oral delivery systems for treatment of inflammatory bowel disease. *Adv Drug Deliv Rev*, 2005, **57**(2):247–265.

67. Maroni, A., et al. Erodible drug delivery systems for time-controlled release into the gastrointestinal tract. *J Drug Deliv Sci Technol*, 2016, **32**(Part B):229–235.

68. Rowe, R., et al. *Handbook of Pharmaceutical Excipients*, 6th ed. Pharmaceutical Press, London, 2009.

69. Thakur, V.K., et al. *Handbook of Polymers for Pharmaceutical Technologies: Structure and Chemistry*. John Wiley & Sons, Hoboken, 2015.

70. Torres, J., et al. Preclinical disease and preventive strategies in IBD: perspectives, challenges and opportunities. Gut, 2016, **65**(7):1061–1069.

71. Taylor, K.M., et al. Optimization of conventional therapy in patients with IBD. *Nat Rev Gastroenterol Hepatol*, 2011, **8**(11):646–656.

72. Lautenschläger, C., et al. Drug delivery strategies in the therapy of inflammatory bowel disease. *Adv Drug Deliv Rev*, 2014, **71**:58–76.

73. Watts, P.J., et al. Colonic drug targeting. *Drug Dev Ind Pharm*, 1997, **23**:893–913.

74. Coco, R., et al. Drug delivery to inflamed colon by nanoparticles: comparison of different strategies. *Int J Pharm*, 2013, **440**(1):3–12.

75. Sosnik, A., et al. Mucoadhesive polymers in the design of nano-drug delivery systems for administration by non-parenteral routes: a review. *Prog Polym Sci*, 2014, **39**(12):2030–2075.

76. Lamprecht, A., et al. Nanoparticles enhance therapeutic efficiency by selectively increased local drug dose in experimental colitis in rats. *J Pharmacol Exp Ther*, 2005, **315**(1):196–202.

77. Lamprecht, A., et al. Size-dependent bioadhesion of micro- and nanoparticulate carriers to the inflamed colonic mucosa. *Pharm Res*, 2001, **18**(6):788–793.

78. Wang, Y.Y., et al. Addressing the PEG mucoadhesivity paradox to engineer nanoparticles that "slip" through the human mucus barrier. *Angew Chem Int Ed Engl*, 2008, **47**(50):9726–9729.

79. Maisel, K., et al. Effect of surface chemistry on nanoparticle interaction with gastrointestinal mucus and distribution in the gastrointestinal tract following oral and rectal administration in the mouse. *J Controlled Release*, 2015, **197**:48–57.

80. Lamprecht, A., et al. Biodegradable nanoparticles for targeted drug delivery in treatment of inflammatory bowel disease. *J Pharmacol Exp Ther*, 2001, **299**(2):775–781.

81. Lamprecht, A. IBD: selective nanoparticle adhesion can enhance colitis therapy. *Nat Rev Gastroenterol Hepatol*, 2010, **7**(6):311–3112.

82. Beloqui, A., et al. Budesonide-loaded nanostructured lipid carriers reduce inflammation in murine DSS-induced colitis. *Int J Pharm*, 2013, **454**(2):775–783.

83. Guada, M., et al. Cyclosporine A-loaded lipid nanoparticles in inflammatory bowel disease. *Int J Pharm*, 2016, **503**(1–2):196–198.

84. Guada, M., et al. Lipid nanoparticles for cyclosporine A administration: development, characterization, and in vitro evaluation of their immunosuppression activity. *Int J Nanomed*, 2015, **10**:6541–6553.

85. Makhlof, A., et al. pH-Sensitive nanospheres for colon-specific drug delivery in experimentally induced colitis rat model. *Eur J Pharm Biopharm*, 2009, **72**(1):1–8.

86. Cheng, R., et al. Dual and multi-stimuli responsive polymeric nanoparticles for programmed site-specific drug delivery. *Biomaterials*, 2013, **34**(14):3647–3657.

87. Kriegel, C., et al. Multi-compartmental oral delivery systems for nucleic acid therapy in the gastrointestinal tract. *Adv Drug Deliv Rev*, 2013, **65**(6):891–901.

88. Kriegel, C., et al. Oral TNF-alpha gene silencing using a polymeric microsphere-based delivery system for the treatment of inflammatory bowel disease. *J Controlled Release*, 2011, **150**(1):77–86.

89. Kriegel, C., et al. Dual TNF-α/cyclin D1 gene silencing with an oral polymeric microparticle system as a novel strategy for the treatment of inflammatory bowel disease. *Clin Transl Gastroenterol*, 2011, **2**(3):e2.

90. Bhavsar, M.D., et al. Oral IL-10 gene delivery in a microsphere-based formulation for local transfection and therapeutic efficacy in inflammatory bowel disease. *Gene Ther*, 2008, **15**(17):1200–1209.

91. Bhavsar, M.D., et al. Gastrointestinal distribution and in vivo gene transfection studies with nanoparticles-in-microsphere oral system (NiMOS). *J Controlled Release*, 2007, **119**(3):339–348.

92. Bhavsar, M.D., et al. Development of novel biodegradable polymeric nanoparticles-in-microsphere formulation for local plasmid DNA delivery in the gastrointestinal tract. *AAPS PharmSciTech*, 2008, **9**(1):288–294.

93. Lamprecht, A., et al. A pH-sensitive microsphere system for the colon delivery of tacrolimus containing nanoparticles. *J Controlled Release*, 2005, **104**(2):337–346.

94. Laroui, H., et al. Drug-loaded nanoparticles targeted to the colon with polysaccharide hydrogel reduce colitis in a mouse model. *Gastroenterology*, 2010, **138**(3):843–853.e1-2.

95. Laroui, H., et al. Targeting intestinal inflammation with CD98 siRNA/PEI-loaded nanoparticles. *Mol Ther*, 2014, **22**(1):69–80.

96. Xiao, B., et al. Nanoparticles with surface antibody against CD98 and carrying CD98 small interfering RNA reduce colitis in mice. *Gastroenterology*, 2014, **146**(5):1289–1300.e1-19.

97. Favoriti, P., et al. Worldwide burden of colorectal cancer: a review. *Updates Surg*, 2016, **68**(1):7–11.

98. Adams, V.R. Adverse events associated with chemotherapy for common cancers. *Pharmacotherapy*, 2000, **20**(7 Pt 2):96s–103s.

99. Ma, Y., et al. Designing colon-specific delivery systems for anticancer drug-loaded nanoparticles: an evaluation of alginate carriers. *J Biomed Mater Res A*, 2014, **102**(9):3167–3176.

100. Ma, Y., et al. The in vivo fate of nanoparticles and nanoparticle-loaded microcapsules after oral administration in mice: evaluation of their potential for colon-specific delivery. *Eur J Pharm Biopharm*, 2015, **94**:393–403.

101. Sjogren, E., et al. In vivo methods for drug absorption - comparative physiologies, model selection, correlations with in vitro methods (IVIVC), and applications for formulation/API/excipient characterization including food effects. *Eur J Pharm Sci* 2014, **57**:99–151.

102. Singh, S., et al. A nanomedicine-promising approach to provide an appropriate colon-targeted drug delivery system for 5-fluorouracil. *Int J Nanomed*, 2015, **10**:7175–7182.

103. Ashwanikumar, N., et al. Methacrylic-based nanogels for the pH-sensitive delivery of 5-fluorouracil in the colon. *Int J Nanomed*, 2012, **7**:5769–5779.

104. Sharma, M., et al. Folic acid conjugated guar gum nanoparticles for targeting methotrexate to colon cancer. *J Biomed Nanotechnol*, 2013, **9**(1):96–106.

105. Zhao, X., et al. Targeted drug delivery via folate receptors. *Expert Opin Drug Deliv*, 2008, **5**(3):309–319.

106. Jain, A., et al. Design and development of ligand-appended polysaccharidic nanoparticles for the delivery of oxaliplatin in colorectal cancer. *Nanomed: Nanotechnol, Biol Med*, 2010, **6**(1):179–190.

107. Lugli, A., et al. Overexpression of the receptor for hyaluronic acid mediated motility is an independent adverse prognostic factor in colorectal cancer. *Mod Pathol*, 2006, **19**(10):1302–1309.

108. Jain, A., et al. In vitro and cell uptake studies for targeting of ligand anchored nanoparticles for colon tumors. *Eur J Pharm Sci*, 2008, **35**(5):404–416.

109. McConnell, E.L., et al. Colonic antigen administration induces significantly higher humoral levels of colonic and vaginal IgA, and serum IgG compared to oral administration. *Vaccine*, 2008, **26**(5):639–646.

110. Zhu, Q., et al. Immunization with adenovirus at the large intestinal mucosa as an effective vaccination strategy against sexually transmitted viral infection. *Mucosal Immunol*, 2008, **1**(1):78–88.

111. Cordeiro, A.S., et al. Nanoengineering of vaccines using natural polysaccharides. *Biotechnol Adv*, 2015, **33**(6 Pt 3):1279–1293.

112. Zhu, Q., et al. Large intestine-targeted, nanoparticle-releasing oral vaccine to control genitorectal viral infection. *Nat Med*, 2012, **18**(8):1291–1296.

113. Cerchiara, T., et al. Chitosan based micro- and nanoparticles for colon-targeted delivery of vancomycin prepared by alternative processing methods. *Eur J Pharm Biopharm*, 2015, **92**:112–119.

114. Mackay, M., et al. Peptide drug delivery: colonic and rectal absorption. *Adv Drug Deliv Rev*, 1997, **28**(2):253–273.

115. Maroni, A., et al. Oral colon delivery of insulin with the aid of functional adjuvants. *Adv Drug Deliv Rev*, 2012, **64**(6):540–556.

116. Guo, F., et al. Modified nanoparticles with cell-penetrating peptide and amphipathic chitosan derivative for enhanced oral colon absorption of insulin: preparation and evaluation. *Drug Deliv*, 2016, **23**(6):2003–2014.

117. Salvioni, L., et al. Oral delivery of insulin via polyethylene imine-based nanoparticles for colonic release allows glycemic control in diabetic rats. *Pharmacol Res*, 2016, **110**:122–130.

Part III

Topical Drug Delivery Approaches

Chapter 7

Nanotechnological Approaches in Drug Absorption through Skin Topical Delivery

Sofia A. Costa Lima and Salette Reis
UCIBIO, REQUIMTE, Department of Chemical Sciences, Faculty of Pharmacy, University of Porto, Portugal
slima@ff.up.pt

In 1959, Feynman gave a lecture called "There's Plenty of Room at the Bottom," in which he explored the possibility of the direct manipulation of individual atoms, which became the inspiration for the application of these concepts decades later. Indeed, contemporary nanotechnology focuses on the exploration of the interesting properties of matter on a scale where the dimension of particles is reduced to that of individual molecules or their aggregates, namely particles whose size is in the nanometer scale. Without a doubt, the development of nanoparticles for dermatologic applications is an area of increasing magnitude and interest, as it has been possible to observe a recent growth in financial investments and an exponential number of registered patents in this context. The usage of nanoparticles in dermatology allows for the creation of newer and better-targeting therapeutic strategies, through surface

Nanoparticles in Life Sciences and Biomedicine
Edited by Ana Rute Neves and Salette Reis

ISBN 978-981-4745-98-7 (Hardcover), 978-1-351-20735-5 (eBook)
www.panstanford.com

modification and binding of particular ligands for specific cell targets; enhancement of the use of already existing (and sometimes discarded) pharmaceutical agents, together with the innovative application of the same drugs but through different administration routes, is also possible by the use of nanoparticles. The following section will unveil prospective approaches in drug absorption through skin topical delivery.

7.1 Introduction

Human skin provides a unique delivery pathway for therapeutic and other active agents. Skin penetration of these agents involves intercellular, intracellular, and transappendageal routes, resulting in topical delivery to the skin or transdermal delivery into the systemic circulation.

Topical/transdermal drug delivery is the most preferred administration route, with higher patient compliance and satisfaction. Advantages also include the avoidance of hepatic first-pass metabolism effects, avoidance of metabolic degradation associated with oral administration, and far fewer systemic side effects in comparison to other administration routes. Nanotechnology has gained interest with the optimization of several methods (e.g., thermo- or radiofrequency) able to control pore opening, and thus improve drug skin penetration. Nanoparticle-based topical delivery systems are promising nanomedicines for the treatment of skin diseases; combining advantages of nanosized carriers with the topical approach. The nanotechnological approaches for improving drug absorption for skin-related diseases will be discussed.

7.2 Skin Structure

The skin is a major human organ, having a surface area that ranges between 1.5 and 2 m^2 for adults. It is much more than a static, impenetrable physical barrier that functions as a first-line defense against the external environment. In fact, the skin is a dynamic and integrated network of cells, tissues, and matrix elements providing multifaceted functions, namely protecting from physical injuries, infectious agents, and ultraviolet (UV) radiation and involved in thermoregulation, wound repair, and regeneration [1]. These various functions are essential to human survival and welfare. The skin is a heterogeneous membrane, lipophilic on its surface and hydrophilic in its deeper layers. In terms of structure, the skin is very complex and composed of three main layers: epidermis, dermis, and hypodermis. These are interdependent, with functional regions connected by surrounding tissue for regulation and modulation of normal structure and function.

7.2.1 Epidermis

The outermost layer is the epidermis, a stratified and cornified region that has a thickness of about 50–150 μm and is characterized by the absence of blood vessels. The most abundant cells are keratinocytes, which are organized into different strata. Along the keratinocytes, throughout the different strata, other resident cells are found, such as melanocytes, Langerhans cells, and Merkel cells. Keratinocytes progressively differentiate from proliferative basal cells at the epidermal basement membrane to the differentiated keratinized stratum corneum (SC), which constitutes the outermost layer of the skin.

In the SC, keratinocytes are completely differentiated into anucleated cells filled with keratin and are designated corneocytes [2, 3]. These cells are embedded in a lamellar matrix of intercellular lipids conferring permeation resistance, which makes the SC responsible for the barrier function of the skin, reducing the passage of molecules larger than 500 Da. In sum, the SC barrier is composed of a two-compartment system of lipid-depleted, protein-enriched corneocytes embedded in an extracellular lipid matrix. The latter is responsible for regulation of permeability, desquamation, toxin exclusion, and selective chemical absorption. On the other hand, corneocytes' role is based on mechanical reinforcement, hydration, cytokine-mediated inflammation, and UV protection.

The SC is lipophilic and contains ca. 10% of water, while a viable epidermis is significantly hydrophilic (>50%) [4].

7.2.2 Dermis

The dermis represents the majority of the skin, with a thickness of approximately 250 μm, and provides pliability, elasticity, and tensile strength. It is an integrated system of fibrous and connective tissue accommodating epidermally derived appendages (e.g., nails and sweat glands) and nerve and vascular networks. Thus metabolic exchanges between the epidermis and the blood systems may occur as also clearance of cell metabolic products and penetrated foreign agents. The dermis contains resident cells (e.g., fibroblasts, macrophages, and mast cells) and transient circulating cells of the immune system. It interacts with the epidermal keratinocytes to maintain the skin structure, enable morphogenesis, and perform wound repair. In the dermis the water content reaches 70%, favoring hydrophilic drug uptake. Below the reticular dermis layer, the fibrous connective tissue transitions to the adipose tissue of the hypodermis. This is mainly constituted of adipocytes interconnected by collagen fibers, forming a thermal barrier able to store energy and protect from physical shock [5].

7.3 Skin Function

7.3.1 Skin Penetration

Skin delivery is of utmost interest and has been extensively used by the pharmaceutical and cosmetic industries [2, 4, 6]. For successful skin delivery the applied substance needs to penetrate into the skin by the transdermal route through the SC or by the transfollicular route using pilosebaceous units, such as hair follicles or sebaceous glands. Different factors influence the cutaneous absorption of a substance (drug molecule or nanoparticle): (i) location and skin environment at the application site, (ii) physicochemical properties of the substance, and (iii) physicochemical properties of the vehicle dispersing the substance [2].

Factors influencing penetration related with skin location and condition include skin integrity, size of pore, density of appendages, SC thickness (varies according to body region), age, skin type, and sex hormones. Thus, skin absorption through the transdermal or transfollicular routes depends on body location. In particular, penetration of nanosized molecules by the transdermal route may be favored in those body locations where the SC is less thick or has been exfoliated [2, 7].

The physicochemical properties of the penetrating substance that influence their skin permeation are solubility, ionization state (unionized molecules are taken to the viable epidermis), molecular weight (preferably lower than 500 Da), and octanol/water partition coefficient of the penetrating substance. For example, a lipophilic molecule will easily penetrate the SC but will leave it with difficulty, whereas a hydrophilic molecule will suffer poor penetration. Additionally, nanocarrier stability affects skin penetration as it involves disassembly or rupture and the ability to form micelles upon contact with skin components.

Skin delivery implies the design of a formulation containing the active agent of interest incorporated in a vehicle of diverse ingredients. Thus, all components of the formulation will be subjected to skin absorption, although to different extents according to their physicochemical properties. As a result, possible synergisms and/or interactions between the vehicle, the agent, and the skin and the application method will definitely affect absorption outcomes. Several situations may occur: after application volatile compounds will evaporate; nonvolatile ingredients will be absorbed to different extents; and skin, sweat, and/or sebum components will diffuse in the applied formulation. Thus, the composition and penetration potential of vehicle components should be considered, as potential enhancing effects could be provided by specific ingredients in the formulations [2].

Regulation of SC hydration depends on the proteolysis of the corneocyte content and prevents SC mechanical failure, ensuring SC metabolic activity and correct self-assembly of intercorneocyte lipids. In general, the SC is considered a lipophilic stratum, composed essentially of intercellular lipids (e.g., cholesterol and derivatives, ceramides, free fatty acids, and triglycerides), in contrast to the higher amount of water in viable epidermis (hydrophilic layer) [2, 7]. The SC presents a barrier to most high-molecular-weight compounds and prevents the ingress of intact nanoparticles [8]. Therefore, corneocytes represent the first macroscopic physical barrier against the penetration of foreign agents, also ensuring impact resistance.

However, therapeutic nanoparticles and microparticles can be delivered in diseased skin and to hair follicles [9].

Sweat glands and hair follicles are appendages that extend from the SC to the dermis or hypodermis (2–5 mm in length) and are involved in thermoregulation and excretion of acids and body wastes. Both skin appendages create openings on the skin surface, providing breaches that may be used as potential ports of ingress.

7.3.2 Major Routes for Entrance into the Skin

The major distinct routes described for drug molecules and nanoparticles to enter into the skin are transcellular, intercellular, and transappendageal transport pathways (Fig. 7.1).

The intercellular pathway involves passage through the lipid matrix that occupies the intercellular spaces of the corneocytes, while the transcellular pathway, also known as the intracellular pathway, is through the successive skin layers and dead cells. The transappendageal pathway uses the different skin appendages (e.g., fair follicles, sebaceous glands, and sweat glands) to enter through the skin.

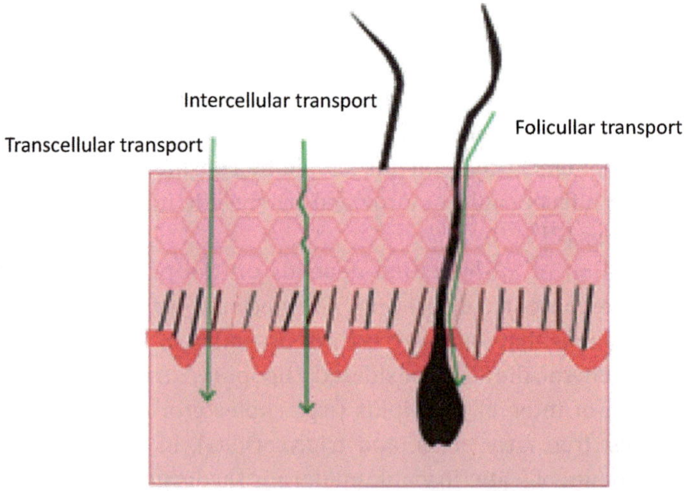

Figure 7.1 Schematic representation of different entry pathways into the skin.

Transdermal delivery involves the transcellular and intercellular pathways, and the transfollicular delivery is restricted to the transfollicular pathway. The intercellular route may dominate during steady-state penetration of compounds, but the skin appendages may offer an alternative pathway for a diffusing molecule, more significant than previously believed [10].

7.4 Nanotechnology for Winning the Skin Barrier

7.4.1 Advantages of Skin Drug Delivery

The main advantages of skin drug delivery include minimal invasiveness or noninvasiveness of an application and improved drug pharmacokinetics and drug targeting.

Minimal invasiveness and pain evasion are relative issues based on personal opinion. Skin perforation with a hypodermic insulin needle, a microneedle, or thermos- or radiofrequency is less but not totally noninvasive. To reduce the danger of cutaneous infection the number and area of transcutaneous passages should be minimal, through the use of a minimally invasive device (e.g., a microneedle or a ballistic skin breaching method). Pore-opening persistency is a parameter that affects transdermal drug transport and the possibility of opportunistic infections, as thinner holes close more rapidly than wide open pores. In some cases occlusions may hamper pore closure and favor epicutaneous bacterial growth. Setting more holes in the skin barriers decreases the overall permeability of the pores, probably due to a local microedema [11]. The skin sonication method promotes skin permeability under occlusion for 24 h, followed by 7–32 h without occlusion, depending on the applied permeant's size [12]. This approach poses a challenge to the skin as infections are more probable than after the insertion of a hypodermic insulin needle.

The improvement of drug pharmacokinetics and drug targeting is complex as it requires surpassing of the primary skin barrier and local control and distribution of the applied drug throughout the blood vessels and lymphatic system. Cutaneous blood vessels are broadly spread over the skin, in the dermis, depending on the skin condition and body activity. Thus clearance of the applied drug varies locally and physiologically [13]. Cutaneous blood capillaries favor small molecules' clearance through the transdermal pathway to systemic distribution but hamper local drug delivery [14]. Transdermal delivery can be improved by several available commercial options for skin barrier elimination, SC/epidermis perforation, and SC penetration, as briefly described in Table 7.1. The multitude of electrical, mechanical, and carrier/enhancer-based methods enable the opening of new or widening of pre-existing entries across the skin barrier of the body.

Table 7.1 Options for improving transdermal delivery of drugs

Skin barrier elimination
Regional SC elimination ▪ Abrasion using sandpaper ▪ Ablation by radiofrequency or ultrasonic field treatment (e.g., RF-MicroChannel®) Localized SC elimination ▪ Microscission (e.g., Airbrasive®) ▪ Thermoporation (e.g., PassPort® patch)
SC/epidermis perforation
▪ Mechanoporation by cutaneous microneedles (e.g., MicroFlux, MicroJet needle, and MicroCor) ▪ Sonoporation (sonic pulses) (e.g., Sonoprep®) ▪ Electroporation (high-voltage pulses) ▪ Mechanoporation by ballistic droplets (high-velocity impact of particles) (e.g. Antares Vision®, Choice®, Mini-Jet®, and Injex®)
Skin penetration with or without depot creation
▪ Penetration enhancers ▪ Soft carriers (e.g., hydrogels, emulsions, and nanoparticles)

Local or regional skin barrier elimination maximizes permeability by "noninvasive" methods, creating deeper pores, which remain open for several days. For SC and/or epidermis perforation several techniques are available, exploiting different sources of energy to break the normally tight cellular junctions in the skin barrier and thus opening transepidermal pores for at least 24 h [15]. The least perturbing approach is SC penetration, comparable to nature's action of pathogens able to enter the body using the noninvasive pericellular route. Similarly, drug nanocarriers can be designed to enter the skin barrier with minimal invasiveness.

Of outmost importance is to ensure that the skin is not significantly compromised, ensuring its protective action even when the skin barrier is locally or temporarily opened to deliver a substance.

7.4.2 Nanocarriers for Skin Delivery

Nanocarriers have unique physical and chemical features related to their nanosized range that confer significant advantage for skin delivery purposes. These colloidal systems can be composed by lipids, carbohydrates, proteins, polymers, and metal compounds. For skin delivery nanocarriers should have a size below 500 nm [8]. Upon application to the skin, the nanocarriers may exert their effect locally (topical or dermal delivery) or systemically through permeation using the transdermal and transfollicular pathways [8].

A significant increase has been observed on the application of nanocarriers to improve skin penetration. Though several types of nanocarriers have been described for skin delivery, the capacity to provide enough dose of a drug to a specific site and SC penetration are still a challenge [2]. Interaction between SC lipids and nanocarriers is a focus of intense research to create the most efficient delivery system able to penetrate without damaging the SC barrier [8].

The application of nanocarriers to transfer therapeutic agents through the skin offers several benefits: protection from the environment, enhancement of product esthetics, sustained release, follicular targeting, and protection against degradation. Nevertheless, the use of delivery systems in the skin remains clinically low, as it is still necessary to unravel drug-vehicle interactions and ensure sufficient drug loading and efficient drug release [9].

Nanocarriers can be generally classified as lipid- or polymeric-based nanoparticles. The most commonly used lipid-based nanoparticles are liposomes, solid lipid nanoparticles (SLNs), nanostructured lipid carriers (NLCs), and emulsions. Liposomes are lipid vesicles formed of one or more lipid bilayers containing an inner aqueous core and according to the composition can be designated as transferosomes, ethosomes, niosomes, and polymersomes [8]. In fact, their composition can be adjusted to improve skin penetration, for example, by inclusion of an ethanol permeating agent in ethosomes. Development of SLNs and NLCs for dermal application takes advantage of their occlusive properties and lipid interaction to improve skin penetration and follicular penetration as well [16]. An emulsion is a colloidal nanocarrier composed of a surfactant, oil, and water and usually has a low viscosity [8]. A major advantage of emulsions is the ability to incorporate hydrophilic and lipophilic drugs. Polymeric-based nanoparticles, made of natural or synthetic polymers, have an advantage in that they slowly release the drug and are often used in dermopharmaceutics and in cosmetics [4, 10, 17].

7.4.3 Lipid-Based Nanocarriers

7.4.3.1 Liposomes

One of the first nanocarriers for transdermal delivery was based on liposomes due to their ability to exert different functions after topical application [18, 19]. Liposome composition determines the penetration effect, most probably due to the phase transition temperature of the phospholipids and correspondent fluidity. Also, lipid-coating or lipophilic properties may favor higher uptake into skin appendages with a lipophilic environment (e.g., hair follicles) [20].

Lipid-based carriers are nontoxic, biocompatible, and biodegradable [21]. These nanocarriers, made of nanosized physiological lipids, can penetrate the skin and break the SC structure, leading to fluidity and efficient drug transport across the skin.

Liposomes comprise one or more lipid bilayers, separated by aqueous compartments. Hence, liposomes can deliver both hydrophilic and hydrophobic bioactive agents. The mechanism of action involves lipid bilayer fusion with the cell membrane, resulting in the delivery of its contents [22]. Modified liposomes with alcohol (ethosomes) or an edge activator (transferosomes or elastic liposomes) are able to deform upon skin application [23].

Ethosomes potentiate the penetration of drugs to reach deeper skin structures, such as pilosebaceous follicles. Their high ethanol concentration, up to 45%, is responsible for disturbing the order of the skin lipid bilayer, resulting in a higher skin penetration and drug delivery into deep skin layers. Therefore, these nanocarriers represent a promising approach for topical delivery [35].

Transferosomes, or ultradeformable nanocarriers, can interact with the skin, exploiting the hydrophilic pathway (spaces between the cells) [36]. Niosomes are made of a lipid mixture and nonionic surfactants, leading to increased membrane fluidity and improved permeation through the SC [37, 38]. Niosomes deposit drugs in the body and allow controlled release in a specific part of the body, thus minimizing the dose required for therapeutic administration [39].

In sum, the high versatility makes liposomes a potential drug nanocarrier that can permeate through the SC. These nanocarriers make efficient noninvasive delivery of a wide range of bioactive agents (e.g., drugs, peptides, growth factors, and antibiotics) with poor skin penetration. Some recent examples of enhanced delivery of bioactive agents through liposome nanocarriers into the skin are listed in Table 7.2.

Table 7.2 Options for improving transdermal delivery of drugs

Delivery system	Bioactive agent	Remark	Ref.
Liposomes	Clotrimazole	Higher skin permeation and enhanced antifungal activity of optimized elastic liposomes	[24]
	Ketoprofen	Surfactant-based-liposome-enhanced ketoprofen transport across human skin	[25]
	Paroxetine	Improved drug bioavailability after liposomal transdermal patch application and sustainment of the therapeutic effects compared to oral administration	[26]
	Recombinant human epidermal growth factor	Enhanced permeation and localization of recombinant human epidermal growth factor in rat skin by facilitated entry through pores of skin	[27]
Ethosomes	Econazole nitrate	Controlled drug release, better antifungal activity, and good storage stability	[28]

(*Continued*)

Table 7.2 *(Continued)*

Delivery system	Bioactive agent	Remark	Ref.
Transferosomes	Insulin	Prolonged hypoglycemic effect in diabetic rats over 24 h after transdermal administration	[29]
	Ketoconazole	Transferosomal formulation improving the release and permeation of ketokonazole	[30]
	Sertraline	Transferosomal drug exhibiting better antidepressant activity as compared to the plain drug	[31]
Niosomes	Fluconazole	Sustained release with enhanced cutaneous retention of the drug	[32]
	Papain	Enhanced transdermal absorption of papain rat skin and the improvement of scar reduction in rabbit ear model	[33]
	Sumatriptan succinate	Successful optimization of sumatriptan succinate niosomes as drug carriers for skin delivery	[34]

7.4.3.2 Lipid nanoparticles

Lipid nanoparticles, such as SLNs and NLCs, have several advantages over other nanocarriers as they combine the main features of emulsions (good tolerability, easy production by high-pressure homogenization, and scale-up) and polymeric nanoparticles (controlled release of drugs, protection from chemical degradation, and slower metabolism) [40]. Lipid-based colloidal nanocarriers have other merits, such as good biocompatibility, low toxicity, and the fact that they can be sterilized. Increasing attention has been paid to lipid nanoparticles for topical administration of pharmaceutical agents or cosmetics [16, 41]. In fact, their unique occlusive properties enhance skin hydration and improve drug penetration through the skin; also lipid nanoparticles interact with skin lipids, allow drug targeting, reduce local irritation, enhance physical stability, and have a low cost when compared to liposomes [40].

Skin administration of SLNs and NLCs is favored by their adhesiveness, occlusion, and skin hydration effect [4, 41]. These lipid nanoparticles form a hydrophobic film monolayer on the skin, known as adhesiveness, and consequently the occlusive effect occurs and prevents moisture loss. These effects lead to reduction of corneocyte packing and opening of intercorneocyte gaps, thus facilitating drug penetration into the skin [41, 42]. In addition, epidermal lipids interact with lipid nanocarriers adhering to the skin and lipid exchange occurs between the outermost layer of the SC and the nanocarriers [42]. Lipophilicity of the entrapped drug determines the interaction rate with the skin lipids. Thus highly lipophilic molecules can penetrate easily and hydrophilic molecules do not diffuse.

Often during storage, changes may occur in SLN lipid conformation, resulting in transformation from polymorphic to perfect crystals and consequent drug expulsion [43]. This drawback was solved with the development of the NLC, as a solid and liquid lipid mixture, which solidifies after cooling, without recrystallization. A NLC remains in the amorphous state during storage, preserving the entrapped drug for a longer period, and its composition allows the increase of drug loading capacity [44].

The lipid nanoparticle composition and the physicochemical characteristics of the entrapped molecule will determine the loading capacity, incorporation efficiency, morphology, size, release profile, and skin absorption properties [45]. In the case of a SLN and to ensure stability the loading capacity is usually limited to ca. 10% of the lipid amount [42]. The high structural similarity between the lipid matrix of the SLN and the epidermal lipids in the skin makes a SLN promising transdermal drug nanocarrier in the nanotechnology field.

Lipid nanoparticles can be produced by different methodologies (e.g., spray chilling, precipitation, and high-pressure homogenization), with the possibility of industrial scale-up and manufacture [45]. Commonly, lipid nanoparticles are prepared by melt emulsification followed by homogenization through a high-pressure homogenizer or an ultrasonic processor [41]. The high-pressure homogenization technique is the most advantageous method due to the ease with which it can be scaled up, avoidance of organic solvents, and short production time, and industrially it is the most feasible one

[41]. It is clear that lipid nanoparticles are able to enhance drug penetration into the skin, allowing increased targeting of the epidermis and consequently increased treatment efficiency and reduction of the systemic absorption of drugs. A few recent reports on SLNs and NLCs as skin delivery systems are listed in Table 7.3.

Table 7.3 Examples of lipid nanoparticle solutions for improving transdermal delivery

Delivery system	Bioactive agent	Remark	Ref.
SLN	Artemisinin	SLNs delivered artemisone into the SC epidermis at a significantly higher concentration.	[46]
	Adapalene and vitamin C	There was enhanced targeting of the skin epidermal layer and reduced systemic penetration; coadministration of vitamin C led to an adjunct effect in acne therapy.	[47]
	Diflucortolone valerate	There was improved drug deposition in the skin with the optimized drug-loaded NLC, in contrast to a commercial formulation.	[48]
	Doxorubicin	In vitro and in vivo results indicated the superiority of the cytotoxic performance of the DOX-loaded SLN compared to the DOX solution.	[17]
	Hydroquinone	It improved hydroquinone localization in the skin and lowered its systemic absorption.	[49]
	Isotretinoin	SLN transported the drug to various skin layers effectively, contrary to the free drug, and showed marked antiacne potential and tolerability.	[50]
	Retinyl palmitate	The negatively charged SLN improved bioavailability and enhanced the skin distribution of retinyl palmitate and delivered it to a greater depth than did neutral SLNs.	[51]

Delivery system	Bioactive agent	Remark	Ref.
NLC	Dexamethasone acetate	Skin deposition and permeation were significantly improved with the NLC.	[52]
	Fluocinolone acetonide	The NLC led to a significant amount of fluocinolone acetonide in the epidermal and dermal layers of the skin in comparison with the free drug.	[53]
	Methotrexate	The NLC led to higher drug skin penetration when compared to free methotrexate.	[54]

The incorporation of a drug into lipid nanoparticles may result in different distributions within the nanocarriers: homogenously within the lipid matrix, enriched in the core or particle shell, or adherent to the surface. Also the chemical nature, concentration of ingredients, and production conditions affect the structure of lipid nanoparticles. Hence, different release profiles may occur for SLNs and NLCs. In particular, for dermal delivery the burst release effect and the sustained release effect might be of interest as burst release can improve drug permeation and sustained release is suitable when an active compound needs to be delivered to the skin for a long period of time [41].

7.4.3.3 Microemulsions

The dispersion of two or more immiscible liquids in the presence of a surfactant stabilizer occurs through the emulsion phenomenon, resulting in macro- or microemulsions. Microemulsions are biphasic dispersions composed of immiscible lipids stabilized by a film of surfactant in conjunction with a cosurfactant. These nanocarriers are easy to prepare and thermodynamically stable, have a high surface area, and can be the oil-in-water (O/W) or water-in-oil (W/O) variety, with a size range of 5–100 nm. Drug release from microemulsions is controlled by drug interactions with the surfactant and the oil/water partition [55] and some examples are described on Table 7.4.

Table 7.4 Possibilities for improving transdermal delivery based on microemulsions

Delivery system	Bioactive agent	Remark	Ref.
	α-Tocopherol and lipoic acid	The microemulsion structure only affected the tocopherol penetration; the delivered levels of both antioxidants were sufficient for a decrease in thiobarbituric acid-reactive substances.	[56]
	Capsaicin	The skin permeation flux of low-dose capsaicin—0.15% (w/w)—was significantly higher for the microemulsion (ca. 4-fold) than for the commercial product and capsaicin in ethanol (control) (ca. 2-fold).	[57]
	Lycopene and ascorbic acid	Skin treatment with monocaprylin containing both lycopene and ascorbic acid increased the tissue antioxidant activity 10.2-fold.	[58]
Microemulsions	Penetratin	The microemulsion containing penetratin promoted a larger increase of transepidermal water loss (2-fold) than the plain or the phytosphingosine-containing microemulsions (1.5-fold).	[59]
	Pentoxifylline	Topical-pentoxifylline-loaded microemulsion application developed superior anti-inflammatory activity when compared to the PTX solution, reducing the paw edema up to 88.83%.	[60]

7.4.4 Polymeric-Based Nanocarriers

Polymeric nanoparticles have tunable chemical and physical features that allow effective drug protection from degradation or denaturation. Other advantages include the ability to enhance skin permeation of drugs that are poorly water soluble by increasing the drug concentration gradient across the skin layers. Also, these nanocarriers exhibit a slow and controlled drug release and can enhance the cutaneous penetration of the drugs across the skin barrier by increasing the concentration gradient. Polymeric-based nanoparticles can be classified as natural or synthetic according to the nature of their composition.

Natural polymeric nanoparticles are based on natural polymers such as chitosan, alginate, and gelatin, purified from nature. These can form hydrogels incorporating peptides or hydrophilic drugs. Chitosan is the most frequently applied for topical skin delivery. It is a cationic natural polymer with antioxidant, anti-inflammatory, and antimicrobial properties suitable for delivery of drugs to dermatological disease sites. Two of the limitations of natural polymers are a lack of batch-to-batch consistency and a lack of the controlled release pattern. When using synthetic polymers it is possible to modulate drug release and to have batch-to-batch reproducibility. The most widely used synthetic polymers are biodegradable and biocompatible aliphatic polyesters, for example, polylactides, poly(lactide-co-glycolide) copolymers, and poly(ε-caprolactone), and the nondegradable polymers poly(methyl methacrylate) and polyacrylates. Usually synthetic polymer nanoparticles are applied for delivery of hydrophobic drugs.

Great efforts have been devoted to the commercialization of nanomedicine technologies. Several nanoparticle-based treatments have already been approved by the Food and Drug Administration (FDA), such as Estrasorb (micellar nanoparticles) for topical menopause therapy and Abraxane (albumin-bounded paclitaxel nanoparticles) for breast cancer treatment. A few examples of polymeric-based delivery systems are listed in Table 7.5, along with their remarks.

Table 7.5 A few examples of topical delivery systems based on natural and synthetic polymeric nanoparticles

Polymer	Bioactive agent	Remark	Ref.
Chitosan	Hydrocortisone and hydroxytyrosol	Cationic polymeric chitosan nanoparticles loaded with anti-inflammatory and antimicrobial drugs penetrated 2.46-fold deeper than the commercial formulation.	[61]
	Curcumin	In vivo studies show that chitosan wound dressing containing chitosan/poly-γ-glutamic acid/pluronic/curcumin nanoparticles promoted neocollagen regeneration and tissue reconstruction.	[62]
Lecithin-chitosan	Clobetasol-17-propionate	Lecithin/chitosan nanoparticles can induce epidermal targeting and improve the risk-benefit ratio for topically applied clobetasol-17-propionate.	[63]
	Diflucortolone valerate	Nanoparticles containing diflucortolone valerate in gel formulations produced significantly higher edema inhibition in rats compared with commercial cream in in-vivo studies.	[64]
PLGA	Indomethacin	A higher amount of indomethacin was delivered through the skin when the drug was loaded in nanoparticles than the free drug.	[65]
PCL	Minoxidil	In vivo and in vitro skin permeation results demonstrated that nanoparticles containing solutes penetrated mainly via shunt routes like hair follicles, resulting in skin absorption of solutes.	[66]
PLGA/ Chitosan	Spantide II and ketoprofen	Deposition of ketoprofen through the KP-SPN, or skin-permeating nanogel, system in epidermis and dermis was 9.75- and 11.55-fold higher, respectively, than plain KP.	[67]

PLGA, poly(lactic-co-glycolic) acid; PCL, poly(ε-caprolactone).

7.5 Conclusions and Future Perspectives

This chapter explains why and how the skin prevents essentially any large molecule's transport into the body: the more hydrophilic the molecule, the more difficult it is for it to cross the barrier pathways. From a global perspective, it is through the exploration of the aforementioned anatomical and functional characteristics of the skin barrier that it becomes possible to develop nanoparticles and therapeutic formulations that are able to penetrate it and adequately deliver drugs for exertion of therapeutic effect.

Transdermal drug delivery systems have existed for several decades now and still are constantly improving. They started as simple patches and have advanced to microneedles and ionophoresis systems. It is possible to breach the skin barrier using enough force and suitable tools, such as microneedles. In any case, typically a wide pore opening occurs through the skin. Crucial for safe transdermal application is the need to maintain the skin's protective role after pore creation. Methods that require an external source of energy often depend on complicated and expensive gadgets.

Topically applied, drug delivery systems for the treatment of skin-related diseases have a number of advantages; however, compared to the formulations developed in the industrial environment, the drug delivery systems have scope for only a limited degree of enhancement in skin permeation. By conjugating a nanotechnological method, a "game change" will enable rapid, more effective, and targeted delivery of therapeutics. This development will definitely go further in coming years to accommodate improved patient convenience.

Future prospects envisage long-term safety and tolerability, large-scale manufacture, and focus on health-related quality-of-life issues, and these tasks involve hard work. Going by the current state-of-the-art technology and the described skin interactions with delivery systems, the lipid-based colloidal nanocarriers are still the mainstream approach and have a great potential for topical delivery and market reach in the foreseeable future.

Acknowledgments

Sofia A. Costa Lima thanks *Operação* NORTE-01-0145-FEDER-000011 (*Qualidade e Segurança Alimentar — uma abordagem (nano) tecnológica*) for her investigator contract. The authors also thank *National Funds from Fundação para a Ciência e a Tecnologia* (FCT) and FEDER for financial support under Program PT2020 (project 007728 -UID/QUI/04378/2013).

References

1. Lee, S.H., et al. An update of the defensive barrier function of skin. *Med J*, 2006, **47**:293–306.
2. Baroli, B. Penetration of nanoparticles and nanomaterials in the skin: fiction or reality? *Structure*, 2010, **99**:21–50.
3. Desai, P., et al. Interaction of nanoparticles and cell-penetrating peptides with skin for transdermal drug delivery. *Mol Membr Biol*, 2010, **27**:247–259.
4. Forster, M., et al. Topical delivery of cosmetics and drugs. Molecular aspects of percutaneous absorption and delivery. *Eur J Dermatol*, 2009, **19**:309–323.
5. Wertz, P.W., et al. The physical, chemical and functional properties of lipids in the skin and other biological barriers. *Chem Phys Lipids*, 1998, **81**:85–96.
6. Chourasia, R., et al. Drug targeting through pilosebaceous route. *Curr Drug Targets*, 2009, **10**:950–967.
7. Pouillot, A., et al. The stratum corneum: a double paradox. *J Cosmet Dermatol*, 2008, **7**:143–148.
8. Neubert, R.H.H. Potentials of new nanocarriers for dermal and transdermal drug delivery. *Eur J Pharm Biopharm*, 2011, **77**:1–2.
9. Barry, B.W. Drug delivery routes in skin: a novel approach. *Adv Drug Deliv Rev*, 2002, S31–S40.
10. Guterres, S.S., et al. Polymeric nanoparticles, nanospheres and nanocapsules, for cutaneous applications. *Drug Target Insights*, 2007, **2**:147–157.
11. Martanto, W., et al. Transdermal delivery of insulin using microneedles in vivo. *Pharm Res*, 2004, **21**:947–952.
12. Gupta, J., et al. Recovery of skin barrier properties after sonication in human subjects. *Ultrasound Med Biol*, 2009, **35**:1405–1408.
13. Cevc, G., et al. Spatial distribution of cutaneous microvasculature and local drug clearance after drug application on the skin. *J Controlled Release*, 2007, **118**:18–26.

14. Kretsos, K., et al. Dermal capillary clearance: physiology and modeling. *Skin Pharmacol Physiol*, 2005, **18**:55–74.

15. Denet, A.R., et al. Skin electroporation for transdermal and topical delivery. *Adv Drug Deliv Rev*, 2004, **56**:659–674.

16. Carbone, C., et al. Pharmaceutical and biomedical applications of lipid-based nanocarriers. *Pharm Patent Anal*, 2014, **3**:199–215.

17. Alvarez-Roman, R., et al. Enhancement of topical delivery from biodegradable nanoparticles. *Pharm Res*, 2004, **21**:1818–1825.

18. Maghraby, G.M.M.E., et al. Can drug bearing liposomes penetrate intact skin? *J Pharm Pharmacol*, 2006, **58**:415–429.

19. Heinrich, K., et al. Influence of different cosmetic formulations on the human skin barrier. *Skin Pharmacol Physiol*, 2014, **27**:141–147.

20. Lu, G.W., et al. Comparison of artificial sebum with human and hamster sebum samples. *Int J Pharm*, 2009, **367**:37–43.

21. Pradhan, M., et al. Novel colloidal carriers for psoriasis: current issues, mechanistic insight and novel delivery approaches. *J Controlled Release*, 2013, **170**:380–395.

22. Elsayed, M.M.A., et al. Deformable liposome and ethosome: mechanism of enhanced skin delivery. *Int J Pharm*, 2006, **322**:60–66.

23. Elsayed, M., et al. Lipid vesicles for skin delivery of drugs: reviewing three decades of research. *Int J Pharm*, 2007, **332**:1–16.

24. Kumar, R., et al. Vesicular system-carrier for drug delivery. *Der Pharmacia Sinica*, 2011, **2**:192– 202.

25. Uchino, T., et al. Characterization and skin permeation of ketoprofen-loaded vesicular systems. *Eur J Pharm Biopharm*, 2013, **86**:156–166.

26. El-Nabarawi, M.A., et al. Transdermal drug delivery of paroxetine through lipid-vesicular formulation to augment its bioavailability. *Int J Pharm*, 2013, **443**:307–317.

27. Jeon, S.O., et al. Enhanced percutaneous delivery of recombinant human epidermal growth factor employing nano-liposome system. *J Microencapsul*, 2012, **29**:234–241.

28. Verma, P., et al. Nanosized ethanolic vesicles loaded with econazole nitrate for the treatment of deep fungal infections through topical gel formulation. *Nanomed Nanotechnol Biol Med*, 2012, **8**:489–496

29. Malakar, J., et al. Formulation, optimization and evaluation of transferosomal gel for transdermal insulin delivery. *Saudi Pharm J*, 2012, **20**:355–363.

30. Rajan, R., et al. Effect of permeation enhancers on the penetration mechanism of transfersomal gel of ketoconazole. *J Adv Pharm Technol Res*, 2012, **3**:112–116.
31. Gupta, A., et al. Transfersomes: a novel vesicular carrier for enhanced transdermal delivery of sertraline: development, characterization and performance evaluation. *Sci Pharm*, 2012, **80**:1061–1080.
32. Gupta, M., et al. Effect of surfactants on the characteristics of fl uconazole niosomes for enhanced cutaneous delivery. *Artif Cells Nanomed Biotechnol*, 2011, **39**:376–384.
33. Manosroi, A., et al. Transdermal absorption enhancement of papain loaded in elastic niosomes incorporated in gel for scar treatment. *Eur J Pharm Sci*, 2013, **48**:474–483.
34. Gonzalez-Rodriguez, M.L., et al. Applying the Taguchi method to optimize sumatriptan succinate niosomes as drug carriers for skin delivery. *J Pharm Sci*, 2012, **101**:3845– 3863.
35. Zhang, Y.T., et al. Evaluation of skin viability effect on ethosome and liposome-mediated psoralen delivery via cell uptake. *J Pharm Sci*, 2014, **103**:3120–3126.
36. Honeywell-Nguyen, P.L., et al. Vesicles as a tool for transdermal and dermal delivery. *Drug Disco Today Technol*, 2005, **2**:67–74.
37. Cevc, G., et al. Ultraflexible vesicles, transfersomes, have an extremely low pore penetration resistance and transport therapeutic amounts of insulin across the intact mammalian skin. *Biochim Biophys Acta Biomembr*, 1998, **1368**:201–215.
38. Sinico, C., et al. Vesicular carriers for dermal drug delivery. *Expert Opin Drug Deliv*, 2009, **6**:813–825.
39. Karim, K.M., et al. Niosome: a future of targeted drug delivery systems. *J Adv Pharm Technol Res*, 2010, **1**:374–380.
40. Cevc, G. Lipid vesicles and other colloids as drug carriers on the skin. *Adv Drug Deliv Rev*, 2004, **56**:675–711.
41. Pardeike, J., et al. Lipid nanoparticles (SLN, NLC) in cosmetic and pharmaceutical dermal products. *Int J Pharm*, 2009, **366**:170–184.
42. Schäfer-Korting, M., et al. Lipid nanoparticles for improved topical application of drugs for skin diseases. *Adv Drug Deliv Rev*, 2007, **59**:427–443.
43. Westesen, K., et al. Physicochemical characterization of lipid nanoparticles and evaluation of their drug loading capacity and sustained release potential. *J Controlled Release*, 1997, **48**:223–236.

44. Müller, R.H., et al. Nanostructured lipid matrices for improved microencapsulation of drugs. *Int J Pharm*, 2002, **242**:121–128.

45. Chambi, H.N.M., et al. Solid lipid microparticles containing water-soluble compounds of different molecular mass: production, characterisation and release profiles. *Food Res Int*, 2008, **41**:229–236.

46. Dwivedi, A., et al. In vitro skin permeation of artemisone and its nano-vesicular formulations. *Int J Pharm*, 2016, **503**:1–7.

47. Jain, A., et al. A synergistic approach of adapalene-loaded nanostructured lipid carriers, and vitamin C co-administration for treating acne. *Drug Dev Ind Pharm*, 2016, **42**(6):897–905.

48. Abdel-Salam, F.S., et al. Nanostructured lipid carriers as semisolid topical delivery formulations for diflucortolone valerate. *J Liposome Res*, 2016, 1–15.

49. Ghanbarzadeh, S., et al. Enhanced stability and dermal delivery of hydroquinone using solid lipid nanoparticles. *Colloids Surf B*, 2015, **136**:1004–1010.

50. Liu, J., et al. Isotretinoin-loaded solid lipid nanoparticles with skin targeting for topical delivery. *Int J Pharm*, 2007, **328**:191–195.

51. Jeon, H.S., et al. A retinyl palmitate-loaded solid lipid nanoparticle system: effect of surface modification with dicetyl phosphate on skin permeation in vitro and anti-wrinkle effect in vivo. *Int J Pharm*, 2013, **452**:311–320.

52. Tung, N.T., et al. Topical delivery of dexamethasone acetate from hydrogel containing nanostructured liquid carriers and the drug. *Arch Pharm Res*, 2015, **38**:1999–2007.

53. Pradhan, M., et al. Design, characterization and skin permeating potential of Fluocinolone acetonide loaded nanostructured lipid carriers for topical treatment of psoriasis. *Steroids*, 2015, **101**:56–63.

54. Pinto, M.F., et al. A new topical formulation for psoriasis: development of methotrexate-loaded nanostructured lipid carriers. *Int J Pharm*, 2014, **477**:519–526.

55. Shaji, J., et al. Microemulsion as drug delivery system. *Pharma Times*, 2004, **36**:17–24.

56. Cichewicz, A., et al. Cutaneous delivery of α-tocopherol and lipoic acid using microemulsions: influence of composition and charge. *J Pharm Pharmacol*, 2013, **65**:817–826.

57. Duangjit, S., et al. Development, characterization and skin interaction of capsaicin-loaded microemulsion-based nonionic surfactant. *Biol Pharm Bull*, 2016, **39**:601–610.

58. Pepe, D., et al. Decylglucoside-based microemulsions for cutaneous localization of lycopene and ascorbic acid. *Int J Pharm*, 2012, **434**:420–428.

59. Pepe, D., et al. Protein transduction domain-containing microemulsions as cutaneous delivery systems for an anticancer agent. *J Pharm Sci*, 2013, **102**:1476–1487.

60. Cavalcanti, A.L., et al. Microemulsion for topical application of pentoxifylline: in vitro release and in vivo evaluation. *Int J Pharm*, 2016, **506**:351–360.

61. Siddique, M.I., et al. In-vivo dermal pharmacokinetics, efficacy, and safety of skin targeting nanoparticles for corticosteroid treatment of atopic dermatitis. *Int J Pharm*, 2016, **507**:72–82.

62. Lin, Y.H., et al. Development of chitosan/poly-γ-glutamic acid/pluronic/curcumin nanoparticles in chitosan dressings for wound regeneration. *J Biomed Mater Res B Appl Biomater*, 2017, **105**:81–90.

63. Senyigit, T., et al. Lecithin/chitosan nanoparticles of clobetasol-17-propionate capable of accumulation in pig skin. *J Controlled Release*, 2010, **142**:368–272.

64. Özcan, I., et al. Enhanced dermal delivery of diflucortolone valerate using lecithin/chitosan nanoparticles: in-vitro and in-vivo evaluations. *Int J Nanomed*, 2013, **8**:461–475.

65. Tomoda, K., et al. Enhanced transdermal delivery of indomethacin-loaded PLGA nanoparticles by iontophoresis. *Colloids Surf B*, 2011, **88**:706–710.

66. Shim, J., et al. Transdermal delivery of mixnoxidil with block copolymer nanoparticles. *J Controlled Release*, 2004, **97**:477–484.

67. Shah, P.P., et al. Skin permeating nanogel for the cutaneous co-delivery of two anti-inflammatory drugs. *Biomaterials*, 2012, **33**:1607–1617.

Part IV
Pulmonary Drug Delivery Approaches

Chapter 8

New Approaches from Nanomedicine and Pulmonary Drug Delivery for the Treatment of Tuberculosis

Joana Magalhães,[a,b] Alexandre C. Vieira,[a] Soraia Pinto,[a] Sara Pinheiro,[a] Andreia Granja,[a] Susana Santos,[b] Marina Pinheiro,[a] and Salette Reis[a]
[a]*UCIBIO, REQUIMTE, Department of Chemical Sciences, Faculty of Pharmacy, University of Porto, Portugal*
[b]*I3S - Institute of Research and Innovation in Health, University of Porto, Portugal*
mpinheiro@ff.up.pt

Tuberculosis (TB) is a highly prevalent global disease resulting in a heavy economic, social, and human burden. Considerable efforts have been made to fight this disease, including the search for new anti-TB drugs and the development of new delivery strategies that improve the efficacy of existing treatments. The development of nanodelivery systems represents an interesting alternative for the delivery of anti-TB drugs to the target site of infection as an attempt to reduce the required dose and to minimize the dose-dependent

Nanoparticles in Life Sciences and Biomedicine
Edited by Ana Rute Neves and Salette Reis

ISBN 978-981-4745-98-7 (Hardcover), 978-1-351-20735-5 (eBook)
www.panstanford.com

side effects. This chapter includes a comprehensive analysis of recent progress in nanodelivery systems for the pulmonary delivery of anti-TB drugs. Research using different types of nanoparticles (NPs) for the pulmonary delivery of anti-TB drugs will be reviewed. Lastly, the future of this growing field and its potential impact will be discussed.

8.1 Tuberculosis: Key Facts

Tuberculosis (TB) is one of the oldest deadly diseases known to humanity. The ancient association of TB is supported by archaeological data, morphological evidence in human fossils, and literary descriptions, including Book I of *The Epidemics of Hippocrates* (460 BC) [1, 2].

In the beginning of the nineteenth century, Rene Laennec examined TB pathology in detail. In his work *D'Auscultation Mediate* (1819), Laennec described the physical signs of the pulmonary manifestation of the disease, unified the concept of the disease (whether pulmonary or extrapulmonary), and introduced many of the terms still in use today [3, 4]. In 1865, Jean-Antoine Villemin demonstrated the transmissibility of the infection from an infected cadaver's tissue to a rabbit. Then, Robert Koch described the etiologic agent of TB, *Mycobacterium tuberculosis* (MTB), in his work *Die Aetiologie der Tuberculose* (1882). During the Industrial Revolution, TB reached epidemic levels in Europe, being responsible for one in four deaths [1]. Mortality rates began to decline in the nineteenth century due to the introduction of the concept of fresh air, good nutrition, and isolation of TB patients in sanatoria, as an attempt to cure TB [4].

In 1908, Albert Calmette and Camille Guerin tested several mediums in Koch's bacillus growth to decrease the virulence and increase the immunity capacity. These studies led to the development of the bacillus Calmette–Guérin (BCG) vaccine, which was introduced in 1921 and was widely employed following World War I [5].

The modern era of TB treatment started after the discovery of streptomycin, in 1944 [4]. A rapid succession of anti-TB drugs appeared in the following years, namely *para*-aminosalicylic acid (1946), isoniazid (INH, 1951), pyrazinamide (PZA, 1952), rifampin

(RIF, 1957) and other rifamycins, and ethambutol (EMB, 1961). In the mid-1960s, the incorporation of RIF in the standard anti-TB regimen allowed the reduction of TB treatment duration to 9 months, and, when used in a regimen that also contained PZA, to 6 months [6, 7].

Nowadays, the first-line anti-TB therapy includes a combination of INH, RIF, PZA, and EMB [6, 8]. Despite the effective therapies available today, TB is still a major global health problem, being the leading cause of death worldwide, alongside the human immunodeficiency virus (HIV). In the last available report, World Health Organization (WHO) estimations indicated 9.6 million new cases of TB in 2014, which resulted in 1.5 million deaths [8].

8.1.1 Epidemiology

The exposure to *M. tuberculosis*, through the inhalation of infectious aerosols, can result in different clinical outcomes. The infection is generally established in 25%–50% of close contacts with the bacilli. The infected individuals can develop a symptomatic and infectious state, known as active TB (5%–10% of the cases), or an asymptomatic and noninfectious state, named latent TB (90%–95%) [8]. Latent individuals constitute a reservoir for new disease and ongoing *M. tuberculosis* transmission, carrying 5%–10% lifetime risk of progression to active disease [9]. With such a reservoir, TB control seems to be a difficult task to achieve.

In the last 20 years, a global TB monitoring system has been created and annual rounds of data collection have provided valuable information regarding TB prevention, diagnosis, and treatment.

In the last available report, WHO estimations indicated 1 million new cases in children, 3.2 million in women, and 5.4 million in men, resulting in 9.6 million new cases of TB in 2014 [8].

Geographically, TB occurs in every part of the world. In 2014, the higher number of new TB cases occurred in Southeast Asia and Western Pacific regions, representing 58% of the new cases. On the other hand, the most severe burden relative to population (281 cases for every 100,000 people) and the higher number of HIV-associated new TB cases (74%) occurred in Africa [8].

TB and HIV infection can form a lethal combination. HIV-infected people are 20 to 30 times more likely to develop active TB than people without HIV. In addition, 1 in every 3 HIV deaths is due to TB.

In 2014, 1.2 million new cases of TB occurred among HIV-positive people and approximately 0.4 million people died of HIV-associated TB [8, 10].

In the last 15 years, TB incidence has fallen by an average of 1.5% per year and more than 43 million lives were saved through effective diagnosis and treatment [8, 11]. Nevertheless, the emergence of multi-drug-resistant TB (resistant to the two main first-line anti-TB drugs, INH and RIF) and extensively drug resistant TB (a multi-drug-resistant TB that responds to even fewer anti-TB drugs, including the most effective second-line drugs) has been a problematic issue in TB control and eradication [12].

In 2014, 480,000 people developed multi-drug-resistant TB and 190,000 died from this disease. Additionally, 20% of previously treated cases had multi-drug-resistant TB and 9.7% of the multi-drug-resistant cases had extensively drug resistant TB [8].

From 2016, the main goal of WHO is to implement the WHO End TB Strategy. This strategy includes the recently adopted sustainable development goals that involve the reduction of new TB cases by 80% and TB deaths by 90% between 2015 and 2030 [8, 13]. For this purpose, current research should focus on the improvement of therapeutic compliance through the development of more effective diagnostics, anti-TB drugs, and vaccines or through the development and application of targeting drug delivery systems that increase the efficacy of available chemotherapy.

8.1.2 Etiology, Transmission, and Physiopathology

TB is caused by the *M. tuberculosis* complex, which comprises a group of phylogenetically closely related bacteria, including *M. africanum, M. bovis, M. caprae, M. canetti, M. microti, M. pinnipedii*, and *M. tuberculosis* [14, 15]. In humans, TB is primarily caused by *M. tuberculosis* and *M. africanum*. Both species are obligate human pathogens that have a limited ability to survive outside the human body [15]. *M. bovis* is a cattle pathogen that may also affect humans. However, with the introduction of pasteurization and meat-control practices, the *M. bovis* infection in humans decreased significantly [16]. The other species of the *M. tuberculosis* complex affect a range of wild and domestic animal species [15].

In humans, the most clinical presentation is pulmonary TB, in which transmission occurs via inhalation of aerosol droplets released from an infected individual, typically through coughing, sneezing, spitting, or speaking [17]. Once in the lung, the bacilli are internalized through phagocytosis by the alveolar macrophages. The initial steps of TB infection involve a pro-inflammatory response, the recruitment of inflammatory cells to the lung, and the formation of granulomas [18]. Most infected individuals are clinically asymptomatic and remain in a latent state of infection. However, if the immune response is subsequently compromised, reactivated pulmonary TB may occur. The individuals with active TB can generate infectious droplets that propagate the infection [19].

In a minority of cases, *M. tuberculosis* spreads via the lymphatic system to other host tissues, thereby becoming disseminated throughout the body, resulting in extrapulmonary forms of the disease. The most common clinical presentation of extrapulmonary TB includes lymph nodes infection; pleural effusions; and genitourinary tract, skeletal, ocular, and abdominal TB [20]. Gastrointestinal TB can occur by the ingestion of infected animal products or by the swallowing of infected sputum, resulting in further transmission of infection via feces and urine [1].

Dissemination of *M. tuberculosis* from the primary site of infection (i.e., lungs) to the lymph nodes and to other organs is one of the key events in TB pathogenesis. Indeed, this event enables the pathogen to spread to new niches, to establish alternative sites of infection, and to develop a protective immune response in the host [21].

In addition, due to its extraordinary stealth and capacity to adapt to environmental changes throughout the course of infection, *M. tuberculosis* is able to survive and proliferate within mononuclear phagocytes and to enter in a quiescent physiological state, maintaining viability for extended periods of time [22]. To secure survival and multiplication within the hostile environment of the host, *M. tuberculosis* has developed defense strategies, such as autophagy, the recruitment of potent hydrolytic lysosomal enzymes, the production of reactive oxygen and nitrogen species, phagosome-lysosome fusion, antigen presentation, and major histocompatibility complex class II expression and trafficking [23, 24]. Moreover, *M. tuberculosis* is a facultative intracellular bacterium shielded by a unique thick cell wall composed of long-chain fatty acids, glycolipids, peptidoglycan, and proteins [25], which forms a diffusion barrier and protects the bacilli from digestion by lysosomal enzymes [26]. The success of *M. tuberculosis* as a pathogen has contributed to the high prevalence of latent TB, which persists for the lifetime of the host and reactivates with sufficient frequency to maintain a worldwide epidemic.

8.1.3 Clinical Manifestations and Diagnosis

In the initial stages of TB infection, the majority of people remain asymptomatic or present with a mild fever [27].

In pulmonary TB, the most common symptoms are fever, chronic cough, weight loss, fatigue, and night sweats. Less common symptoms are chest pain, dyspnea, hemoptysis, production of blood-tinged sputum, pleuritic pain, and severe respiratory failure [28]. Clinically, pulmonary TB is associated with extensive lung damage (i.e., pulmonary atelectasis, fibrosis, and scarring), which results in decreased pulmonary compliance and impaired gas exchange [27].

Extrapulmonary TB has a wider range of symptoms, depending on which organ is affected, and in many cases, infection produces systemic effects, rather than local ones [25, 29]. Moreover, these effects are usually associated with other ailments, such as HIV infection, diabetes mellitus, and neoplastic diseases, which considerably delays diagnosis and increases misdiagnoses, especially in patients coinfected with HIV [29].

An accurate TB diagnosis is essential for a timely detection of infection. The most common methods for TB diagnosis are chest radiology, sputum smear microscopy, culture, and tuberculin skin test [30].

A sputum smear test is the most common method for diagnosing TB. This simple and inexpensive technique involves the quantification of mycobacteria in stained sputum preparations by microscopic examination. This test is very useful but has a low sensitivity and does not distinguish between different species of mycobacteria [29, 31]. Culture-based methods using selective media for mycobacteria are very reliable and sensitive but require more than a week to obtain conclusive results [8]. A chest radiography can identify lung damage but has limited specificity and reader inconsistency [32].

The tuberculin skin test was the first method capable of diagnosing latent TB [33]. This test measures the immune response against soluble antigens of *M. tuberculosis* as a delayed-type hypersensitivity reaction at the site of injection [34].

In recent years, novel experimental approaches and detection technologies have provided more sensitive alternatives for TB diagnosis. Recently developed molecular tests can be applied directly to the sputum sample, giving rapid, highly specific, and sensitive diagnosis without the need for culture-based techniques [28]. The most common methods are Xpert® MTB/RIF assay (Cepheid, US), line probe assays (LPAs), and the urine lateral flow lipoarabinomannan (LF-LAM) assay (Alere Determine™ TB LAM Ag test, Alere Inc., US) [8]. In addition, a diagnostic platform called the GeneXpert Omni® and a next-generation cartridge called Xpert Ultra® are in development. These new diagnostic tools may replace conventional culture as the primary diagnostic method for TB [8]. Despite all the advantages of these molecular tools for TB diagnostics, conventional microscopy and culture-based techniques remain necessary for monitoring patients' response to treatment.

8.1.4 Vaccines and Treatment

As stated before, the only approved TB vaccine is BCG, an attenuated form of *M. bovis* that has been used in humans since 1921 [4]. BCG vaccination is recommended by WHO and has been successfully carried out in children for over a century [5, 8]. BCG is relatively cheap, safe, and administered as a single intradermal injection. Despite its efficacy against severe forms of childhood TB, BCG does not prevent the most prevalent form of TB, adult pulmonary TB [35]. In this context, a widely applicable and effective vaccine is urgently needed for TB control.

In recent years, considerable effort has been made to develop new vaccines. New strategies include viral vectors that express *M. tuberculosis* antigens, *M. tuberculosis* proteins with improved adjuvants, recombinant BCG strains, and live attenuated *M. tuberculosis* vaccines [19, 35]. Currently, there are almost 20 vaccine candidates in clinical trials, and several new candidates are ready to enter the clinical trial pipeline [35].

Vaccine candidates for TB can be classified according to their target populations (therapeutic and preventive vaccines), their composition (killed mycobacteria, viable recombinant mycobacteria, viral-vectored, and adjuvant subunit vaccines), time of administration regarding *M. tuberculosis* infection (pre-exposure and postexposure vaccines), and relation to conventional BCG vaccine (BCG replacement and heterologous prime-boost vaccines) [35].

Recent advances in vaccine development have offered promising perspectives to explore immunotherapy as a treatment option, especially for multi-drug-resistant and extensively drug resistant TB [35, 36].

TB treatment requires a long period of chemotherapy (6–9 months) with a combination of multiple agents to achieve bacterial clearance, to reduce the risk of transmission, and to minimize the emergence of drug resistance [26, 27]. Indeed, the current recommended treatment includes an initial phase of 2 months of daily administration of a four-drug combination (INH, PZA, RIF, and EMB), followed by a continuous phase of 4–7 months of daily administration of INH and RIF [37]. These four drugs, together with streptomycin, constitute the first-line therapy for TB [7, 26]. Second-line agents (e.g., rifabutin, ethionamide, amikacin, kanamycin A, and levofloxacin) are used when treatment with first-line drugs fails or in the presence of multi-drug-resistant TB cases. These second-line drugs are less effective, more toxic, and more expensive than first-line drugs [7, 37]. In some cases, more severe drug resistance can occur, responding to even fewer available antibiotics, including the most effective second-line anti-TB drugs [37]. The third-line anti-TB drugs have more side effects and are even more expensive than first- and second-line anti-TB drugs. In addition, in the majority of cases, the third-line drugs are under clinical trials and have unclear efficacy [6, 37].

Besides drug resistance, the interaction of anti-TB drugs with antiretroviral agents and other anti-infective drugs prescribed for opportunistic infections can also make TB treatment difficult [27]. New drugs with interest for TB treatment are being developed and tested and are currently at different steps of preclinical or clinical trials [8]. Despite the drawbacks of existing drugs, mechanisms to improve their efficacy may present a faster strategy to fight TB [26].

Improvement of the therapeutic index of existing anti-TB drugs through encapsulation into nanodelivery systems should be considered as a means of increasing drug concentration at infected sites, reducing the severe toxic side effects, and decreasing treatment duration [26, 38]. Moreover, another important advantage of nanodelivery systems is the feasibility of more versatile routes of administration, including the inhalatory route [39]. Inhalable drug-loaded nanoparticles (NPs) can improve drug effectiveness and decrease side effects, due to their ability to carry drugs specifically to the infected area, as well as ensure their prolonged and sustained release [40, 41]. Indeed, pulmonary drug delivery may play a crucial role in the improvement of patient compliance to therapy and, consequently, contribute to TB control.

8.2 Respiratory System as a Route for Drug Delivery

Anatomically, the human respiratory system is divided into the upper respiratory tract and the lower respiratory tract. The nose, pharynx, and larynx are the components of the upper respiratory tract. The lower respiratory tract consists of the trachea, bronchial tree, and lungs. The thoracic cavity is delimited by the chest wall and the diaphragm [42].

The main respiratory system functions are related to ventilation, gas exchange, and oxygen utilization, which involve obtaining oxygen from the external environment and distributing it to the cells, as well as eliminating carbon dioxide produced by cellular metabolism [42].

The lungs are composed of two treelike structures, the vascular tree and the airway tree, that are embedded in the lung tissue [43]. The vascular tree is composed of arteries and veins connected by capillaries. The airway tree consists of a series of branching tubes, in which each level of branching results in another generation of airways, each generation smaller in diameter than the previous one [42, 43].

The trachea is the main airway and branches into two bronchi, each supplying a lung. Within each lung, the bronchus branches many times, into progressively smaller bronchi, the bronchioles [43]. The trachea and the first 16 generations of airways make up the conducting zone. The conducting airways have three major functions: to warm and humidify inspired air, to distribute inspired air to the more distal gas-exchanging regions of the lungs, and to serve as a part of the body defense system [42]. The last referred function is linked to the mucociliary transport system and will be discussed in the following section.

The last seven generations of airways form the site of gas exchange, called respiratory zone. The respiratory zone is composed of respiratory bronchioles, alveolar ducts, and alveoli [43].

The lungs have the most extensive capillary network, surrounding each alveolus and bringing blood into close proximity with the gas inside the alveolus [43].

8.2.1 Barriers of the Respiratory System

The respiratory system has physiological barriers that affect the uptake and translocation of molecules through the respiratory tract (Fig. 8.1).

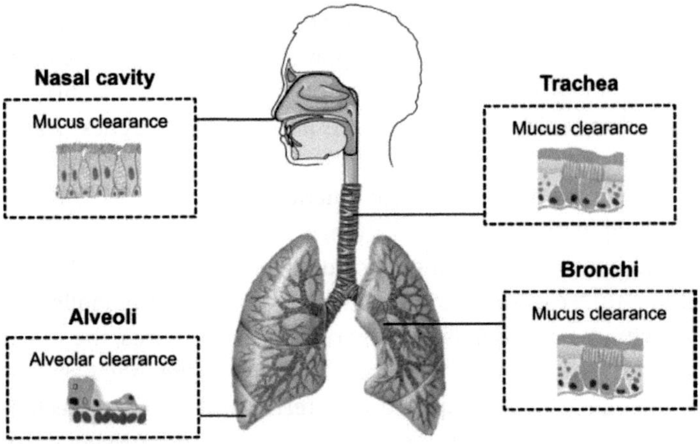

Figure 8.1 Schematic of the physiological barriers (i.e., mucus and alveolar clearance) that affect the uptake and translocation of molecules through the respiratory tract.

The conducting airways (i.e., trachea, bronchi, and bronchioles) are lined with different cell types, such as basal, ciliated, epithelial, goblet, and secretory cells, which serve the specialized functions of each compartment [39]. Epithelial cells are the major cell type of the airways, forming a ciliated columnar epithelium with a height of 58 μm in the bronchi and a cellular monolayer of 0.1–0.2 μm in the alveoli [39].

Mucus-secreting goblet cells are interspersed among the epithelial cells, forming a protective viscous mucus layer called the mucociliary escalator [42]. The mucociliary escalator acts as a primary barrier in the bronchial epithelium, trapping inhaled particles and removing them by coughing them up and out of the lungs [44, 45]. Due to the high concentration of mucopolysaccharides and macromolecules tethered to the cilia, microvilli, and surface of bronchial epithelial cells, the underlying periciliary liquid layer acts as a second barrier to particle dissolution and diffusion toward the epithelium [39, 46].

The mucus layer is composed of inorganic salts, proteins, glycoproteins (mucins), lipids, and water; is 8 μm thick in the bronchi; and transitions through the airways into a 0.07 μm thick surfactant in the alveoli [39]. The pulmonary surfactant (PS) layer is composed of phospholipids, cholesterol, and various proteins that are released from type II epithelial cell lamellar bodies [44, 47]. Besides affecting particle dissolution and diffusion toward the epithelium, this protective coating of the lungs also affects the interactions between drugs and cell surfaces and/or receptors [39].

Apart from coughing, mucociliary clearance and macrophage clearance affect the residence time of inhaled drugs in the lungs [47]. Mucociliary clearance involves the movement of mucus under the influence of coordinated ciliary beat at the surface of airways [47, 48]. Particles can migrate through the mucus to the epithelium for absorption or, in the extreme insoluble particles, are transported by mucociliary-mediated transport [48]. Mucociliary clearance is a fast and unspecific mechanism, with 80%–90% of inhaled particles being excreted from the upper and central lung within 24 h [47]. The inhaled particles that exhibit delays in dissolution are usually taken up by macrophages and transported to lymph nodes [48]. Alveolar macrophages are able to phagocyte inhaled organic or inorganic particles. Nevertheless, in contrast to mucociliary clearance, alveolar macrophages show size-dependent uptake, being more effective for particles with a geometric diameter of 0.5–5 μm [47].

Reformulating drugs to be nanosized offers a number of potential benefits related to penetrating into the protective barriers of the lungs. Moreover, according to the proposed action, pulmonary drug delivery can be designed to avoid phagocytic cells or to take advantage of the phagocytic nature of the alveolar macrophages [44].

8.2.2 Pulmonary Administration of Drugs

The lung is the ideal site of anti-TB drug delivery and an important route of access in case of pulmonary TB [49]. Indeed, pulmonary drug delivery offers advantages over both oral and parenteral routes of administration [40].

The inhalatory route is a noninvasive delivery portal that requires lower administration doses for efficacy and exhibits reduced toxicity and fewer side effects in comparison with the oral route [49]. In addition, lung mucosa has a large surface of absorption and extensive vascularization, which may allow drugs to be systemically distributed by the bloodstream, bypassing the hepatic first-pass metabolism that occurs upon oral administration [49, 50]. In contrast to the parenteral route, pulmonary drug delivery is a "free of needles" treatment that can be self-administered easily [40, 50].

In the last decade, the interest in the inhalatory route for TB treatment has been steadily growing. Regarding the characteristics of second- and third-line anti-TB drugs, the inhalatory route has several advantages when compared to the oral and parenteral administrations [51]. Pulmonary delivery of anti-TB drugs constitutes an interesting approach for the treatment of both pulmonary and extrapulmonary TB [50]. In this regard, the delivery of drugs directly to the lungs allows a high drug concentration in the lungs and a better targeting of the infected alveolar macrophages in the case of pulmonary TB [50, 51]. In addition, the direct absorption of active agents into systemic circulation bypasses hepatic first-pass metabolism and avoids degradation of drugs in the acidic environment of the stomach, which is advantageous for treating extrapulmonary TB [52].

8.2.3 Lung-Targeting and Inhalation Devices

The deposition of inhaled therapeutic agents in the respiratory airways is crucial to yielding local and systemic therapeutic effect. Impaction, sedimentation, and/or diffusion are the main mechanisms by which particles are deposited in the respiratory tract (Fig. 8.2) [45, 48]. To ensure deep lung deposition, parameters such as aerosol particle diameter, shape, density, electrical charge, and hygroscopicity are the determinants [40, 45]. In addition to the need for particles and formulation to have adequate characteristics, inhalatory administration of drugs must be done using a suitable device that produces an appropriate aerosol [41, 50, 52].

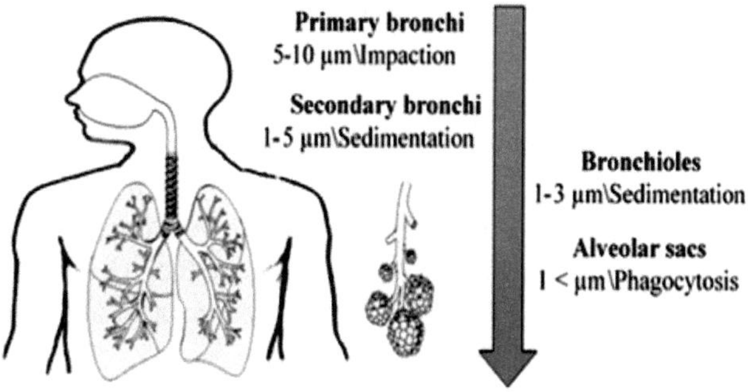

Figure 8.2 Correlation between area of the lung and the size of the deposition particles.

The aerosolization of drugs as fine powders using dry powder inhalers (DPIs) is a convenient way of delivering drugs to the lungs. Liquid formulations can also be administered through a nebulizer. Alternatively, drugs can be directly delivered to the lungs, without prior aerosolization, using an insufflator [52].

Different inhalatory devices have distinct properties and specifications, making them more appropriate than others for ensuring the efficacy of a specific drug formulation [41]. Nebulization is the most common method of aerosol delivery of antibiotics. In addition, adjustments in drug dosing are easier to achieve using nebulizers [52]. Compared to nebulizers, DPIs offer many advantages, such as greater stability, improved tolerability, and extended release profile. In addition, no liquid propellant is involved, DPIs are easy to use and less time consuming, and they do not require storage at a low temperature [45]. These characteristics make DPIs a popular choice for the treatment of pulmonary chronic diseases [50].

Besides the characteristics discussed above, inhaler devices should also be accurate, small, easy to handle, and user friendly to be accepted by both clinicians and patients [41].

8.3 Nanotechnology as a Tool for Drug Delivery

Nanotechnology is a field of science that involves the design and study of structures called NPs, in which at least one of the dimensions is measured in the nanoscale range (1 to 1000 nm) [50].

The application of nanotechnology in medicine, and more specifically in drug delivery, has been investigated mainly to reduce toxicity and side effects of drugs [53].

NPs are attractive due to their unique features, such as their surface-to-mass ratio, which is much higher than that of other particles; their quantum properties; and their capacity to adsorb and carry other compounds [54]. In addition, NPs can be designed to target specific tissues, cells, and cellular receptors, enabling spatial controlled drug release [55].

The application of NPs for drug delivery provides technological advantages, including high stability (i.e., a long shelf life), high carrier capacity, feasibility of incorporation of both hydrophilic and hydrophobic substances, and possibility of variable routes of administration [56]. Besides the application of NPs as drug carriers, the drug itself can be formulated at a nanoscale level and function as its own carrier [54].

The interaction of NPs with membranes, tissues, and cells greatly depends on their composition. For that purpose, the surface of NPs can be functionalized with polymers or appropriate ligands to target specific cells and/or locations within the human body [55]. NPs have also the ability to be activated by chemical stimuli, by changes in environmental pH, by application of external heat source, and by an oscillating magnetic field [57–59].

On the basis of the type of material from which their matrix is made, NPs can be classified as organic or inorganic. Organic NPs are the most studied type, accounting for more than two-thirds of the systems approved for therapeutic use in humans [60]. Organic NPs include liposomes, polymeric NPs, polymeric micelles, solid lipid nanoparticles (SLNs), and nanostructured lipid carriers (NLCs). Inorganic NPs comprise metallic NPs made of gold, silver, platinum, aluminum, zinc, titanium, palladium, iron, or copper.

Different types of nanocarriers can target the site of the disease, either through passive or active targeting of drugs [61]. In passive targeting the uptake occurs through phagocytosis, pinocytosis, or endocytosis. This pathway mainly depends on size, but the physicochemical characteristics of the NP's surface also play a crucial role in the uptake [60]. On the other hand, active targeting involves the conjugation of receptor-specific ligands to the NP surface. Such ligands enhance the recognition of NPs by specific receptors, promoting site-specific targeting [61]. Alveolar macrophages have drawn special attention as a relevant clinical target in the treatment of pulmonary diseases, including TB. In fact, alveolar macrophages exhibit a number of receptors that can be exploited by nanocarriers with appropriate ligands to actively target the site of infection [50]. Sugars such as mannose [62] and lactose [63] are the most commonly used ligands since the receptors to these ligands are highly expressed in alveolar macrophages. Other common ligands include maleylated bovine serum albumin (MBSA); O-steroyl amylopectin (O-SAP); tetrapeptid tuftsin [26, 41]; and anionic lipids, such as dicetylphosphate (DCP) [50].

In the past few decades, a wide variety of nanosystems have been shown to be more beneficial than conventional formulations. Despite the above-mentioned advantages, many challenges have to be addressed regarding the potential toxicity and adverse effects of NP formulations [54]. Therefore, the application of nanotechnology in formulation design should be fully explored.

8.3.1 Nanoparticles for Pulmonary Drug Delivery

One of the main challenges in drug delivery is to target the place where the drug is needed, thereby avoiding potential side effects in nondiseased sites. In recent years, NPs have been explored to improve efficacy of existing drugs, reduce dosage schedules, and provide more comfortable administration routes for the treatment of pulmonary infectious diseases [40, 41, 60].

The application of nanotechnology in pulmonary drug delivery is a promising area of research for several reasons. First, the size of the particles can be fine-tuned to reach different areas of the lung through successful passive targeting strategies [50]. Second, the NP's surface can be modified and functionalized with ligands to actively target sites of interest, such as the alveolar macrophages [64]. Third, studies have demonstrated that pulmonary delivery of nanosuspensions favors higher lung tissue concentrations and markedly raises the lung to the serum ratio of drugs, compared with other routes of administration [50, 65].

Inhalation is the best route for the drugs to target the lungs. As mentioned above, the respiratory system provides a noninvasive route for delivery, with a large surface area and extensive blood supply. In addition, the drug avoids first-pass metabolism and the respiratory tract has a lower enzymatic activity compared to other organs (e.g., the gastrointestinal tract). After inhalation, NPs are transported with the inhaled air through the respiratory tract until they can be deposited or exhaled [41].

Particle size, density, and geometry are the determining factors to prevent exhalation and ensure that NPs will be deposited in the region of interest [60]. The optimal mean aerodynamic diameter range for particle deposition in the lower airways is 1 to 5 μm [41, 60]. Particles larger than 5 μm are deposited in the oropharynx and are consequently swallowed; particles smaller than 1 μm reach the tertiary bronchi and bronchioles; while articles smaller than 0.5 μm are deposited in the alveoli, but if they have a very low mass they may be exhaled during administration [60, 66].

The surface charge of NPs, which is commonly characterized by zeta potential, affects the distribution and uptake by cells. Macrophages display a negatively charged surface. Thus cationic NPs undergo better uptake than negatively charged NPs [67]. In the case of infections such as HIV infection, anionic NPs provide better interaction since infected cells express positively charged proteins [68].

To improve drug delivery to the lungs, NPs should be able to avoid mucociliary clearance from the airways. For this purpose, NPs can be formulated using mucoadhesive polymers, such as alginate or chitosan [60, 69].

Regarding other NP characteristics, the increment of hydrophobicity decreases the diffusion kinetic in mucus [70], increases cellular uptake [67], and influences output from endosome and drug release [71].

Due to the above-mentioned characteristics, diverse types of NPs are suitable for the treatment of pulmonary infections through inhalation [40, 41, 60]. Apart from technological innovation and the possibility to counter pulmonary drug delivery limitations, nanosystems have advantages and disadvantages and there is no single carrier that is perfect for all drugs. The choice of the carrier system for pulmonary drug delivery should always take into account the physicochemical properties of the therapeutic agent, the targeted site, the challenges of NPs' translocation through the respiratory tract, and the possibility of large-scale production.

8.3.2 Pulmonary Anti-TB Drug Delivery Nanotechnologies

The most commonly used nanocarriers to achieve the pulmonary delivery of anti-TB drugs are polymer-based NPs and also lipid-based NPs. Besides, other NPs are currently under research in the TB field, including drug nanocrystals, magnetic NPs, NPs with effervescent activity, and gold NPs for the study of internalization of NPs by alveolar macrophages (Fig. 8.3).

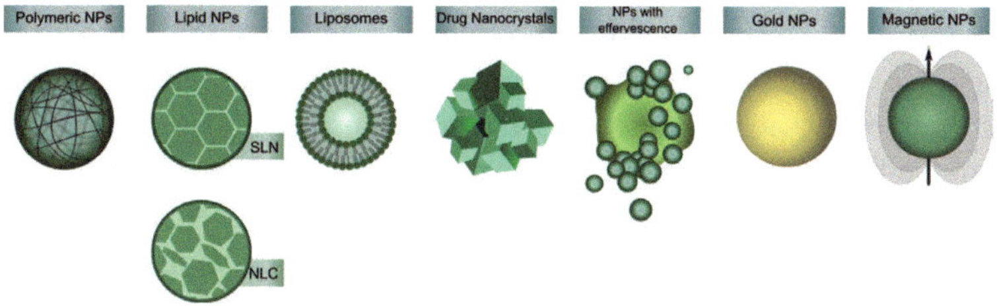

Figure 8.3 Schematic representation of the different types of NPs currently used to achieve pulmonary delivery of anti-TB drugs.

8.3.2.1 Polymeric NPs

Natural and synthetic polymers are commonly used for producing polymeric NPs as nanocarriers for anti-TB drug delivery. Research into the rational delivery of and targeting by pharmaceutical, therapeutic, and diagnostic agents is at the forefront of projects in nanomedicine [55]. Natural polymers commonly chosen for pulmonary delivery are alginate, chitosan, and gelatin. Synthetic polymers include poly(lactide-co-glycolide) acid (PLGA), polylactic acid, polyanhydride, and polyacrylate [72].

Currently, polymeric NPs are the most widely researched systems for pulmonary delivery since these NPs fulfill most requirements needed for pulmonary delivery, such as sufficient association of the therapeutic agent with the carrier particles, targeting of specific sites or cell populations in the lung, protection of the therapeutic agent against degradation, release of the therapeutic agent at a therapeutically optimal rate, ability to be transferred into an aerosol, low toxicity, and stability against forces generated during aerosolization. However, the respiratory tract provides substantial barriers that need to be overcome for successful pulmonary application. In this regard, micro- and nanosized particles offer novel concepts for the development of optimized therapeutic tools in pulmonary research. Polymeric nanocarriers are generally preferred as controlled pulmonary delivery systems due to prolonged retention in the lung. Specific manipulation of nanocarrier characteristics enables the design of "intelligent" carriers with modified duration and intensity of pharmacological effects. New formulations should be tested for pulmonary absorption and distribution using more advanced ex vivo and in vivo models. The delivery of nanocarriers to the airspace enables a detailed characterization of the interaction between the carrier vehicle and the natural pulmonary environment. In summary, polymeric NPs seem to be especially promising as controlled delivery systems and represent a solid basis for future advancement of pulmonary delivery applications [73]. The selection of natural or synthetic materials is important in designing particle or NP clusters with the desired characteristics, such as biocompatibility, size, charge, drug release, and polymer degradation rate [66].

Polymeric NPs as drug deliverers of anti-TB drugs have already been studied. Thus, Jain et al. developed polymeric NPs for ciprofloxacin delivery made of albumin, gelatin, and also chitosan [74]. The authors studied the drug release profile of the formulation and concluded that albumin and chitosan polymeric NPs were more capable of drug incorporation (48% and 35% loading efficiency

[LE], respectively) and sustained drug release (up to 120 and 96 h, respectively). Another study involving natural polymeric NPs was performed by Ahmad et al. for the incorporation of RIF, INH, and PZA in alginate polymeric NPs [75]. The NPs developed presented aerodynamic diameters in the respirable range and high drug LE (between 70%–90%). In addition, the bioavailability of the formulations was studied, showing better results in comparison with the oral administration of the free drugs. In a study conducted by Saraogi et al. polymeric NPs made of gelatin were used for the incorporation of INH. These NPs were functionalized with mannose to increase the uptake in the target cells (i.e., infected macrophages) [76]. The authors obtained LE of around 50% and reported higher accumulation of INH in the lungs when using mannosylated NPs in comparison with nonfunctionalized NPs, rendering them suitable for pulmonary delivery of anti-TB drugs.

Abdulla et al. used poly-(ethylene oxide)-block-distearoyl phosphatidyl-ethanolamine (mPEG2000–DSPE and mPEG5000–DSPE) polymers to produce polymeric NPs for the pulmonary delivery of RIF. The authors reported high LE (84%–104%) and reported that these values were influenced by the drug-polymer ratio, but not by the mPEG–DSPE molecular weight [77]. Particle size and aerodynamic characterization showed that the formulations developed are suitable for lung deposition through inhalation.

Chitosan has some important reported properties to act as an inert carrier, such as biocompatibility, biodegradability, and low toxicity, being also mucoadhesive, promoting macromolecule permeation through well-organized epithelia [78]. In this context, Pourshahab et al. used chitosan/tripolyphosphate NPs to incorporate INH. The NPs developed showed a size of 449 nm and LE of 17%. Moreover, the formulation demonstrated an initial drug release burst, followed by slow and sustained drug release in the following 6 days [79]. Garg et al. prepared spray-dried inhalable chitosan NPs for the delivery of INH and RIF [80]. The results obtained showed NPs with diameters of 230 nm and LE of 69% for INH and 71% for RIF. In addition, compared with the free drug, the formulations demonstrated lower cytotoxicity and a reduction in the number of bacilli in the lungs of *M. tuberculosis* $H_{37}Rv$ infected BALB/c mice [80].

Polymeric NPs made of PLGA are extremely common and have been used to encapsulate some anti-TB drugs. Sung et al. showed that

PLGA NPs loaded with RIF can be formulated, resulting in "porous NP- aggregate particles" with aerosol properties appropriate for lung delivery [81]. The authors conducted in vitro and in vivo studies and found evidence of delayed drug release, a delay of up to 8 h. Moreover, the presence of RIF in the lungs of guinea pigs was detected up to 8 h after the delivery [81]. Jain et al. reported better results when using PLGA NPs conjugated with lactose and loaded with RIF (LE of approximately 38%–42%) [82]. The authors demonstrated that RIF remains in the plasma for 4–6 days and INH/PZA for 8–9 days. Pandey et al. and Sharma et al. developed PLGA NPs for the incorporation of RIF, INH, and PZA (LE of 50%, 60%, and 70%, respectively) [83, 84]. The authors reported the presence of RIF in the plasma for 4–6 days and the presence of INH and PZA for 8–9 days. Pandey et al. showed that 5 doses of nebulized anti-TB PLGA NPs achieved the equivalent therapeutic benefits of 46 daily doses of orally administered free drugs [83]. Later, the same research group coated similar NPs with wheat germ agglutinin and reported an increased period during which all drugs were detectable in plasma, namely 6–7 days for RIF and 13–14 days for INH/PZA [84].

The incorporation of hydrophilic drugs in polymeric nanosystems is challenging. Cheow and Hadinoto developed modified PLGA to achieve higher LE of water-soluble antibiotics, using levofloxacin as the model drug [85]. The formulation exhibited LE between 4% and 23%, depending of the method used for NP production, drug release being maintained. Additionally, the formulation developed showed antibacterial activity even after the spray drying. Later, the same authors developed lipid-polymer hybrid NPs to incorporate levofloxacin, ciprofloxacin, and ofloxacin [86]. The formulations developed obtained LE of 10%–19% for levofloxacin, 4%–6% for ciprofloxacin, and 10%–25% for ofloxacin. After the initial burst drug release, hybrid NPs showed a slower (or very slow) drug release than their nonhybrid counterparts. More recently, Varma et al. developed polymeric NPs using a copolymer of ethyl acrylate, methyl methacrylate, and a low content of methacrylic acid ester with quaternary ammonium groups [87]. The NPs developed showed a size range from 45 to 300 nm and LE of 80.9%. The formulation was showed to be stable during two months and was demonstrated to be efficiently internalized by alveolar macrophages.

8.3.2.2 Liposomes

Liposomes are vesicular structures consisting of at least one phospholipid bilayer enclosing an aqueous medium [88]. Their size can be fine-tuned to target different regions of the lung by passive targeting, and their structure and composition can be changed to actively target specific cells, namely alveolar macrophages. The functionalization of these NPs with mannose is one of the most successful examples of this strategy, as it has been shown to increase uptake of NPs by alveolar macrophages [89, 90]. Moreover, liposomes are particularly interesting for pulmonary delivery, since they can be produced using endogenous compounds, such as the components of the PS [91]. For all the above-mentioned reasons, in the literature it is possible to find a vast number of studies involving pulmonary delivery of drugs with liposomal formulations, many of them focusing on the delivery of antibiotics and particularly anti-TB drugs. Chattopadhyay et al. demonstrated that changing liposome composition by incorporation of charged lipids and cholesterol molecules into the bilayer prevented particle aggregation and preserved bilayer integrity after air-jet nebulization [92]. This is an important aspect since many aerosolization techniques can compromise liposome structure and integrity. A strategy to avoid this is to protect the surface of the vesicles by using polyethylene glycol (PEG) and plain tuftsin-conjugated liposomes, which was demonstrated to protect and maintain the stability of the liposomes during the nebulization process [93]. Another approach is based on the fact that a drug–lipid mixture solubilized in chlorofluorocarbon (CFC) will form liposomes upon hydration in small airways [94]. Gaur et al. explored the hypothesis by producing the liposomes in situ, since the lung has a wet surface and for that reason can provide an aqueous phase for local formation [94]. The developed formulation was formed by egg phosphatidylcholines (EPC), cholesterol, and DCP and was encapsulated with RIF with LE of 29%–38%. The authors reported that no vesicle rupture was observed and the liposomes were formed in situ and demonstrated prolonged drug release. Ciprofloxacin was one of the first anti-TB drugs encapsulated in liposomes. Wong, Finlay, and coworkers developed liposomes with phosphatidylcholine (PC) and cholesterol encapsulated with ciprofloxacin with high LE of 90% before nebulization, of which

2%–30% remained encapsulated after the nebulization process [95]. The authors studied liposome disruption during aerosolization, using 25 different nebulizers [95]. Later, the same authors published results related with the in situ formation of liposomes on dispersion of phospholipid-based powder formulations [96, 97]. The formulations consisting of PCs (dipalmitoylphosphatidylcholine [DPPC] or EPC) and cholesterol demonstrated impressive LE of 97% before lyophilization, of which up to 90% was retained by the lyophilized cake and up to 40% after jet milling. In another study, among all tested formulations (DPPC, dimyristoylphosphatidylglycerol [DMPG], EPC:DMPG, dimyristoylphosphatidylcholine [DMPC], and DMPC:DMPG), the best results in terms of LE, with a value near 100%, were obtained with formulations composed of lipids with a negative charge, namely DMPG. Additionally, the nebulization of the particles causes an overall decrease of the drug's LE. Bhavane et al. developed liposome agglomerates with ciprofloxacin that could be triggered by the instillation of a biologically acceptable agent [98]. The formulations were produced using DPPC, cholesterol and mPEG-DSPE or DPPC, and cholesterol and distearoylphosphoethan olamine[amino(polyethylene glycol)] conjugate (DSPE-PEG-NH2). Cysteine was used to facilitate postadministration modulation of the drug release rate since it forms cross links between nanosized liposomes to form the agglomerates. It could allow treatment regimens where the administration of a single dose would be sufficient for an extended period of time, since drug release could be periodically accelerated. The authors also found that progressive release of the drug does not cause significant inflammation, unlike the administration of the free drug ciprofloxacin.

Other anti-TB drugs have already been encapsulated in liposomes. Liposomes made of DPPC for the delivery of INH have been developed and evaluated in vitro by Chimote and Banerjee [99]. The authors obtained LE of approximately 37% with sustained drug release, held over 24 h after a burst release of 50% of the drug in the first 5 h. They have also performed biocompatibility and stability studies and found that the formulations were hemocompatible, cytocompatible, and stable for at least a month. Justo and Moraes studied the possibility of coencapsulating several anti-TB drugs in the same formulation made of distearoylphosphatidylcholine (DSPC) and cholesterol [91]. The anti-TB drugs encapsulated chosen were the first-line anti-TB drugs INH, PZA, and RIF and also ethionamide and streptomycin. The results demonstrated that RIF and ethionamide were not successfully encapsulated. Low LE was obtained in the case of INH (3%) and PZA (2%), with the LE of streptomycin the highest (42%). Gaur et al. published a feasibility study, using RIF as a model drug [94]. In this study, liposomes made of PC and cholesterol were formed in situ and the results showed a better-sustained drug release profile than the preformed liposomes. The liposomal aerosols showed improved delivery of RIF over plain drug aerosols, with LE of approximately 30%. Patil et al. developed RIF-loaded liposomes with LE of 79.25%. The in vitro results showed increased solubility of the drug and higher anti-TB activity in comparison with the free drug [100]. An aerodynamic characterization result demonstrated that the NPs developed were within a respirable size range; and in vivo results pointed to the enhancement of drug permeation in the alveolar epithelium [100].

The possibility of surface coating to achieve active targeting with liposomes has also been a subject of interest. Vyas et al. developed liposomes formed by PC, cholesterol, and encapsulated RIF with LE of roughly 50% [101]. The authors tested different macrophage-specific ligands (i.e., DCP, MBSA, or O-SAP) and reported a preferential accumulation of ligand-coated formulations in the lung macrophages. Deol and Khuller developed coated liposomes formed by EPC and cholesterol and coated with O-SAP for the encapsulation of the first-line anti-TB drug RIF (LE 44%–49%) and INH (LE 8%–10%) [102]. The authors reported that encapsulating drugs within liposomes reduced the toxicity of the free drugs and that an O-SAP coating increased the accumulation of the NPs in the lungs. Agarwal and coworkers studied tuftsin functionalization of EPC liposomes encapsulated with RIF (LE 30%) [103]. The authors reported that regular administration of the functionalized formulation over 2 weeks was more efficient in controlling TB than the nonfunctionalized one. Patil-Gadhe et al. developed rifapentine-loaded proliposomes for the inhalation by spray drying [104]. The results showed sustained drug release, a higher accumulation of the drug in the lungs having been achieved. Later, the same authors confirmed the potential of the developed formulation for the direct delivery of the drug to the lungs in spray-dried DPI, achieving higher drug concentrations in the lungs and reducing systemic toxicity [100].

8.3.2.3 Lipid NPs

Lipid NPs show a high drug loading capacity and high stability and do not require the use of organic solvents during production [105]. Moreover, these nanocarriers are biocompatible, can be produced with the appropriate size and morphology for lung targeting and deposition [106], and have been studied as a suitable pulmonary drug delivery strategy. The surface of lipid NPs can be easily modified to achieve active targeting. SLNs made of only solid lipids and NLCs with solid as well as liquid lipids are the most common lipid NPs used.

There are several studies using lipid NP formulations for the delivery of anti-TB drugs. Jain and coworkers developed SLNs for ciprofloxacin prolonged drug release [74]. Their work is one of only four reports found so far in the literature concerning the pulmonary delivery of SLNs loaded with drugs for the treatment of TB, namely rifabutin, INH, RIF, and PZA. Nimje et al. prepared rifabutin loaded in SLNs made of tristearin and compared uncoated formulations with formulations coated with mannose sugar [62]. The results demonstrated that the developed formulation is suitable for sustained drug delivery, the cellular uptake by alveolar macrophages being enhanced by roughly six times due to the mannose coating. The in vivo results confirmed the higher drug accumulation in the lungs for coated SLNs, as well as a less immunogenic formulation. Pandey and Kuller developed SLNs made of stearic acid for pulmonary delivery through nebulization [107]. The authors encapsulated INH, RIF, and PZA with LE of roughly 50% for each drug. The formulations were capable of sustained drug release, with a burst drug release <20% in the first 6 h and of 11%–15% in the 6–72 h period in the case of INH and PZA; for RIF it was of 9% in the first 6 h and of 11% in the 6–72 h period. The nebulized SLNs were successfully deposited in the lungs and were detected in other organs up to 7 days after administration, while the free drug was cleared from the system within 24–48 h. Jain and Banerjee developed SLNs made of stearic acid to deliver ciprofloxacin (with LE of 39%) and concluded that these NPs were able to sustain drug release for up to 80 h [74]. Chuan et al. developed RIF-loaded SLNs to target alveolar macrophages [108]. The loaded SLNs presented an average size of around 800 nm and showed relatively low cytotoxicity when their concentration was

>20 μg/ml. In addition, their results demonstrated that RIF-loaded SLNs were internalized more selectively in alveolar macrophages than in alveolar epithelial type II cells. Gaspar et al. developed SLNs using glyceryl dibehenate or glyceryl tristearate as lipid components for the delivery of rifabutin [109]. The SLNs produced showed a size between 99 and 210 nm, with high LE of above 80%. The release studies demonstrated an almost complete drug release, with low NP cytotoxicity [109].

The second generation of lipid NPs, NLCs, due to their composition (a mixture of solid and liquid lipids), possess an imperfect crystal matrix that enhances the LE and minimizes drug expulsion during long-term storage [84]. To the best of our knowledge, there are only two studies reporting the development of NLCs loaded with anti-TB drugs. Song et al. developed RIF-loaded cationic mannosylated NLCs with an average size of 160 nm and LE higher than 90%, with minimum cytotoxicity and no inflammatory response [110]. Additionally, the formulation showed tissue selectivity and significantly improved accumulation of RIF in the lungs when compared to the free drug. More recently, Pinheiro et al. prepared mannose-functionalized rifabutin-loaded NLCs made of Precirol® ATO 5, Tween 60, and Miglyol-812 [26]. The size of the NPs was found to be approximately 200 nm, and they had LE higher than 80%. Moreover, the NPs were found to be highly stable for at least 6 months. Moreover, the drug release was found to be pH sensitive, with a faster release at acidic pH than at neutral pH, which may be useful for a higher amount of release of the drug mainly under the acidic conditions seen in phagosomes and phagolysosomes, where the etiologic agent of TB is located.

8.3.2.4 Other NPs

Nanocrystals have been proposed as systems for drug delivery. They are used as dispersions of pure drug NPs, kept stable through the presence of a minimum amount of a surfactant—nanosuspensions. Drug nanocrystals dissolve rapidly in the lung lining fluid, leading to a local high concentration, which is interesting for the localized treatment of respiratory diseases, such as pulmonary TB. The results point out that these NPs could be used in drug delivery formulations to improve pharmacokinetic and pharmacodynamic properties of poorly soluble drugs.

Gao et al. reported two different pulmonary formulations, one containing drug nanocrystals [65] (i.e., aqueous nanosuspensions) packaged and administered using a nebulizer and the other containing drug nanocrystals collected and transported into the lung by the small aerosol droplets generated by the nebulizer. Spore-like drug particles for deep lung deposition have also been proposed as an innovative system [111]. Thus, hollow and spore-like nanoagglomerates can be formed by mixing the drug solution with an antisolvent in a high-gravity environment. The production of drug particles similar to spores may improve the pulmonary drug delivery efficiency in DPIs, this being a more efficient, cost-effective, and easy-to-scale-up method to prepare nanosuspensions than milling, homogenization, spray freezing into liquid, and supercritical antisolvent precipitation. According to the authors, monodispersed particle size and controlled morphology can be achieved using this technique.

Presently, only one study was found related with nanocrystals or nanoagglomerates for anti-TB drug delivery. El-Gendy et al. developed nanosuspensions loaded with ciprofloxacin with high LE (81%–96%), which were then flocculated to form NP agglomerates [112]. The authors performed dissolution studies and compared the results with the stock drug. The results demonstrated that the dissolution rate was improved, reinforcing that these techniques may help to overcome the insolubility issues presented by new anti-TB drugs, especially the ones that have shown high potential as anti-TB drugs but have not yet passed the clinical trials due to low solubility.

NPs with effervescent activity have been recently suggested for pulmonary delivery. Effervescent activity of the NPs occurs when the carrier particles are exposed to humidity, adding an active release mechanism to the pulmonary route of administration. In addition, effervescent NPs can be synthesized with the adequate size for deep lung deposition and this technology seems to be safe for pulmonary delivery [113]. Although effervescent NPs have been studied as a promising strategy for the pulmonary delivery of different drugs, only one report was found regarding their use for the delivery of anti-TB drugs [114]. Ely and coworkers have developed and studied different powder compositions with effervescent activity and have found two formulations suitable for pulmonary delivery of ciprofloxacin

[115]. The formulations made of polybutylcyanoacrylate (PBCA) had 1-leucine and PEG 6000, which improved the aerodynamic characteristics of the powder particles. Effervescent activity of the prepared formulations resulted in the drug release of 56% into the solution in comparison with 32% into the carrier particles made just of lactose.

Gold NPs have recently been used to study internalization and intracellular translocation of inhaled NPs in alveolar macrophages of rats [116], and it was demonstrated that ciprofloxacin binds to gold NPs [117]. However, the studies do not focus on pulmonary delivery and no other reports have been found regarding the use of these particles for pulmonary delivery.

The use of magnetic aerosols using superparamagnetic iron oxide NPs has also been suggested as a tool to improve drug delivery to the lungs [118], and ciprofloxacin was once again used as a drug model in the development of superparamagnetic nanocomposites with magnetically mediated release of the loaded anti-TB drug [119]. However, no study has been found combining these two strategies to achieve magnetically mediated pulmonary delivery of anti-TB drugs.

8.4 Challenges and Future Directions

In the last few decades, there has been considerable effort to speed up detection and improve treatment outcomes, with the main aim being to reduce the number of new TB cases and deaths. Despite the novel rapid diagnostic tests and effective therapies available today, low patient compliance and the emergence of multi-drug-resistant TB cases constitute problematic issues, impairing TB control and eradication [13].

To improve patient compliance to the therapy, treatment support and direct-observation programs have been extensively applied. In spite of these efforts, 16% to 49% of the patients do not complete their treatment regimen [8]. Thus, new delivery strategies to improve the efficacy of existing treatments may contribute to reduce these numbers.

Inhalation is the best route to ensure the drugs target the lungs. Indeed, lungs are the primary site of TB infection, inhalation being an interesting alternative for the delivery of anti-TB drugs [50].

Although promising and vastly researched, pulmonary delivery strategies face obstacles difficult to overcome. Regarding anti-TB drugs, most of the first-line drugs have been studied in vitro for use in powder formulations for inhalation [120–124].

In recent years, nanotechnology has been explored not only to carry and protect drugs but also to ensure targeted delivery and reduce dosing frequencies, which may contribute to prevent toxicity related to therapy [125]. For instance, a wide variety of nanosystems have been shown to be more beneficial than conventional formulations. Despite the above-mentioned advantages of nanotechnology approaches, a clearer understanding of the pharmacokinetic profile of anti-TB delivery nanosystems is required and in vivo studies in complex biological systems should be performed to understand whether these novel formulations have the desirable characteristics to reach the market. For this purpose, more resources to support the translation of research from bench to clinics are needed.

8.5 Conclusions

Despite the considerable research efforts that have been made in the last few decades, TB treatment continues to be a great challenge due to the long-term multidrug schedule that is required and due to the associated side effects. The classical anti-TB medicines are mainly administered through the oral route, but an effort has been made to develop new systems targeting the main site of TB infection (i.e., the lung). With the advent of innovative NP-based formulations, a wide variety of pulmonary nanodelivery systems have been tested and a newer hope has emerged. Efforts to define more effective treatment approaches, including the development of pulmonary nanosystems with individualized doses and regimens, may reduce treatment duration, which ultimately will lead to improve patients' compliance to therapy. The success of these nanodelivery systems remains to be demonstrated and will probably depend on the design of biosafe and cost-effective formulations that address different limitations of classical anti-TB pharmacotherapies. Technologies involving the use of lipid NPs seem to offer a suitable economical solution for the pulmonary administration of anti-TB drugs.

To conclude, NP-based formulations to improve TB treatment should be pursued to meet all current challenges. In fact, some of the above-discussed results are very promising and so should encourage new research on NP-based pulmonary delivery systems.

Acknowledgments

This work received financial support from the European Union (FEDER funds POCI/01/0145/FEDER/007728) and National Funds (*Fundação para a Ciência e Tecnologia* and *Ministério da Educação e Ciência* [FCT/MEC]) under the partnership agreement PT2020 UID/MULTI/04378/2013. Joana Magalhães and Marina Pinheiro thank FCT and *Programa Operacional Potencial Humano* (POPH) for the PhD grant (SFRH/BD/110683/2015) and the postdoctoral grant (SFRH/BPD/99124/2013), respectively. Alexandre Vieira thanks the CNPq, Ministry of Education of Brazil, for the fellowship 246514/2012-4. This work was also supported by FCT through the FCT PhD programs, specifically by the BiotechHealth Programme (Doctoral Programme on Cellular and Molecular Biotechnology Applied to Health Sciences). The authors thank Ricardo Ribeiro for his contribution in Fig. 8.3.

References

1. Donoghue, H.D. Human tuberculosis--an ancient disease, as elucidated by ancient microbial biomolecules. *Microbes Infect*, 2009, **11**(14–15):1156–1162.
2. Galagan, J.E. Genomic insights into tuberculosis. *Nat Rev Genet*, 2014, **15**(5):307–320.
3. Davies, M.K., et al. Rene Theophile-Hyacinthe Laennec (1781–1826). *Heart*, 1996, **76**(3):196–196.
4. Daniel, T.M. The history of tuberculosis. *Respir Med*, 2006, **100**(11):1862–1870.
5. Roche, P.W., et al. BCG vaccination against tuberculosis: past disappointments and future hopes. *Trends Microbiol*, 1995, **3**(10):397–401.
6. Zumla, A., et al. Advances in the development of new tuberculosis drugs and treatment regimens. *Nat Rev Drug Discov*, 2013, **12**(5):388–404.

7. Zumla, A., et al. Tuberculosis treatment and management: an update on treatment regimens, trials, new drugs, and adjunct therapies. *Lancet Respir Med*, 2015, **3**(3):220–234.
8. WHO, Global Tuberculosis Report. WHO Library Cataloguing-in-Publication Data, 2015.
9. Hartman-Adams, H., et al. Update on latent tuberculosis infection. *Am Fam Physician*, 2014, **89**(11):889–896.
10. Narendran, G., et al. TB-HIV co-infection: a catastrophic comradeship. *Oral Dis*, 2016, **22**(Suppl 1):46–52.
11. Ahmad, S., et al. Current status and future trends in the diagnosis and treatment of drug-susceptible and multidrug-resistant tuberculosis. *J Infect Public Health*, 2014, **7**(2):75–91.
12. Chiang, C.Y., et al. Drug-resistant tuberculosis: past, present, future. *Respirology*, 2010, **15**(3):413–432.
13. WHO, Global strategy and targets for tuberculosis prevention, care and control after 2015, Executive Board, 2015(134/12):1–23.
14. Niemann, S., et al. Differentiation among members of the Mycobacterium tuberculosis complex by molecular and biochemical features: evidence for two pyrazinamide-susceptible subtypes of M. bovis. *J Clin Microbiol*, 2000, **38**:152–157.
15. Gagneux, S. Host-pathogen coevolution in human tuberculosis. *Philos Trans R Soc Lond B Biol Sci*, 2012, **367**(1590):850–859.
16. Smith, N.H., et al. Bottlenecks and broomsticks: the molecular evolution of Mycobacterium bovis. *Nat Rev Microbiol*, 2006, **4**(9):670–681.
17. Dannenberg, A. Immune mechanisms in the pathogenesis of pulmonary tuberculosis. *Rev Infect Dis*, 1989, **11**(2):S369–S378.
18. Flynn, J.L., et al. Macrophages and control of granulomatous inflammation in tuberculosis. *Mucosal Immunol*, 2011, **4**(3):271–278.
19. Nunes-Alves, C., et al. In search of a new paradigm for protective immunity to TB. *Nat Rev Microbiol*, 2014, **12**(4):289–299.
20. Lee, J.Y. Diagnosis and treatment of extrapulmonary tuberculosis. *Tuberc Respir Dis (Seoul)*, 2015, **78**(2):47–55.
21. Krishnan, N., et al. The mechanisms and consequences of the extra-pulmonary dissemination of Mycobacterium tuberculosis. *Tuberculosis (Edinb)*, 2010, **90**(6):361–366.
22. Boon, C., et al. How Mycobacterium tuberculosis goes to sleep: the dormancy survival regulator DosR a decade later. *Future Microbiol*, 2012, **7**(4):513–518.

23. Weiss, G., et al. Macrophage defense mechanisms against intracellular bacteria. *Immunol Rev*, 2015, **264**:182–203.

24. Hmama, Z., et al. Immunoevasion and immunosuppression of the macrophage by Mycobacterium tuberculosis. *Immunol Rev*, 2015, **264**:220–232.

25. Kaufmann, S.H. Tuberculosis: back on the immunologists' agenda. *Immunity*, 2006, **24**(4):351–357.

26. Pinheiro, M., et al. Liposomes as drug delivery systems for the treatment of TB. *Nanomedicine*, 2011, **6**(8):1413–1428.

27. Sia, I.G., et al. Current concepts in the management of tuberculosis. Mayo Clin Proc, 2011, **86**(4):348–361.

28. Yan, L., et al. Systematic review: comparison of Xpert MTB/RIF, LAMP and SAT methods for the diagnosis of pulmonary tuberculosis. *Tuberculosis (Edinb)*, 2016, **96**:75–86.

29. Dunlap, N.E., et al. Diagnostic standards and classification of tuberculosis in adults and children. *Am J Respir Crit Care Med*, 2000, **161**:1376–1395.

30. Swick, B.L. Polymerase chain reaction-based molecular diagnosis of cutaneous infections in dermatopathology. *Semin Cutan Med Surg*, 2012, **31**(4):241–246.

31. Wlodarska, M., et al. A microbiological revolution meets an ancient disease: improving the management of tuberculosis with genomics. *Clin Microbiol Rev*, 2015, **28**(2):523–539.

32. Tsara, V., et al. Problems in diagnosis and treatment of tuberculosis infection. *Hippokratia*, 2009, **13**(1):20–22.

33. Druszczynska, M., et al. Latent M. tuberculosis infection--pathogenesis, diagnosis, treatment and prevention strategies. *Pol J Microbiol*, 2012, **61**(1):3–10.

34. Maertzdorf, J., et al. Enabling biomarkers for tuberculosis control. *Int J Tuberc Lung Dis*, 2012, **16**(9):1140–1148.

35. Kaufmann, S., et al. Progress in tuberculosis vaccine development and host-directed therapies: a state of the art review. *Lancet Respir Med*, 2014, **2**(4):301–320.

36. Groschel, M.I., et al. Therapeutic vaccines for tuberculosis--a systematic review. *Vaccine*, 2014, **32**(26):3162–3168.

37. WHO, Treatment of tuberculosis: guidelines, 4th ed. WHO Library Cataloguing-in-Publication Data, 2010.

38. Choudhary, S., et al. Potential of nanotechnology as a delivery platform against tuberculosis: current research review. *J Controlled Release*, 2015, **202**:65–75.

39. Mortensen, N.P., et al. The role of particle physico-chemical properties in pulmonary drug delivery for tuberculosis therapy. *J Microencapsul*, 2014, **31**(8):785–795.

40. Mehanna, M.M., et al. Respirable nanocarriers as a promising strategy for antitubercular drug delivery. *J Controlled Release*, 2014, **187**:183–197.

41. Andrade, F., et al. Nanotechnology and pulmonary delivery to overcome resistance in infectious diseases. *Adv Drug Deliv Rev*, 2013, **65**(13–14):1816–1827.

42. Kelly, L. *Essentials of Human Physiology for Pharmacy*. CRC Press Pharmacy Education Series, 2004, pp. 239–277.

43. Rhoades, R., et al. *Human Physiology*, 4th ed. Thomson, 2003, pp. 631–635.

44. Thorley, A.J., et al. New perspectives in nanomedicine. *Pharmacol Ther*, 2013, **140**(2):176–185.

45. Muralidharan, P., et al. Inhalable nanoparticulate powders for respiratory delivery. *Nanomedicine*, 2015, **11**(5):1189–1199.

46. Button, B., et al. A periciliary brush promotes the lung health by separating the mucus layer from airway epithelia. *Science*, 2012, **337**(6097):937–941.

47. Ruge, C., et al. Pulmonary drug delivery: from generating aerosols to overcoming biological barriers-therapeutic possibilities and technological challenges. *Lancet Respir Med*, 2013, **1**(5):402–413.

48. Hickey, A.J. Controlled delivery of inhaled therapeutic agents. *J Controlled Release*, 2014, **190**:182–188.

49. Misra, A., et al. Inhaled drug therapy for treatment of tuberculosis. *Tuberculosis (Edinb)*, 2011, **91**(1):71–81.

50. Costa, A., et al. The formulation of nanomedicines for treating tuberculosis. *Adv Drug Deliv Rev*, 2016, **102**:102–115.

51. Parumasivam, T., et al. Dry powder inhalable formulations for anti-tubercular therapy. *Adv Drug Deliv Rev*, 2016, **102**:83–101.

52. Pandey, R., et al. Antitubercular inhaled therapy: opportunities, progress and challenges. *J Antimicrob Chemother*, 2005, **55**(4):430–435.

53. Gupta, A., et al. Nanotechnology and its applications in drug delivery: a review. *Int J Med Mol Med*, 2012, **3**(1):WMC002867.

54. De Jong, W.H., et al. Drug delivery and nanoparticles: applications and hazards. *Int J Nanomed*, 2008, **3**(2):133–149.

55. Moghimi, S.M., et al. Nanomedicine: current status and future prospects. *FASEB J*, 2005, **19**(3):311–330.

56. Gelperina, S., et al. The potential advantages of nanoparticle drug delivery systems in chemotherapy of tuberculosis. *Am J Respir Crit Care Med*, 2005, **172**(12):1487–1490.

57. Clark, H.A., et al. Optical nanosensors for chemical analysis inside single living cells. 1. Fabrication, characterization, and methods for intracellular delivery of PEBBLE sensors. *Anal Chem*, 1999, **71**(21):4831-4836.

58. Drummond, D.C., et al. Current status of pH-sensitive liposomes in drug delivery. *Prog Lipid Res*, 2000, **39**(5):409–460.

59. Moghimi, S.M., et al. Long-circulating and target-specific nanoparticles: theory to practice. *Pharmacol Rev*, 2001, **53**(2):283–318.

60. Zazo, H., et al. Current applications of nanoparticles in infectious diseases. *J Controlled Release*, 2016, **224**:86–102.

61. Parveen, S., et al. Nanoparticles: a boon to drug delivery, therapeutics, diagnostics and imaging. *Nanomedicine*, 2012, **8**(2):147–166.

62. Nimje, N., et al. Mannosylated nanoparticulate carriers of rifabutin for alveolar targeting. *J Drug Target*, 2009, **17**(10):777–787.

63. Jain, A., et al. Mannosylated solid lipid nanoparticles as vectors for site-specific delivery of an anti-cancer drug. *J Controlled Release*, 2010, **148**(3):359–367.

64. Gill, S., et al. Nanoparticles: characteristics, mechanisms of action, and toxicity in pulmonary drug delivery: a review. *J Biomed Nanotechnol*, 2007, **3**(2):107–119.

65. Gao, L., et al. Drug nanocrystals: in vivo performances. *J Controlled Release*, 2012, **160**(3):418–430.

66. Rytting, E., et al. Biodegradable polymeric nanocarriers for pulmonary drug delivery. *Expert Opin Drug Deliv*, 2008, **5**(6):629–639.

67. Chellat, F., et al. Therapeutic potential of nanoparticulate systems for macrophage targeting. *Biomaterials*, 2005, **26**(35):7260–7275.

68. Gunaseelan, S., et al. Surface modifications of nanocarriers for effective intracellular delivery of anti-HIV drugs. *Adv Drug Deliv Rev*, 2010, **62**(4–5):518–531.

69. Netsomboon, K., et al. Mucoadhesive vs. mucopenetrating particulate drug delivery. *Eur J Pharm Biopharm*, 2016, **98**:76–89.

70. Abdulkarim, M., et al. Nanoparticle diffusion within intestinal mucus: three-dimensional response analysis dissecting the impact of particle surface charge, size and heterogeneity across polyelectrolyte, pegylated and viral particles. *Eur J Pharm Biopharm*, 2015, **97**(Pt A):230–238.

71. Lorenz, S., et al. The softer and more hydrophobic the better: influence of the side chain of polymethacrylate nanoparticles for cellular uptake. *Macromol Biosci*, 2010, **10**(9):1034–1042.

72. Pandey, R., et al. Nanomedicine and experimental tuberculosis: facts, flaws, and future. *Nanomedicine*, 2011, **7**(3):259–272.

73. Beck-Broichsitter, M., et al. Controlled pulmonary drug and gene delivery using polymeric nano-carriers. *J Controlled Release*, 2012, **161**(2):214–224.

74. Jain, D., et al. Comparison of ciprofloxacin hydrochloride-loaded protein, lipid, and chitosan nanoparticles for drug delivery. *J Biomed Mater Res B Appl Biomater*, 2008, **86**(1):105–112.

75. Ahmad, Z., et al. Inhalable alginate nanoparticles as antitubercular drug carriers against experimental tuberculosis. *Int J Antimicrob Agents*, 2005, **26**(4):298–303.

76. Saraogi, G.K., et al. Mannosylated gelatin nanoparticles bearing isoniazid for effective management of tuberculosis. *J Drug Target*, 2011, **19**(3):219–227.

77. Abdulla, J.M., et al. Rehydrated lyophilized rifampicin-loaded mPEG-DSPE formulations for nebulization. *AAPS PharmSciTech*, 2010, **11**(2):663–671.

78. Grenha, A., et al. Microencapsulated chitosan nanoparticles for lung protein delivery. *Eur J Pharm Sci*, 2005, **25**(4–5):427–437.

79. Pourshahab, P.S., et al. Preparation and characterization of spray dried inhalable powders containing chitosan nanoparticles for pulmonary delivery of isoniazid. *J Microencapsul*, 2011, **28**(7):605–613.

80. Garg, T., et al. Inhalable chitosan nanoparticles as antitubercular drug carriers for an effective treatment of tuberculosis. *Artif Cells Nanomed Biotechnol*, 2015, 1–5.

81. Sung, J.C., et al. Formulation and pharmacokinetics of self-assembled rifampicin nanoparticle systems for pulmonary delivery. *Pharm Res*, 2009, **26**(8):1847–1855.

82. Jain, S.K., et al. Lactose-conjugated PLGA nanoparticles for enhanced delivery of rifampicin to the lung for effective treatment of pulmonary tuberculosis. *PDA J Pharm Sci Technol*, 2010, **64**(3):278–287.

83. Pandey, R., et al. Poly (DL-lactide-co-glycolide) nanoparticle-based inhalable sustained drug delivery system for experimental tuberculosis. *J Antimicrob Chemother*, 2003, **52**(6):981–986.

84. Sharma, A., et al. Lectin-functionalized poly (lactide-co-glycolide) nanoparticles as oral/aerosolized antitubercular drug carriers for treatment of tuberculosis. *J Antimicrob Chemother*, 2004b, **54**(4):761–766.

85. Cheow, W.S., et al. Enhancing encapsulation efficiency of highly water-soluble antibiotic in poly(lactic-co-glycolic acid) nanoparticles: modifications of standard nanoparticle preparation methods. *Colloids Surf, A*, 2010, **370**(1–3):79–86.

86. Cheow, W.S., et al. Factors affecting drug encapsulation and stability of lipid-polymer hybrid nanoparticles. *Colloids Surf B*, 2011, **85**(2):214–220.

87. Varma, J.N., et al. Formulation and characterization of pyrazinamide polymeric nanoparticles for pulmonary tuberculosis: efficiency for alveolar macrophage targeting. *Indian J Pharm Sci*, 2015, **77**(3):258–266.

88. Bangham, A.D. Liposomes: the Babraham connection. *Chem Phys Lipids*, 1993, **64**:275–285.

89. Chono, S., et al. Effect of surface-mannose modification on aerosolized liposomal delivery to alveolar macrophages. *Drug Dev Ind Pharm*, 2010, **36**(1):102–107.

90. Kong, F., et al. Mannosylated liposomes for targeted gene delivery. *Int J Nanomed*, 2012, **7**:1079–1089.

91. Justo, O.R., et al. Incorporation of antibiotics in liposomes designed for tuberculosis therapy by inhalation. *Drug Deliv*, 2003, **10**(3):201–207.

92. Chattopadhyay, S., et al. Morphology and bilayer integrity of small liposomes during aerosol generation by air-jet nebulisation. *J Nanopart Res*, 2012, **14**(4).

93. Anabousi, S., et al. Effect of PEGylation on the stability of liposomes during nebulisation and in lung surfactant. *J Nanosci Nanotechnol*, 2006, **6**(9):3010–3016.

94. Gaur, P.K., et al. Targeted drug delivery of Rifampicin to the lungs: formulation, characterization, and stability studies of preformed aerosolized liposome and in situ formed aerosolized liposome. *Drug Dev Ind Pharm*, 2010, **36**(6):638–646.

95. Finlay, W.H., et al. Regional lung deposition of nebulized liposome-encapsulated ciprofloxacin. *Int J Pharm*, 1998, **167**(1–2):121–127.

96. Desai, T.R., et al. A facile method of delivery of liposomes by nebulization. *J Controlled Release*, 2002, **84**(1–2):69–78.

97. Desai, T.R., et al. A novel approach to the pulmonary delivery of liposomes in dry powder form to eliminate the deleterious effects of milling. *J Pharm Sci*, 2002, **91**(2):482–491.

98. Bhavane, R., et al. Triggered release of ciprofloxacin from nanostructured agglomerated vesicles. *Int J Nanomed*, 2007, **2**(3):407–418.

99. Chimote, G., et al. Evaluation of antitubercular drug-loaded surfactants as inhalable drug-delivery systems for pulmonary tuberculosis. *J Biomed Mater Res A*, 2009, **89**(2):281–292.

100. Patil, J.S., et al. A novel approach for lung delivery of rifampicin-loaded liposomes in dry powder form for the treatment of tuberculosis. *Lung India*, 2015, **32**(4):331–338.

101. Vyas, S.P., et al. Design of liposomal aerosols for improved delivery of rifampicin to alveolar macrophages. *Int J Pharm*, 2004, **269**(1):37–49.

102. Deol, P., et al. Therapeutic efficacies of isoniazid and rifampin encapsulated in lung-specific stealth liposomes against Mycobacterium tuberculosis infection induced in mice. *Antimicrob Agents Chemother*, 1997, **41**(6):1211–1214.

103. Agarwal, A., et al. Tuftsin-bearing liposomes as rifampin vehicles in treatment of tuberculosis in mice. *Antimicrob Agents Chemother*, 1994, **38**(3):588–593.

104. Patil-Gadhe, A., et al. Single step spray drying method to develop proliposomes for inhalation: a systematic study based on quality by design approach. *Pulm Pharmacol Ther*, 2014, **27**(2):197–207.

105. Sosnik, A., et al. New old challenges in tuberculosis: potentially effective nanotechnologies in drug delivery. *Adv Drug Deliv Rev*, 2010, **62**(4–5):547–559.

106. Videira, M.A., et al. Lymphatic uptake of pulmonary delivered radiolabelled solid lipid nanoparticles. *J Drug Target*, 2002, **10**(8):607–613.

107. Pandey, R., et al. Solid lipid particle-based inhalable sustained drug delivery system against experimental tuberculosis. *Tuberculosis (Edinb)*, 2005, **85**(4):227–234.

108. Chuan, J., et al. Enhanced rifampicin delivery to alveolar macrophages by solid lipid nanoparticles. *J Nanopart Res*, 2013, **15**(5):1–9.

109. Gaspar, D.P., et al. Rifabutin-loaded solid lipid nanoparticles for inhaled antitubercular therapy: physicochemical and in vitro studies. *Int J Pharm*, 2016, **497**(1–2):199–209.

110. Song, X., et al. Rifampicin loaded mannosylated cationic nanostructured lipid carriers for alveolar macrophage-specific delivery. *Pharm Res*, 2015, **32**(5):1741–1751.

111. Shen, Z.G., et al. Fabrication of inhalable spore like pharmaceutical particles for deep lung deposition. *Int J Pharm*, 2012, **430**(1–2):98–103.

112. El-Gendy, N., et al. Agglomerates of ciprofloxacin nanoparticles yield fine dry powder aerosols. *J Pharm Innovation*, 2010, **5**(3):79–87.

113. Azarmi, S., et al. Formulation and in vivo evaluation of effervescent inhalable carrier particles for pulmonary delivery of nanoparticles. *Drug Dev Ind Pharm*, 2008, **34**(9):943–947.

114. Roa, W.H., et al. Inhalable nanoparticles, a non-invasive approach to treat lung cancer in a mouse model. *J Controlled Release*, 2011, **150**(1):49–55.

115. Ely, L., et al. Effervescent dry powder for respiratory drug delivery. *Eur J Pharm Biopharm*, 2007, **65**(3):346–353.

116. Takenaka, S., et al. Efficient internalization and intracellular translocation of inhaled gold nanoparticles in rat alveolar macrophages. *Nanomedicine (Lond)*, 2012, **7**(6):855–865.

117. Saha, B., et al. In vitro structural and functional evaluation of gold nanoparticles conjugated antibiotics. *Nanoscale Res Lett Nanoscale Res Lett*, 2007, **2**(12):614–622.

118. Dames, P., et al. Targeted delivery of magnetic aerosol droplets to the lung. *Nat Nanotechnol*, 2007, **2**(8):495–499.

119. Bajpai, A.K., et al. Magnetically mediated release of ciprofloxacin from polyvinyl alcohol based superparamagnetic nanocomposites. *J Mater Sci Mater Med*, 2011, **22**(2):357–369.

120. Son, Y.J., et al. A new respirable form of rifampicin. *Eur J Pharm Biopharm*, 2011, **78**(3):366–376.

121. Hanif, S.N., et al. Pharmaceutical aerosols for the treatment and prevention of tuberculosis. *Front Cell Infect Microbiol*, 2012, **2**:118.

122. Manion, J.a.R., et al. Inhalable antibiotics manufactured through use of near-critical or supercritical fluids. *Aerosol Sci Technol*, 2012, **46**(4):403–410.

123. Chan, J.G., et al. A novel dry powder inhalable formulation incorporating three first-line anti-tubercular antibiotics. *Eur J Pharm Biopharm*, 2013, **83**(2):285–292.

124. Das, S., et al. Inhaled dry powder formulations for treating tuberculosis. *Curr Drug Deliv*, 2015, **12**:26–39.

125. Abed, N., et al. Nanocarriers for antibiotics: a promising solution to treat intracellular bacterial infections. *Int J Antimicrob Agents*, 2014, **43**(6):485–496.

Part V

Brain Drug Delivery Approaches

Chapter 9

Nanoparticles and New Challenges in Site-Specific Brain Drug Delivery

Ana C. R. Joyce Coutinho, Rúben G. R. Pinheiro, and Ana Rute Neves
UCIBIO, REQUIMTE, Department of Chemical Sciences, Faculty of Pharmacy, University of Porto, Portugal
ananeves@ff.up.pt

Neurological disorders are rapidly increasing as the population is getting older. Early diagnosis would enable improved disease outcomes, but the entry of drugs into the brain is commonly restricted across the blood–brain barrier (BBB) due to the presence of tight junctions and lack of fenestration. In fact, this barrier separates the blood from the brain tissue and is composed of endothelial cells, astrocytes, pericytes, neurons, and microglia, which makes the brain delivery of drugs a great challenge. This creates a protection barrier to potential threats to the brain, which at the same time turns extremely difficult the passage of drugs across the blood–brain barrier. To circumvent this problem, new strategies have been developed based on nanotechnology to produce promising brain

Nanoparticles in Life Sciences and Biomedicine
Edited by Ana Rute Neves and Salette Reis

ISBN 978-981-4745-98-7 (Hardcover), 978-1-351-20735-5 (eBook)
www.panstanford.com

nanodelivery systems that may improve diagnostic and therapeutic outcomes. These strategies aim to optimize the amount of drug-loaded nanoparticles that reaches and accumulates in the target brain cells, in order to improve therapeutic effects, while decreasing toxic effects by reducing their concentration in the healthy tissues. This chapter aims to provide an overview of recent advances and achievements in brain drug delivery systems, providing a wide coverage of strategies and possible applications in the field of nanomedicine.

9.1 Introduction

9.1.1 Concerns about Neurological Diseases

According to the World Health Organization (WHO), neurological disorders are a growing twenty-first-century global health crisis that demands our urgent attention. In fact, Alzheimer's disease (AD), Parkinson's disease (PD), Huntington's disease, amyotrophic lateral sclerosis, multiple sclerosis, epilepsy, and stroke are rapidly increasing as the population is getting older and also as a consequence of wrong lifestyle and increased risk factors. In Europe, about 35% of the total diseases are caused by brain disorders, and about 1.5 billion people worldwide suffer from central nervous system (CNS) diseases [1, 2].

Neurodegenerative diseases are a group of chronic and progressive pathologies characterized by the loss of neurons in the CNS and the formation of misfolded proteins, such as amyloid β-fibrils and tau tangles in AD and α-synuclein protein in PD [3, 4]. Several studies also support the idea that oxidative stress and inflammatory processes may significantly affect the synaptic transmission [5, 6]. The socioeconomic impact of these age-related neurological disorders will be one of the biggest problems of the future society, given the ageing population, the increased life expectancy, and the fact that the incidence of dementia increases exponentially with age.

Brain cancers are very aggressive, progressive, and difficult to treat due to their anatomical location, which often precludes the removal of the tumor by surgery. Gliomas are the most common brain cancers, originating from glial cells, which constitute the tissue that surrounds and supports neurons in the CNS. Glioblastoma multiforme is a malignant astrocytoma and the most common and aggressive form of primary brain tumors among adults, making the treatment a challenge [7, 8]. In fact, after resection, radiotherapy, and systemic chemotherapy, approximately 90% of the patients have a recurrence and the life span after diagnosis is only around 21 months [8, 9].

Therefore, there are presently no definitive treatments for patients affected by brain tumors and a successful medication to delay the progression of neurodegenerative diseases is also lacking. Indeed, the efficient delivery of many potentially therapeutic and diagnostic compounds to specific areas of the brain has been constrained due to the existence of a protective barrier that restricts the number of drugs that can enter the brain upon systemic administration, the blood–brain barrier (BBB) [10].

9.1.2 Importance and Challenges of the Blood–Brain Barrier

The BBB is a dynamic and extremely complex interface that separates the brain from the rest of the body. This barrier has an important role in controlling the rate of influx and efflux of biological substances needed in the CNS, hindering the delivery of diagnostic and therapeutic agents to the brain. In fact, it protects the brain from most substances in the blood, supplying brain tissues with nutrients and filtering out harmful compounds by a selective transport mechanism. Therefore, the BBB is crucial to maintaining the homeostasis of the brain microenvironment, being essential for proper neuronal activity and functioning of the CNS [11–13].

The existence of the BBB was discovered by Paul Ehrlich, who observed that intravenous administration of trypan blue stained all organs except the brain and the spinal cord. Later on, Edwin Ellen Goldmann also discovered that the same dye applied to the cerebrospinal fluid exclusively stained the brain tissue. These findings led to the concept of a barrier between blood and the brain [14].

The BBB is formed by specialized brain endothelial cells, which form the cerebral microvascular endothelium. This endothelium, together with astrocytes, pericytes, neurons, microglia, and the extracellular matrix, constitutes a "neurovascular unit" that is essential for the health and function of the CNS (Fig. 9.1) [15].

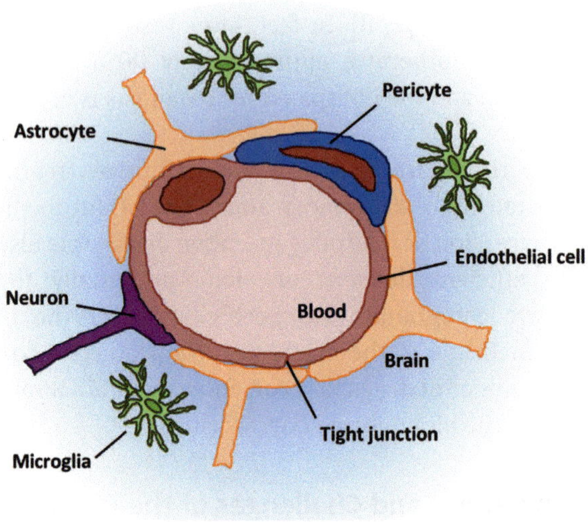

Figure 9.1 Schematic representation of a neurovascular unit cross section in the CNS.

Endothelial cells of the BBB differ from endothelial cells in the rest of the body due to the lack of fenestrations, the presence of more extensive adherens junctions (AJs) and tight junctions (TJs), a low level of pinocytotic vesicles, a more negatively charged membrane, and the expression of various transporters [16]. The extreme tightness of TJs is a key feature that regulates permeability, thereby limiting paracellular flux of ions, polar solutes, and hydrophilic molecules across the BBB [17].

Concerning astrocytes, they are responsible for physical support and maintenance of BBB integrity, since astrocytic end feet cover the brain capillaries almost completely. Moreover, astrocytes secrete factors that may influence endothelial cells, such as transforming growth factor-β (TGF-β), glial-derived neurotrophic factor (GDNF), basic fibroblast growth factor (bFGF), interleukin-6 (IL-6), and angiopoetin-1 [18].

Pericytes wrap around the endothelial cells and provide structural support to the vessel, maintain the integrity of the vessel, as well as provide vasodynamic capacity to the microvasculature

[19]. Like astrocytes, pericytes also secrete factors that may influence endothelial function, such as TGF-β, angiopoetin-1, and vascular endothelial growth factor (VEGF) [20]. Moreover, pericytes can also induce the expression of the TJ protein occludin in brain capillary endothelial cells, which suggests its involvement in regulating BBB permeability [21].

Regarding neurons and microglia, little is known about their precise role influencing BBB properties. However, there is some evidence that neurons affect cerebral blood flow and can regulate the function of blood vessels and that microglia have a role in the immune response of the CNS, and consequently in the BBB integrity [22].

As a result of these specific structural and biochemical properties, the BBB strictly constrains the transport of substances across this barrier. Only those molecules that are small, electrically neutral, and lipophilic can cross the barrier via passive transcellular diffusion. Thus, for a drug molecule to cross the BBB, among other requirements it must have high lipid solubility, a neutral charge, and a molecular mass lower than 400 Da and must not be a substrate for active efflux transporters [23]. Therefore, the BBB represents a critical obstacle in the treatment of neurological diseases since more than 98% of the new promising and potential neurotherapeutic molecules are unable to cross this barrier [24].

9.1.3 Strategies to Overcome the Hurdles of Brain Delivery

Nowadays, there is a huge interest in the development of effective solutions to overcome the limitations of brain delivery through the BBB. Several strategies have been developed for the sole purpose of effective brain targeting and drug delivery into the CNS for the treatment of neurological diseases. These include the enhancement of penetration through the BBB by opening of TJs, hypertonic BBB disruption, intracarotid administration, and intranasal delivery directly from the nasal cavity to the brain [25, 26]. However, the risk of embolism and hemorrhage leads to the avoidance of these strategies.

Currently, nanotechnology has been developing quickly and has been widely studied for diagnosis and treatment of brain diseases.

In this chapter, we will cover the efficient design of noninvasive nanocarrier systems that can facilitate drug delivery to specific regions of the brain, as well as the factors that influence their transport through the bloodstream to the therapeutic target place.

9.2 Nanotechnology as a Tool for Brain Delivery

Nanoparticles (NPs) have been regarded as having great potential for the delivery of drugs into the CNS. According to *Encyclopedia of Pharmaceutical Technology* and *Encyclopedia of Nanoscience and Nanotechnology*, NPs are well-defined solid colloidal particles ranging in size approximatively from 1 to 1000 nm, in which the active principle may be entrapped inside or adsorbed to the surface. It is important to keep in mind a range of useful properties and characteristics required for an optimal nanocarrier design and efficient application in drug delivery across the BBB.

9.2.1 Factors Affecting Nanocarriers' Brain Delivery

Once NPs are designed for human application, the materials used for their preparation must be nontoxic, biocompatible, and excretable or biodegradable. For this reason, toxicity studies have been performed to understand the implications of NPs for biocompatibility and the occurrence of possible side effects [27]. In fact, this issue is essential for the Food and Drug Administration's (FDA) approval of new nanocarriers. Another factor strongly affecting NP characteristics is the drug loading capacity, and it is possible to optimize the components in a way that maximizes the drug encapsulation depending on their hydrophilic/hydrophobic characteristics. Besides, it may also allow the codelivery of more than one therapeutic agent simultaneously to enhance therapeutic efficacy through controlled delivery of diverse therapeutic cargoes with synergistic activities [28]. At the same time, it is required that NPs provide drug stability and protection, ensuring that early chemical and physical degradation of drugs does not take place before or after administration [29]. Additionally, these nanosystems increase drug solubility mainly of very hydrophobic drugs, which can contribute to the stability of compounds in solution. Hence, drug delivery systems show promise in terms of providing desirable delivery properties by altering the biopharmaceutic and pharmacokinetic properties of drugs, contributing toward extending their bioavailability [30]. However, the major limiting factor for the systemic use of NPs is their rapid clearance from blood circulation by the reticuloendothelial system (RES), which depends mainly on the NP's surface modifications. Therefore, a proper modification of a NP surface can represent a highly effective strategy for NP biodistribution, resulting in a longer circulating half-life [31]. In fact, surface covalent modification with the highly hydrophilic polyethylene glycol (PEG) has been explored as a possibility for decreasing opsonization with protein adsorption, slowing down the clearance of stealth NPs [32]. Moreover, due to their improved blood circulation time, PEGylated NPs accumulate more efficiently in the brain [33]. In fact, a desirable drug delivery system should remain most drug bound during blood circulation and quickly release the drugs on arrival at the brain. This would allow the release of drugs in a controlled manner, for example, through stimuli such as pH or temperature-sensitive controlled release systems [34]. Therefore, nanocarriers may provide a more prolonged, controlled, and slow release of the encapsulated drugs over time, directing the entrapped substance to a specific target.

Furthermore, the nanosize plays an important and helpful role while developing systems for brain drug delivery, considerably affecting the in vivo behavior and distribution of NPs. First of all, the ideal size to escape RES recognition is below 200 nm. However, NPs should also not be too small to prevent renal clearance [35]. Additionally, several studies have shown a clear inverse correlation among NP size and BBB penetration [36]. In fact, most of the successfully used NPs for the transport of drugs across the BBB present a size ranging from 100 to 300 nm [37]. Another important aspect relates with the surface charge. Indeed, the NP charge has a very important effect in its clearance. Neutral and zwitterionic NPs have a longer circulation time after intravenous administration, in contrast to negatively and positively charged NPs. Besides, since the BBB has a negative charge, positively charged carriers would be expected to be more efficient in drug delivery to the brain. However, the strong cell binding may prevent permeation across the barrier, resulting in bioaccumulation and cytotoxic effects [38]. On the other hand, anionic NPs are able to permeate the BBB without damage to the endothelium, with efficient uptake rates [37]. Therefore, most of the NPs described in the literature for brain delivery have moderate (between −1 to −15 mV) or high (between −15 to −45 mV) negative zeta potentials.

Additionally, NPs can enhance drug transport to the CNS by a number of different mechanisms. Since TJs play an important role in keeping the integrity of the BBB, temporary opening of the BBB is an applicable method to improve the permeability, allowing NPs to directly diffuse into the brain. Hence, modification of the tightness of TJs can be achieved by several physical and pharmacological methods, such as the use of borneol or alkylglycerols, which reversibly opens TJs, enhancing permeability and facilitating brain delivery of drugs [39]. An alternative is the use of NPs that can inhibit the transmembrane P-glycoprotein (Pgp) efflux system, for instance, by coating with surfactants such as polysorbate 80, allowing the drug to penetrate within the carrier into the endothelial cells by endocytosis, bypassing the drug efflux mechanism [40]. However, the passage of drugs through the BBB is extremely limited and the traditional systems fail in the delivery of drugs in therapeutic concentrations into the brain. Therefore, NPs can be further coated with ligands in order to be recognized actively by specific sites and cells through ligand–receptor interactions. To maximize the specificity, a surface marker or receptor should be overexpressed on brain target cells relative to nontarget ones. In general, NPs have large surface-to-volume ratios that contribute to their easy functionalization and, consequently, their intracellular penetration [41].

As a result, NPs have emerged as exciting systems for brain drug delivery due to the possibility that they can be modulated in terms of shape, size, surface charge, hydrophobicity, and coating, which may enhance the ability of NPs to improve the therapeutic agent stability in circulation, to escape the RES, to control the cargo release into the desired target site, and to enhance BBB penetration and cellular internalization [42]. The increase of drug concentration at the target site allows the reduction of nonspecific biodistribution, decreasing its toxic effects while potentiating the desired therapeutic effect in the brain [43].

Table 9.1 Examples of nanocarrier applications on brain drug delivery

NP type	Therapeutic agents	Administration route	In vitro/in vivo results	Refs.
Liposomes	5-HT, OVAL, TMP, PTX, AM	Intravenous, intranasal	- Brain uptake of liposomal 5-HT higher than its free form in rats - Liposomes loaded with OVAL widely distributed in rat brains - TMP-loaded liposomes able to effectively permeate in vitro and ex vivo BBB models - Display of the most promising results for the treatment of brain glioma tumors in rats by PTX plus AM-loaded liposomes	[44–47]
Lipid nanoparticles	RSV, STRS, PTX, bFGF, CUR	Intraperitoneal, intranasal	- Higher accumulation of RSV-loaded SLNs in the brain tissue of Wistar rats - Effective treatment for cerebral tuberculosis due to higher distribution of STRS-SLNs in the brain compared to free STRS - NPs able to cross the BBB in vitro model (hCMEC/D3) and the PTX-blocked cellular proliferation of glioblastoma cell lines - Efficient show of therapeutic effects by bFGF-loaded carriers on hemiparkinsonian rats - Enhanced permeation of CUR into the CNS when encapsulated in NLCs, suggesting better pharmacokinetic properties in mice	[48–52]

(Continued)

Table 9.1 (*Continued*)

NP type	Therapeutic agents	Administration route	In vitro/in vivo results	Refs.
Polymeric nanoparticles	DVF, MTX	Intravenous, intranasal	- Successful BBB passage for NPs fabricated with nonionic and cationic surfactants but not with anionic ones - Enhanced DVF concentration in the brain when PLGA-chitosan NPs administered intranasally, reducing depression symptoms - Higher circulation half-life and brain delivery of MTX when administered in encapsulated form	[53–55]
Cyclodextrins	LIG	Oral	- Permeability across bovine brain endothelial cell monolayers cocultured with rat glial cells: α-CDs > β-CDs >> γ-CDs - Higher bioavailability of LIG-loaded HP-β-CDs in rats compared to free LIG	[56, 57]
Dendrimers	n.i.	Intravenous	- Extended blood circulation time and enhanced accumulation in microglia and injured neuronal cells in canine brain injury model - Influence of cationic dendrimers on mitochondrial activity, apoptosis, and neuronal differentiation in human neural progenitor cells	[58, 59]
Silica nanoparticles	CUR, CHR	Intraperitoneal	- Accumulation of phytochemicals within cells after 2 h incubation of SiO_2 NPs with porcine olfactory bulb neuroblastoma cell line - SiO_2 NPs able to pass through in vitro BBB models; permeability mainly dependent on the size - Inflammation, oxidative stress, and DNA damage in mice and neurotoxicity in neuron cell lines caused by SiO_2 NPs	[60–63]

NP type	Therapeutic agents	Administration route	In vitro/in vivo results	Refs.
Magnetic nanoparticles	PTX, CA	Intravenous	- Prolonged blood circulation time, increased PTX concentration in rat brain tissues, and enhanced theranostic efficacy of MNPs - Effective nanoprobes for antioxidative and anticancer theranostics using CA-MNPs toward brain tumors in mice	[64, 65]
Gold nanoparticles	n.i.	Intracerebral	- Maximal uptake for 70 nm Au NPs in the mouse brain endothelial cell line - Au NP toxicity dependent on the dose and size of NPs administered in the cerebral cortex of female Sprague–Dawley rats	[66, 67]
Carbon dots	n.i.	Intravenous	- High distribution and accumulation in the glioma site of C6 glioma-bearing mice	[68]
Carbon nanotubes	n.i.	n.i.	- CNTs able to penetrate murine microvascular cerebral endothelial monolayers without damaging the cells - GO able to induce autophagy in neuroblastoma cell lines, conferring theranostic applications	[69, 70]

n.i., not indicated; 5-HT, serotonin; OVAL, ovalbumin; TMP, tetramethylpyrazine; PTX, paclitaxel; AM, artemether; RSV, resveratrol; STRS, streptomycin sulfate; bFGF, basic fibroblast growth factor; CUR, curcumin; SLNs, solid lipid nanoparticles; NLCs, nanostructured lipid carriers; DVF, desvenlafaxine; MTX, methotrexate; PLGA, poly(lactic-co-glycolic) acid; LIG, Z-ligustilide; CDs, cyclodextrins; HP-β-CDs, hydroxypropyl-β-CDs; CHR, chrysin; SiO_2, silicon dioxide or silica; CA, caffeic acid; MNPs, magnetic nanoparticles; Au NPs, gold nanoparticles; CNTs, carbon nanotubes; GO, graphene oxide nanoribbons

9.2.2 Nanocarriers for Brain Delivery

Several nanosystems have been exploited in order to accomplish the desired brain drug delivery, namely liposomes, lipid NPs, polymeric NPs, CDs, dendrimers, silica NPs, magnetic NPs, gold NPs, quantum dots (QDs), and carbon nanotubes (Table 9.1).

9.2.2.1 Liposomes

Liposomes are vesicles composed of unilamellar or multilamellar phospholipid bilayers surrounding aqueous compartments that can efficiently incorporate both lipophilic and hydrophilic compounds, representing a suitable system for enhancing the solubility and bioavailability of active compounds [71]. Liposomes represent vehicles particularly suitable for targeted drug delivery due to their biocompatibility, biodegradability, and nonimmunogenic properties. One of the disadvantages of these nanosystems is that they are rapidly captured by the RES and removed from blood circulation after intravenous injection, reducing their bioavailability [72]. One of the strategies introduced to overcome this problem consists in modifying the liposomal surface by introducing PEG in order to avoid uptake by macrophages, leading to extended blood circulation time in vivo [73]. Furthermore, an optimal liposomal formulation should have maximum drug encapsulation, limited leakage during storage time, and an effective release in the target place.

Afergan et al. developed negatively charged liposomes encapsulated with serotonin (5-HT), a BBB-impermeable neurological drug, and studied the distribution and brain uptake of liposome formulations [44]. The results were obtained 4 and 24 h after intravenous administration in rats and showed that brain uptake of liposomal serotonin was 1.7- to 2.0-fold higher than serotonin in solution. Xia et al. developed and studied liposomes encapsulated in tetramethylpyrazine (TMP), which is a natural product suitable for the treatment of cerebral ischemic diseases [46]. The study was performed in models that mimic the phospholipid bilayer membrane of the BBB and ex vivo subcutaneous-mucous-membranes. The TMP-loaded liposomes were around 150 nm in size and had highly negative charges of −30 mV. Cell viability studies indicate that formulations were biocompatible and permeability results indicate that TMP-loaded liposomes could effectively permeate both BBB models, being suitable for brain deliver of TMP. Li et al. studied liposomes loaded with paclitaxel (PTX) and artemether (AM) for the treatment of invasive brain glioma [47]. The studies were done on brain glioma cells in vitro and in brain glioma-bearing rats, using PTX as the anticancer drug and AM as the regulator of apoptosis and inhibitor of brain cancer vasculogenic mimicry channels. The liposomes were around 80 nm in size and showed entrapment efficiency of above 80% for both active compounds. The penetrating ability and inhibitory effect against brain glioma tumors suggested that functional targeting with PTX plus AM within liposomes displayed the most promising results for the treatment of brain tumors in rats. Migliore et al. used another strategy with cationic liposomes for brain delivery of ovalbumin (OVAL) by intranasal route in rats [45]. Liposomes were loaded with Alexa 488-OVAL, and the delivery was measured by fluorescence microscopy. The results revealed that Alexa 488-OVAL became widely distributed throughout the brain and peripheral tissues and the highest brain concentration was achieved at the shortest time point (1 h), providing a novel noninvasive strategy for the delivery of neuroactive proteins to the brain for the treatment of CNS disorders.

9.2.2.2 Lipid nanoparticles

Solid lipid nanoparticles (SLNs) and nanostructured lipid carriers (NLCs) are drug delivery systems based on a solid lipid matrix at room and body temperature. While SLNs are made solely of solid lipids, NLCs are composed of a mixture of solid and liquid lipids that creates a less ordered lipid matrix and reduces drug leakage during storage [74]. Lipid nanoparticles have a large number of advantages as drug carriers due to their good biocompatible properties, high drug loading capacity, and sustained drug release behavior. Gastaldi et al. suggested that drug brain bioavailability may be improved by the use of polysorbates in the preparation of SLNs, which may induce endocytosis and inhibition of efflux systems [75]. The authors have also reported that modification of SLNs with surfactants may prevent their rapid recognition by the RES [75]. Jose et al. developed glyceryl behenate-based SLNs incorporating resveratrol (RSV) toward brain targeting [48]. The reported RSV-loaded SLNs demonstrated a size of around 250 nm, a zeta potential of 25 mV, and encapsulation efficiency of 30%. X-ray diffraction and differential scanning calorimetry analyses revealed that lipid matrix of drug-loaded SLNs was in a disordered crystalline phase as supposed. The in vivo assay measured the biodistribution in Wistar rats and demonstrated a higher accumulation of RSV-loaded SLNs in the brain tissue, which may be a promising strategy for treating brain neoplastic diseases. Kumar et al. designed streptomycin sulfate (STRS) loaded in SLNs (STRS-SLNs) for noninvasive intranasal delivery into the brain [49]. STRS-SLNs were developed using a nanocolloidal aqueous dispersion technique, achieving particles of 140 nm and with significant entrapment efficiency of 50%. Noninvasive intranasal administration of STRS-SLNs in mice revealed a 3-fold enhancement of STRS in the brain. The authors reported that the presence of Tween 80 may be responsible for the inhibition of Pgp efflux, resulting in an effective treatment for cerebral tuberculosis due to a high relative distribution in the brain tissues in comparison to free STRS. Chirio et al. developed positively charged SLNs encapsulated with PTX for the treatment of glioblastoma [50]. The NPs had mean diameters in the range of 300 to 600 nm, zeta potentials between 8 and 20 mV, and encapsulation efficiency of 25% to 90%. To obtain NPs with a positive charge they introduced charged molecules into the formulation, such as stearylamine and glycol chitosan. In vitro assays were performed to evaluate drug release, permeability across the BBB, and cytotoxicity and the results suggested that PTX had a slow release from SLNs and NPs were able to cross the BBB in a time-dependent manner. The results also showed that PTX was able to block cellular proliferation, in a dose-dependent manner, demonstrating a high cytotoxic activity of all PTX-loaded SLNs when compared with the free drug. Zhao et al. developed new type of phospholipid-based gelatin NLCs encapsulating bFGF to target the brain via nasal administration [51]. These gelatin vehicles are effective carriers with a size around 170 nm, –20 mV zeta potential, and entrapment efficiency higher than 85%, and they have revealed therapeutic effects on hemiparkinsonian rats. Puglia et al. designed curcumin (CUR)-loaded NLCs for inducing histone hypoacetylation in the CNS after intraperitoneal administration in mice [52]. NPs presented an average particle size of around 160 nm, a zeta potential of 15 mV, and encapsulation efficiency of 87%. A better permeation of CUR when encapsulated in NLCs suggested a novel approach to upgrading the pharmacokinetics of this compound, allowing a better release into the CNS.

9.2.2.3 Polymeric nanoparticles

Polymeric nanoparticles are composed of amphiphilic block copolymers that have hydrophobic cores and hydrophilic shells in their structure. The core allows the entrapment of the bioactive compound, and the shell increases the stability of the nanoparticle [76]. The major requirement for NPs to deliver drugs inside the brain is their rapid biodegradability. For this reason, poly(butyl cyanoacrylate) (PBCA), poly(lactic acid) (PLA), poly(lactic-co-glycolic) acid (PLGA), and chitosan appear to be the materials of choice for the preparation of polymeric nanoparticles. In fact, these polymers create NPs that are biodegradable and biocompatible, have good stability and low toxicity, and can be manufactured by simple methods [77]. Voigt et al. designed several fluorescent PBCA NPs and evaluated their passage over the blood–retina barrier, which is a model of the BBB [53]. These formulations were injected into rats and real-time imaging of retinal blood vessels and retinal tissue was carried out with in vivo confocal neuroimaging (ICON). It was possible to obtain successful BBB passage with subsequent cellular labeling with NPs fabricated with nonionic surfactants and cationic stabilizers but not with anionic compounds. Therefore, the authors concluded that NPs can be designed to specifically enhance drug delivery to the brain or prevent brain penetration in order to reduce unwanted psychoactive effects of drugs in the CNS. Tong et al. developed desvenlafaxine (DVF)-loaded PLGA-chitosan NPs and evaluated their efficacy and distribution of the drug in the brain [54]. The main purpose of this study was to understand the potential of mucoadhesive PLGA-chitosan NPs for brain delivery in the treatment of depression in rodents. The optimized NPs presented a size of around 170 nm and a zeta potential of 35 mV and were able to encapsulate 76% of DVF. Pharmacokinetic results proved a 2.5-fold higher DVF concentration in the brain when PLGA-chitosan NPs were administered intranasally compared to intravenous administration. Jain et al. developed PLA-thermosensitive NPs incorporating MTX for targeting brain tumor via intranasal administration [55]. The formulated NPs' size and zeta potential was found to be around 350 nm and 25 mV, respectively. The entrapment efficiency for MTX NPs was found to be 59%. These vehicles were capable of delivering MTX to the brain, reaching a maximum concentration 2 h after the administration through the nasal route. The pharmacokinetic parameters demonstrated an enhancement in the circulation half-life of MTX when administered intranasally in an encapsulated form.

9.2.2.4 Cyclodextrins

Cyclodextrins (CDs) are composed of glucose units connected by glycosidic linkages to create a cylinder-shaped structure that contains a hydrophilic external surface and a hydrophobic cavity. These ring-shaped molecules have numerous hydroxyl moieties, all facing outside, which makes them highly hydrophilic. At the same time, their hydrophobic cavity enables the encapsulation of hydrophobic compounds, enhancing their solubility in water [78]. Monnaert et al. studied CDs' permeability across bovine brain endothelial cell monolayers cocultured with rat glial cells and concluded that permeation was higher for α-CDs, followed by β-CDs, and finally γ-CDs [56]. Lu et al. developed hydroxypropyl-β-CD (HP-β-CD) complexes encapsulated with Z-ligustilide (LIG) in order to improve its stability and oral bioavailability [57]. LIG is a lactone used for the treatment of cerebrovascular diseases, preventing brain damage and presenting neuroprotective effects in AD. However, its oral formulation is strongly restricted due to poor water solubility and stability. The prepared formulations showed a size range between 300 and 400 nm and were able to entrap approximately 78% of LIG. Bioavailability studies showed a higher plasma concentration of LIG-loaded HP-β-CDs in rats when compared to LIG alone, proving the efficacy of the complexation with CDs for improving LIG oral bioavailability.

9.2.2.5 Dendrimers

Each dendrimer has a highly branched 3D architecture formed by a series of chemical shells, enclosing a small core [79]. These nanovehicles possess the ability to facilitate the transport of bioactive compounds across various cell membranes or biological barriers, which depends on the surface groups and molecular mass of the dendrimers. The two main advantages in using dendrimers for the delivery of drugs are that they have high biocompatibility and great water solubility. Different varieties of dendrimers have been exploited regarding drug delivery and imaging, including polyamidoamine (PAMAM), polyetherhydroxylamine (PEHAM), and poly(propylene imine) (PPI). Among these, PAMAM dendrimers have been the most studied due to their unique structures and properties [80]. Zhang et al. designed generation-6 (G6) hydroxyl PAMAM dendrimers for the evaluation of the effect of dendrimer size on CNS penetration after intravenous administration in a canine brain injury model [58]. G6 dendrimers showed sizes of around 7 nm and demonstrated extended blood circulation times and enhanced accumulation in the injured brain when compared to G4 dendrimers, which were undetectable in the brain until 48 h after administration. The results also suggest that G6 dendrimers specifically target microglia and injured neuronal cells, which are the major cell types associated with neuroinflammation. Zheng et al. designed amino-PAMAM dendrimers and evaluated their effect on cytotoxicity and neuronal differentiation using human neural progenitor cells [59]. The in vitro assays showed that these cationic PAMAM dendrimers might affect mitochondrial activity, apoptosis, and neuronal differentiation.

9.2.2.6 Silica nanoparticles

Silica is a biocompatible material often used for producing inorganic nanoparticles. Silica nanoparticles (SiO_2 NPs) are highly porous in nature, with high structural stability and easy surface functionalization, which makes them highly adaptable for creating excellent tools for application in CNS drug delivery and imaging [81, 82]. Recent data have shown that SiO_2 NPs can pass through in vitro BBB models and proved that the permeability of these NPs was mainly dependent on size [61]. Lungare et al. studied the effect of mesoporous silica NPs, loaded with phytochemicals like CUR and chrysin, for nose-to-brain olfactory drug delivery [60]. This study was based on in vitro assays using NPs of approximately 200 nm. The results showed the accumulation of fluorescein isothiocyanate (FITC)-loaded NPs within cells after 2 h incubation, making them useful carriers for poorly soluble phytochemicals. Successful intranasal delivery of mesoporous silica NPs into the olfactory mucosa significantly enhances their application in nose-to-brain delivery of phytochemicals. Nevertheless, some studies reveal that SiO_2 NPs cause inflammation, oxidative stress, DNA damage, and neurotoxicity in brain, highlighting the need for exhaustive evaluation of silica NPs before their use in humans [62, 63].

9.2.2.7 Magnetic nanoparticles

Magnetic nanoparticles (MNPs) have unique magnetic properties that create an attractive platform for several applications, for example, as carriers for disease diagnosis, for magnetic resonance imaging (MRI), and for drug delivery [83]. As a therapeutic tool, MNPs have been studied as prospective carriers in order to overcome limitations associated with systemic distribution of conventional chemotherapeutic agents [84]. The high magnetic susceptibility of a MNP's core facilitates noninvasive manipulation through magnetic fields and external magnet assistance to enhance brain targeting. Hence, MNPs have been implemented for disease diagnosis and therapy combined, the so-called nanotheranostic applications [85]. Dilnawaz et al. investigated the transport of non-surfactant-based PTX-loaded MNPs through the BBB in rat models [64]. The aim of this project was to develop a potential therapeutic method for the treatment of glioma. The nanosystems showed a size of around 200 nm, a zeta potential of 20 mV, and encapsulation efficacy of 95%. The results revealed prolonged blood circulation time in vivo, as well as a significant increase of the drug amount in rat brain tissues compared to native PTX. The ability of MNPs to be an effective MRI contrast enhancement agent was illustrated using in vivo rat brain and liver MRI images. The enhanced theranostic efficacy of PTX MNPs in glioblastoma cell lines revealed a useful candidate for the treatment of brain tumors. Richard et al. have also developed iron oxide NPs as nanoprobes for antioxidative theranostics toward brain tumors [65]. The magnetic Fe_2O_3 NPs were used to encapsulate caffeic acid (CA). The preliminary in vitro studies showed high potential of CA as an anticancer drug when associated with MNPs. In fact, these nanosystems were able to permeate the cells in a dose- and time-dependent manner and were able to buffer the reactive oxygen species within the cells. The in vivo MRI acquisitions showed a negative contrast enhancement in mice after intravenous injection, demonstrating passive targeting of brain tumors by these nanosystems.

9.2.2.8 Gold nanoparticles

Gold NPs (Au NPs) exhibit several advantageous properties since gold material is highly unreactive and chemically inert, displaying high potential for drug delivery due to its nontoxic, nonimmunogenic, and biocompatible characteristics [86]. Therefore, this nanomaterial is used in several applications, for example, in cell labeling and antitumor therapy, for drug delivery, and as a photothermal conducting agent [87]. Moreover, Au NPs are able to cross the BBB for reaching the brain [88]. Au NPs' properties influence particle biodistribution and accumulation in vivo, and many authors have reported that particle size strongly influences the in vivo fate of Au NPs. Shilo et al. performed a study with a mouse brain endothelial cell line, which indicated maximal uptake for 70 nm NPs, while the smallest ones used in the study, of 20 nm, had the highest carrying capacity [66]. This research highlights the fact that the number of cargo molecules is the most important fact regarding CNS permeation, rather than the amount of gold, unless the aim is to use the gold itself for imaging. Lee et al. injected Au NPs into the cerebral cortex of rats in order to evaluate the expression of nestin, which is mainly activated in astrocytes after CNS damage [67]. The authors found that Au NP toxicity is dependent on the dose and size of NPs administrated; they came to this conclusion because they found a larger quantity of small Au NPs at the nestin expression site than large-sized particles on short-term implantation. Several studies have also reported that neurotoxicity depends on the administration route. For instance, tail vein injection showed lower toxicity compared to oral administration and intraperitoneal injection [89]. However, care should be taken prior to use since there is some evidence that Au NPs with different characteristics are able to promote neurotoxic effects and neuroinflammation in different CNS regions [90].

9.2.2.9 Quantum dots

QDs, also known as fluorescent semiconductor NPs, represent versatile platforms for designing and engineering nanoparticle-based drug delivery systems [91]. QDs can be used not only as vehicles to deliver therapeutic agents but also as imaging agents, called theranostics [92]. There is some evidence that QDs provide an unprecedented opportunity for transmigrating across the BBB and reaching the brain parenchyma depending on their nanoscale sizes, which has contributed to the development of theranostic nanosystems for various neurological disorders [93]. The major drawback in designing QD-based imaging techniques is their poor stability in an aqueous environment and in biological systems [94, 95]. Some authors have proven that encapsulation of QDs with amphiphilic copolymers and phospholipids represents a solution for increasing their solubility and stability [96, 97]. Another drawback is the toxicity associated with QDs to cells and organisms, for instance, due to Cd^{2+} release from the CdSe core [98, 99]. Hence, carbon-based QDs, also known as carbon dots (CDs), have attracted increasing attention due to their unique properties, such as low toxicity and great biocompatibility [100, 101]. Ruan et al. developed a simple one-step method for producing fluorescent CDs and tested their potential application in noninvasive glioma imaging [68]. Several in vitro assays have demonstrated that CDs possess great serum stability and low cytotoxicity and are taken up by C6 glioma cells in a time- and concentration-dependent manner. The in vivo assay proved that CDs show high distribution and accumulation in the glioma site of the brain with a high intensity, while in the normal brain tissue CD distribution occurs with a low intensity.

9.2.2.10 Carbon nanotubes

Carbon nanotubes (CNTs) are cylindrical nanosystems composed of a continuous, unbroken, and hexagonal mesh of carbon atoms [102, 103]. In particular, CNTs display singular physical and chemical properties, which allow conjugation with an extensive amount of therapeutic and imaging agents [104–106]. Furthermore, these nanosystems have a high length-to-diameter ratio, which enables them to efficiently permeate biological membranes and accumulate into their intracellular compartments [107]. Despite CNTs' great potential, some authors have reported some associated toxicity [108]. Therefore, their lengths and diameters can be adjusted in order to increase their biocompatibility and bioavailability [109]. CNTs have the capacity to cross the BBB by transcytosis and can be loaded and deliver high doses of drugs into the therapeutic site. Moreover, their intrinsic thermal properties show a great potential for real-time tracking and photothermal applications [110]. Shityakov et al. demonstrated that FITC-labeled CNTs were able to penetrate murine microvascular cerebral endothelial monolayers over a 48 h period without bringing any damage to the cells [69]. Mari et al. developed graphene oxide (GO) nanoribbons and demonstrated that small doses of GO were able to induce autophagy within the first 48 h of exposure in neuroblastoma cell lines [70]. Therefore, the authors believe that GO nanoribbons can act with great potential as drug delivery systems for theranostic applications in the treatment of neuroblastoma.

9.3 Concluding Remarks

Brain drug delivery still represents a great challenge to be pursued. Many aging disorders and brain tumors require drugs acting on the CNS, and the number of patients looking for efficient treatments is constantly increasing. In fact, current therapies are far from being satisfactory, mainly due to the difficulty for the drugs to cross the most complex, selective, and impermeable barrier, the BBB. In this context, a range of nanosystems based on liposomes, lipid NPs, polymeric NPs, CDs, dendrimers, silica NPs, magnetic NPs, gold NPs, QDs, and CNTs have been considered as promising systems for brain drug delivery and in vivo imaging. Several examples show the applicability of NPs to enhance the therapeutic agent stability in circulation, to escape the RES, to control the cargo release into the desired target site, and to enhance BBB penetration and cellular internalization, which may improve current therapies for CNS disorders.

Acknowledgments

This work received financial support from the European Union (FEDER funds) and National Funds (*Fundação para a Ciência e a Tecnologia* and *Ministério da Educação e Ciência* [FCT/MEC]) under the partnership agreement PT2020 UID/MULTI/04378/2013 - POCI/01/0145/FEDER / 007728. Ana Rute Neves also thanks ICETA for her postdoctoral grant (FOOD_RL3_PHD_GABAI_02) under the project NORTE-01-0145-FEDER-000011.

References

1. Pardridge, W.M. Why is the global CNS pharmaceutical market so under-penetrated? *Drug Discov Today*, 2002, **7**(1):5–7.
2. Croquelois, A., et al. Diseases of the nervous system: patients' aetiological beliefs. *J Neurol Neurosurg Psychiatry*, 2005, **76**(4):582–584.
3. Sezgin, Z., et al. Alzheimer's disease and epigenetic diet. *Neurochem Int*, 2014, **78**:105–116.
4. Hoozemans, J.J., et al. Activation of the unfolded protein response is an early event in Alzheimer's and Parkinson's disease. *Neurodegener Dis*, 2012, **10**(1–4):212–215.
5. Meraz-Rios, M.A., et al. Inflammatory process in Alzheimer's disease. *Front Integr Neurosci*, 2013, **7**:59.
6. Taylor, J.M., et al. Neuroinflammation and oxidative stress: co-conspirators in the pathology of Parkinson's disease. *Neurochem Int*, 2013, **62**(5):803–819.
7. Ostrom, Q.T., et al. The epidemiology of glioma in adults: a "state of the science" review. *Neuro Oncol*, 2014, **16**(7):896–913.
8. Mrugala, M.M. Advances and challenges in the treatment of glioblastoma: a clinician's perspective. *Discov Med*, 2013, **15**(83):221–230.
9. Rape, A., et al. Engineering strategies to mimic the glioblastoma microenvironment. *Adv Drug Deliv Rev*, 2014, **79–80**:172–183.
10. Neuwelt, E., et al. Strategies to advance translational research into brain barriers. *Lancet Neurol*, 2008, **7**(1):84–96.
11. Abbott, N.J. Blood-brain barrier structure and function and the challenges for CNS drug delivery. *J Inherit Metab Dis*, 2013, **36**(3):437–449.
12. Pardridge, W. Targeting neurotherapeutic agents through the blood-brain barrier. *Arch Neurol*, 2002, **59**:35–40.
13. Persidsky, Y., et al. Blood-brain barrier: structural components and function under physiologic and pathologic conditions. *J Neuroimmune Pharmacol*, 2006, **1**(3):223–236.
14. Liebner, S., et al. *The Blood-Brain Barrier and Its Microenvironment: Basic Physiology to Neurological Diseases*, Vol. 1. Taylor and Francis Group, Boca Raton, FL, 2005.
15. Hawkins, B.T., et al. The blood-brain barrier/neurovascular unit in health and disease. *Pharmacol Rev*, 2005, **57**(2):173–185.

16. Wolburg, H., et al. Brain endothelial cells and the glio-vascular complex. *Cell Tissue Res*, 2009, **335**(1):75–96.

17. Forster, C. Tight junctions and the modulation of barrier function in disease. *Histochem Cell Biol*, 2008, **130**(1):55–70.

18. Abbott, N.J., et al. Astrocyte-endothelial interactions at the blood-brain barrier. *Nat Rev Neurosci*, 2006, **7**(1):41–53.

19. Cardoso, F.L., et al. Looking at the blood-brain barrier: molecular anatomy and possible investigation approaches. *Brain Res Rev*, 2010, **64**(2):328–363.

20. Thanabalasundaram, G., et al. The impact of pericytes on the blood-brain barrier integrity depends critically on the pericyte differentiation stage. *Int J Biochem Cell Biol*, 2011, **43**(9):1284–1293.

21. Hori, S., et al. A pericyte-derived angiopoietin-1 multimeric complex induces occludin gene expression in brain capillary endothelial cells through Tie-2 activation in vitro. *J Neurochem*, 2004, **89**(2):503–513.

22. Choi, Y.K., et al. Blood-neural barrier: its diversity and coordinated cell-to-cell communication. *BMB Rep*, 2008, **41**(5):345–352.

23. Nagpal, K., et al. Drug targeting to brain: a systematic approach to study the factors, parameters and approaches for prediction of permeability of drugs across BBB. *Expert Opin Drug Deliv*, 2013, **10**(7):927–955.

24. Pardridge, W.M. The blood-brain barrier: bottleneck in brain drug development. *NeuroRx*, 2005, **2**(1):3–14.

25. Badruddoja, M.A., et al. Improving the delivery of therapeutic agents to CNS neoplasms: a clinical review. *Front Biosci*, 2006, **11**:1466–1478.

26. Vyas, T.K., et al. Intranasal drug delivery for brain targeting. *Curr Drug Deliv*, 2005, **2**(2):165–175.

27. Bahadar, H., et al. Toxicity of nanoparticles and an overview of current experimental models. *Iran Biomed J*, 2016, **20**(1):1–11.

28. Godsey, M.E., et al. Materials innovation for co-delivery of diverse therapeutic cargos. *RSC Adv*, 2013, **3**(47):24794–24811.

29. Devapally, H., et al. Role of nanotechnology in pharmaceutical development. *J Pharm Sci*, 2007, **96**:2547–2565.

30. Alyautdin, R., et al. Nanoscale drug delivery systems and the blood–brain barrier. *Int J Nanomed*, 2014, **9**:795–811.

31. Kreuter, J. Nanoparticulate systems for brain delivery of drugs. *Adv Drug Deliv Rev*, 2001, **47**(1):65–81.

32. Walkey, C.D., et al. Nanoparticle size and surface chemistry determine serum protein adsorption and macrophage uptake. *J Am Chem Soc*, 2012, **134**(4):2139–2147.

33. Nance, E.A., et al. A dense poly(ethylene glycol) coating improves penetration of large polymeric nanoparticles within brain tissue. *Sci Transl Med*, 2012, **4**(149):149ra119.

34. Doane, T.L., et al. The unique role of nanoparticles in nanomedicine: imaging, drug delivery and therapy. *Chem Soc Rev*, 2012, **41**(7):2885–2911.

35. Choi, H.S., et al. Renal clearance of nanoparticles. *Nat Biotechnol*, 2007, **25**:1165–1170.

36. Hanada, S., et al. Cell-based in vitro blood-brain barrier model can rapidly evaluate nanoparticles' brain permeability in association with particle size and surface modification. *Int J Mol Sci*, 2014, **15**(2):1812–1825.

37. Wohlfart, S., et al. Transport of drugs across the blood-brain barrier by nanoparticles. *J Controlled Release*, 2012, **161**(2):264–273.

38. Lockman, P.R., et al. Nanoparticle surface charges alter blood-brain barrier integrity and permeability. *J Drug Target*, 2004, **12**(9–10):635–641.

39. Gao, H. Progress and perspectives on targeting nanoparticles for brain drug delivery. *Acta Pharm Sin B*, 2016, **6**(4):268–286.

40. Miller, D.S., et al. Modulation of P-glycoprotein at the blood-brain barrier: opportunities to improve central nervous system pharmacotherapy. *Pharmacol Rev*, 2008, **60**(2):196–209.

41. Juillerat, J. The targeted delivery of cancer drugs across the blood brain barrier: Chemical modifications of drugs or drug nanoparticles. *Drug Discov Today*, 2008, **13**:1099–1106.

42. Singh, R., et al. Nanoparticle-based targeted drug delivery. *Exp Mol Pathol*, 2009, **86**(3):215–223.

43. Lockman, P.R., et al. Nanoparticle technology for drug delivery across the blood-brain barrier. *Drug Dev Ind Pharm*, 2002, **28**(1):1–13.

44. Afergan, E., et al. Delivery of serotonin to the brain by monocytes following phagocytosis of liposomes. *J Controlled Release*, 2008, **132**(2):84–90.

45. Migliore, M.M., et al. Brain delivery of proteins by the intranasal route of administration: a comparison of cationic liposomes versus aqueous solution formulations. *J Pharm Sci*, 2010, **99**(4):1745–1761.

46. Xia, H., et al. Investigating the passage of tetramethylpyrazine-loaded liposomes across blood-brain barrier models in vitro and ex vivo. *Mater Sci Eng C Mater Biol Appl*, 2016, **69**:1010–1017.

47. Li, X.Y., et al. Multifunctional liposomes loaded with paclitaxel and artemether for treatment of invasive brain glioma. *Biomaterials*, 2014, **35**(21):5591–5604.

48. Jose, S., et al. In vivo pharmacokinetics and biodistribution of resveratrol-loaded solid lipid nanoparticles for brain delivery. *Int J Pharm*, 2014, **474**(1–2):6–13.

49. Kumar, M., et al. Intranasal delivery of streptomycin sulfate (STRS) loaded solid lipid nanoparticles to brain and blood. *Int J Pharm*, 2014, **461**(1–2):223–233.

50. Chirio, D., et al. Positive-charged solid lipid nanoparticles as paclitaxel drug delivery system in glioblastoma treatment. *Eur J Pharm Biopharm*, 2014, **88**(3):746–758.

51. Zhao, Y.Z., et al. Gelatin nanostructured lipid carriers-mediated intranasal delivery of basic fibroblast growth factor enhances functional recovery in hemiparkinsonian rats. *Nanomedicine*, 2014, **10**(4):755–764.

52. Puglia, C., et al. Curcumin loaded NLC induces histone hypoacetylation in the CNS after intraperitoneal administration in mice. *Eur J Pharm Biopharm*, 2012, **81**(2):288–293.

53. Voigt, N., et al. Surfactants, not size or zeta-potential influence blood-brain barrier passage of polymeric nanoparticles. *Eur J Pharm Biopharm*, 2014, **87**(1):19–29.

54. Tong, G.-F., et al. Development and evaluation of Desvenlafaxine loaded PLGA-chitosan nanoparticles for brain delivery. *Saudi Pharm J*, 2016, in press.

55. Jain, D.S., et al. Thermosensitive PLA based nanodispersion for targeting brain tumor via intranasal route. *Mater Sci Eng C Mater Biol Appl*, 2016, **63**:411–421.

56. Monnaert, V., et al. Behavior of alpha-, beta-, and gamma-cyclodextrins and their derivatives on an in vitro model of blood-brain barrier. *J Pharmacol Exp Ther*, 2004, **310**(2):745–751.

57. Lu, Y., et al. Complexation of Z-ligustilide with hydroxypropyl-beta-cyclodextrin to improve stability and oral bioavailability. *Acta Pharm*, 2014, **64**(2):211–222.

58. Zhang, F., et al. Generation-6 hydroxyl PAMAM dendrimers improve CNS penetration from intravenous administration in a large animal brain injury model. *J Controlled Release*, 2017, **249**:173–182.

59. Zeng, Y., et al. Effects of PAMAM dendrimers with various surface functional groups and multiple generations on cytotoxicity and

neuronal differentiation using human neural progenitor cells. *J Toxicol Sci*, 2016, **41**(3):351–370.

60. Lungare, S., et al. Phytochemical-loaded mesoporous silica nanoparticles for nose-to-brain olfactory drug delivery. *Int J Pharm*, 2016, **513**(1–2):280–293.
61. Nemmar, A., et al. Interaction of amorphous silica nanoparticles with erythrocytes in vitro: role of oxidative stress. *Cell Physiol Biochem*, 2014, **34**(2):255–265.
62. Nemmar, A., et al. Oxidative stress, inflammation, and DNA damage in multiple organs of mice acutely exposed to amorphous silica nanoparticles. *Int J Nanomed*, 2016, **11**:919–928.
63. Xie, H., et al. Silica nanoparticles induce alpha-synuclein induction and aggregation in PC12-cells. *Chem Biol Interact*, 2016, **258**:197–204.
64. Dilnawaz, F., et al. The transport of non-surfactant based paclitaxel loaded magnetic nanoparticles across the blood brain barrier in a rat model. *Biomaterials*, 2012, **33**(10):2936–2951.
65. Richard, S., et al. Antioxidative theranostic iron oxide nanoparticles toward brain tumors imaging and ROS production. *ACS Chem Biol*, 2016, **11**(10):2812–2819.
66. Shilo, M., et al. The effect of nanoparticle size on the probability to cross the blood-brain barrier: an in-vitro endothelial cell model. *J Nanobiotechnol*, 2015, **13**:19.
67. Lee, U., et al. Cytotoxicity of gold nanoparticles in human neural precursor cells and rat cerebral cortex. *J Biosci Bioeng*, 2016, **121**(3):341–344.
68. Ruan, S., et al. A simple one-step method to prepare fluorescent carbon dots and their potential application in non-invasive glioma imaging. *Nanoscale*, 2014, **6**(17):10040–10047.
69. Shityakov, S., et al. Blood-brain barrier transport studies, aggregation, and molecular dynamics simulation of multiwalled carbon nanotube functionalized with fluorescein isothiocyanate. *Int J Nanomed*, 2015, **10**:1703–1713.
70. Mari, E., et al. Graphene oxide nanoribbons induce autophagic vacuoles in neuroblastoma cell lines. *Int J Mol Sci*, 2016, **17**(12):1995.
71. Eloy, J.O., et al. Liposomes as carriers of hydrophilic small molecule drugs: Strategies to enhance encapsulation and delivery. *Colloids Surf B*, 2014, **123**:345–363.

72. Chen, J., et al. Thermosensitive liposomes with higher phase transition temperature for targeted drug delivery to tumor. *Int J Pharm*, 2014, **475**(1–2):408–415.

73. Immordino, M.L., et al. Stealth liposomes: review of the basic science, rationale, and clinical applications, existing and potential. *Int J Nanomed*, 2006, **1**(3):297–315.

74. Naseri, N., et al. Solid lipid nanoparticles and nanostructured lipid carriers: structure, preparation and application. *Adv Pharm Bull*, 2015, **5**(3):305–313.

75. Gastaldi, L., et al. Solid lipid nanoparticles as vehicles of drugs to the brain: current state of the art. *Eur J Pharm Biopharm*, 2014, **87**(3):433–444.

76. Zhang, T., et al. Micellar emulsions composed of mPEG-PCL/MCT as novel nanocarriers for systemic delivery of genistein: a comparative study with micelles. *Int J Nanomed*, 2015, **10**:6175–6184.

77. Nagpal, K., et al. Chitosan nanoparticles: a promising system in novel drug delivery. *Chem Pharm Bull (Tokyo)*, 2010, **58**(11):1423–1430.

78. Bonnet, V., et al. Cyclodextrin nanoassemblies: a promising tool for drug delivery. *Drug Discov Today*, 2015, **20**(9):1120–1126.

79. Xu, L., et al. Dendrimer advances for the central nervous system delivery of therapeutics. *ACS Chem Neurosci*, 2014, **5**(1):2–13.

80. Xu, L., et al. Dendrimer-based RNA interference delivery for cancer therapy, in *Tailored Polymer Architectures for Pharmaceutical and Biomedical Applications*. American Chemical Society, 2013, pp. 197–213.

81. Nehoff, H., et al. Nanomedicine for drug targeting: strategies beyond the enhanced permeability and retention effect. *Int J Nanomed*, 2014, **9**:2539–2555.

82. Wang, Y., et al. Mesoporous silica nanoparticles in drug delivery and biomedical applications. *Nanomed Nanotechnol Biol Med*, 2015, **11**(2):313–327.

83. Gamarra, L.F., et al. Biocompatible superparamagnetic iron oxide nanoparticles used for contrast agents: a structural and magnetic study. *J Magn Magn Mater*, 2005, **289**:439–441.

84. Dilnawaz, F., et al. Dual drug loaded superparamagnetic iron oxide nanoparticles for targeted cancer therapy. *Biomaterials*, 2010, **31**(13):3694–3706.

85. Arruebo, M., et al. Magnetic nanoparticles for drug delivery. *Nano Today*, 2007, **2**(3):22–32.

86. Mieszawska, A.J., et al. Multifunctional gold nanoparticles for diagnosis and therapy of disease. *Mol Pharm*, 2013, **10**(3):831–847.

87. Male, D., et al. Gold nanoparticles for imaging and drug transport to the CNS. *Int Rev Neurobiol*, 2016, **130**:155–198.

88. Kim, J.H., et al. Intravenously administered gold nanoparticles pass through the blood-retinal barrier depending on the particle size, and induce no retinal toxicity. *Nanotechnology*, 2009, **20**(50):505101.

89. Zhang, X.D., et al. Toxicologic effects of gold nanoparticles in vivo by different administration routes. *Int J Nanomed*, 2010, **5**:771–781.

90. Leite, P.E.C., et al. Hazard effects of nanoparticles in central nervous system: Searching for biocompatible nanomaterials for drug delivery. *Toxicol in Vitro*, 2015, **29**(7):1653–1660.

91. Barreto, J.A., et al. Nanomaterials: applications in cancer imaging and therapy. *Adv Mater*, 2011, **23**(12):H18–H40.

92. Xie, J., et al. Nanoparticle-based theranostic agents. *Adv Drug Deliv Rev*, 2010, **62**(11):1064–1079.

93. Xu, G., et al. Theranostic quantum dots for crossing blood-brain barrier in vitro and providing therapy of HIV-associated encephalopathy. *Front Pharmacol*, 2013, **4**:140.

94. Xu, G., et al. Bioconjugated quantum rods as targeted probes for efficient transmigration across an in vitro blood-brain barrier. *Bioconjug Chem*, 2008, **19**(6):1179–1185.

95. Kloepfer, J.A., et al. Photophysical properties of biologically compatible CdSe quantum dot structures. *J Phys Chem B*, 2005, **109**(20):9996–10003.

96. Gao, X., et al. In vivo cancer targeting and imaging with semiconductor quantum dots. *Nat Biotechnol*, 2004, **22**(8):969–976.

97. Dubertret, B., et al. In vivo imaging of quantum dots encapsulated in phospholipid micelles. *Science*, 2002, **298**(5599):1759–1762.

98. Bruneau, A., et al. In vitro immunotoxicology of quantum dots and comparison with dissolved cadmium and tellurium. *Environ Toxicol*, 2015, **30**(1):9–25.

99. Tang, Y., et al. The role of surface chemistry in determining in vivo biodistribution and toxicity of CdSe/ZnS core-shell quantum dots. *Biomaterials*, 2013, **34**(34):8741–8755.

100. Cao, L., et al. Carbon dots for multiphoton bioimaging. *J Am Chem Soc*, 2007, **129**(37):11318–11319.

101. Li, W., et al. Simple and green synthesis of nitrogen-doped photoluminescent carbonaceous nanospheres for bioimaging. *Angew Chem Int Ed Engl*, 2013, **52**(31):8151–8155.

102. Wenrong, Y., et al. Carbon nanotubes for biological and biomedical applications. *Nanotechnology*, 2007, **18**(41):412001.

103. Bianco, A., et al. Applications of carbon nanotubes in drug delivery. *Curr Opin Chem Biol*, 2005, **9**(6):674–679.

104. Heister, E., et al. Triple functionalisation of single-walled carbon nanotubes with doxorubicin, a monoclonal antibody, and a fluorescent marker for targeted cancer therapy. *Carbon*, 2009, **47**(9):2152–2160.

105. Lay, C.L., et al. Delivery of paclitaxel by physically loading onto poly(ethylene glycol) (PEG)-graft-carbon nanotubes for potent cancer therapeutics. *Nanotechnology*, 2010, **21**(6):065101.

106. Peigney, A., et al. Specific surface area of carbon nanotubes and bundles of carbon nanotubes. *Carbon*, 2001, **39**(4):507–514.

107. Pantarotto, D., et al. Functionalized carbon nanotubes for plasmid DNA gene delivery. *Angew Chem Int Ed*, 2004, **43**(39):5242–5246.

108. Poland, C.A., et al. Carbon nanotubes introduced into the abdominal cavity of mice show asbestos-like pathogenicity in a pilot study. *Nat Nanotechnol*, 2008, **3**(7):423–428.

109. Al-Jamal, K.T., et al. Degree of chemical functionalization of carbon nanotubes determines tissue distribution and excretion profile. *Angew Chem Int Ed Engl*, 2012, **51**(26):6389–6393.

110. Costa, P.M., et al. Functionalised carbon nanotubes: from intracellular uptake and cell-related toxicity to systemic brain delivery. *J Controlled Release*, 2016, **241**:200–219.

Part VI
Cancer Drug Delivery Approaches

Chapter 10

The Emerging Role of Nanomedicine in the Advances of Oncological Treatment

Petra Gener, Diana Rafael, Simó Schwartz, and Fernanda Andrade
[a]*Molecular Biology & Biochemistry Research Centre for Nanomedicine (CIBBIM-Nanomedicine), Vall d'Hebron Institut de Recerca (VHIR), Barcelona, Spain*
[b]*Networking Research Centre for Bioengineering, Biomaterials and Nanomedicine (CIBER-BBN), Instituto de Salud Carlos III, 50018, Zaragoza, Spain*
fernanda.silva@vhir.org, jose.p.l.araujo@gmail.com

Cancer is one of the most prevalent and fearsome diseases of our times, mainly due to the development of resistant and advanced forms of the disease. The tremendously negative impact of cancer on public health and the economy of countries boosted the research for timely diagnosis and improved treatments. In the last few decades important advances have been observed and clinical outcomes reached, many of them supported by the application of biotechnology and nanotechnology to health with the development of biopharmaceuticals and nanotechnology-based drug delivery systems (nanoDDSs), for example.

Nanoparticles in Life Sciences and Biomedicine
Edited by Ana Rute Neves and Salette Reis

ISBN 978-981-4745-98-7 (Hardcover), 978-1-351-20735-5 (eBook)
www.panstanford.com

A wide range of different nanoDDSs have been designed in the last few years with different drugs or biomolecules. The active targeting of specific cells with a nanoDDS was a step toward achieving even more precise therapeutic treatment without the common severe secondary effects. Due to the problem of cancer resistance and recurrence the targeting of cancer stem cell (CSC) subpopulations is a major concern nowadays.

Following the growing tendency for specific therapies, personalized medicines have become the current focus of research, where the exosomes gain protagonism as a personalized therapeutic carrier. Despite the difficulties regarding the scaling up of nanoDDS production to the industrial scale, we expect an increase in the coming years of nanoDDS-based products able to jump from the bench to the bedside.

10.1 Introduction

In the past decades important advances were observed in the molecular biology, biotechnology, and pharmaceutical fields, allowing a more efficient understanding, control, and treatment of oncological diseases. Nanomedicines have been one of the driving motors of the most recent progresses, being responsible for the development of diagnostic tools and key medicines like Doxil®/Caelix® and Abraxane®, which led to a decrease of cancer-related mortality. However, and despite all the advances, cancer is still a highly prevalent and deadly disease, partially due to its high complexity and variability among patients, treatment inefficiency/resistance, and tumor recurrence.

Nowadays, collaborating efforts between research centers, hospitals, government authorities, pharmaceutical industry, and patients' associations have been made to search for and develop new treatment alternatives and improved therapeutic regimens. Great efforts have been made in order to identify and characterize new biological players that can be used as disease biomarkers or as new specific targets for anticancer drugs or gene-based therapies. Additionally, improved drug and gene delivery systems have been developed (e.g., liposomes, lipidic and polymeric nanoparticles [NPs], drug-polymer conjugates, nanovesicles, and exosomes).

All efforts together have been paving the way to a more specific and personalized medicine, with the ultimate goal of cancer becoming a curable disease.

This chapter will focus on the role of nanomedicine in cancer treatment and the state of the art of marketed and under-development medicines. Special attention will be paid to the importance of targeting therapy and personalized medicine, as well as the difficulties observed in the translation of these therapies from research to clinic.

10.2 Why Nanotechnology Is Important for Cancer Therapy

The efforts applied by research groups and the pharmaceutical industry in the last decades regarding cancer diagnosis and therapy have resulted in encouraging clinical outcomes, such as, importantly, a decrease in cancer-related deaths (by more than 20% since the 1990s) and an increase in patient survival rates [1, 2]. However, especially in developed countries, cancer is, according to the World Health Organization (WHO), a leading cause of morbidity and death, accounting for 14 million new cases (expected to increase to 22 million within two decades) and 8 million cancer-related deaths in 2012 [3]. Just taking the US as an example, it is estimated that there were more than 1,600,000 new cases in 2016, with the number of predicted deaths thought to be more than 500,000, mainly among men (Fig. 10.1). Unfortunately, despite the overall reduction in cancer-related deaths, some cancers, such as cancers of the liver, pancreas, and uterine corpus, have been presenting increasing mortality rates [4].

As a consequence of the high incidence and prevalence of oncological diseases, the sales of anticancer drugs have been growing in the last decades, these drugs becoming the leaders of the global drug market share. An increase of 8% in anticancer drug sales was observed in 2014 compared to 2013, accounting for more than 10% of the total drug sales that year [5]. Regarding the statistical estimations of the disease profile, it is predicted that in the upcoming years there will be an increase in the prescription of oncological treatments, with their global market share growing by almost 15% by 2020 [5]. This scenario indicates great costs to the health care systems related to both cancer treatment and care. For example, in the US alone, cancer care is projected to cost around 174 billion US dollars in 2020 [6].

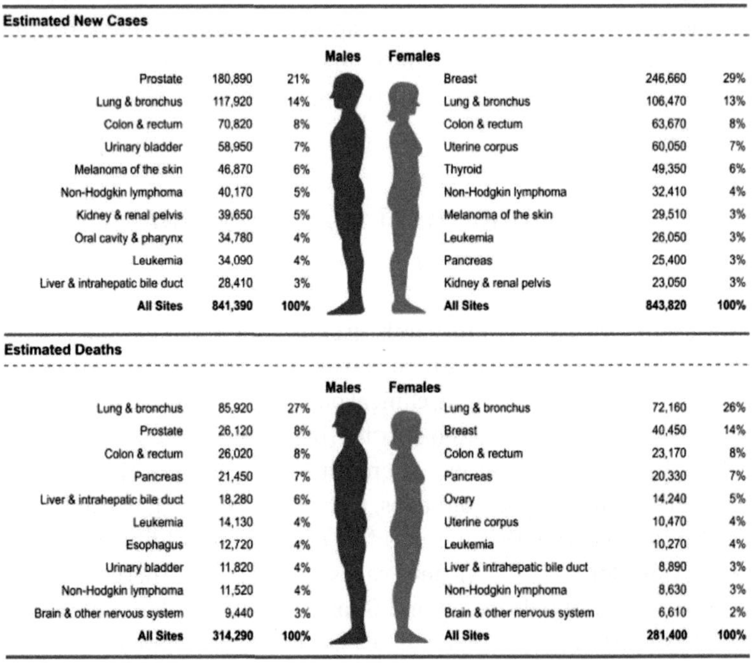

Estimated New Cases

Males			Females		
Prostate	180,890	21%	Breast	246,660	29%
Lung & bronchus	117,920	14%	Lung & bronchus	106,470	13%
Colon & rectum	70,820	8%	Colon & rectum	63,670	8%
Urinary bladder	58,950	7%	Uterine corpus	60,050	7%
Melanoma of the skin	46,870	6%	Thyroid	49,350	6%
Non-Hodgkin lymphoma	40,170	5%	Non-Hodgkin lymphoma	32,410	4%
Kidney & renal pelvis	39,650	5%	Melanoma of the skin	29,510	3%
Oral cavity & pharynx	34,780	4%	Leukemia	26,050	3%
Leukemia	34,090	4%	Pancreas	25,400	3%
Liver & intrahepatic bile duct	28,410	3%	Kidney & renal pelvis	23,050	3%
All Sites	**841,390**	**100%**	**All Sites**	**843,820**	**100%**

Estimated Deaths

Males			Females		
Lung & bronchus	85,920	27%	Lung & bronchus	72,160	26%
Prostate	26,120	8%	Breast	40,450	14%
Colon & rectum	26,020	8%	Colon & rectum	23,170	8%
Pancreas	21,450	7%	Pancreas	20,330	7%
Liver & intrahepatic bile duct	18,280	6%	Ovary	14,240	5%
Leukemia	14,130	4%	Uterine corpus	10,470	4%
Esophagus	12,720	4%	Leukemia	10,270	4%
Urinary bladder	11,820	4%	Liver & intrahepatic bile duct	8,890	3%
Non-Hodgkin lymphoma	11,520	4%	Non-Hodgkin lymphoma	8,630	3%
Brain & other nervous system	9,440	3%	Brain & other nervous system	6,610	2%
All Sites	**314,290**	**100%**	**All Sites**	**281,400**	**100%**

Figure 10.1 Leading cancer types for the estimated new cancer cases and deaths by gender for 2016 in the US. Reprinted from Ref. [4] with permission from Wiley.

Late or improper diagnosis is one of the main reasons for the elevated number of advanced-stage cancers and, consequently, for the observed high mortality rates and care costs [6]. Also, the first generation of cancer treatments has shown to be unable to properly eradicate the disease, usually causing cancer relapse. This incomplete and insufficient treatment is mainly due to the lack of selectivity of these anticancer drugs to the affected cells/tissues, which results in dose-limiting toxicity and acquired therapeutic resistance [7]. Thus, the fight against cancer still needs to pave its way to more timely diagnosis and efficient and cost-effective treatments, with continued clinical and basic research by researchers, health care agencies, and pharmaceutical companies leading to a better ability to control the disease.

These days, more than 800 medicines and vaccines to treat cancer are under development and/or clinical assessment [1], more than 73% of the cases being based on biomarkers, presenting potential for personalized treatment [8].

Some of the medicines under developed are based on biopharmaceuticals, namely monoclonal antibodies, vaccines, and cell or gene therapy. In 2013, the majority of total biopharmaceuticals under development were intended to treat cancer and related diseases (more than 300), most of these based on monoclonal antibodies (170 medicines), and a few on gene therapy [9].

Some of the most important advances in cancer diagnosis and therapy observed in the last few years and expected in the near future arise from the application of nanotechnology to medicine. According to BCC Research, the global nanomedicine market has been growing steadily, reaching a value of $248.3 billion in 2014. It is expected to increase at an annual growth rate of 16.3% until 2019, reaching $528 billion [10].

The pharmacokinetics and pharmacodynamics of drugs are highly dependent on their physical and chemical characteristics, which are influenced by the system used to deliver them. NanoDDSs like polymeric NPs, solid lipid nanoparticles (SLNs), liposomes, drug-polymer conjugates, and micelles (Fig. 10.2), due to their small sizes and high surface areas, can modulate and improve the performance of many drugs to an extent not achievable by conventional formulations. By encapsulating drugs, a nanoDDS can (i) increase their solubility, (ii) protect them from degradation, (iii) enhance their transepithelial absorption, (iv) enhance their uptake by cells, (v) protect them from the in vivo defensive systems, thus increasing their blood circulation time, (vi) release them in a controlled manner in response to a specific stimulus, and/or (vii) target specific cells, tissues, and organs with the drugs [11–13]. Additionally, a multifunctional nanoDDS can simultaneously diagnose and treat a tumor by encompassing both imaging and therapeutic compounds, an emerging field known as theranostics [14].

Among the different nanoDDSs available, liposomes were the first to be studied, originating the first nanomedicines approved by the Food and Drug Administration (FDA) and reaching the market (Doxil/Caelyx®) in 1995. The success of Doxil/Caelyx®, stealth liposomes with a size of 100 nm encapsulating doxorubicin, boosted the development of new nanoDDSs. This system was able to alter the biodistribution of the drug by increasing its half-life and promote passive targeting of tumors, increasing therefore its therapeutic efficacy and reducing its well-known side effects, namely cardiotoxicity and neutropenia [15, 16]. Stealth liposomes like Doxil/Caelyx®, Novantrone®, and Lipoplatin® are a good example of how PEGylation allows the improvement of blood circulation time and the therapeutic efficacy of many drugs [17].

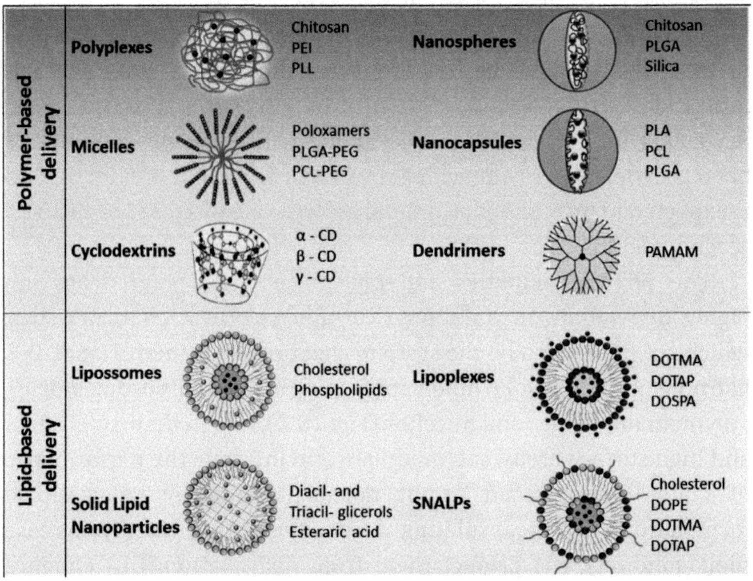

Figure 10.2 Examples of nanoDDSs and their main components. CD, cyclodextrin; DOPE, 1,2-di-(9Z-octadecenoyl)-sn-glycero-3-phosphoethanolamine; DOSPA, 2,3-dioleyloxy-*N*-[2(sperminecarboxamido) ethyl]-*N*,*N*-dimethyl-1-propanaminium trifluoroacetate; DOTAP, 1,2-dioleoyl-3-trimethylammonium-propane; DOTMA, 1,2-di-O-octadecenyl-3-trimethylammonium propane; PAMAM, polyamidoamine; PCL, poly(ε-caprolactone); PEG, poly(ethylene glycol); PEI, poly(ethylene imine); PLA, poly(d,l-lactide); PLGA, poly(d,l-lactide-co-glycolide); PLL, poly(L-lysine); SNALPs, stable nucleic acid lipid particles.

Abraxane® is another good example of how oncological treatment can benefit from the application of nanomedicine. This albumin-paclitaxel conjugate forming particles of 130 nm was approved in 2005 for the treatment of metastatic breast cancer and lately for advanced non-small-cell lung cancer (2012) and late-stage pancreatic cancer (2013), enjoying high clinical applicability.

The success of Abraxane® is mainly due to its ability to solubilize the poorly soluble drug paclitaxel in an aqueous environment without the need of ethanol and Cremophor® EL, the main components of Taxol and associated with serious and dose-limiting toxicities like sensory neuropathy, neutropenia, and strong hypersensitivity reactions [18]. Also, its passive targeting and improved cellular uptake and tumor penetration account for its enhanced therapeutic efficacy compared to Taxol [18].

In 2013, LipoDox® was approved by the FDA, becoming the first nanosimilar, that is, generic of a nanomedicine product (in this case Doxil/Caelyx®), reaching the market.

Nowadays, some nanomedicines are key medicines in many therapeutic regimens of a variety of cancer types and used on a daily basis all over the world. In Table 10.1 are presented some examples of nanoDDSs approved for the treatment of cancer.

10.3 Targeted Therapeutic: Does It Matter?

Most nanomedicines passively accumulate inside solid tumors due to the enhanced permeability and retention effect (EPR effect) that is characteristic of an imperfect intratumoral vasculature and a deficient lymphatic drainage system of the tumors [19]. However, the passive accumulation of NPs is not possible in small tumors with no internal vasculature or in metastasis that does not present an EPR effect. Therefore, functionalization with targeting moieties (e.g., targeting peptides and antibodies) to increase the specific accumulation of NPs in metastatic sites or in resistant cells is currently being investigated. This chapter will address the benefits and drawbacks of this approach.

10.3.1 The Importance of Active Targeting

In general, solid tumors are metabolically very active and thus require constant oxygen supply for survival and further proliferation. When

Table 10.1 Examples of approved nanomedicines with oncological applications

Type of nanocarrier	Drug	Indication	Trade name
Polymeric nanoparticles and polymer conjugates	Paclitaxel	Breast cancer, pancreatic cancer, non-small-cell lung cancer	Abraxane
	Pegaspargase	Leukemia	Onscaspar
	Pegfilgrastim	Neutropenia after chemotherapy	Neulasta
	Neocarzinostatin (Smancs)	Liver cancer, renal cancer	Zinostatin Stimalmer
	Paclitaxel	Breast cancer, lung cancer	Genexol-PM
Liposomes and vesicles	Cisplatin	Pancreatic cancer	Lipoplatin
	Doxorubicin	Kaposi's sarcoma, breast and ovarian cancer	LipoDox
	Cytarabine	Neoplastic meningitis	DepoCyt
	Daunorubicin	Advanced HIV-related Kaposi's sarcoma	Daunoxome
	Doxorubicin	Metastatic breast cancer	Myocet
	Doxorubicin	Ovarian cancer, breast cancer, HIV-related Kaposi's sarcoma	Doxil/Caelyx
	Mitoxantrone	Acute nonlymphocytic leukemia, advanced hormone-refractory prostate cancer	Novantrone
	Vincristine	Acute lymphoblastic leukemia	Marqibo
	Paclitaxel	Advancer solid tumors	Lipusu
	Mifamurtide	Osteosarcoma	Mepact
	Irinotecan	Metastatic adenocarcinoma of the pancreas	Onivyde

Type of nanocarrier	Drug	Indication	Trade name
Nanocrystals	Aprepitant	Nausea after chemotherapy	Emend
Antibody or protein-drug conjugate	Gemtuzumab-ozogamicin	Acute myeloid leukemia	Mylotarg
	Tositumomab-iodine I131	Non-Hodgkin lymphoma	Bexxar
	Ibritumomab-tiuxetan	Non-Hodgkin lymphoma	Zevalin
	Denileukin-diftitox	Recurrent cutaneous T cell lymphoma	Ontak
	Trastuzumab-emtansine	Breast cancer	Kadcyla
	Brentuximab vedotin	Non-Hodgkin lymphoma	Adcetris
Inorganic particles	Superparamagnetic iron oxide nanoparticle	Thermal ablation glioblastoma	NanoTherm

the normal vasculature present in the tumor vicinity is not sufficient to provide enough oxygen, dying tumor cells start to secrete growth factors that trigger the budding of new blood vessels from the surrounding capillaries [20]. This promotes the rapid generation of new blood vessels that have discontinuous epithelium and lack the basal membrane, which is typical for normal vascular structures [21]. The resulting blood capillaries present defects that can reach sizes ranging from 200 to 2000 nm, depending on the tumor type and its environment and localization [22]. The defective tumor vessels offer just a little resistance to extravasation to the tumor interstitium, favoring the accumulation of the macromolecules within the tumor. This phenomenon, known as the EPR effect, was first observed more than 30 years ago by Matsumura and Maeda [23] (Fig. 10.3). They showed that proteins larger than 30 kDa (i.e., labeled murine and bovine albumins and IgGs) get preferentially distributed in the tumor interstitium and remain there for prolonged periods of time [23].

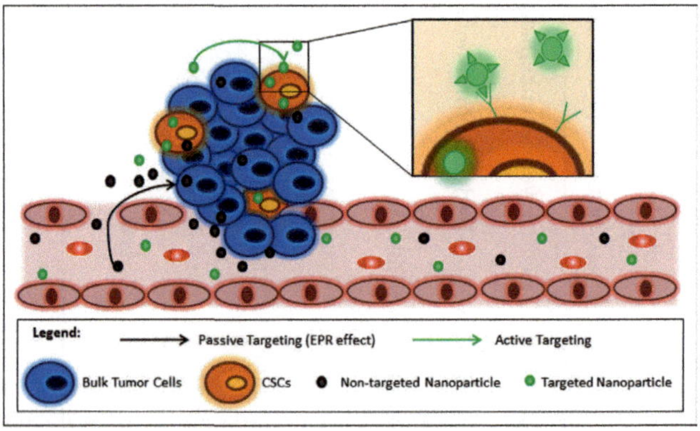

Figure 10.3 Mechanisms of nanoDDS tumor targeting. EPR, enhanced permeability and retention; CSCs, cancer stem cells.

Nevertheless, the passive targeting associated with successful extravasation from blood vessels is an important factor, but not the only one, associated with the distribution of NPs within tumor cells. Very important for the therapeutic success of nanomedicines is the further diffusion of NPs through the extravascular tumor tissue and their interaction with intracellular and/or extracellular targets within the tumor environment (Fig. 10.3). To increase NP retention and uptake by the cells of interest it is necessary to add affinity ligands (e.g., antibodies and small peptides) to their surface, specifically selected to bind surface molecules or receptors overexpressed in diseased organs, tissues, cells, or subcellular domains [24–28]. Folate and transferrin are some of the commonly used ligands [24] and for antibodies, examples include human epidermal growth factor receptor 2 (HER-2), Pgp (MRK2 and MH162), Cripto, and CD19 [29, 30]. An actively targeted material needs to be in the proximity of the target to benefit from this increased affinity. Targeting moieties should increase interactions between NPs and cells of interest without altering the overall biodistribution [24, 31, 32]. Thus, even though some in vivo biodistribution studies showed no change in the accumulation of NPs within the tumor by targeting moieties, the individual cellular uptake might be significantly improved by active targeting. This could be very important for successful elimination of small metastasis or resistant cancer stem cells (CSCs) within the tumor.

10.3.2 The Problematic CSCs

CSCs represent a small, but essential, cell population with the ability to generate and maintain the tumor [33–35]. Due to the self-renewal capacity, CSCs have almost unlimited propagating capacity, while any other cell will stop dividing due to clonal exhaustion [36, 37]. In addition CSCs are substantially insensitive to most conventional anticancer therapies, antimitotic agents, and/or radiation [38]. Such aggressive behavior is a consequence of a unique CSCs phenotype—its gene expression and epigenetic modifications—that differs from more differentiated "bulk" tumor cells. In particular, CSCs have increased capacity to migrate, invade, and possess a potent machinery of antiapoptotic proteins and are more permissive to DNA damage, also showing high DNA repair capabilities [38, 39]. Accordingly, the percentage of CSCs within a tumor often increases after anticancer treatment like chemotherapy or radiation [38, 40–42]. This often leads to cancer recurrence and metastatic growth since only a few CSCs are necessary and sufficient for tumor regeneration. It has been shown that as few as 100 CSCs displaying stem cell properties such as self-renewal and differentiation can promote tumorigenesis in vivo [43].

Even though long circulation times favor the extravasation of nanomedicines at the tumor site, it is nonetheless necessary to actively target CSCs to improve efficacy. Because CSCs express specific cell surface biomarkers, active CSC targeting has been pursued by using either small molecules or antibodies against CSC surface receptors.

In a recent paper, our group has reported that actively targeted nanomedicines increase CSCs' sensitivity to paclitaxel treatment. We observed a stronger cytotoxic effect of paclitaxel-loaded poly(d,l-lactide-co-glycolide)–poly(ethylene glycol) (PLGA-PEG) NPs against breast CSCs when using CD44 antibodies on the surface of NPs for active targeting. Similarly, improved cytotoxicity was also seen in colorectal CSC models using Cetuximab as the director moiety against epidermal growth factor receptor (EGFR), which is overexpressed in colorectal CSCs [44]. Limitations of this approach include toxicity associated with the use of monoclonal antibodies and the heterogenicity of intratumoral CSC populations. This makes even more difficult the development of antibodies or other ligand groups that can selectively recognize and target CSCs but not healthy stem cells or other off-target cells.

10.3.3 Actively Targeted Nanomedicines in Clinical Trials

Currently, there is no commercially available targeted nanomedicine in the market and only a few of them have reached clinical trials. This is mainly due to the very complex and difficult development of targeted nanomedicines compared to nontargeted systems. When actively targeted NP drug carriers have to be developed, the choice of a specific ligand first needs to be considered. The ligand should be as specific as possible to not affect the healthy cells of tissues. In the case of specific targeting of CSCs that share several targets with somatic stem cells it is also important to not attack the healthy stem cells within the tissue. The possible targets are screened by time- and money-consuming techniques like proteomics or microarray and RNA sequencing. Next, the types of ligand available and effective ligand conjugation chemistry need to be examined, since it can considerably affect the size of the NPs [45] and/or ligand density and conformation [46], and thus the overall efficacy of NPs. The choice of the targeting ligand also affects the efficacy of the active targeting strategy in vitro and, most importantly in vivo, where there is always risk of nonspecific binding to the proteins in the bloodstream or unwanted immune reaction [47, 48]. The collaboration of multidisciplinary teams formed by pharmacists, chemists, biologists, and clinicians is thus essential to successfully create an effective and safe targeted nanomedicine.

Even though the use of targeted nanomedicines to improve the efficacy of the treatment is the future of personalized medicine, it is worth mentioning that there are several drawbacks that have to be considered regarding the use of actively targeted NPs in human patients. Measuring the impact of an active targeting medicine in humans is not a simple task. The biggest problem is that it is not possible to evaluate in parallel targeted and matching nontargeted NPs in clinical studies like it is normally done in preclinical models. One way of addressing this problem is to evaluate the early efficacy of the targeted nanomedicine in homogenous target-positive populations to ensure its benefit compared to a standard treatment. However, since targeted NPs rely on sufficient surface expression of specific markers to exert their effect, resistance could theoretically arise from the natural selection of target-negative cell populations less exposed to the drug. Thus, in next step the evaluation of patient cohorts, who differ in the expression of the target biomarker, should be considered. Nevertheless, it is likely that the impact of an active targeting might only be assessed in postapproval, phase 4 studies. However, only after the commercialization of the first targeted systems to treat a larger cohort of patients will it be possible to obtain data regarding their real therapeutic potential in clinical use.

10.4 Recent Advances in the Field of Nanomedicines for Druggable and Nondruggable Targets: A State of Art

10.4.1 Drug Delivery

As mentioned, among all therapeutic areas, the oncological field is undoubtedly the one that has been benefiting the most from nanomedicine. A huge variety of systems have been developed and proposed for delivery of anticancer agents, with some systems already used in clinical practice (Table 10.1) and many others in the different phases of preclinical and clinical development (Tables 10.2 and 10.3).

NanoDDSs have shown to be especially useful for the encapsulation and delivery of poorly water soluble anticancer drugs (biopharmaceutical classification system [BCS] class II and IV) like paclitaxel, docetaxel, etoposide, campthotecin, irinotecan, and methotrexate and to protect and improve the bioavailability of BCS class I and IV drugs such as doxorubicin and 5-fluorouracil, allowing their proper formulation and administration, otherwise insufficient or impossible with conventional drug delivery systems. Additionally, as previously mentioned, due to the pathophysiology of cancer, nanoDDSs take advantage of passive targeting, improving by that the therapeutic index of the drugs [49, 50].

In Table 10.2 are presented some of the systems currently under development and enrolled in clinical evaluation. As is possible to observe, different drugs and systems are being tested for a variety of oncological diseases, some of them presenting active targeting properties, such as MM-302 (Merrimack), MBP-426 (Mebiopharm), 2B3-101 (to-BBB), anti-EGFR immunoliposomes (University of Basel), and BIND-014 (Bind Therapeutics).

A brief reference should be made to sustained-release inhaled liposome targeting (SLIT) cisplatin, developed in partnership between Transave and Insmed, and L9NC, developed by the University of New Mexico, both lipidic-based products in the form of aerosol for local delivery of cisplatin and 9-nitro-20 (S)-camptothecin, respectively, for the treatment of lung cancer [51, 52].

Also, systems like Lipotecan from Taiwan Liposome Company (hepatocellular carcinoma), Paclical from Oasmia Pharmaceutical (ovarian cancer), Livatag from BioAlliance Pharma (primary liver cancer), EndoTAG-1 from SynCore Biotechnology (pancreatic cancer), and SP1049C from Supratek Pharma Inc. (advanced esophagus and gastric cancers) were granted orphan designation. It is worth mentioning that SP1049C, a polymeric micellar system based in poloxamers, has shown to be particularly effective in multi-drug-resistant cancers due to its capacity to inhibit multidrug resistance and efflux systems [53, 54], opening the possibility of treating more efficiently late-stage or recurrent cancers.

Table 10.2 Examples of nanoDDSs for drug delivery to treat cancer under clinical trials

System name/code	Targeting	Encapsulated agent	Indication	Clinical state	Reference
		Liposomes/lipidic particles			
2B3-101	Glutathione	Doxorubicin	Brain metastases, breast cancer	2	NCT01386580
Alocrest	NA	Vinorelbine	Advanced solid tumors, non-Hodgkin's and Hodgkin's lymphoma	1	NCT00364676
C225-ILS-DOX (anti-EGFR immunoliposomes)	Anti-EGFR	Doxorubicin	Advanced triple-negative EGFR-positive breast cancer, solid tumors	2	NCT02833766 NCT01702129
Aroplatin	NA	DACH-platin	Advanced solid tumors	2	NCT00043199 NCT00057395
ATI-1123	NA	Docetaxel	Advanced solid tumors	1	NCT01041235
ATRA-IV	NA	Tretinoin	Solid tumors malignancies	1	NCT00195156
Brakiva	NA	Topotecan	Advanced cervical cancer	1	NCT00054444
CPX-1	NA	Irinotecan:Floxuridine	Advanced colorectal cancer	2	NCT00361842
E7389 Liposome	NA	Eribulin mesylate	Solid tumors	1	NCT01945710

(*Continued*)

Table 10.2 (*Continued*)

System name/code	Targeting	Encapsulated agent	Indication	Clinical state	Reference
EndoTag-1	NA	Paclitaxel	Breast, pancreas, and liver tumors	2	NCT00448305 NCT00542048 NCT00377936
IHL-305	NA	Irinotecan	Solid tumors	1	NCT00364143
L9NC	NA	9-nitro-20-(S)-camptothecin	Non-small-cell lung cancer, metastatic endometrial cancer	2	NCT00249990 NCT00250068
L-Annamycin	NA	Annamycin	Acute lymphocytic leukemia, acute myelogenous leukemia	2	NCT00271063 NCT00430443
LE-DT	NA	Docetaxel	Advanced or metastatic pancreatic cancer and prostate cancer	2	NCT01186731 NCT01188408
LEM-ETU	NA	Mitoxantrone	Solid tumors	1	NCT00024492
LEP-ETU	NA	Paclitaxel	Advanced breast cancer	2	NCT01190982
LE-SN38	NA	SN38	Colorectal cancer	1	NCT00046540
LiPlaCis	NA	Cisplatin	Advanced or refractory solid tumors	1	NCT01861496

System name/code	Targeting	Encapsulated agent	Indication	Clinical state	Reference
Lipocurc	NA	Curcumin	Advanced solid tumors	2	NCT02138955
MCC-465	F(ab') 2 fragment Ab GAH	Doxorubicin	Metastatic stomach cancer	1	[55]
MBP-426	Transferrin	Oxaliplatin	Gastric cancer	2	NCT00964080
MM-302	HER2	Doxorubicin	Advanced/metastatic breast cancer	3	NCT02213744
MM-398 (PEP02)	NA	Irinotecan	Pancreatic cancer, unresectable advanced cancer	3	NCT01494506
NL CPT-11	NA	Irinotecan	Recurrent glioma	1	NCT00734682
OSI-211	NA	Lurtotecan	Recurrent small-cell lung cancer, relapsed epithelial ovarian cancer	2	NCT00046787 NCT00046800
OSI-7904L	NA	Thymidylate synthase inhibitor	Metastatic cancer of the head and neck, gastroesophageal adenocarcinoma	2	NCT00073502 NCT00116909
Promitil	NA	Mitomycin-C lipid-based prodrug	Solid tumors	1	NCT01705002

(*Continued*)

Table 10.2 (*Continued*)

System name/code	Targeting	Encapsulated agent	Indication	Clinical state	Reference
S-CKD602	NA	Belotecan	Advanced malignancies	1	NCT00177281
SLIT cisplatin	NA	Cisplatin	Lung cancer	2	NCT00102531
SPI-077	NA	Cisplatin	Ovarian cancer	2	NCT00004083
ThermoDox	NA	Doxorubicin	Liver and breast cancer	3	NCT02112656 NCT00826085
Vyxeos	NA	Cytarabine:Daunorubicin	Acute myeloid leukemia	3	NCT01696084
Polymeric particles					
ABI-008	NA	Docetaxel	Metastatic breast cancer, prostate cancer	2	NCT00477529 NCT00531271
ABI-009	NA	Rapamycin	Solid tumors, bladder cancer	2	NCT02009332 NCT02494570
ABI-011	NA	Thiocolchicine dímer	Solid tumors, lymphoma	1	NCT02582827
AP5280	NA	Platinum	Solid tumors	1	[56]
BIND-014	PSMA	Docetaxel	Non-small-cell lung cancer, prostate cancer	2	NCT01792479 NCT01812746

System name/code	Targeting	Encapsulated agent	Indication	Clinical state	Reference
CT-2106	NA	Camptothecin	Colon cancer, ovarian cancer	2	NCT00291837 NCT00291785
Delimotecan	NA	T-2513	Solid tumors	1	[57]
DHAD-PBCA-N	NA	Mitoxantrone	Hepatocellular carcinoma	2	[58]
Docetaxel-PNP	NA	Docetaxel	Solid tumors	1	NCT02274610
DX-8951f	NA	Exatecan mesylate	Solid tumors	2	NCT00004108 NCT00005938
EZN-2208	NA	SN38	Solid tumors, breast cancer, lymphoma, colorectal cancer	2	NCT01036113 NCT00520637 NCT00931840
IT-101 (CRLX101)	NA	Camptothecin	Advanced solid tumors, ovarian cancer	2	NCT00753740 NCT00333502
Lipotecan	NA	TLC388	Liver cancer, renal cancer	2	NCT02267213 NCT01831973
Livatag	NA	Doxorubicin	Hepatocellular carcinoma	3	NCT01655693
MAG-CPT	NA	Camptothecin	Solid tumors	1	[59]
MTX-HSA	NA	Methotrexate	Kidney and bladder cancer	2	[60]
Nanotax	NA	Paclitaxel	Peritoneal neoplasms	1	NCT00666991

(Continued)

Table 10.2 *(Continued)*

System name/code	Targeting	Encapsulated agent	Indication	Clinical state	Reference
Nanoxel	NA	Paclitaxel	Advanced breast cancer	1	NCT00915369
NC-4016	NA	Oxaliplatin	Advanced solid tumors or lymphoma	1	NCT01999491
NC-6004 (Nanoplatin)	NA	Cisplatin	Solid tumors; non-small-cell lung, biliary, and bladder cancer; pancreatic cancer	3	NCT02043288 NCT02240238
NK012	NA	SN38	Advanced breast cancer, small-cell lung cancer	2	NCT00951613 NCT00951054
NK105	NA	Paclitaxel	Advanced or recurrent breast cancer	3	NCT01644890
NK-911	NA	Doxorubicin	Solid tumors	1	[61]
NKTR-102	NA	Irinotecan	Breast cancer, ovarian cancer, small-cell lung cancer, colorectal cancer	3	NCT01492101 NCT01876446 NCT00806156 NCT00598975
NKTR-105	NA	Docetaxel	Solid tumors, ovarian cancer	1	[62]
Paclical	NA	Paclitaxel	Ovarian cancer	3	NCT00989131

System name/code	Targeting	Encapsulated agent	Indication	Clinical state	Reference
Pegamotecan	NA	Camptothecin	Gastric cancer, soft-tissue sarcoma	2	NCT00080002 NCT00079950
PK1	NA	Doxorubicin	Breast cancer	2	NCT00003165
PK2	NA	Doxorubicin	Hepatocellular carcinoma	1	[63]
PNU166945	NA	Paclitaxel	Solid tumors	1	[64]
ProLindac	NA	DACH-oxaliplatin	Advanced ovarian cancer	2	[65]
SP1049C	NA	Doxorubicin	Esophagus and gastric carcinoma	3	[66]
Taxoprexin	NA	Paclitaxel	Melanoma, liver cancer, non-small-cell lung cancer	3	NCT00243867 NCT00249262 NCT00422877
XMT-1001	NA	Camptothecin	Solid tumors	1	NCT00455052
Xyotax/Opaxio	NA	Paclitaxel	Lung cancer, ovarian cancer, fallopian tube cancer, esophageal cancer, breast cancer	3	NCT00576225 NCT00265733 NCT00069901 NCT00522795 NCT00108745

Note: In the reference section, the NCT number corresponds to the trial identifier number in the ClinicalTrials.gov database (www. clinicaltrials.gov). ATRA, all *trans*-retinoic acid; DACH, diaminocyclohexane; EGFR, epidermal growth factor receptor; HER2, human epidermal growth factor receptor 2; NA, not applicable; PSMA, prostate-specific membrane antigen; SN38, 7-ethyl-10-hydroxycamptothecin; T-2513, 7-ethyl-10-aminopropyloxy-camptothecin; TLC388, camptothecin analog.

10.4.2 Biopharmaceuticals and Gene Delivery

As mentioned previously, the majority of the biopharmaceuticals under development are intended to treat oncological diseases. The advent of molecular biology, biotechnology, bioengineering, and recombinant DNA technology brings the opportunity to better understand the pathophysiological processes and to identify possible therapeutic targets. Also importantly, it allows the production of biopharmaceuticals on a large scale, making them available for the development of medicines [67, 68].

The high interest of research groups and pharmaceutical companies in the development of biopharmaceuticals is explained by the great selectivity and specificity of biopharmaceuticals resulting in their high therapeutic efficacy against complex and challenging diseases and the overall improvement in the quality of life of patients to an extent sometimes not reached by conventional drugs [69]. Consequently, many biopharmaceuticals have been granted market authorization over the years [70], presenting an increased share of the global pharmaceutical market year by year [71].

The discovery of the RNA interference mechanism for gene silencing, which granted the Nobel Prize in Physiology or Medicine in 2006 to Andrew Zachary Fire and Craig Cameron Mello, opened a door of opportunities for new and improved treatments for cancer based on the precise control of gene expression and disease state [72]. However, the delivery of genes into cells is a hard, complex, and challenging task, which has been taking advantage of nanotechnology to mimic the commonly used viral gene delivery systems but without presenting the safety concerns usually associated with virus-based transfection agents. Despite the toxicological potential, some viral gene delivery systems, such as GliAtak® and ProsAtak® from Advantagene Inc., consisting of a adenoviral vector, are under clinical evaluation to treat malignant glioma (Phase 2) [73] and prostate cancer (Phase 3) [74], respectively.

At the moment some nanoDDSs for biopharmaceuticals and gene delivery are enrolled in clinical trials (Table 10.3).

Table 10.3 Examples of nanoDDSs for biopharmaceuticals and gene delivery to treat cancer under clinical trials

System name/ code	Targeting	Encapsulated agent	Indication	Clinical state	Reference
		Liposomes/lipidic particles			
ALN-VSP		siRNA KSP and VEGF	Liver cancer, liver metastases	1	NCT00882180
Atu027	Protein kinase N3	siRNA PKN3	Advanced pancreatic cancer	2	NCT01808638
BP1001		ASO Grb2	Leukemia	2	NCT02781883 NCT01159028
C-VISA BikDD		Bik	Pancreatic cancer	1	NCT00968604
DCR-MYC		siRNA MYC	Solid tumors, multiple myeloma, lymphoma	2	NCT02314052
DPX-0907 (vaccine)		HLA-A2-restricted peptides, T helper peptide, polynucleotide adjuvant	Advanced-stage ovarian, breast, and prostate cancer	1	NCT01095848
DPX-Survivac (vaccine)		Survivin	Advanced-stage ovarian, fallopian, or peritoneal cancer	2	NCT01416038
EGFR antisense DNA liposomes		EGFR antisense DNA	Head and neck cancer	1	NCT00009841
EphA2 targeting liposomes		siRNA EphA2	Solid tumors	1	NCT01591356

(*Continued*)

Table 10.3 (*Continued*)

System name/ code	Targeting	Encapsulated agent	Indication	Clinical state	Reference
INGN-401		FUS1 gene	Lung cancer		
Interleukin-2 gene therapy		Interleukin-2 gene	Head and neck cancer	2	NCT00006033
LErafAON		ASO *c-raf*	Advanced solid tumors	1	NCT00024661
Lipovaxin-MM (vaccine)		Melanoma antigens and IFNγ	Melanoma	1	NCT01052142
MRX34		miRNA RX34	Solid tumors, hematologic malignancies	1	NCT01829971
PNT2258		BCL-2 antisense DNA	Relapsed or refractory non-Hodgkin's lymphoma, Richter's transformation, relapsed or refractory diffuse large B-cell lymphoma	2	NCT01733238 NCT02378038 NCT02226965
SGT53-01	Antitransferrin receptor single-chain Ab fragment (TfRscFv)	P53 gene	Solid tumors	1	NCT00470613
SGT-94		RB94 gene	Solid tumors	1	NCT01517464

System name/ code	Targeting	Encapsulated agent	Indication	Clinical state	Reference
Stimuvax (vaccine)		Tecemotide	Non-small-cell lung cancer, prostate cancer, multiple myeloma	3	NCT01094548 NCT01496131 NCT01015443
STMN-1		Pbi-shRNA STMN1	Solid tumors, metastatic cancer	1	NCT01505153
TKM-PLK1		siRNA PLK1	Solid tumors or lymphomas that are refractory to standard therapy	2	NCT01262235
		Polymeric particles			
ADI-PEG 20		ADI	Hepatocellular carcinoma, melanoma, pleural mesothelioma, leukemia	3	NCT02709512 NCT01287585 NCT00450372 NCT01910012
BC-819		H19 gene	Bladder cancer, ovarian cancer	2	NCT00595088 NCT00826150
CALAA-01	Transferrin	siRNA RRM2	Solid tumors	1	NCT00689065
GEN-1		IL-12 gene	Ovarian cancer	1	NCT02480374

Note: In the reference section, the NCT number corresponds to the trial identifier number in the ClinicalTrials.gov database (www. clinicaltrials. gov). ADI, arginine deiminase; ASO, antisense oligonucleotide; BCL-2, B-cell lymphoma 2; BIK, BCL-2 interacting killer; c-raf, proto-oncogene serine/ threonine-protein kinase; EGFR, epidermal growth factor receptor; EphA2, ephrin type-A receptor 2; FUS1, fusion protein 1; Grb2, growth factor receptor binding protein 2; HLA-A2, human leukocyte antigen A2; IFNγ, interferon γ; IL-12, interleukin 12; KSP, kinesin spindle protein; miRNA, micro-RNA; MYC, bHLH transcription factor; P53, tumor protein 53; PKN3, protein kinase N3; PLK1, polo-like kinase 1; RB94, retinoblastoma 94; RRM2, ribonucleoside-diphosphate reductase subunit M2; shRNA, short hairpin RNA; siRNA, small interfering RNA; STMN1, human stathmin 1; VEGF, vascular endothelial growth factor.

10.5 Difficulties of Passing from the Bench to the Bedside

Despite the promising advances of nanomedicine in cancer treatment and the high amount of research performed in the last few years (Fig. 10.4), the number of clinical trials accounts for only about 2% of the published works [75]. This is a consequence of the troubles that the majority of the proposed nanomedicines face when intending to jump from the researcher bench to the clinical bedside.

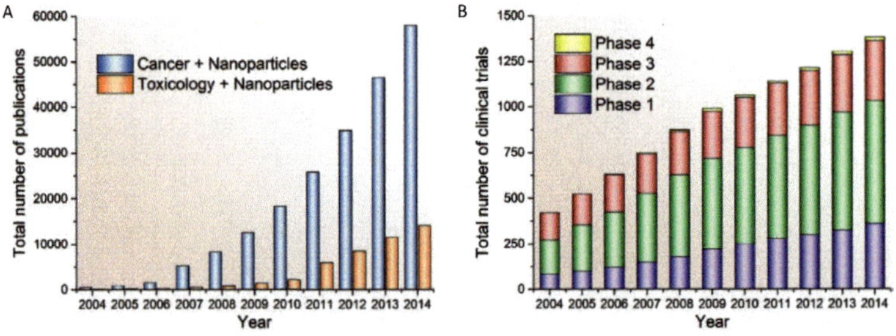

Figure 10.4 Total number of publications (Web of Science®) (A) and clinical trials (www. clinicaltrials.gov) (B) between 2004 and 2014. Reprinted from Ref. [75], Copyright (2015), with permission from Elsevier.

Many factors and limitations make this transition hard, including difficulties in proper characterization of the systems, manufacture on the industrial scale needing to comply with the quality-control guidelines of good manufacturing practices (GMPs), and demanding regulatory and approval requirements by the regulatory agencies [75]. Also, the high number of studies and systems developed led to patent disputes over potentially partial overlapping of content, delaying or making impossible complete product development and approval [76].

The biological behavior of nanomedicines is highly dependent on their properties, namely structure, size, surface charge, and composition, and small variations in these features could result in significant differences in the clinical outcome both regarding efficacy and safety. These variations can not only modify the interaction of the NPs with the biological environment and components, altering their biodistribution, cell uptake, or toxicological profile, but also alter the storage stability of the product [47, 77]. Because of that, batch-to-batch characterization of the system is generally required. Also, improved techniques to better represent the biological interfaces to truly assess and characterize the nanoparticulate systems in order to predict more accurately their biological behavior are needed [77]. Also important for the correct translation of preclinical assessment to clinical assessment is the development of clinically relevant in vitro and in vivo models for the disease of interest. The generally used xenograft models lack proper resemblance to the complexity of the human pathophysiology and patient tumor features, resulting many times in overestimated therapeutic efficacy [78].

The industrial production of nanomedicines under ' is many times an obstacle on the way of such products to the bedside. The scaling up of the multistep methodologies generally applied to produce the nanomedicines on a laboratory scale is often difficult, complex, expensive, and time-consuming, and sometimes not possible. Also, due to the high dependence of a nanomedicine's therapeutic behavior on its physical-chemical characteristics, a tight control of the production parameters and the final product properties, to an extent superior to that requested for conventional drugs, is imperative to avoid batch-to-batch variations [79].

Worthy of note is the patients' and health care professionals' distrust of and lack of confidence in nanomedicine, especially due to safety concerns. Unfortunately, the majority of published studies regarding nanotoxicology are relative to inorganic nanomaterials and environmental exposure, lacking information about the toxicology of nanobased drugs. As for particles' physical and chemical characterization, the development of cutting-edge methods and technologies for proper assessment of the nanoDDS toxicological profile and its impact on human health and environment is of utmost importance [80, 81].

The concerns regarding the safety of nanoDDSs are also one of the reasons for the long, complex, and challenging approval procedure and regulations imposed by regulatory agencies [82]. One of the main difficulties for regulatory agencies is the classification of products into devices (carriers) or drugs (effectors) [83]. Specific guidelines and regulations regarding nanotechnology-based products are lacking, the assessment of the approved products being studied case by case [75]. This procedure is complex and time-consuming, delaying the entry of the products into the market. Most importantly, a unanimously accepted definition of nanomaterials (up to 100 nm or below 1 μm) among researchers, companies, and regulatory agencies is missing. Up to now, only two guidelines were released by the FDA (Considering Whether an FDA-Regulated Product Involves the Application of Nanotechnology—2014) and EMA (Reflection Paper on Nanotechnology-Based Products for Human Use—2006) regarding this subject.

With the continuous progress of nanomedicine, with the development of multifunctional carriers—systems for personalized medicine and nanosimilars—the complexity and difficulties associated with the approval processes are expected to increase [75]. Thus, specific guidelines are urgent in order to develop new tools, standards, and approaches to assess the safety, efficacy, and quality of nanotechnology-based products [84, 85], allowing their faster access to the market and clinical practice.

Finally, but not less important, the high costs associated with the development, clinical assessment, production, and approval of nanomedicines needs to be outshined by a clinical benefit highly superior to that of the conventional drugs, in order to persuade pharmaceutical companies to invest in the product development and the consumers to pay for their consequential high cost [78, 86]. To provide more cost-effective and successful nanomedicines to patients, it is necessary to improve their preclinical and clinical assessment with more precise and disease-focused methodologies [78].

10.6 Future Perspectives: The Importance of Personalized Medicine

Every person on earth is different. We differ in our genetic and epigenetic backgrounds, lifestyles, alimentation habits, etc. This means that our organisms are reacting very differently in stress conditions, leading to the development of differentiated oncological disease process. Moreover, nowadays we know that tumors are also very individual entities, with different behaviors and prognosis, depending on many factors, like the place of their origin, genetic background, and the patient´s age. All the differences lead to more individualized treatment responses. However, the current medicine is based on the broad population average, which means that patients carrying similar types of tumors are likely to receive the same medical treatment. This traditional practice often misses being effective because each person's genetic makeup is slightly different from everyone else's, often in very important ways that affect health. Our growing understanding of genetics and genomics, and how they drive cancer and drug responses in each person, is enabling us to provide more accurate diagnoses and more effective treatments.

The introduction of personalized medicine is moving us closer to more precise, predictable, and powerful health care that is customized for each individual patient.

There is still a long way to go, but in the future we will be able to offer to the patient a real personalized nanomedicine, which will be designed and tailored according to the person's unique genetic makeup and necessity.

10.6.1 Exosomes as the Future of Nanomedicine

The best treatment is always offered and designed by nature. Since time immemorial, in the field of bioengineering the scientist has tried to mimic what is already present naturally in our body. This way, the problems related to adverse immune reactions and efficacy of the treatment are easier to solve.

In this perspective, the most promising candidates for personalized, individualized nanomedicine are modified exosomes. Exosomes are nature's nanocarriers that transport biological information in humans. Moreover, their structural properties, origin, and functions make them interesting, innovative tools for drug delivery.

Exosomes are nanosized vesicles released in the extracellular space from the endosomal compartment. These structures are produced by most cell types and act as transporters of different proteins (hormones, receptors, and signaling molecules), lipids, and genetic materials (messenger RNA [mRNA], short hairpin RNA [shRNA], and miRNA), among others (Fig. 10.5). Exosomes are defined by a size ranging of 30–100 nm [87]. The exact size differs according to the origin of the exosome. The structure of exosomes depends on the cell type they originate from, as well as on the function they play in intercellular communication [88].

Although exosomes were described more than two decades ago, new evidence suggests that exosomes are involved in fundamental physiological processes such as adhesion, migration, invasion, angiogenesis, embryonic development, and tumor progression [88, 89].

It has been demonstrated that tumor cells show a high production of exosomes. Exosomes produced by tumor cells can stay in the extracellular space at the site of origin or can enter biological fluids, reach distal sites, and participate in the creation of premetastatic niches [90, 91].

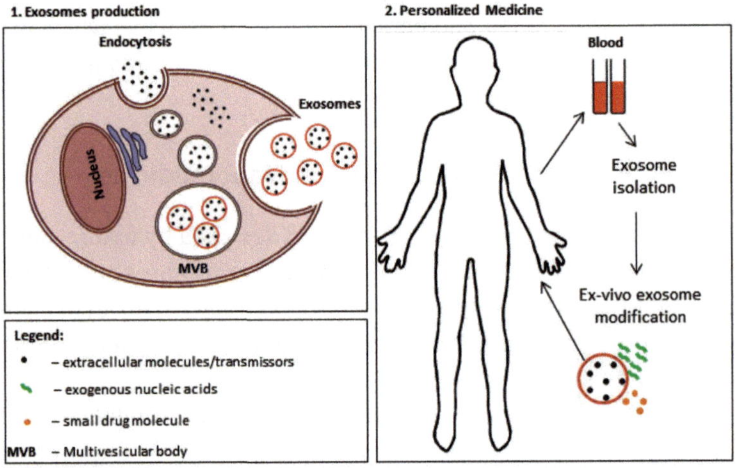

Figure 10.5 Conceptual overview of exosome-based therapeutics: (1) Exosome production and (2) exosome ex vivo modification and clinical therapeutic use in personalized medicine.

In fact, it has been shown that released exosomes can directly stimulate surrounding cells by interaction with specific surface receptors or being transferred from one cell to another through various bioactive molecules, including membrane receptors, proteins, mRNA, microRNA, and even organelles [92]. This "information exchange" represents one of the critical components of conditioning premetastatic niches and is involved in the control of CSC self-renewal and tumor expansion [88, 92]. Exosomes are proposed to be key regulators of the variable phenotype of stem cells, thanks to its ability to mediate the exchange of genetic information and/or protein between cells [93]. It was reported that exosomes released by embryonic stem cells can reprogram hematopoietic progenitors by horizontal transfer of specific mRNA and proteins [94]. Moreover, it seems that resistance of tumor cells to chemotherapeutic treatment also involves cellular intercommunication mediated by exosomes. Indeed, resistant CSCs release significantly more exosomes than cells responsive to antitumor therapy [95, 96].

Furthermore, exosomes also appear to favor CSCs' survival by reducing the responsiveness of T cells or directly induce apoptosis in activated antitumor T cells [97, 98].

There are still a lot of unanswered questions regarding exosomes, and a lot of investigation should be done, but we can already now imagine several ways to treat cancer via exosomes. One possibility is to use exosomes derived from dendritic cells to induce antitumor immunity [99]. Effectively, exosome-based therapy was approved by the FDA in 2010 as the first therapeutic vaccine with antigen-presenting dendritic-cell-derived exosomes by the name of Provenge® for the personalized treatment of metastatic prostate cancer [100]. So far it is the only FDA-approved exosome-based cancer immune therapy.

Another option for cancer therapy involves the general removal of exosomes from circulation [87]. This idea comes from the fact that exosomes play an important role in the pathomechanisms of tumor progression and can also interfere with antitumor chemotherapy, leading to the development of drug resistance. However, there is no currently available technique by which we could effectively remove just exosomes derived from malignant cells, leaving the "good" exosomes, produced by healthy cells. Since exosomes are important for cell-to-cell communication. Their removal from circulation may lead to serious adverse effects. Our knowledge about the different types of exosomes is increasing, and it is possible that in the future we will be able to identify specific molecules present only on the surface of cancer cells derived exosomes. Thus we will be able to distinguish and to specifically remove from circulation the "malignant" exosomes. In this scenario we would not provide nanomedicine to the patient but we would remove the malignant nanocarriers from circulation to block the tumor communication system and prevent its spread to distal organs.

A more futuristic idea is to use the patient´s exosomes for personalized drug delivery. It is clear that production of a stable formulation that mimics the structure and function of exosomes, with all the necessary proteins, receptors, and lipids involved in the lipid bilayer, is a complex task. As emphasized before, each individual and each tumor is different and also the exosomes originating from it are different and unique. Exosomes derived from tumor cells are synthesized on the basis of the tumor cells' requirements and antibody spectrum, having imprinted the specific destination to deliver their cargo. Hereby exosomes loaded with different agents (such as drugs, siRNA, or diagnostic particles) will specifically bind to the target cells, creating a therapeutic accumulation in a specific area like the metastatic niche. This type of therapy was proven to work on a large-scale basis [101]. Exosomes extracted from a pancreatic tumor were combined with staphylococcal enterotoxin B. On the basis of their specific antigen biomarker these modified exosomes reached tumor sites, where the enterotoxin was unloaded and induced apoptosis of tumor cells [102].

The biggest challenge regarding drug delivery by tumor-derived exosomes is their effective drug loading. Currently, exosomes are mostly investigated for delivery of therapeutic RNA. One of the techniques used for their loading is electroporation. Some types of siRNA were successfully loaded to exosomes by electroporation while loading of other types of RNA, such as mRNA, miRNA, and shRNA, was not feasible. Another strategy that was tested was the overexpression of RNA by the exosome-producing cells, which allows even the incorporation of fully functional miRNA and mRNA [103]. A similar strategy could also be used for loading exosomes with drug molecules or therapeutic proteins, peptides, and hormones since these molecules are also a natural cargo of exosomes. Cancer cells could be genetically modified to overexpress functional therapeutic proteins that will be highly present in the exosomes derived from these cells. Moreover, on the basis of the exosomes' specific surface proteins, they could be directed into the cancer sites. The problem that remains is the impossibility of the exact quantification of exosomes' cargo and the prediction of therapeutic efficacy. The amount of exosomes injected into circulation should be as minimal as possible to prevent any adverse effect. But if we are not able to predict the minimal-efficacy dose, since we don't know the amount of therapeutic molecules in the exosomes, it is very difficult to set up the correct dose and treatment regimen.

In spite of all these drawbacks substantial progress was already made in the field of cancer therapy using exosomes as nanoDDSs. In Table 10.4 are summarized some recent therapeutic applications of exosomes, like Alzheimer's disease, HIV infection, atherosclerosis, and immunosuppression [103]. Definitely in the future we will learn more about how to effectively use these natural nanocarriers for patient treatment.

Table 10.4 Examples of application of exosome nanoDDSs [103]

Exosome source	Recipient cell type	Cargo delivered	Functional consequences
	Immunosuppressive effects		
(H) EBV transformed B cells	(H) Monocyte-derived DC	Viral miRNA	Downregulation of antiviral response
(H) Serum of pregnant patients	(H) Jurkat T cells	FasL	Suppression of CD3 signaling and IL-2 production
Murine BMDC overexpressing IL-10	Murine T cells	Antigen, presented on MHC	Suppression of T cell proliferation
	Immunostimulatory effects		
Murine BMDC	Murine $CD8^+$ and $CD4^+$ T cells (in vitro and in vivo)	Antigen, presented on MHC	Induction of T cell proliferation
(H) CD28 stimulated $CD3^+$ T cells	(H) Unstimulated $CD3^+$ T cells	Unidentified	T cell activation, induction of proliferation, and cytokine production when codelivered with IL-2
Murine BMDC	Murine BMDC (allogeneic)	Antigen	Transfer of foreign antigen, followed by foreign antigen presentation to and activation of T cells

(Continued)

Table 10.4 (*Continued*)

	Therapeutic effects		
(H) H9 $CD4^+$ T cells	(H) Jurkat T cells, human PMBC	APOBEC3 protein (HIV replication inhibitor)	Reduction of HIV replication
(H) Endothelial cells	(H) Aortic smooth muscle cells	miR-143, miR-145	Reduction of atherosclerotic lesions
Murine MSC	Murine primary neurons	miR-133b	Neurite outgrowth after injury
	Pathogenic effects		
(H) B cell lymphoma cell lines	None		Binding and sequestering of rituximab (antibody used in B cell lymphoma immunotherapy)
(H) CSF	None	Phosphorylated tau	Transport of neurotoxic protein in Alzheimer's disease
(H) PMBC derived DC incubated with HIV	(H) Jurkat T cell line expressing CCR5	HIV viral particles	Delivery of functional HIV viral particles encapsulated in exosomes, leading to HIV infection of recipient cells

(H), human cell; APOBEC3, apolipoprotein B mRNA-editing enzyme catalytic polypeptide-like 3; BMDC, bone marrow-derived dendritic cells; CCR5, C-C chemokine receptor type 5; CD28, cluster of differentiation 28; CD3, cluster of differentiation 3; CD4, cluster of differentiation 4; CD8, cluster of differentiation 8; CSF, cerebrospinal fluid; DC, dendritic cells; EBV, Epstein–Barr virus; FasL, Fas ligand; HIV, human immunodeficiency virus; IL-10, interleukin 10; IL-2, interleukin 2, MHC, major histocompatibility complex; miR-133b, microRNA 133b; miR-143, microRNA 143; miR-145, microRNA 145; miRNA, micro-RNA; MSC, mesenchymal stem cells; PMBC, peripheral blood mononuclear cells.

10.7 Conclusions

Since the development of the first anticancer drugs and the progresses observed in the area of diagnostic tools, huge advances have been made in the field of cancer therapy. Nevertheless, the conventional therapies, despite the success observed in the treatment of primary cancers, fails many times in preventing and treating recurrent and advanced cancers.

Nanomedicines have revolutionized the oncological treatment scenario since the first medicines were approved and entered clinical practice, being key players today in the therapeutic schemes. Resorting to nanocarriers it was possible to increase the therapeutic index of many anticancer drugs and thus improve the clinical outcomes. It also allowed the delivery of biopharmaceutical drugs with high specificity to the pathophysiology. However, cancer recurrence is still a drawback and a major challenge.

Nowadays, several research groups have been focusing on the development of targeted nanomedicines to further improve treatment specificity. In this, CSCs arise as an important and promising therapeutic target to treat advanced stages of the disease. It is expected that in the near future, some of the targeted systems currently under clinical evaluation will enter the market and become important players in cancer therapy. Also, CSCs should become a more intensive focus of research for the big biotechnology and pharmaceutical companies.

In the last few years, much attention has been given to personalized medicine as a means to improve the treatment of cancer. The development of biomarkers has made it possible to better predict the therapeutic response of patients and to guide the practitioners in the design of therapeutic schemes. However, the concept of personalized medicine used nowadays consists more of the treatment applied to a group of people with the same characteristics than to a single person. The recently explored exosomes, which are nanoDDSs based on the own cells of the patients, could be the solution to reach the so-desired individualized medicine.

References

1. PhRMA. Medicines in development for cancer, 2015, www.phrma.org.

2. American Cancer Society. Cancer facts & figures 2016, www.cancer.org.
3. IARC. World cancer report 2014, Stewart, B.W., Wild, C.P., eds. 2014.
4. Siegel, R.L., et al. Cancer statistics 2016, *CA Cancer J Clin*, 2016, **66**(1):7–30.
5. EvaluatePharma. World preview 2015, Outlook to 2020, 2015, www.evaluatepharma.com.
6. Yabroff, K.R., et al. Economic burden of cancer in the United States: estimates, projections, and future research. *Cancer Epidemiol Biomarkers Prev*, 2011, **20**(10):2006–2014.
7. Holohan, C., et al. Cancer drug resistance: an evolving paradigm. *Nat Rev Cancer*, 2013, **13**(10):714–726.
8. Tufts Center for the Study of Drug Development. Personalized Medicine Gains Traction but Still Faces Multiple Challenges. 2015, www.csdd.tufts.edu.
9. PhRMA. Medicines in development - biologics, 2013, www.phrma.org.
10. Evers, P. Nanotechnology in Medical Applications: The Global Market, 2015.
11. Andrade, F., et al. Micelle-based systems for pulmonary drug delivery and targeting. *Drug Deliv Lett*, 2011, **1**(2):171–185.
12. Bailey, M., et al. Nanoparticle formulations in pulmonary drug delivery. *Med Res Rev*, 2009, **29**(1):196–212.
13. Mansour, H.M., et al. Nanomedicine in pulmonary delivery. *Int J Nanomed*, 2009, **4**:299–319.
14. Bawarski, W.E., et al. Emerging nanopharmaceuticals. *Nanomedicine*, 2008, **4**(4):273–282.
15. Barenholz, Y. Doxil(R)-the first FDA-approved nano-drug: lessons learned. *J Controlled Release*, 2012, **160**(2):117–134.
16. Chang, H.I., et al. Clinical development of liposome-based drugs: formulation, characterization, and therapeutic efficacy. *Int J Nanomed*, 2012, **7**:49–60.
17. Immordino, M.L., et al. Stealth liposomes: review of the basic science, rationale, and clinical applications, existing and potential. *Int J Nanomed*, 2006, **1**(3):297–315.
18. Miele, E., et al. Albumin-bound formulation of paclitaxel (Abraxane ABI-007) in the treatment of breast cancer. *Int J Nanomed*, 2009, **4**:99–105.

19. Padera, T.P., et al. Pathology: cancer cells compress intratumour vessels. *Nature*, 2004, **427**(6976):695.

20. Bates, D.O., et al. Regulation of microvascular permeability by vascular endothelial growth factors. *J Anat*, 2002, **200**(6):581–597.

21. Jain, R.K., et al. Delivering nanomedicine to solid tumors. *Nat Rev Clin Oncol*, 2010, **7**(11):653–664.

22. Hobbs, S.K., et al. Regulation of transport pathways in tumor vessels: role of tumor type and microenvironment. *Proc Natl Acad Sci USA*, 1998, **95**(8):4607–4612.

23. Matsumura, Y., et al. A new concept for macromolecular therapeutics in cancer chemotherapy: mechanism of tumoritropic accumulation of proteins and the antitumor agent smancs. *Cancer Res*, 1986, **46**(12 Pt 1):6387–6392.

24. Peer, D., et al. Nanocarriers as an emerging platform for cancer therapy. *Nat Nanotechnol*, 2007, **2**(12):751–760.

25. Kamaly, N., et al. Targeted polymeric therapeutic nanoparticles: design, development and clinical translation. *Chem Soc Rev*, 2012, **41**(7):2971–3010.

26. Shi, J., et al. Self-assembled targeted nanoparticles: evolution of technologies and bench to bedside translation. *Acc Chem Res*, 2011, **44**(10):1123–1134.

27. Cheng, Z., et al. Multifunctional nanoparticles: cost versus benefit of adding targeting and imaging capabilities. *Science*, 2012, **338**(6109):903–910.

28. Koshkaryev, A., et al. Immunoconjugates and long circulating systems: origins, current state of the art and future directions. *Adv Drug Deliv Rev*, 2013, **65**(1):24–35.

29. Milane, L., et al. Multi-modal strategies for overcoming tumor drug resistance: hypoxia, the Warburg effect, stem cells, and multifunctional nanotechnology. *J Controlled Release*, 2011, **155**(2):237–247.

30. Zhang, Y., et al. The eradication of breast cancer and cancer stem cells using octreotide modified paclitaxel active targeting micelles and salinomycin passive targeting micelles. *Biomaterials*, 2012, **33**(2):679–691.

31. Byrne, J.D., et al. Active targeting schemes for nanoparticle systems in cancer therapeutics. *Adv Drug Deliv Rev*, 2008, **60**(15):1615–1626.

32. Alexis, F., et al. Factors affecting the clearance and biodistribution of polymeric nanoparticles. *Mol Pharm*, 2008, **5**(4):505–515.

33. Wicha, M.S., et al. Cancer stem cells: an old idea-a paradigm shift. *Cancer Res*, 2006, **66**(4):1883–1890; discussion 1895–1896.

34. Shackleton, M., et al. Heterogeneity in cancer: cancer stem cells versus clonal evolution. *Cell*, 2009, **138**(5):822–829.

35. Reya, T., et al. Stem cells, cancer, and cancer stem cells. *Nature*, 2001, **414**(6859):105–111.

36. Shackleton, M. Normal stem cells and cancer stem cells: similar and different. *Semin Cancer Biol*, 2010, **20**(2):85–92.

37. Kreso, A., et al. Evolution of the cancer stem cell model. *Cell Stem Cell*, 2014, **14**(3):275–291.

38. Dean, M., et al. Tumour stem cells and drug resistance. *Nat Rev Cancer*, 2005, **5**(4):275–284.

39. He, Y.C., et al. Apoptotic death of cancer stem cells for cancer therapy. *Int J Mol Sci*, 2014, **15**(5):8335–8351.

40. Cho, Y.M., et al. Long-term recovery of irradiated prostate cancer increases cancer stem cells. *Prostate*, 2012, **72**(16):1746–1756.

41. Kurtova, A.V., et al. Blocking PGE2-induced tumour repopulation abrogates bladder cancer chemoresistance. *Nature*, 2015, **517**(7533):209–213.

42. Lagadec, C., et al. Survival and self-renewing capacity of breast cancer initiating cells during fractionated radiation treatment. *Breast Cancer Res*, 2010, **12**(1):R13.

43. Du, L., et al. CD44 is of functional importance for colorectal cancer stem cells. *Clin Cancer Res*, 2008, **14**(21):6751–6760.

44. Gener, P., et al. Fluorescent CSC models evidence that targeted nanomedicines improve treatment sensitivity of breast and colon cancer stem cells. *Nanomedicine*, 2015, **11**(8):1883–1892.

45. Jiang, W., et al. Nanoparticle-mediated cellular response is size-dependent. *Nat Nanotechnol*, 2008, **3**(3):145–150.

46. Gu, F., et al. Precise engineering of targeted nanoparticles by using self-assembled biointegrated block copolymers. *Proc Natl Acad Sci USA*, 2008, **105**(7):2586–2591.

47. Bertrand, N., et al. The journey of a drug-carrier in the body: an anatomo-physiological perspective. *J Controlled Release*, 2012, **161**(2):152–163.

48. Monopoli, M.P., et al. Biomolecular coronas provide the biological identity of nanosized materials. *Nat Nanotechnol*, 2012, **7**:779–786.

49. Pérez-Herrero, E., et al. Advanced targeted therapies in cancer: drug nanocarriers, the future of chemotherapy. *Eur J Pharm Biopharm*, 2015, **93**:52–79.

50. Egusquiaguirre, S.P., et al. Nanoparticle delivery systems for cancer therapy: advances in clinical and preclinical research. *Clin Transl Oncol*, 2012, **14**(2):83–93.

51. Wittgen, B.P., et al. Phase I study of aerosolized SLIT cisplatin in the treatment of patients with carcinoma of the lung. *Clin Cancer Res*, 2007, **13**(8):2414–2421.

52. Mexico, U.o.N. Study of aerosolized liposomal 9-nitro-20 (S)-camptothecin (L9NC) - full text view, ClinicalTrials.gov.

53. Alakhov, V., et al. Block copolymer-based formulation of doxorubicin. From cell screen to clinical trials. *Colloids Surf B*, 1999, **16**:113–134.

54. Alakhova, D.Y., et al. Effect of doxorubicin/pluronic SP1049C on tumorigenicity, aggressiveness, DNA methylation and stem cell markers in murine leukemia. *PLoS One*, 2013, **8**(8):e72238.

55. Matsumura, Y., et al. Phase I and pharmacokinetic study of MCC-465, a doxorubicin (DXR) encapsulated in PEG immunoliposome, in patients with metastatic stomach cancer. *Ann Oncol*, 2004, **15**(3):517–525.

56. Rademaker-Lakhai, J.M., et al. A Phase I and pharmacological study of the platinum polymer AP5280 given as an intravenous infusion once every 3 weeks in patients with solid tumors. *Clin Cancer Res*, 2004, **10**(10):3386–3395.

57. Veltkamp, S.A., et al. Clinical and pharmacologic study of the novel prodrug delimotecan (MEN 4901/T-0128) in patients with solid tumors. *Clin Cancer Res*, 2008, **14**(22):7535–7544.

58. Zhou, Q., et al. A randomized multicenter phase II clinical trial of mitoxantrone-loaded nanoparticles in the treatment of 108 patients with unresected hepatocellular carcinoma. *Nanomedicine*, 2009, **5**(4):419–423.

59. Bissett, D., et al. Phase I and pharmacokinetic (PK) study of MAG-CPT (PNU 166148):a polymeric derivative of camptothecin (CPT). *Br J Cancer*, 2004, **91**(1):50–55.

60. Bolling, C., et al. Phase II study of MTX-HSA in combination with cisplatin as first line treatment in patients with advanced or metastatic transitional cell carcinoma. *Invest New Drugs*, 2006, **24**(6):521–527.

61. Matsumura, Y., et al. Phase I clinical trial and pharmacokinetic evaluation of NK911, a micelle-encapsulated doxorubicin. *Br J Cancer*, 2004, **91**(10):1775–1781.

62. Calvo, E., et al. Dose-escalation phase I study of NKTR-105, a novel pegylated form of docetaxel. *J Clin Oncol*, 2010, **28**:TPS160.

63. Seymour, L.W., et al. Hepatic drug targeting: phase I evaluation of polymer-bound doxorubicin. *J Clin Oncol*, 2002, **20**(6):1668–1676.

64. Meerum Terwogt, J.M., et al. Phase I clinical and pharmacokinetic study of PNU166945, a novel water-soluble polymer-conjugated prodrug of paclitaxel. *Anticancer Drugs*, 2001, **12**(4):315–323.

65. Nowotnik, D.P., et al. ProLindac (AP5346): a review of the development of an HPMA DACH platinum polymer therapeutic. *Adv Drug Deliv Rev*, 2009, **61**(13):1214–1219.

66. Sawant, R., et al. Micellar nanopreparations for medicine, in *Handbook of Nanobiomedical Research: Fundamentals, Applications and Recent Developments*, Torchilin, V., ed. World Scientific, Hackensack, New Jersey, 2014.

67. Andrade, F., et al. Nanocarriers for pulmonary administration of peptides and therapeutic proteins. *Nanomedicine (Lond)*, 2011, **6**(1):123–141.

68. Grenha, A. Systemic delivery of biopharmaceuticals: parenteral forever? *J Pharm Bioallied Sci*, 2012, **4**(2):95.

69. Morishita, M., et al. Is the oral route possible for peptide and protein drug delivery? *Drug Discov Today*, 2006, **11**(19–20):905–910.

70. Rader, R.A. FDA biopharmaceutical product approvals and trends in 2012. *BioProcess Int*, 2013, **11**(3):18–27.

71. EvaluatePharma. World Preview 2018, 2012, www.evaluatepharma.com.

72. Xu, X., et al. Cancer nanomedicine: from targeted delivery to combination therapy. *Trends Mol Med*, 2015, **21**(4):223–232.

73. Advantagene, I. Phase 2a study of AdV-tk with standard radiation therapy for malignant glioma (BrTK02) - full text view, ClinicalTrials.gov.

74. Advantagene, I. Randomized controlled trial of ProstAtak® immunotherapy during active surveillance for prostate cancer (ULYSSES) - full text view, ClinicalTrials.gov.

75. Wicki, A., et al. Nanomedicine in cancer therapy: challenges, opportunities, and clinical applications. *J Controlled Release*, 2015, **200**:138–157.

76. Harris, D., et al. Strategies for resolving patent disputes over nanoparticle drug delivery systems. *Nanotechnol Law Bus*, 2004, **1**(4):1–18.

77. Nel, A.E., et al. Understanding biophysicochemical interactions at the nano-bio interface. *Nat Mater*, 2009, **8**(7):543–557.

78. Harea, J.I., et al. Challenges and strategies in anti-cancer nanomedicine development: an industry perspective. *Adv Drug Deliv Rev*, 2017, **108**:25–38.

79. Zamboni, W.C., et al. Best practices in cancer nanotechnology: perspective from NCI nanotechnology alliance. *Clin Cancer Res*, 2012, **18**(12):3229–3241.

80. Stone, V., et al. Development of in vitro systems for nanotoxicology: methodological considerations. *Crit Rev Toxicol*, 2009, **39**(7):613–626.

81. Bosetti, R., et al. Future of nanomedicine: obstacles and remedies. *Nanomedicine (Lond)*, 2011, **6**(4):747–755.

82. Morigi, V., et al. Nanotechnology in medicine: from inception to market domination. *J Drug Deliv*, 2012, **2012**(Article ID 389485):1–7.

83. Miller, J. Beyond biotechnology: FDA regulation of nanomedicine. *Columbia Sci Technol Law Rev*, 2003, **4**:E5.

84. Hamburg, M. FDA's approach to regulation of products of nanotechnology. *Science*, 2012, **336**(6079):299–300.

85. U.S. Department of Health and Human Services, Food and Drug Administration. Guidance for industry: considering whether an FDA-regulated product involves the application of nanotechnology, 2011.

86. Allen, T.M., et al. Liposomal drug delivery systems: from concept to clinical applications. *Adv Drug Deliv Rev*, 2013, **65**(1):36–48.

87. Marleau, A.M., et al. Exosome removal as a therapeutic adjuvant in cancer. *J Transl Med*, 2012, **10**:134.

88. Camussi, G., et al. Exosome/microvesicle-mediated epigenetic reprogramming of cells. *Am J Cancer Res*, 2011, **1**(1):98–110.

89. Atay, S., et al. Tumor-derived exosomes: a message delivery system for tumor progression. *Commun Integr Biol*, 2014, **7**(1):e28231.

90. Jung, T., et al. CD44v6 dependence of premetastatic niche preparation by exosomes. *Neoplasia*, 2009, **11**(10):1093–1105.

91. Costa-Silva, B., et al. Pancreatic cancer exosomes initiate pre-metastatic niche formation in the liver. *Nat Cell Biol*, 2015, **17**(6):816–826.

92. Camussi, G., et al. Role of stem-cell-derived microvesicles in the paracrine action of stem cells. *Biochem Soc Trans*, 2013, **41**(1):283–287.

93. Quesenberry, P.J., et al. The paradoxical dynamism of marrow stem cells: considerations of stem cells, niches, and microvesicles. *Stem Cell Rev*, 2008, **4**(3):137–147.

94. Ratajczak, J., et al. Membrane-derived microvesicles: important and underappreciated mediators of cell-to-cell communication. *Leukemia*, 2006, **20**(9):1487–1495.

95. Safaei, R., et al. Abnormal lysosomal trafficking and enhanced exosomal export of cisplatin in drug-resistant human ovarian carcinoma cells. *Mol Cancer Ther*, 2005, **4**(10):1595–1604.

96. Shedden, K., et al. Expulsion of small molecules in vesicles shed by cancer cells: association with gene expression and chemosensitivity profiles. *Cancer Res*, 2003, **63**(15):4331–4337.

97. Huber, V., et al. Human colorectal cancer cells induce T-cell death through release of proapoptotic microvesicles: role in immune escape. *Gastroenterology*, 2005, **128**(7):1796–1804.

98. Iero, M., et al. Tumour-released exosomes and their implications in cancer immunity. *Cell Death Differ*, 2008, **15**(1):80–88.

99. Bu, N., et al. Exosome-loaded dendritic cells elicit tumor-specific CD8+ cytotoxic T cells in patients with glioma. *J Neurooncol*, 2011, **104**(3):659–667.

100. Tan, A., et al. The application of exosomes as a nanoscale cancer vaccine. *Int J Nanomed*, 2010, **5**:889–900.

101. Jo, W., et al. Large-scale generation of cell-derived nanovesicles. *Nanoscale*, 2014, **6**:12056–12064.

102. Mahmoodzadeh Hosseini, H., et al. Exosome/staphylococcal enterotoxin B, an anti tumor compound against pancreatic cancer. *J BUON*, 2014, **19**(2):440–448.

103. Marcus, M.E., et al. FedExosomes: engineering therapeutic biological nanoparticles that truly deliver. *Pharmaceuticals (Basel)*, 2013, **6**(5):659–680.

Chapter 11

On the Trail of Oral Delivery of Anticancer Drugs via Nanosystems

José Lopes-de-Araújo and Cláudia Nunes
UCIBIO, REQUIMTE, Department of Chemical Sciences, Faculty of Pharmacy, University of Porto, Portugal
jose.p.l.araujo@gmail.com

Oral chemotherapy is a desired route for cancer treatment. Oral administration of anticancer drugs will increase the patients' quality of life, facilitate the administration of combined therapies, and prolong therapeutic exposure to the anticancer drugs. Nevertheless, the choice of this administration route is associated with several limitations and there are still many challenges that need to be surpassed. In this sense, nanotechnology-based drug delivery systems appear as a potential solution to overcoming many of the drawbacks of oral administration.

11.1 Introduction

Intravenous administration of anticancer drugs is the most-used administration route in the vast majority of the cancer treatments.

Nanoparticles in Life Sciences and Biomedicine
Edited by Ana Rute Neves and Salette Reis

ISBN 978-981-4745-98-7 (Hardcover), 978-1-351-20735-5 (eBook)
www.panstanford.com

Nevertheless, being more convenient, noninvasive, and painless and because it promotes an increase in the patient's quality of life, oral administration of the anticancer drugs is becoming a more attractive administration route.

Oral chemotherapy will, therapeutically, facilitate the administration of combined therapies and prolong therapeutic exposure to the anticancer drugs. However, oral administration of anticancer drugs faces several limitations and among them are the harsh gastrointestinal environment, the gastrointestinal tract barriers, poor oral bioavailability, the fact that most of the anticancer drugs are substrates of cytochrome P450, the multidrug efflux pumps present in the cells along the gastrointestinal tract, and interpatient variability.

In this sense, although oral administration of anticancer drugs represents a very attractive approach, there are still many challenges that need to be overcome. Thereby, nanotechnology-based drug delivery systems may provide a solution capable of surpassing many of the drawbacks of oral administration. In fact, several types of NPs can be used to achieve these goals, which is prompting the interest in and increasing the importance of these systems.

11.2 Cancer

Cancer is a pathology in which the cells acquire new characteristics due to mutations. The new capabilities acquired by these cells lead to their abnormal growth and development. Some of these distinctive capabilities that promote tumor growth and metastatic dissemination are resistance to cell death, sustenance of proliferative signaling, evasion of growth suppressors, avoidance of immune destruction, replicative immortality, tumor-promoting inflammation, activation of invasion and metastasis, and induction of angiogenesis and genome instability and mutations [1].

According to the World Health Organization and the International Agency for Research on Cancer, in 2012 there were more than 14 million new cancer cases and over 8 million deaths due to cancer worldwide. With the continuous growth and aging of the world population, it is estimated that in 2035 there will be around 24 million new cancer cases and almost 15 million deaths [2].

The main current cancer therapies are limited to a few alternatives that can be conjugated, namely surgery, radiotherapy, chemotherapy (anticancer drug therapy), and immunotherapy. There is still a high amount of treatment failure [3]. For that reason, the discovery of highly effective anticancer drugs with few side effects continues to be a major quest. The ineffectiveness in attaining this goal leads to a shift in the paradigms of anticancer drug discovery toward:

- Molecularly targeted therapeutics
- A better understanding of the molecular drug interactions
- The use of NPs as drug delivery systems (DDSs) to develop targeted therapies

11.3 Anticancer Drugs

In this context, a huge effort has been put in the development of more effective anticancer drugs. Matthias Dobbelstein and Ute Moll [4] suggested that the quest to obtain highly effective and low-toxicity anticancer drugs could be roughly divided into three "waves" that have emerged sequentially (Fig. 11.1).

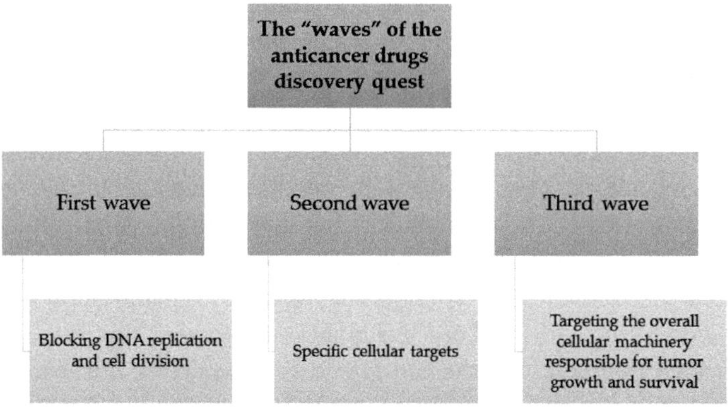

Figure 11.1 The "waves" of anticancer drugs' discovery quest.

The first wave of anticancer drugs encompasses drugs that mainly block DNA replication and cell division. Due to this nature, first-wave anticancer drugs are not selective, affecting also the normal cells, resulting in high toxicity. Chemotherapeutics are an example of first-wave anticancer drugs, which include, for example, methotrexate, prednisone, doxorubicin, and topotecan.

The second wave of anticancer drugs are intended to decrease the side effects. The focus is on the development of anticancer drugs with higher specificity to several cellular targets. These targets are mainly signaling intermediates that are genetically altered in cancer cells and that are essential to the survival, growth, and development of the tumor. Nevertheless, cancer cells can develop several resistance mechanisms for these anticancer drugs. Second-wave anticancer drugs include imatinib, trastuzumab, and sunitinib.

The third wave corresponds to the actual phase of anticancer drug development, in which anticancer drugs are being developed to target the cellular machinery that is essential for the tumor growth and survival, instead of single molecules, taking also advantage of the chronic stress conditions that cancer cells undergo. Third-wave anticancer drugs include drugs that are already well established, like paclitaxel and cisplatin; anticancer drugs recently developed, like vorinostat and bortezomib; and anticancer drugs in the investigational stage and established drugs whose anticancer activity was recently discovered, like actinomycin D, piperlongumine, and cationic amphiphilic drugs (such as terfenadine and siramesine).

A combination of different drugs from different waves targeting different cell mechanisms is also a strategy garnering increasing interest since it may result in the enhancement of the efficacy of the cancer treatments.

Currently, the vast majority of the anticancer drugs are administered by the intravenous route. Actually, this is a rare event in medicine and oncology is one of the rare areas in which intravenous administration is dominant over oral administration [5]. In effect, a report from 2008 by the National Comprehensive Cancer Network stated that at that time only 10% of the cancer chemotherapy was provided in the form of oral formulations and that this value was expected to increase in the coming years. Supporting this expected increase in the oral chemotropic formulations was the fact that in 2008 25%–30% of the drugs in the oncology pipeline were oral agents [6].

11.4 Intravenous Administration

Intravenous administration of anticancer drugs promotes a more direct, quick, and complete bioavailability of the drugs, without great variability among patients, bypassing the tissue absorption barriers promoting therefore an accurate dose control [7, 8].

Preparation of the injectable drug formulations often requires the use of several excipients in order to surpass the low solubility or stability of the anticancer drugs, which may not be well tolerated by patients, causing some toxicity in issues and an increase in the toxicity of the treatment [7].

This therapy is designed to administer in a short period of time the maximum-tolerated dose to kill the cancer cells followed by a "rest" period without the administration of the drug. This process has an associated severe risk since a high drug concentration is also delivered to health cells [7]. However, acute toxic events are easily detected since the administration requires the presence of trained personnel and intravenous administration allows quick dose modification or easy treatment suspension [8].

One should also notice that intravenous administration of drugs requires a hospital visit and nursing or, as an alternative, the use of ambulatory pump or indwelling catheters. Nevertheless, these procedures are painful; cause discomfort to the patient; and may result in bleedings, infections, and venous thrombosis [7].

11.5 Oral Administration

As an alternative, the use of the oral route to administer anticancer drugs is gaining more attention. It is not yet clear whether this interest represents a change in the chemotherapy paradigm or whether oral chemotherapy is an ephemeral hot topic. The reality is that there is an increase in the number of emerging oral formulations and future developments in oral chemotherapy are expectable [5].

From the point of view of the patients, oral chemotherapy is preferred in comparison with intravenous chemotherapy [9]. Oral chemotherapy patients reported feeling less sick, with more freedom and more control of the disease, better quality of life, and more time to spend at home [7, 9, 10]. Furthermore, oral administration is often associated with greatest safety and convenience [10].

Since oral administration requires fewer hospitalizations and less nursing, patient compliance becomes a great concern due to the fact that the responsibility of monitoring the doses and toxicity migrates to the patient or caregivers [11]. Suboptimal compliance may affect the activity of the drug, hindering treatment efficiency or resulting in increased toxicity [7]. On the other hand, above-optimal compliance may result in increased toxicity, with patients having greater doses than the recommended or continuing the treatment despite facing severe side effects [5, 11]. Bad compliance from patients may mainly reflect the complexity of the treatments, forgetfulness, and willful or manipulative behavior from patients [6]. Therefore, not all patients are suitable for oral chemotherapy. So carefully selecting the most suitable patients and promoting the correct education of the patients, by instructing them on the risks and benefits of the treatments and encouraging the patients to follow the instructions, are of extreme importance for the success of oral chemotherapy [5, 7, 11, 12].

The expectations regarding the frequent clinical application of oral chemotherapy are diverse among all the stakeholders [5, 13]:

- The patients, whose interests are efficiency, low side effects, and convenience
- The health care professional, who look for effective treatments, with low toxicity, low bioavailability variability, compliance from the patients, and control of the treatment
- The pharmaceutical companies, which are concerned with the higher cost of developing oral formulations and technical problems that emerge from the use of oral formulations, such as solubility and permeability of the drugs
- The health care funders, whose income from hospitalizations (medical, nursing, disposable items, pharmacy, and administrative costs) will be diminished

Contrarily to intravenous chemotherapy, which is designed to create a high peak above the maximum tolerable concentration of the drug followed by its fast excretion from circulation, resulting in a limited area under the curve (AUC), oral chemotherapy can maintain a sustained moderate concentration of the drug for a greater period of time, which may result in higher efficiency and less toxicity [14]. In fact, changing the treatment from an intravenous to an oral therapy will allow many clinically relevant advantageous alterations, such as [9, 11, 13, 15, 16]:

- Maintenance of an adequate plasma drug concentration, allowing longer drug exposure, which will promote an increase in the efficacy and a reduction of side effects
- Capability to modulate drug release
- Application of chronic treatment regimens, with prolonged exposure to the drug preventing the intermittence of the intravenous therapy
- Circumvention of the pain and discomfort associated with injection and the risks of infections and extravasation associated with intravenous infusions
- Reduction of the costs associated with the therapy by avoidance of hospitalizations, nursing, and trained personal assistance; management of the side effects; and sterile manufacturing
- Convenience, less psychological distress, and enhancement of the patient's quality of life, which may result in enhanced patient cooperation
- Opening of the possibility to explore the prophylactic segment

Despite these reasons, the intravenous route is still the most used route to administer chemotherapy. The factors underlying this predominance in intravenous chemotherapy include the properties of the drugs, since most anticancer drugs are not suitable for oral administration because they belong to class IV of the Biopharmaceutical Classification System (BCS) (low solubility and low apparent permeability drugs) and physiological barriers [7, 10, 14]. Therefore, complex problems arise from the oral use of these drugs, namely [7, 8, 10, 11, 14, 17–20]:

- Low solubility in aqueous media, which makes the creation of efficient formulations difficult.
- Low intestinal permeability, which among other factors may be related to insufficient transit time in the gastrointestinal tract and results in incomplete drug absorption.
- Degradation in gastrointestinal fluids, since the highly acidic conditions in the stomach and the wide variety of enzymes present in gastrointestinal fluids may potentiate the degradation of the drugs along the gastrointestinal tract.
- Incorrect adhesion and diffusion in the mucosal environment along the gastrointestinal tract and difficulties in transposing

of other extra- and intracellular barriers that will allow the drugs to reach their final destination, that is, the specific tissue, cell, or molecule.

- Affinity for liver and intestinal cytochrome P450 (CYP3A4), which leads to the elimination of the drugs by metabolic processes dependent of the enzymes.
- Affinity to P-glycoprotein (Pgp) in the intestinal barrier (this efflux pump hinders the permeability of the drugs through these physiological barriers).
- Presystemic metabolism (in the gastrointestinal tract and liver—first passage effect after being absorbed).
- Drug–drug interactions, which may result in unpredicted toxicity and alterations in efficiency. These interactions may also occur with food, nutritional supplements, and herbal remedies among other elements.
- Incorrect dissolution rate from the dosage form.
- Alterations in the dose due to vomiting, especially immediately after drug intake, which will result in drug expulsion and thus reduced drug absorption.
- Inter- and intrapatient variability in the above-mentioned factors, which results in different drug absorption profiles, drug metabolism, and drug excretion and, therefore, variability in the bioavailability.
- Specific pharmacokinetic and pharmacodynamic profiles and narrow therapeutic windows, which can lead to alterations in the benefits or adverse effects of the drugs when small alterations in the dose occur.

Therefore one should consider that the drug bioavailability is consequently affected and expect that upon oral administration of anticancer drugs only a part of the administered drugs will be systemically available for a therapeutic effect [10].

There are other factors impairing the use of oral chemotherapy, like toxicity issues since high concentrations of anticancer drugs near the gastrointestinal walls may result in damage of these tissues, limitations on selection of the excipients, and a narrow clinical setting in which they may be applied [5, 7, 10, 21].

One should also notice that the majority of the available oral formulations are essentially new formulations of already existing ones, a majority of which are low-priced generics with a low interest for the pharmaceutical companies. This lack of interest from this industry is also extended to clinicians and drug developers since there is still the belief that intravenous chemotherapy is more effective [5].

11.6 Strategies to Overcome the Hurdles of Oral Chemotherapy

To surpass the problems associated with the oral delivery of anticancer drugs and the achievement of oral chemotherapy, different strategies can be followed, for example, the synthesis of prodrugs that improve the solubility of the precursor drug and coadministration of inhibitors of efflux pumps of the intestinal wall in order to increase the absorption of the drugs.

Another strategy of great interest is the development of nanotechnology-based drug delivery systems to promote the oral administration of anticancer drugs. This last strategy will be further discussed herein [11, 22].

11.7 Nanotechnology-Based Drug Delivery Systems

Nanotechnology is, perhaps, the most important and versatile strategy to overcome the limitations that hinder the establishment of oral chemotherapy. In fact, nanotechnology-based drug delivery systems have many characteristics that can help overcome these limitations. NPs can be used to effectively encapsulate the anticancer drugs, increasing the drug solubility and absorption and therefore the bioavailability. NPs may also be useful for modulating the pharmacokinetic profile and tissue distribution of the drugs, preventing the degradation of the drugs and the local toxic effects in the gastrointestinal tract and circumventing drug–drug interactions and other hurdles that hinder the drugs' bioavailability.

Nanotechnology-based drug delivery systems come in a wide and diverse catalog that includes lipid NPs, liposomes, lipid micelles, lipid–drug conjugates, self-nanoemulsifying drug delivery systems, nanocrystals, polymeric NPs, and polymeric micelles, carbon nanotubes among others. Furthermore, characteristics that are of major importance in the success of these drug delivery systems (such as size, shape, surface charge, release profile, and mucoadhesion) can be easily manipulated and custom-built specific surface functionalization can also be achieved to promote the targeting of specific agents.

The use of nanotechnology to develop drug delivery systems is very interesting not only as a platform to create formulations that can deliver already-in-use drugs but also as a platform to implement the use of drugs with promising activity but with bad characteristics and that therefore were abandoned. Also the use of phytopharmaceutical drugs with anticancer activity can be prompted by the use of nanotechnology.

Similarly, it is also worth mentioning the potential to design systems capable of encapsulating two or more drugs, very useful for combined therapy or for taking advantage of the synergistic effect between drugs.

There are some NP-formulation-loading anticancer drugs already in the clinic [23]. Nevertheless the use of oral NP anticancer formulations in the clinical scenario is still in its infancy, with very few examples, like Megace® ES (Par Pharmaceutical) (nanocrystals) [24].

Despite the absence of a large number of nanotechnology-based drug delivery systems clinically available, these drug delivery systems have already proved to overcome the adversities associated with the oral administration of anticancer drugs.

The goal of regular application of oral chemotherapy is a few steps away, but this goal will certainly be achieved by the use of nanotechnology-based drug delivery systems. However, and once again, a lot of steps need to be climbed to prove the efficiency and safety of these systems in the oral administration of anticancer drugs.

Published works combining nanotechnology and chemotherapy are many, but works focused on oral chemotherapy are relatively scarce since this area is yet in its infancy and is somehow only now becoming a field of emerging interest. Furthermore, the existing works are mainly focused on a few types of NPs, despite the broad range of possibilities. Herein, we will only focus on works published in the last three years, because since that time some reviews have been published about the use of nanotechnology-based drug delivery systems, such the ones by Mazzaferro et al. [25], Thanki et al. [10], and Mei et al. [14] in 2013, and Luo et al. [19] in 2014.

11.7.1 Drug Nanocrystals

Drug nanocrystals have the capability to increase the overall bioavailability and efficacy of the anticancer drugs by increasing their solubility, dissolution, absorption, and stability. It is also possible to incorporate other functional excipients in order to improve the general activity of the final formulation. High payload, easy scale-up, and widespread industrial adaptability are other factors that potentiate the interest in these particles in the establishment of oral chemotherapy. Despite all these optimal characteristics, the downsides of these systems include the potential to promote gastrointestinal toxicity and "small" intellectual property privileges [10].

To surpass the bad solubility and rapid metabolism of paclitaxel, Sharma et al. [26] developed paclitaxel nanocrystals with the capacity to increase the bioavailability of the drug in male Wistar rats, when compared with a drug suspension (plain paclitaxel crystals). In fact, the developed nanocrystals prompted an increase of 9- to 10-fold in the AUC when compared to the plain paclitaxel crystals and were also capable of increasing the efficacy of the drug in in vitro studies using breast cancer cells (MCF-7 and MDA-MB cell lines). During the development stage several stabilizers were tested and the selected one, that presented better stability, was the synthetic polyelectrolyte sodium polystyrene sulfonate (PSS). PSS in a nonbiodegradable polymer, and therefore its use in drug delivery systems should be avoided. Due to this fact, the authors have synthesized a designed hydrophobically modified water-soluble Pluronic 127–grafted chitosan (Pl-g-CH) to be used as a functional stabilizer for nanocrystals [27]. A Pl-g-CH stabilizer has the capability to enhance even further the stability of the nanocrystals, to facilitate their absorption, and to have an inhibitory effect on the Pgp efflux pump. In effect, the synthesized nanocrystals improved both the relative bioavailability (a 12.6-fold increase), also in male Wistar rats, and the efficacy of paclitaxel when compared to the commercial intravenous formulation Taxol™ (from Bristol-Myers Squibb).

Another interesting strategy regarding the use of these systems was followed by Patel et al. [28] that orally coadministered paclitaxel nanocrystals with clarithromycin (a dual Pgp and CYP3A4 inhibitor). The strategy led to an enhancement of the oral bioavailability of the drug, in Sprague–Dawley® rats, and an improvement in the efficacy, resulting in an enhanced survival rate of the melanoma-tumor-bearing BDF mice and a reduction in the tumor volume.

The use of nanocrystals is not limited to paclitaxel and has been attempted for other drugs, like, for example, bexarotene, which resulted in a decrease in the side effects of the drug and an increase in its bioavailability in Wistar rats when compared to the intravenous administration of the drug [29].

11.7.2 Polymeric Nanosystems

11.7.2.1 Polymeric nanoparticles

Polymeric NPs are one of the most studied systems in the development of drug delivery systems for oral chemotherapy. These NPs generally are biocompatible and biodegradable and can be produced using an extensive number of polymers with different properties. Polymeric NPs can be used to encapsulate a wide variety of drugs by adjusting the hydrophilicity and hydrophobicity of the structure. Besides, these particles are robust, can be designed to be very stable in the gastrointestinal tract, and have a high and rapid absorption in the gastrointestinal tract and the sustained-drug-release profiles can be easily tuned according to the desired specifications. The principal disadvantages of the polymeric NPs are the relative high costs associated with most used polymers and concerns about the toxicity of some polymers. So, the quest for alternative inexpensive polymers is a huge demand for the real establishment of these systems as a completely effective and satisfying alternative for promoting oral chemotherapy [10].

With the aim of enhancing the bioavailability of docetaxel by taking advantage of a polymeric system with mucoadhesive properties, Saremi et al. [30] developed a particle composed of a polymethyl methacrylate core and a thiolated chitosan shell. The in vitro results demonstrated an increased permeability of these drugs across the Caco-2 cell monolayer compared to that of the free drug. The in vivo results in male Wistar rats showed that the properties of the particles due to the presence of the thiolated chitosan (mucoadhesion, Pgp inhibition, and permeability enhancing effects) promoted an increase in the half-life of the particles and an increase in the oral bioavailability compared to that of the free drug. Nevertheless, no studies related to the anticancer activity of the formulation were performed to prove the efficacy of the system. Wu et al. [31] also developed chitosan NPs for the oral administration of docetaxel. The developed chitosan NPs included cyclodextrin complexes for an improved aqueous solubility of the drug. The optimized NPs revealed enhanced permeability and increased oral bioavailability in male Wistar rats after oral administration when compared to those of the free drug, but once again no studies regarding the anticancer activity of the system were performed. Still in the line of chitosan NPs for oral delivery of anticancer drugs, Feng et al. [32] developed a pH-sensitive nanosystem using a mixture of chitosan and o-carboxymethyl chitosan. The pH-sensitive NPs diminished doxorubicin release in acidic conditions while enhancing the intestinal absorption of the drug. The result, in Sprague–Dawley® rats, was an increase in the drug present in the blood and therefore an increase in the absolute bioavailability of the drug. Besides, tissue biodistribution demonstrated decreased doxorubicin accumulation in both the heart and kidney, which led to a decrease in cardiac and renal toxicities. Nevertheless, no studies were performed regarding drug accumulation in the tumor tissue and the system efficacy.

Oral administration of shRNA for anticancer treatment is a complex process that needs to overcome several adversities, like nuclease degradation, cell uptake, nuclear location, intestinal permeation, and degradation of the gastrointestinal environment. To do so, Zheng et al. [33] synthesized an amino-acid-modified (histidine-cysteine) chitosan NP for the delivery of survinin shRNA-expressing plasmid DNA. The developed polymeric NPs improved the stability of the shRNA and promoted its absorption and internalization, which as a consequence increased the survinin downregulation in the human liver cancer cell line QGJ7703, leading to its death and therefore decreasing the tumor proliferation.

PLGA is also a FDA-approved polymer frequently used in several works for the oral delivery of anticancer drugs [34–40]. One interesting example of the use of these polymeric NPs is the work performed by Jain et al. [41]. The authors produced a PLGA NP for a dual therapy, taking advantage of the synergistic effect between tamoxifen and quercetin. Quercetin is an antioxidant that reduces the hepatotoxicity and resistance to tamoxifen and also increases its bioavailability by inhibiting its Pgp efflux. Additionally, quercetin presents antiproliferative effects. The optimized formulation revealed an enhanced in vitro permeability (in Caco-2 monolayers) that may help to explain the increased in vivo bioavailability of the drug, in Sprague–Dawley® rats, when compared to that of the free drugs. In vitro studies using human breast adenocarcinoma cell line MCF-7 also revealed increased cellular uptake and cytotoxicity, which was corroborated with the in vivo studies using 7,12-dimethyl [α] benzanthracene (DMBA)-induced breast-tumor-bearing female Sprague–Dawley® rats, which revealed higher tumor suppression when compared with the free separated and conjugated drugs. Also, the use of this system prevented any hepatotoxicity and oxidative stress on the studied animals.

In addition to the above-mentioned examples several other works have been performed using polymeric NPs to achieve the oral administration of anticancer drugs [42–46].

11.7.2.2 Polymeric micelles

Polymeric micelles have innumerous characteristics that show them as promising players in the future of oral chemotherapy. These micelles have enhanced stability, high solubilizing capability, a small size, and a long circulation time due to their hydrophilic surface. Additionally, they can be functionalized in order to achieve specific targeting or controlled release dependent on the temperature or pH, for example [10]. Once again, the main disadvantages of these micelles are the cost of the used polymers and safety concerns. In vivo studies proving the ability and efficacy of these systems for oral delivery of anticancer drugs are yet scarce [14, 25].

Styrene maleic acid was used by Parayath et al. [47] in the development of micelles bearing epirubicin. The performed studies demonstrated an efficient uptake of the micelles mediated by the

M-cells and enterocytes without the micelles interfering with the integrity of the intestinal tissue. In vivo studies in severe combined immunodeficiency (SCID) mice injected with MDA-MB-231 breast cancer cell line were performed with fluorescent-labeled micelles and revealed an increase in the overall distribution of the fluorophore in tissues with high vasculature and mainly in tumor site. In vitro cytotoxicity studies in MDA-MB-231 cells revealed that the activity of the drug was not affected by its encapsulation in the micelles, but no further in vitro or in vivo studies with micelles bearing the drug were performed in order to prove the efficacy of the system.

Another appealing strategy using polymeric micelles is by coupling the drug with a polymer, forming a prodrug that is used to produce micelles. An example of this strategy is the work of Li et al. [48] that coupled poly(ethylene glycol) (PEG) with different chain lengths to all *trans*-retinoic acid (ATRA), developing a prodrug (ATRA-PEG) that was then used to produce micelles. The optimized particles revealed high stability, increased permeability when compared to that of the free drug, and the capability to be quickly hydrolyzed in hepatic homogenate, releasing the drug. The overall results demonstrated an increase in the bioavailability of the drug (in male Sprague–Dawley® rats) when compared to the bioavailability of the free drug. Nevertheless, no studies regarding the anticancer activity of the system or toxic effects were performed.

Wang et al. [29] have conducted a very interesting work in which polymeric micelles loaded with docetaxel composed of PEG and poly(ε-caprolactone) (PCL) were embedded in a pH-sensitive hydrogel. The hydrogel released the polymeric micelles preferably in intestinal pH and the overall result was an increase in the oral bioavailability of the drug (in Sprague–Dawley® rats) when compared to free micelles. Furthermore, the system was effective in inhibiting tumor growth in a subcutaneous 4T1 mouse model and revealed less toxicity when compared with intravenous treatment.

11.7.2.3 Dendrimers

Dendrimers can be designed in order to incorporate a wide variety of drugs with high payload, either by being entrapped in the structure or by being chemically conjugated with the branched structure.

Therefore, dendrimers can protect the drugs from gastrointestinal degradation and increase the solubility, permeability, and overall bioavailability of the drugs. Nevertheless, a few works are there concerning the use of dendrimers for oral delivery of anticancer drugs, and the known works were conducted only in vitro [10, 14].

The work of Sadekar et al. [49] is an example of the use of dendrimers for the oral administration of anticancer drugs. The authors produced polyamidoamine (PAMAM) dendrimers to orally deliver camptothecin. The developed dendrimers were capable of controlling the solubilization of the drug in gastrointestinal fluids and to increase the bioavailability of the drug without causing any visible histological alterations on the epithelial layer of the intestine in CD-1 mice. Nevertheless, no further studies proving the safety of this dendrimers and their anticancer activity were done.

11.7.3 Lipid Nanosystems

11.7.3.1 Nanoemulsions

A nanoemulsion is composed of a very small system that comprises mainly a liquid lipid, a surfactant, and a cosurfactant in an aqueous medium that can be refined according to the characteristics of the drugs, being especially useful in the delivery of lipophilic drugs. The greater advantage and interest of these particles in oral chemotherapy is their ability to be absorbed through the gastrointestinal barrier. Moreover, the constituents of these particles are inexpensive and their production is cheap and easy to do on an industrial scale. However, one should consider that the high amount of surfactant used in the production of a nanoemulsion may lead to side effects mainly in the gastrointestinal tract [10, 14].

One of the works conducted using nanoemulsions as a vehicle for the delivery of anticancer drugs was performed by Lu et al. [50]. The authors developed a nanoemulsion with a small droplet size for the oral delivery of ent-11α-hydroxy-15-oxo-kaur-16-en-19-oic-acid (5F). The developed nanoemulsion promoted controlled and sustained drug release and increased bioavailability, and there were no hepatotoxic effects in comparison with the suspension of the free drug in the Kunming mice model. No further studies were performed regarding the efficacy of the system.

11.7.3.2 Self-nanoemulsifying drug delivery systems

Self-nanoemulsifying drug delivery systems (SNEDDSs) are a different approach on emulsions that count on the capability of their constituents to self-assemble and form nanoemulsions in aqueous media, for example, like physiologic fluids. The exact mechanism that is at the base of this self-assembling process is not yet understood, but it's believed to be affected by the characteristics of the drug and of the constituents. The nanosytem is composed of liquid lipids (like long- and medium-chain triglycerides with different degrees of saturation, modified or hydrolyzed vegetable oils, and novel semisynthetic medium-chain derivatives), which are stabilized by the use of surfactants and cosurfactants (namely nonionic surfactants with a higher hydrophilic-lipophilic balance).

A SNEDDS has better stability than the traditional nanoemulsions and also promotes an increase in the solubilization of the drugs and their absorption across the gastrointestinal tract. A SNEDDS therefore increases the bioavailability of the encapsulated drugs and can also be functionalized in order to promote the targeted delivery of a drug. Once again, one of the advantages of these particles is the price of the excipients used. A SNEDDS can also be absorbed in inert absorbents such as silicates, dextran, carbon nanotubes, and charcoal, solidifying the system in order to further increase its stability. These processes, however, may lead to alterations in the characteristics of the SNEDDS and obviously increase the production costs. More recent studies are focused on using excipients that at the same time emulsify and solidify these emulsions [10].

Jain et al. [51, 52] developed solid SNEDDSs loaded with both tamoxifen and quercetin, to take advantage of the above-described synergistic effect, using a lyophilization process with Aerosil 200 as a solidification method. The results revealed an increase in the bioavailability of both drugs, in female Sprague–Dawley® rats, when compared to that of the free drugs and an increase in the cytotoxicity of both drugs when compared with the free drugs separated or in conjugation in human breast adenocarcinoma MCF-7 cell line and in DMBA-induced-tumor-bearing mice. These events suggest the reduction of the dose index for both tamoxifen and quercetin. Furthermore, tamoxifen hepatotoxicity was not visible after oral administration of the system.

Other substances can be used to solidify the liquid SNEDDS. For example, Ahmad et al. [53] used polyoxyethylene as an inert solid carrier for a paclitaxel SNEDDS and Quan et al. [54] used colloidal silica as a solid carrier for a docetaxel-loaded SNEDDS. Both systems have shown to be good candidates for improving the oral bioavailability of these drugs.

Another work worth mentioning is the work performed by Kamel et al. [55]. The authors produced solid SNEDDSs loaded with rosuvastatin using mannitol as a solid carrier. The used systems revealed no cytotoxicity by themselves but increased the cytotoxicity of the drugs in the Caco-2 cell line by increasing their uptake. The solid SNEDDSs were then introduced in a tablet and the oral bioavailability compared to the commercial product in healthy human volunteers. The pharmacokinetic studies demonstrated an increase in the rosuvastatin oral bioavailability most probability due to an increase in the mucosal permeability and enhancement in the absorption. Therefore, despite no further studies, the enhanced bioavailability may result in increased efficacy and consequently in dose reduction and decreased side effects.

11.7.3.3 Solid lipid nanoparticles

Solid lipid nanoparticles (SLNs) are composed by solid lipids and surfactants that are usually inexpensive excipients with high biocompatibility. These particles can be produced using methods that avoid the use of organic solvents and that can be easily scaled up. SLNs have the capability to entrap and transport difficult-to-deliver drugs, improving their bioavailability. Furthermore, the characteristics of the particles can be tuned according to the drug that is intended to be encapsulated and the overall release profile specifically controlled. SLNs can also be functionalized in order to target a specific agent and therefore obtain a localized release [10].

Cho et al. [56] optimized the production of two SLNs, both capable of encapsulating docetaxel for its oral administration. The optimized formulations promoted in vitro controlled sustained docetaxel release and an increase in the intestinal uptake of the drug and bioavailability, in male Sprague–Dawley® rats, when compared to Taxotere® (Sanofi Aventis), an intravenous docetaxel formulation. The SLNs emulsified with D-α-tocopheryl PEG_{1000} succinate (TPGS

1000) proved to have better characteristics (improved intestinal absorption and oral bioavailability) that those emulsified with Tween-80, which may be related to their capability to better inhibit the drug efflux. Nevertheless, despite the good results obtained, no studies were performed regarding the toxicity and anticancer activity of the formulations.

SLNs were also used by Battani et al. [57] to develop a system capable of orally administering raloxifene. The developed SLNs had high drug encapsulation efficiency, were highly stable in gastrointestinal-simulated fluids, and were capable of promoting prolonged, sustained in vitro raloxifene release. In vivo studies in female Sprague–Dawley® rats revealed an increase in the oral bioavailability and enhanced efficacy in DMBA-induced-breast-cancer-tumor-bearing rats. Furthermore, no alterations in the weights of the animals were observed.

Besides these examples that use SLNs for oral administration of anticancer drugs, other works following this line have also been published in the last years [58–60].

11.7.3.4 Nanostructured lipid carriers

Nanostructured lipid carriers (NLCs) are very similar to SLNs, the difference being that they are composed not only of solid lipids but also of liquid ones. The main characteristics that make them a suitable system to orally deliver anticancer drugs are the same as those of the SLNs, but with increased advantages such as increased loading capacity and shelf life stability [10].

An example of the use of NLCs to promote the oral administration of anticancer drugs is the work of Neupane et al. [61]. The authors produced NLCs loaded with decitabine with high drug encapsulation efficiency; sustained in vitro drug release, with an initial burst release; and a high shelf life stability. Furthermore, the NLCs revealed increased drug permeation in the ex vivo male albino rat gut and excellent in vitro anticancer activity in the human non-small-cell lung cancer A549 cell line. In vivo biodistribution studies in Ehrlich ascites tumor–bearing mice proved the accumulation of the NPs at the tumor site 4 h after oral administration. Nevertheless, no further studies were performed regarding the oral bioavailability and toxic effects of the formulation.

11.7.3.5 Lipid nanocapsules

Lipid nanocapsules (LNCs) encompass core-shell particles whose liquid core can be oil or water based. So, LNCs can be used to efficiently encapsulate hydrophobic or hydrophilic drugs with excellent drug payloads. The most important properties of these systems related to the oral delivery of anticancer drugs are their ability to be absorbed in the gastrointestinal tract and their proficiency in inhibiting the membrane efflux transporters. In addition, they can be designed in order to achieve controlled and localized release of the drugs [10].

Groo et al. studied the characteristics that paclitaxel-loaded LNCs with PEG coating should have in order to achieve optimal mucus diffusion and thereby to promote the gastrointestinal uptake of the drug [62]. From the different LNCs tested the authors concluded that the diffusion of paclitaxel in mucus was improved using neutral or positive LNCs with short PEG chain lengths [62]. Following this study, the authors produced LNCs bearing these characteristics and interestingly despite showing an increase in the bioavailability of the drugs (in syngeneic Fisher F344 female rats), the nanocapsules failed to demonstrate good anticancer efficacy after oral administration, most likely due to a loss of integrity [63]. Nevertheless, no further improvement studies have been performed.

11.7.3.6 Liposomes

Liposomes are well-known vesicular structures composed of phospholipids with aqueous cores, making them suitable for the encapsulation of both hydrophobic and hydrophilic drugs. However, their use in oral administration is very limited since liposomes cannot withstand the gastrointestinal environment, being easily degraded. The moderate drug loading capacity, encapsulation efficiency, shelf life stability, and price of production are other arguments that also hinder the use of liposomes in oral chemotherapy [10].

Nevertheless, attempts are being made in order to protect liposomes from the harsh gastrointestinal environment and the work performed by Joshi et al. [64] is an example of one of these attempts. The authors developed a hybrid lipopolymeric system composed of a liposome coupled to a chitosan blanket that shielded the vesicle, offering protection against the gastrointestinal environment. The developed NPs increased the stability of the vesicles through the gastrointestinal tract and increased their mucoadhesion when compared to nonmodified vesicles. The use of this lipopolymeric system led to an increase in the bioavailability of paclitaxel against the commercial formulation Taxol™ (Brystol-Myers Squibb) in female Wistar rats. Furthermore, an equivalent therapeutic efficacy to Taxol™ (Brystol-Myers Squibb) and Abraxane® (Celgene) was also observed as an improved survival and reduced toxicity.

11.7.4 Other Nanotechnology-Based Drug Delivery Systems Strategies

Several other strategies have been applied in the construction of nanotechnology-based drug delivery systems for oral administration of anticancer drugs. Munagala et al. [65], for example, used bovine milk as an easily scalable source of exosomes and these exosomes were separately loaded with several drugs with anticancer activity. The exosomes revealed a high shelf life stability and high cross-species tolerance. The drug-loaded exosomes promoted an increase in the in vitro (in human lung cancer A549 and H1299 cell lines, breast cancer MDA-MB-231 and T47D cancer cell lines, and human normal bronchial epithelial Beas-2B cell lines) and in vivo (in nude mice carrying the A549 lung tumor xenograft) anticancer activity of the drugs. Furthermore, the addition of folate to the exosomes promoted the targeting of the cancer cells, increasing even further the anticancer activity of the drugs.

The strategy followed by Zhang et al. [66] is another interesting approach to this problematic. The authors developed lipid-coated nanodiamonds (particles from the carbon NPs family) loaded with sorafenib. Despite not being absorbed into circulation the nanodiamonds increased the concentration of the drug in the gastrointestinal tract, leading to an increase in the sorafenib bioavailability in male Sprague–Dawley® rats. Moreover, the concentration of the drug at the tumor site was also increased, leading to an improvement in the tumor growth inhibition and a diminution of metastasis in a tumor induced in BALB/c nude mice by subcutaneous injection of cells from the gastric BCG-823 cell line. Nevertheless, despite no observable alterations in the weight of the animals no further studies assessing the safety of this formulation were performed.

Keeping to the theme of NPs from the carbon NP family, carbon nanopowder was also used by Ding et al. [67] in order to increase the bioavailability of apigenin, a bioactive flavonoid with anticancer activity. The carbon nanopowder successfully increased the oral bioavailability of apigenin without increasing the intestinal toxicity of the drug in male Sprague–Dawley® rats. However, despite these results no further toxicity studies, biodistribution, or anticancer activity studies were performed.

Another interesting strategy is the use of the glycoprotein lactoferrin to produce NPs. In fact, Zhang et al. [68] produced lactoferrin NPs capable of encapsulate gambogic acid, whose anticancer activity has been reported. The lactoferrin NPs promoted active intestinal absorption of gambogic acid. Furthermore, the anticancer activity was therefore increased when compared to an intravenous formulation. Nevertheless, no additional pharmacokinetic studies have been performed.

Liquid crystalline NPs are self-assembled structures that after the self-assembling process may produce a wide variety of secondary structures, like cubic or hexagonal NPs, vesicular carriers, and micelles. Swarnakar et al. developed cubic liquid crystalline NPs loaded with doxorubicin in order to improve their therapeutic efficacy [69]. In fact, doxorubicin incorporation into these structures improved the oral bioavailability of the drug upon oral administration and the in vitro (in the human adenocarcinoma breast cancer MCF-7 cell line) and in vivo (in DMBA-induced breast cancer in female Sprague–Dawley® rats) anticancer activity of the drug while decreasing the cardiotoxicity when compared to the commercial intravenous formulations Adriamycin® (Bedford Laboratories) and Lipodox® (Sun Pharma) [69]. To further increase the efficacy and safety of this system these particles were used in combination with cubic liquid crystalline NPs loaded with coenzyme Q10, which is a potent antioxidant with reported anticancer activity. The combination therapy resulted in an ablation of the cardiotoxicity effects of doxorubicin and in anticancer activity enhancement when compared to doxorubicin cubic liquid crystalline NPs by themselves and the commercial formulation Adriamycin® (Bedford Laboratories) [70].

Vong et al. [71] also developed a fascinating piece of work. The authors used the reactive-oxygen-species-scavenging properties of redox NPs (by self-assembly of methoxy-poly(ethylene glycol)-b-poly(4-[2,2,6,6-tetramethylpiperidine-1-oxyl]oxymethylstyrene), or MeO-PEG-b-PMOT) to diminish the side effects associated with colon cancer treatment. The developed NPs by themselves revealed no long-term toxicity, capacity to scavenge reactive oxygen species, and suppression of tumor growth in azoxymethane- and dextran-sodium-sulfate-induced-colitis-associated colorectal cancer in male imprinting control region (ICR) mice. A combination of these particles with the conventional drug irinotecan improved the therapeutic efficacy while decreasing the side effects. This work opens the door to another strategy that may pass by the conjugation of both drug and redox NPs in an orally administered system.

In addition to these examples, several other strategies have been followed and other different types of NPs have been developed in order to promote the oral administration of anticancer drug [72–78].

11.8 Conclusion

The vast majority of cancer treatments rely on the intravenous administration of anticancer drugs since it leads to immediate availability of the drugs. Nevertheless, it is an invasive route that requires frequent visits to hospitals or clinics for the administration of the anticancer drugs, increasing therefore the costs associated with the treatment.

Oral administration of anticancer drugs is becoming a more attractive therapy due to its convenience and noninvasiveness and the fact that it is painless. Furthermore, it will turn the treatment from a hospital-based to an ambulatory/home-based treatment, which will increase the quality of life of the patients, create a great sense of patient autonomy, and may also reduce the costs associated with the treatment. Nevertheless, patient adherence must be carefully controlled to ensure the success of the treatment and patient safety.

Therapeutically, oral administration will facilitate the administration of combined therapies and promote a prolonged therapeutic exposure to the anticancer drugs. However, the gastrointestinal tract environment and barriers, poor oral bioavailability since most of the anticancer drugs are substrates of cytochrome P450, the multidrug efflux pumps present in the cells along the gastrointestinal tract, and interpatient variability are among the limitations of oral administration of anticancer drugs.

In this sense, although orally administered anticancer drugs seem to be very attractive, there are still many challenges that need to be overcome to achieve this approach. Nanotechnology-based drug delivery systems may provide a solution capable of surpassing many of the drawbacks of oral administration, improving the physicochemical properties of the anticancer drugs, resulting in increased bioavailability. Furthermore, these drug delivery systems can be designed to promote controlled and localized drug release, which may reduce the systemic side effects, leading to safer and efficient treatments. In fact, several types of NPs can be used to achieve these goals, which is prompting the interest in and promoting the importance of these systems. Nevertheless, a lot still needs to be done until nanotechnology-based drug delivery systems capable of promoting efficient and safe oral administration of anticancer drugs become widely available in the market. The ultimate goal of oral administration as the principal route of administration of anticancer drugs is still a few steps away, but nanotechnology-based drug delivery systems will certainly play a key role in the achievement of this goal.

Acknowledgments

This work was supported by *Fundação para a Ciência e Tecnologia* (FCT) through the FCT PhD programs and by *Programa Operacional Capital Humano* (POCH), specifically by the BiotechHealth Programme (Doctoral Programme on Cellular and Molecular Biotechnology Applied to Health Sciences), reference PD/00016/2012. The authors also thank FEDER funds (POCI/01/0145/FEDER/007728) and National Funds (*Fundação para a Ciência e a Tecnologia* and *Ministério da Educação e Ciência* [FCT/MEC]) for financial support under the partnership agreement PT2020 UID/MULTI/04378/2013. José Lopes-de-Araújo and Cláudia Nunes acknowledge the FCT for financial support through the PhD grant PD/BD/114012/2015 and investigator grant IF/00293/2015, respectively.

References

1. Hanahan, D., et al. Hallmarks of cancer: the next generation. *Cell*, 2011, **144**(5):646–674.
2. Ferlay, J., et al. GLOBOCAN 2012 v1.0, Cancer Incidence and Mortality Worldwide: IARC CancerBase No. 11 [Internet]. Lyon, France: International Agency for Research on Cancer; 2013. Available from: http://globocan.iarc.fr, accessed on 26th May, 2015.
3. Aslan, B., et al. Nanotechnology in cancer therapy. *J Drug Target*, 2013, **21**(10):904–913.
4. Dobbelstein, M., et al. Targeting tumour-supportive cellular machineries in anticancer drug development. *Nat Rev Drug Discov*, 2014, **13**(3):179–196.
5. O'Neill, V.J., et al. Oral cancer treatment: developments in chemotherapy and beyond. *Br J Cancer*, 2002, **87**(9):933–937.
6. Weingart, S.N., et al. NCCN task force report: oral chemotherapy. *J Natl Compr Canc Netw*, 2008, **6**(Suppl 3):S1–S14.
7. Mazzaferro, S., et al. Oral delivery of anticancer drugs I: general considerations. *Drug Discov Today*, 2013, **18**(1–2):25–34.
8. Terwogt, J.M., et al. Clinical pharmacology of anticancer agents in relation to formulations and administration routes. *Cancer Treat Rev*, 1999, **25**(2):83–101.
9. Liu, G., et al. Patient preferences for oral versus intravenous palliative chemotherapy. *J Clin Oncol*, 1997, **15**(1):110–115.
10. Thanki, K., et al. Oral delivery of anticancer drugs: challenges and opportunities. *J Controlled Release*, 2013, **170**(1):15–40.
11. Irshad, S., et al. Considerations when choosing oral chemotherapy: identifying and responding to patient need. *Eur J Cancer Care*, 2010, **19**:5–11.
12. Regnier Denois, V., et al. Adherence with oral chemotherapy: results from a qualitative study of the behaviour and representations of patients and oncologists. *Eur J Cancer Care (Engl)*, 2011, **20**(4):520–527.
13. Shen, C., et al. A review of economic impact of targeted oral anticancer medications. *Expert Rev Pharmacoecon Outcomes Res*, 2014, **14**(1):45–69.

14. Mei, L., et al. Pharmaceutical nanotechnology for oral delivery of anticancer drugs. *Adv Drug Deliv Rev*, 2013, **65**(6):880–890.

15. Gornas, M., et al. Oral treatment of metastatic breast cancer with capecitabine: what influences the decision-making process? *Eur J Cancer Care (Engl)*, 2010, **19**(1):131–136.

16. Banna, G.L., et al. Anticancer oral therapy: emerging related issues. *Cancer Treat Rev*, 2010, **36**(8):595–605.

17. Schellens, J.H., et al. Modulation of oral bioavailability of anticancer drugs: from mouse to man. *Eur J Pharm Sci*, 2000, **12**(2):103–110.

18. Halfdanarson, T.R., et al. Oral cancer chemotherapy: the critical interplay between patient education and patient safety. *Curr Oncol Rep*, 2010, **12**(4):247–252.

19. Luo, C., et al. Emerging integrated nanohybrid drug delivery systems to facilitate the intravenous-to-oral switch in cancer chemotherapy. *J Controlled Release*, 2014, **176**:94–103.

20. DeMario, M.D., et al. Oral chemotherapy: rationale and future directions. *J Clin Oncol*, 1998, **16**(7):2557–2567.

21. Partridge, A.H., et al. Adherence to therapy with oral antineoplastic agents. *J Natl Cancer Inst*, 2002, **94**(9):652–661.

22. Mazzaferro, S., et al. Oral delivery of anticancer drugs II: the prodrug strategy. *Drug Discov Today*, 2013, **18**(1–2):93–98.

23. Anselmo, A.C., et al. Nanoparticles in the clinic. *Bioeng Transl Med*, 2016, **1**(1):10–29.

24. Singare, D.S., et al. Optimization of formulation and process variable of nanosuspension: an industrial perspective. *Int J Pharm*, 2010, **402**(1–2):213–220.

25. Mazzaferro, S., et al. Oral delivery of anticancer drugs III: formulation using drug delivery systems. *Drug Discov Today*, 2013, **18**(1–2):99–104.

26. Sharma, S., et al. Development of stabilized Paclitaxel nanocrystals: in-vitro and in-vivo efficacy studies. *Eur J Pharm Sci*, 2015, **69**:51–60.

27. Sharma, S., et al. Investigating the role of Pluronic-g-Cationic polyelectrolyte as functional stabilizer for nanocrystals: impact on Paclitaxel oral bioavailability and tumor growth. *Acta Biomater*, 2015, **26**:169–183.

28. Patel, K., et al. Oral delivery of paclitaxel nanocrystal (PNC) with a dual Pgp-CYP3A4 inhibitor: preparation, characterization and antitumor activity. *Int J Pharm*, 2014, **472**(1–2):214–223.

29. Wang, Y., et al. PEG-PCL based micelle hydrogels as oral docetaxel delivery systems for breast cancer therapy. *Biomaterials*, 2014, **35**(25):6972–6985.

30. Saremi, S., et al. Enhanced oral delivery of docetaxel using thiolated chitosan nanoparticles: preparation, in vitro and in vivo studies. *Biomed Res Int*, 2013, **2013**:150478.

31. Wu, J., et al. Sulfobutylether-beta-cyclodextrin/chitosan nanoparticles enhance the oral permeability and bioavailability of docetaxel. *Drug Dev Ind Pharm*, 2013, **39**(7):1010–1019.

32. Feng, C., et al. Chitosan/o-carboxymethyl chitosan nanoparticles for efficient and safe oral anticancer drug delivery: in vitro and in vivo evaluation. *Int J Pharm*, 2013, **457**(1):158–167.

33. Zheng, H., et al. Oral delivery of shRNA based on amino acid modified chitosan for improved antitumor efficacy. *Biomaterials*, 2015, **70**:126–137.

34. Yuan, Z., et al. Preparation, characterization, and in vivo study of rhein-loaded poly(lactic-co-glycolic acid) nanoparticles for oral delivery. *Drug Des Devel Ther*, 2015, **9**:2301–2309.

35. Voruganti, S., et al. Oral nano-delivery of anticancer ginsenoside 25-OCH3-PPD, a natural inhibitor of the MDM2 oncogene: nanoparticle preparation, characterization, in vitro and in vivo anti-prostate cancer activity, and mechanisms of action. *Oncotarget*, 2015, **6**(25):21379–21394.

36. Tariq, M., et al. Biodegradable polymeric nanoparticles for oral delivery of epirubicin: In vitro, ex vivo, and in vivo investigations. *Colloids Surf B*, 2015, **128**:448–456.

37. Graves, R.A., et al. Formulation and evaluation of biodegradable nanoparticles for the oral delivery of fenretinide. *Eur J Pharm Sci*, 2015, **76**:1–9.

38. Bu, X., et al. Co-administration with cell penetrating peptide enhances the oral bioavailability of docetaxel-loaded nanoparticles. *Drug Dev Ind Pharm*, 2015, **41**(5):764–771.

39. Bhatnagar, P., et al. Anti-cancer activity of bromelain nanoparticles by oral administration. *J Biomed Nanotechnol*, 2014, **10**(12):3558–3575.

40. Zhang, H.Y., et al. Ergosterol-loaded poly(lactide-co-glycolide) nanoparticles with enhanced in vitro antitumor activity and oral bioavailability. *Acta Pharmacol Sin*, 2016, **37**(6):834–844.

41. Jain, A.K., et al. Co-encapsulation of tamoxifen and quercetin in polymeric nanoparticles: implications on oral bioavailability,

antitumor efficacy, and drug-induced toxicity. *Mol Pharm*, 2013, **10**(9):3459–3474.

42. Pooja, D., et al. Fabrication, characterization and bioevaluation of silibinin loaded chitosan nanoparticles. *Int J Biol Macromol*, 2014, **69**:267–273.

43. Ray, L., et al. The activity against Ehrlich's ascites tumors of doxorubicin contained in self assembled, cell receptor targeted nanoparticle with simultaneous oral delivery of the green tea polyphenol epigallocatechin-3-gallate. *Biomaterials*, 2013, **34**(12):3064–3076.

44. Elbaz, N.M., et al. Chitosan-based nano-in-microparticle carriers for enhanced oral delivery and anticancer activity of propolis. *Int J Biol Macromol*, 2016, **92**:254–269.

45. Qin, J.J., et al. Oral delivery of anti-MDM2 inhibitor SP141-loaded FcRn-targeted nanoparticles to treat breast cancer and metastasis. *J Controlled Release*, 2016, **237**:101–114.

46. Kiani, M., et al. Thiolated carboxymethyl dextran as a nanocarrier for colon delivery of hSET1 antisense: In vitro stability and efficiency study. *Mater Sci Eng C Mater Biol Appl*, 2016, **62**:771–778.

47. Parayath, N.N., et al. Styrene maleic acid micelles as a nanocarrier system for oral anticancer drug delivery - dual uptake through enterocytes and M-cells. *Int J Nanomed*, 2015, **10**:4653–4667.

48. Li, Z., et al. Critical determinant of intestinal permeability and oral bioavailability of pegylated all trans-retinoic acid prodrug-based nanomicelles: Chain length of poly (ethylene glycol) corona. *Colloids Surf B*, 2015, **130**:133–140.

49. Sadekar, S., et al. Poly(amido amine) dendrimers as absorption enhancers for oral delivery of camptothecin. *Int J Pharm*, 2013, **456**(1):175–185.

50. Lu, Y., et al. Characterization and evaluation of an oral microemulsion containing the antitumor diterpenoid compound ent-11alpha-hydroxy-15-oxo-kaur-16-en-19-oic-acid. *Int J Nanomed*, 2013, **8**:1879–1886.

51. Jain, A.K., et al. Solidified self-nanoemulsifying formulation for oral delivery of combinatorial therapeutic regimen: part II in vivo pharmacokinetics, antitumor efficacy and hepatotoxicity. *Pharm Res*, 2014, **31**(4):946–958.

52. Jain, A.K., et al. Solidified self-nanoemulsifying formulation for oral delivery of combinatorial therapeutic regimen: part I. Formulation

development, statistical optimization, and in vitro characterization. *Pharm Res*, 2014, **31**(4):923–945.

53. Ahmad, J., et al. Solid-nanoemulsion preconcentrate for oral delivery of paclitaxel: formulation design, biodistribution, and gamma scintigraphy imaging. *Biomed Res Int*, 2014, **2014**:984756.
54. Quan, Q., et al. Physicochemical characterization and in vivo evaluation of solid self-nanoemulsifying drug delivery system for oral administration of docetaxel. *J Microencapsul*, 2013, **30**(4):307–314.
55. Kamel, A.O., et al. Enhancement of human oral bioavailability and in vitro antitumor activity of rosuvastatin via spray dried self-nanoemulsifying drug delivery system. *J Biomed Nanotechnol*, 2013, **9**(1):26–39.
56. Cho, H.J., et al. Surface-modified solid lipid nanoparticles for oral delivery of docetaxel: enhanced intestinal absorption and lymphatic uptake. *Int J Nanomed*, 2014, **9**:495–504.
57. Battani, S., et al. Evaluation of oral bioavailability and anticancer potential of raloxifene solid lipid nanoparticles. *J Nanosci Nanotechnol*, 2014, **14**(8):5638–5645.
58. Andey, T., et al. Lipid nanocarriers of a lipid-conjugated estrogenic derivative inhibit tumor growth and enhance cisplatin activity against triple-negative breast cancer: pharmacokinetic and efficacy evaluation. *Mol Pharm*, 2015, **12**(4):1105–1120.
59. Hashem, F.M., et al. In vitro cytotoxicity and bioavailability of solid lipid nanoparticles containing tamoxifen citrate. *Pharm Dev Technol*, 2014, **19**(7):824–832.
60. Ngwuluka, N.C., et al. Design and characterization of metformin-loaded solid lipid nanoparticles for colon cancer. *AAPS PharmSciTech*, 2017, **18**(2):358–368.
61. Neupane, Y.R., et al. Lipid based nanocarrier system for the potential oral delivery of decitabine: formulation design, characterization, ex vivo, and in vivo assessment. *Int J Pharm*, 2014, **477**(1–2):601–612.
62. Groo, A.C., et al. Development of 2D and 3D mucus models and their interactions with mucus-penetrating paclitaxel-loaded lipid nanocapsules. *Pharm Res*, 2014, **31**(7):1753–1765.
63. Groo, A.C., et al. In vivo evaluation of paclitaxel-loaded lipid nanocapsules after intravenous and oral administration on resistant tumor. *Nanomedicine (Lond)*, 2015, **10**(4):589–601.
64. Joshi, N., et al. Carboxymethyl-chitosan-tethered lipid vesicles: hybrid nanoblanket for oral delivery of paclitaxel. *Biomacromolecules*, 2013, **14**(7):2272–2282.

65. Munagala, R., et al. Bovine milk-derived exosomes for drug delivery. *Cancer Lett*, 2016, **371**(1):48–61.

66. Zhang, Z., et al. The use of lipid-coated nanodiamond to improve bioavailability and efficacy of sorafenib in resisting metastasis of gastric cancer. *Biomaterials*, 2014, **35**(15):4565–4572.

67. Ding, S.M., et al. Enhanced bioavailability of apigenin via preparation of a carbon nanopowder solid dispersion. *Int J Nanomed*, 2014, **9**:2327–2333.

68. Zhang, Z.H., et al. Studies on lactoferrin nanoparticles of gambogic acid for oral delivery. *Drug Deliv*, 2013, **20**(2):86–93.

69. Swarnakar, N.K., et al. Bicontinuous cubic liquid crystalline nanoparticles for oral delivery of Doxorubicin: implications on bioavailability, therapeutic efficacy, and cardiotoxicity. *Pharm Res*, 2014, **31**(5):1219–1238.

70. Swarnakar, N.K., et al. Enhanced antitumor efficacy and counterfeited cardiotoxicity of combinatorial oral therapy using Doxorubicin- and Coenzyme Q10-liquid crystalline nanoparticles in comparison with intravenous Adriamycin. *Nanomedicine*, 2014, **10**(6):1231–1241.

71. Vong, L.B., et al. Development of an oral nanotherapeutics using redox nanoparticles for treatment of colitis-associated colon cancer. *Biomaterials*, 2015, **55**:54–63.

72. Zhang, Z., et al. A self-assembled nanocarrier loading teniposide improves the oral delivery and drug concentration in tumor. *J Controlled Release*, 2013, **166**(1):30–37.

73. Razmi, M., et al. Beta-casein and its complexes with chitosan as nanovehicles for delivery of a platinum anticancer drug. *Colloids Surf B*, 2013, **112**:362–367.

74. Varthya, M., et al. Development of novel polymer-lipid hybrid nanoparticles of tamoxifen: in vitro and in vivo evaluation. *J Nanosci Nanotechnol*, 2016, **16**(1):253–260.

75. Yogesh, B., et al. Biosynthesized platinum nanoparticles inhibit the proliferation of human lung-cancer cells in vitro and delay the growth of a human lung-tumor xenograft in vivo: in vitro and in vivo anticancer activity of bio-Pt NPs. *J Pharmacopuncture*, 2016, **19**(2):114–121.

76. Su, C.W., et al. Sodium dodecyl sulfate-modified doxorubicin-loaded chitosan-lipid nanocarrier with multi polysaccharide-lecithin nanoarchitecture for augmented bioavailability and stability of oral administration in vitro and in vivo. *J Biomed Nanotechnol*, 2016, **12**(5):962–972.

77. Izadi, Z., et al. β-lactoglobulin-pectin nanoparticle-based oral drug delivery system for potential treatment of colon cancer. *Chem Biol Drug Des*, 2016, **88**(2):209–216.

78. Roy, K., et al. Biodegradable Eri silk nanoparticles as a delivery vehicle for bovine lactoferrin against MDA-MB-231 and MCF-7 breast cancer cells. *Int J Nanomed*, 2016, **11**:25–44.

Part VII

Anti-Inflammatory Drug Delivery Approaches

Chapter 12

Nanodelivery Systems for NSAIDs: Challenges and Breakthroughs

José Lopes-de-Araújo,[a,*] Catarina Pereira-Leite,[a,b,*] Iolanda M. Cuccovia,[b] Salette Reis,[a] and Cláudia Nunes[a]

[a]*UCIBIO, REQUIMTE, Department of Chemical Sciences, Faculty of Pharmacy, University of Porto, Portugal*

[b]*Department of Biochemistry, Institute of Chemistry, University of São Paulo, Brazil*

cdnunes@ff.up.pt

The toxicity of nonsteroidal anti-inflammatory drugs (NSAIDs) is a major limitation to their long-term use in the treatment of inflammatory conditions. Therefore, research focused on reducing toxicity and improving the efficacy of NSAIDs is far from being complete and encompasses two main fields: the design of novel drugs and the development of delivery systems. Indeed, nanoparticles (NPs) may be used to safely deliver NSAIDs, preventing the occurrence of adverse effects while maintaining or even improving the therapeutic efficacy of these drugs. Herein, the breakthroughs in the development of nanodelivery systems for NSAIDs will be addressed, focusing on

*José Lopes-de-Araújo and Catarina Pereira-Leite contributed equally to the chapter.

Nanoparticles in Life Sciences and Biomedicine
Edited by Ana Rute Neves and Salette Reis

ISBN 978-981-4745-98-7 (Hardcover), 978-1-351-20735-5 (eBook)
www.panstanford.com

the different approaches used to overcome the toxicity of NSAIDs, such as increasing the bioavailability of NSAIDs, targeting inflamed sites, avoiding gastric release, and enhancing skin permeation.

12.1 Introduction

Nonsteroidal anti-inflammatory drugs (NSAIDs) are among the most used medicines worldwide. For instance, approximately 30 million of US adults were regular users of NSAIDs in 2010 [1]. Such global use is supported by the remarkable efficacy of NSAIDs against inflammation, pain, and fever, which mainly arises from the inhibition of prostanoid biosynthesis through the cyclooxygenase (COX) pathway [2]. These pharmaceuticals are indicated for the relief of painful conditions (headache, menstrual pain, etc.) and for the treatment of chronic inflammatory diseases (osteoarthritis, rheumatoid arthritis [RA], etc.). Moreover, NSAIDs are available as prescription-only medicines and as over-the-counter preparations, being considered for short- or long-term therapies [3].

The long-term use of NSAIDs was soon associated with the occurrence of gastrointestinal (GI) adverse events. About 40% of regular users of NSAIDs report upper GI symptoms, from gastroesophageal reflux and dyspeptic symptoms to bleeding, perforation, and obstruction. The latter are known as GI complications, which may occur in the upper or lower GI tract, and 1%–2% of the users of NSAIDs face dangerous complications during therapy [4]. The discovery of an inducible isoform of COX (COX-2) led to the quick development of COX-2 selective inhibitors (coxibs) to overcome NSAID-induced GI toxicity. The development of coxibs was based on the hypothesis that the therapeutic actions of NSAIDs are related to COX-2, while their GI toxicity is associated with the inhibition of COX-1, a constitutive isoform. Nevertheless, the therapy with coxibs was shortly associated with the occurrence of cardiovascular (CV) adverse effects, such as myocardial infarction, heart failure, and stroke. Truly, these events raised questions regarding the CV tolerability of traditional NSAIDs and led to the end of commercialization of some coxibs [5, 6]. Coxibs-related disappointment has driven the further characterization of pharmacological actions of NSAIDs and triggered the advent of several strategies to circumvent the toxicity of NSAIDs, from the design of novel drugs to the development of drug delivery systems.

The emergence of nanomedicine in the last few decades has been providing a broad range of solutions to improve the safety of NSAIDs. Indeed, nanoparticles (NPs) are versatile platforms to achieve better drug pharmacokinetic and pharmacodynamic profiles, thereby avoiding the drug's toxicity and/or improving the drug's therapeutic efficacy. These nanodelivery systems may be engineered to improve drug bioavailability; to protect drugs from degradation; and to control drug release and/or to target specific cells, tissues, or areas. A wide variety of organic and inorganic materials have been used to synthesize NPs for drug delivery, including carbon, lipids, synthetic polymers, metals, proteins, and carbohydrates. Such variety of materials enables the production of NPs with different sizes, shapes, textures, and surface functionalization, which may be adapted to a specific goal [7, 8].

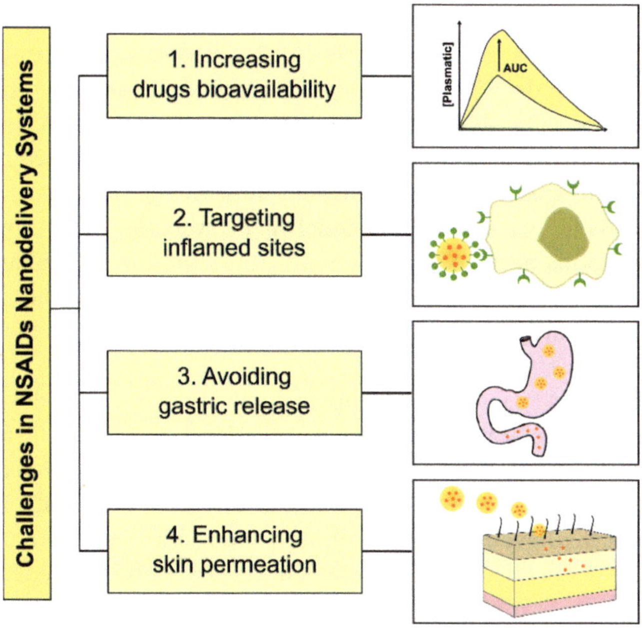

Figure 12.1 Challenges in nanodelivery systems for NSAIDs.

In this context, this section aims to provide an overview of some recent strategies in nanomedicine employed to overcome the toxicity of NSAIDs. In fact, diverse approaches have been considered to improve the safety of NSAID therapy, namely (i) increasing the drug's bioavailability, (ii) targeting inflamed sites, (ii) avoiding gastric release, and (iv) enhancing skin permeation (Fig. 12.1). Thus, the breakthroughs concerning each of these challenges will be discussed in the following sections.

12.2 Challenge 1: Increasing Drugs' Bioavailability

Most NSAIDs are acidic drugs and are classified as class II (high permeability, low solubility) compounds based on the US Food and Drug Administration (FDA) guidance on the Biopharmaceutics Classification System [9]. It means that most NSAIDs exhibit low water solubility, in particular at lower pH, when the compounds are protonated and neutral. In addition, the occurrence of adverse effects caused by NSAIDs is dose dependent, since the incidence of toxic events is greater for higher doses [6]. Therefore, the encapsulation of NSAIDs in drug delivery systems seems to be a useful strategy to increase their solubility and/or to control their in vivo release, thereby increasing the drugs' bioavailability and reducing the needed dosages, which may ultimately improve the safety of NSAIDs.

A wide variety of nanodelivery systems with diverse compositions have been developed with the purpose of enhancing the bioavailability of NSAIDs, such as micellar NPs, liposomes, solid lipid nanoparticles (SLNs), dendrimers, silica NPs, and polymeric NPs.

Micellar NPs made of amphiphilic polymers have been designed to improve the oral bioavailability of NSAIDs [10–14]. For instance, the use of a chitosan derivative to obtain micellar formulations allowed a 60-fold enhancement of celecoxib aqueous solubility. These micellar NPs have exhibited faster, more intense, and more prolonged analgesic effect than celecoxib itself or a suspension obtained from a commercial formulation (Celebrex® from Pfizer) in an in vivo study [11]. Furthermore, Pluronic® was used to efficiently incorporate dexibuprofen in polymeric micelles. The optimized NPs were then lyophilized and used to produce tablets. Such tablets showed a faster release of drug than a commercial tablet of dexibuprofen (Seractil®

from Gebro Pharma GmbH). In vivo absorption studies in healthy human volunteers have also shown a quicker and greater absorption of dexibuprofen from the polymeric-micelle-based tablets [12].

Liposomes have also been used to enhance the solubility and absorption of NSAIDs [15, 16]. For example, chitosan-coated liposomes were developed by Sugihara et al. [16], relying on the mucoadhesive properties of chitosan to facilitate the contact between indomethacin and the absorption site (GI mucosa). Indeed, chitosan-coated liposomes displayed higher retention times in the GI tract than noncoated liposomes and improved the oral absorption of indomethacin in rats. In addition, Zhang et al. [17] have designed a novel nanodelivery system called drug-in-cyclodextrin-in-liposomes, which may circumvent the disruption of liposome properties by direct drug encapsulation and also some problems of cyclodextrin administration, such as renal toxicity and fast drug release. In fact, the optimized nanocarrier significantly increased the bioavailability of flurbiprofen in an in vivo pharmacokinetic study.

The improvement of NSAIDs' solubility and bioavailability has also been accomplished through their encapsulation in SLNs [18–21]. For instance, SLNs composed of shea butter, a natural lipid from *Butyrospermum parkii*, were successfully prepared with high entrapment efficiency for nimesulide. In vivo studies have further shown the significant antinociceptive and anti-inflammatory efficacy of these nimesulide-loaded nanocarriers [18]. Interestingly, Din et al. [22] have developed a dual-reverse thermosensitive SLN-loaded hydrogel for rectal administration. This formulation promoted a sustained drug release, avoiding the initial burst effect, as well as an improvement in the in vivo bioavailability of flurbiprofen, without damaging the rectal mucosa of rats.

Dendrimers have also been used to control the in vivo release and bioavailability of NSAIDs [23–25]. Polyamidoamine (PAMAM) dendrimers have shown to be valuable to increase the solubility of ketoprofen, diflunisal, ibuprofen, and naproxen, as well as to provide a sustained release of naproxen in simulated intestinal conditions [23, 24]. In another study, poly(propyl ether imine) (PETIM) dendrimers have exhibited less cellular toxicity than PAMAM and induced no significant alterations in acute and subacute toxicity experiments in mice. Moreover, PETIM dendrimers have shown to promote a sustained release of ketoprofen in in vitro studies at physiological pH [25].

Similarly, mesoporous silica nanoparticles (MSNs) have been successfully used to encapsulate NSAIDs and to control their release rate [26–29]. The loading capacity of these nanocarriers seems to depend on pore size, impregnation temperature, surface area, and crystal growth, whereas the drug release profile may be conditioned by the diameter and curvature of pores, the drug content, the range of silica order, and pH [27–29]. In addition, the functionalization of MSNs with amino groups increased the drug loading efficiency and further decreased the rate of drug release [26]. Mesoporous silica nanotubes functionalized with amino groups and blue fluorescent CdS quantum dots have also shown to be efficient nanocarriers for ibuprofen. These nanodelivery systems have the additional advantage of using quantum dots as biolabels, which allows the tracking of their location in biological systems [30].

Polymeric NPs have been by far the most explored nanodelivery system to overcome the bioavailability issues of NSAIDs. A wide diversity of synthetic polymers [31–49], proteins [46, 50–53], and polysaccharides [54–59] have been used as matrix ingredients to produce these nanocarriers.

Among synthetic polymers, Eudragit®, poly(d,l-lactide-co-glycolide) (PLGA) and poly(*N*-vinyl pyrrolidone) (PVP) have been extensively used to increase the solubility of NSAIDs and to control their in vivo release. For instance, Eudragit® EPO, a gastrosoluble polymer, was successfully applied to encapsulate meloxicam and to improve its dissolution in a simulated gastric medium. These meloxicam-loaded Eudragit® EPO NPs have also shown higher anti-inflammatory actions and lower ulcerogenic effects than a meloxicam suspension in animal studies [35]. Moreover, several NSAID-loaded PLGA NPs with diverse physicochemical properties have been formulated by varying some parameters of the preparation methodology, such as drug/polymer ratio, type and quantity of the surfactant, and procedure parameters [31–34]. The encapsulation of naproxen in PLGA NPs resulted in a prolonged drug liberation with a reduced initial burst release compared with the intact drug powder and the drug/polymer physical mixtures [33]. Furthermore, PVP NPs were shown to be valuable in improving the dissolution and oral bioavailability of celecoxib and flurbiprofen in in vitro and in vivo studies, respectively [46, 48].

The use of natural polymers, such as proteins and polysaccharides, to generate nanocarriers seems also to be suitable to improve the bioavailability of NSAIDs. First, the incorporation of indomethacin in gelatin NPs ensured a controlled drug release and the enhancement of indomethacin oral bioavailability in animal studies. Additionally, the anti-inflammatory action of this nanodelivery system was confirmed in a rat paw edema model [50]. Second, a PEGylated gelatin nanocarrier was developed to reduce the dosage frequency of ibuprofen sodium when administered parenterally. This formulation has proved to be nontoxic, hemocompatible, nonimmunogenic, and biocompatible in in vitro and in vivo studies. The sustained drug release from this nanodelivery system resulted in a longer circulation time and a better plasma concentration of ibuprofen, which may reduce the number of needed doses [53]. Third, in vivo pharmacokinetic studies in dogs and humans have shown that the encapsulation of celecoxib in ethyl cellulose/casein NPs enabled a higher and faster oral absorption of the drug compared with commercial capsules (Celebrex® from Pfizer). Such improvement may lead to a more rapid onset of the celecoxib therapeutic actions [57].

Other strategies in nanomedicine to improve the bioavailability of NSAIDs comprise the formulation of cubosomes, aquasomes, and drug-in-cyclodextrin-in-NPs [60–63]. Lastly, redox nanoparticles (RNPs) were designed as a combined approach to enhance the bioavailability of NSAIDs and to counteract the NSAID-induced inflammation in the small intestine [64, 65]. These nanotherapeutics consist in polymeric micelles containing nitroxide radicals that are able to scavenge reactive oxygen species (ROS), thereby circumventing inflammation. These RNPs were shown to accumulate in the small intestinal mucosa and to promote a controlled indomethacin release, avoiding the initial burst liberation. Moreover, animal studies with indomethacin-loaded RNPs indicated that this nanodelivery system is valuable in increasing the drug absorption after oral administration and suppressing the small intestinal inflammation related to indomethacin therapy.

12.3 Challenge 2: Targeting Inflamed Sites

Nanodelivery systems may also be engineered to specifically release NSAIDs at the inflammatory sites, avoiding the systemic distribution of the drugs, which may minimize the occurrence of adverse effects. The targeted delivery of NSAIDs to inflammatory sites has been accomplished by three main strategies: (i) the design of nanocarriers coupled with ligands, (ii) the design of pH-sensitive nanocarriers, and (iii) the design of external stimuli-responsive nanocarriers (Fig. 12.2).

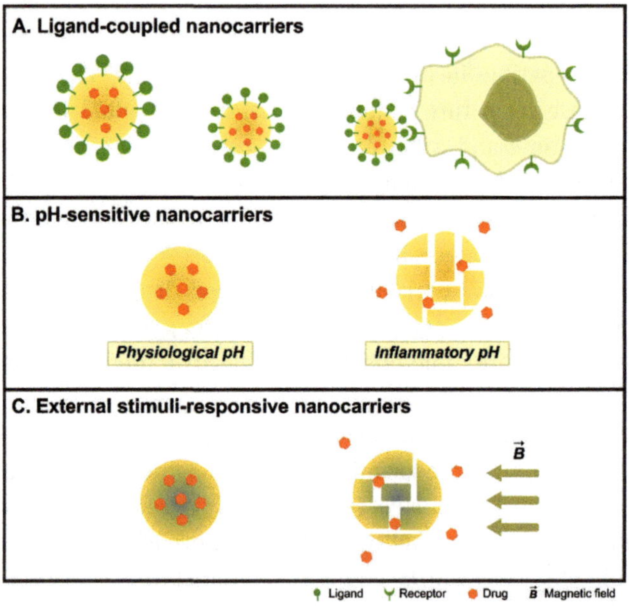

Figure 12.2 Strategies for the target-specific delivery of NSAIDs to inflamed sites.

Diverse ligands have been used to target the nanodelivery systems for NSAIDs at inflamed areas. For instance, folic acid (FA) was exploited to direct nanocarriers to activated macrophages, which overexpress folate receptors (FRs) in the cellular membrane and seem to be crucial effector cells in RA. This targeting strategy was used with dendrimers and bovine serum albumin (BSA) NPs to specifically deliver indomethacin and etoricoxib, respectively

[66–68]. In vivo pharmacokinetic and biodistribution studies demonstrated that these folate-nanocarrier conjugates have high efficiency in targeting the inflammatory region [66–68]. Moreover, albumin-coupled lipid nanoemulsions were successfully prepared to site-specifically deliver diclofenac, since inflammatory tissues are characterized by hypoalbuminemia and expression of albondin receptors [69]. As inflammation triggers the expression of cell adhesion molecules by vascular endothelial cells (VECs), including the vascular cell adhesion molecule (VCAM-1), liposomes were coupled to anti-VCAM-1-FAB' fragments to specifically release celecoxib in activated endothelial cells [70]. In addition, antioxidant-surface-loaded liposomes were also developed to target inflamed sites by taking advantage of the overproduction of ROS in these regions. In detail, coenzyme Q10 and ascorbyl palmitate were shown to be promising targeting ligands when diclofenac-loaded liposomes were intravenously administered [71].

The various microenvironments that exist in the human body have also been exploited for site-specific delivery of NSAIDs. For example, Sousa et al. [72] developed pH-sensitive silica nanotubes to deliver naproxen only in the acidic conditions of inflamed areas. In vitro release studies demonstrated that at physiologic pH virtually no naproxen is released from the nanotubes, while a drug release up to 35% was observed at inflammatory pH (5.0). Naproxen release at inflammatory pH seems to occur due to the partial protonation of the drug, decreasing the strength of the electrostatic interactions between naproxen and the inner surface of nanotubes. Therefore, this may be a promising strategy for target-specific delivery of naproxen after intravenous administration. Moreover, a pH-responsive nanocarrier with a magnetite core and a triblock copolymer shell was designed to direct the delivery of indomethacin to intracellular organelles with characteristic low pH (endosomes and lysosomes). Indeed, this nanodelivery system allowed a faster release of indomethacin in acidic media by the protonation of drug molecules, which decreases the ionic bond between indomethacin and copolymer, and by the swelling of the inner shell due to the protonation of amine groups, which facilitates the drug diffusion [73].

The design of magnetic nanocarriers, which apply an external magnetic field, has also been considered to direct NSAIDs to specific inflamed areas. For instance, magnetite NPs coated with chitosan and oleic acid as well as magnetic NPs composed of an iron core and an ethylcellulose shell were successfully prepared to encapsulate diclofenac for intravenous administration [74, 75]. Moreover, magnetic *N*-benzyl-*O*-carboxymethyl-chitosan NPs were designed to efficiently incorporate indomethacin [76]. Furthermore, biohybrid magnetic matrices composed of alginate and graphite NPs were engineered to load ibuprofen. These bionanocomposite systems when stimulated by an external magnetic field released ibuprofen to a higher extent, showing that the drug release may be modulated by an external stimulus [77]. Hydrogel beads derived from κ-carrageenan and carboxymethyl-chitosan with incorporated magnetic NPs were developed to load diclofenac. The in vitro drug release from this formulation was dependent not only on the environment pH but also on the application of an alternative magnetic field. Thus, this dual-responsive system seems to be a good candidate for site-specific delivery by NSAIDs [78].

12.4 Challenge 3: Avoiding Gastric Release

The gastric side effects of NSAIDs are the most frequent and are in part associated with their topical actions on gastric mucosa, which are closely related to the acidic nature of NSAIDs. Therefore, avoiding the gastric release of these drugs will inhibit their contact with the gastric mucosa and may reduce the occurrence of NSAID-induced gastric side effects. In this context, the design of nanodelivery systems sensitive to pH seems to be a valuable strategy to avoid the release of NSAIDs in gastric media, while promoting a controlled release in other parts of the body.

Polymers have been the most used material to develop these types of pH-sensitive nanocarriers. PEG-b-PCL, for example, is a block polymer composed by polyethylene glycol (PEG) and poly(ε-caprolactone) (PCL) that tends to self-organize into pH-responsive polymeric micelles in water. Ibuprofen encapsulation in these micelles was achieved as well as a controlled release throughout the GI tract after oral administration [79]. In fact, under simulated gastric conditions, these micelles demonstrated to be very stable, releasing less than 20% of the encapsulated ibuprofen. On the other hand, in simulated intestinal conditions, these micelles disrupted, releasing almost all the encapsulated ibuprofen. pH-sensitive polymeric NPs have also been achieved by the combination of a pH-dependent polymer (Eudragit® L100, soluble at intestinal pH) with PLGA [80]. By controlling the polymer ratios, it was possible to obtain a system capable of encapsulating diclofenac and also capable of promoting a controlled drug release at intestinal pH. More recently, a similar strategy was also used by Dupeyron et al. [81], using Eudragit® L100 and PEG to increase indomethacin bioavailability, while promoting its localized release. Once again, the NP characteristics were dependent of the polymer ratios. Moreover, indomethacin was not released at gastric pH, but drug release was observed at intestinal pH with an initial burst liberation. Eudragit® L100 has also been used to produce nanofibers loading diclofenac [82]. In vitro release studies showed that after an oral administration, diclofenac would not be released at gastric pH. However, at intestinal pH the polymer solubilizes, leading to a sustained drug release. Other polymers have also been used to produce pH-responsive NPs [83, 84]. For instance, Zhu et al. [83] produced a self-assembling poly(ethylene imine) (PEI) and β-cyclodextrin nanocarrier. The carrier, based on noncovalent interactions between the drug and the polymers, allowed encapsulation efficiency (EE) rates of over 90% for indomethacin. In vitro release studies showed that at gastric pH, drug release was suppressed. However, at intestinal pH, a fast drug release was observed, with almost all the encapsulated indomethacin released within 24 h. In addition, high indomethacin concentrations were observed in vivo, after oral administration, for up to 4 days with minimal gastric irritation. This nanocarrier also increased indomethacin bioavailability, leading to an area under the curve (AUC) 9 times greater than that of free indomethacin.

Lipid NPs were also designed to hamper the contact of NSAIDs with the gastric mucosa. Indeed, Lopes-de-Araújo et al. [85] synthesized oxaprozin-loaded nanostructured lipid carriers (NLCs). In vitro release studies revealed a pH-dependent release profile, in which the drug release in simulated gastric fluid was avoided. At intestinal pH, high oxaprozin release was observed (up to 70%), which may allow the systemic distribution of the drug. To increase the bioavailability of oxaprozin to inflamed areas, NLCs were also functionalized with FA to target activated macrophages. In vitro cell studies proved the efficient macrophage-targeting capability and low cytotoxicity of these NPs.

The pH-selective release of NSAIDs seems to be a valuable strategy, since it avoids the drug's gastric release and, consequently, may reduce the NSAID-induced gastric side effects, while also controlling the bioavailability of the drugs.

A different approach to avoid gastric release of NSAIDs is the design of colon-targeted NPs. The production of these colonic nanodelivery systems has been accomplished by taking advantage not only of the pH-gradient existent in the GI tract but also of the colonic microflora metabolism. For instance, solid NPs made of hydroxypropyl-β-cyclodextrin and Eudragit® L100 were designed for the colonic transmucosal delivery of diclofenac [86]. As hydroxypropyl-β-cyclodextrin is degradable by the colonic microflora and Eudragit® L100 is not soluble in gastric pH, these nanocarriers only released 25% of diclofenac at pH 1.2, while about 94% of the drug was released at pH 6.8. Moreover, in vitro permeation studies showed that this formulation enables diclofenac permeation through Caco-2 cells and porcine colonic mucosa. Two other nanocarriers were successfully developed for the colonic delivery of diclofenac, namely SLNs made of Compritol® ATO888 and hydroxypropyl-β-cyclodextrin and those made of poly(methyl acrylates)-coated chitosan NPs [87, 88]. Indeed, the optimized formulation of the latter NPs showed a selective drug release at the simulated colonic fluid, together with low cytotoxicity toward Caco-2 cells. These nanocarriers were also capable of inducing a higher uptake of diclofenac into Caco-2 cells [87]. Interestingly, colonic nanodelivery systems have the additional advantage of allowing the topical delivery of NSAIDs for the treatment of colonic inflammatory diseases, such as Crohn's disease and ulcerative colitis.

12.5 Challenge 4: Enhancing Skin Permeation

Transdermal administration and topical administration represent alternative strategies to reduce the adverse effects of NSAIDs, since they can be used, respectively, for systemic drug release and localized drug release. These routes of administration prevent contact between NSAIDs and the gastric mucosa and, in the latter case, the systemic distribution of NSAIDs. Moreover, the transdermal and topical delivery of NSAIDs may be advantageous for the treatment of musculoskeletal and cutaneous inflammatory conditions.

To achieve transdermal administration, several types of nanocarriers have been used. So, to explore the use of lipid nanocapsules (LNCs) for transdermal application, Abdel-Mottaleb et al. [89] performed a comparative ibuprofen transdermal delivery study using LNCs, different types of lipid NPs (NLCs and SLNs), and polymeric NPs. All the lipid carriers revealed to be suitable for transdermal application with a high permeation enhancing effect. However, LNCs showed low accumulation in the skin and higher ibuprofen loading efficiency when compared with SLNs and NLCs. Polymeric NPs revealed a lower permeation enhancing effect and higher drug accumulation on the skin, making them more suitable for dermal applications than for transdermal delivery. Therefore, in this study, LNCs revealed to be a more suitable nanodelivery system for the transdermal application of ibuprofen. Another comparative study with different types of NPs for transdermal delivery was performed by Gonüllü et al. [90]. In this case, SLNs, NLCs, and nanoemulsions containing lornoxicam were prepared and the last nanocarrier provided the highest rate of drug permeation through rat skin, even though all formulations were shown to be stable for 6 months.

Vesicular systems have gained great interest for the transdermal and topical delivery of NSAIDs, resulting in a considerable number of works published in this field [91–108]. Using lipids that are present in the stratum corneum (SC) barrier of the skin, liposomes were developed for the transdermal delivery of diclofenac [91, 92]. For instance, the synthesized vesicles were used in the preparation of gels to improve their application and were compared with a commercial transdermal diclofenac formulation (Voltaren Emulgel® from Novartis) [92]. The final vesicle formulation presented high stability (up to 6 months), drug EE of around 70%–80%, and a controlled diclofenac release (up to 24 h) at physiological pH. The liposomes also presented good compatibility with the skin. Incorporation of the vesicles in a gel promoted its permeation through the thick layer of the skin and an increase in the AUC of almost 6 times, as well as greater edema inhibition compared to the commercial formulation. Therefore, the use of lipids from the SC promoted the liposomes' permeability through the skin, which allowed diclofenac to access the systemic circulation by transdermal delivery.

Still related to the use of lipid vesicular carriers, Caddeo et al. [93] synthesized different types of these carriers (liposomes, ethosomes, and penetration enhancer–containing vesicles [PEVs]) to evaluate their efficiency and safety and to compare their therapeutic efficiency with an aqueous solution of diclofenac and a commercial formulation (Voltaren Emulgel® from Novartis). Regarding the diclofenac solution and the commercial formulation, both revealed continuous diffusion through the skin. However, all the vesicular carriers revealed drug accumulation in the SC, improving diclofenac deposition on the skin and reducing its flux and permeation, thereby preventing its diffusion into the systemic circulation. Moreover, these vesicular carriers successfully reduced inflammation both in ex vivo and in vivo tests. Therefore, liposomes, ethosomes, and PEVs have proved to be useful tools to promote a localized topical drug delivery, avoiding diclofenac diffusion into systemic circulation. Another type of vesicular carrier, transferosomes, was developed to enhance the skin delivery of diclofenac [94]. Transferosomes are optimized ultradeformable lipid vesicles that easily cross skin pores, and this study showed that these nanocarriers can increase the skin permeation and deposition of diclofenac to a higher extent than a marketed product (Olfen® gel from Mepha).

Proniosomes are also interesting vesicular carriers for this purpose, since transdermal application will result in the hydration of these NPs, leading to the formation of niosomes, with great potential for transdermal drug delivery. To reduce the adverse side effects of tenoxicam, Ammar et al. [95] produced a tenoxicam-loaded proniosome gel formulation with a shelf life of up to 3 months. No skin irritation was observed in vivo after applying the formulation for a week, showing that it is a nonirritant. Furthermore, comparing the formulation with an oral marketed tenoxicam tablet (Tilcotil® from Roche), the novel formulation had an increased therapeutic efficacy, with greater anti-inflammatory and antinociceptive effects. Niosomes were also engineered to load diclofenac for transdermal delivery [96]. Indeed, these nanodelivery systems were shown to be percutaneous permeation enhancers in ex vivo studies, since all niosomal formulations induced higher skin permeation of diclofenac than a drug solution.

Transdermal administration of NSAIDs has also been achieved by the use of nanoemulsions [109–112]. For instance, lornoxicam was included in nanoemulsions lately incorporated into a hydrogel [109]. The nanoemulsion and the hydrogel slightly reduced the drug's in vitro release, increased the drug's permeation through the skin, and improved the anti-inflammatory activity of lornoxicam for a prolonged period of time, when compared with a simple lornoxicam hydrogel in in vivo studies. Moreover, to enhance diclofenac transport into the skin and the formation of a drug deposit in the skin for a sustained transdermal drug delivery, Fouad et al. [112] optimized the synthesis of a microemulsion and a poloxamer nanoemulsion-based gel. The comparison of the formulations with the marketed Flector® gel (Institut Biochimique SA) suggested that the nanoemulsion formulation had the most promising properties for a sustained transdermal diclofenac release. In detail, the nanoemulsion showed the highest amount of drug permeation after 8 h and the capability to provide a continuous drug release from the skin, as well as to exhibit an anti-inflammatory activity even after the removal of the microemulsion from contact with the skin.

Lipid NPs have also been designed for the transdermal and topical delivery of NSAIDs [113–118]. Gonzalez-Mira et al. [113], intending to improve the percutaneous absorption and bioavailability of flurbiprofen, formulated two different NLCs based on different solid lipids (stearic acid and Compritol® 888 ATO). Comparing the NLCs with a conventional flurbiprofen solution, the authors verified that both NLCs showed a higher penetration-enhancing ratio, resulting in a greater amount of retained drug in the skin. Ex vivo studies also showed that these formulations provided delayed and sustained drug permeation. So, it is expected that these formulations may promote a sustained local anti-inflammatory effect. Furthermore, both formulations proved to be nonirritants and well tolerated. Similarly, Han et al. [116] formulated NLCs that were inserted in a Carbopol® gel to promote the topical administration of flurbiprofen. The encapsulation of the drug in NLCs increased the in vitro permeation of the drug through the skin by about 1.79-fold in comparison with a gel with a free drug. In contrast to an oral flurbiprofen formulation, the gel led to a smaller plasma concentration and a higher amount of drug accumulated in the skin. Therefore, the gel could be a promising system to promote a localized drug release and to avoid NSAID-related side effects by diminishing the drug's systemic circulation. Interestingly, when compared with a marketed diclofenac topical formulation (Votalin® from Novartis), this flurbiprofen gel demonstrated an equally dose-dependent anti-inflammatory activity by significantly reducing ovalbumin-induced rat paw edema. In addition, the formulation was considered safe since it did not induce irritation in in vivo experiments.

Despite the vast majority of the works for transdermal and topical administration relying on vesicular nanocarriers and lipid NPs, other types of NPs have also been applied for this approach, such as polymeric NPs, quantum dots, lyotrophic liquid crystals, and carbon nanotubes.

Polymeric NPs, such as PLGA-PEG nanospheres with cyclodextrins for the delivery of flurbiprofen, are an example of other types of NPs that have been designed [119]. Vucen et al. [120] also produced polymeric NPs and combined them with the use of solid silicon microneedles for percutaneous ketoprofen delivery. The application of the microneedle arrays associated with the use of polymeric NPs promoted an increased ketoprofen flux across the skin for a long period of time.

Quantum dots have also been engineered to deliver NSAIDs across the skin, as suggested by Degim et al. [121]. The authors successfully increased the transdermal penetration of ketoprofen and dexketoprofen by the use of drug-coated ZnO quantum dots.

Lyotrophic liquid crystals are also interesting systems for these applications [122–125]. Cohen-Avrahami et al. [123–125] developed a series of works with a reversed hexagonal (H_{II}) mesophase made of glycerol mono-oleate. In one of these works, the authors incorporated diclofenac into the H_{II} mesophase to achieve a transdermal system. The capacity of different cell-penetrating peptides (CPPs) to further enhance the transdermal penetration of these systems was also evaluated. Among the studied CPPs, penetratin and oligoarginine revealed to be the best, increasing the drug penetration by 100% due to their interaction with the skin [123].

Carbon nanotubes seem also to be suitable nanocarriers for these administration pathways [126, 127]. Giri et al. [127] suggested a different and innovative strategy that relied on the use of carbon nanotubes incorporated in a hydrogel for the sustained transdermal release of diclofenac to circumvent the drug's adverse effects.

Other types of NPs have also been successfully incorporated in gels to facilitate the application of these nanocarriers on the skin, such as liposomes [128], niosomes [129], lipid NPs [130], and polymeric NPs [131].

Nanodelivery systems for transdermal and topical administration have also been designed to be external-stimuli responsive. First, Spizzirri et al. [132] produced an electroresponsive drug delivery system to modulate the release of diclofenac in response to an external direct current voltage. Carbon nanotubes were used as an electroconducting element and incorporated in a gelatin hydrogel in which diclofenac was incorporated. This drug delivery system promoted a sustained diclofenac release, due to the swelling behavior of the microgels, and the electrical stimulation of the NPs, which resulted in an increased drug release by up to 20%. Second, the application of ultrasounds—sonophoresis—can also be used as an external stimulus to enhance the drug penetration across the skin. In fact, sonophoresis and a hydrogel with diclofenac-loaded dendrimers have already been used as a mechanism to increase the skin penetration of diclofenac [133]. Third, iontophoresis, which uses electropotential energy to increase the skin permeation, was also used to enhance the penetration of negatively charged indomethacin when loaded in PLGA NPs [134, 135].

The use of patches as a reservoir for a sustained release of NSAIDs is already a clinical option. Nevertheless, the compatibility and the overall performance of these patches may be further improved by the use of nanofiber mats as a base material to develop plasters for the transdermal delivery of NSAIDs [136].

12.6 Conclusions

The development of nanodelivery systems seems to be a valuable approach to accomplish an efficacious and safe NSAID therapy. In fact, a wide range of NPs have been designed to increase the bioavailability of NSAIDs, to target the inflamed areas, to avoid their gastric release, and to enhance the skin permeation of NSAIDs. Nevertheless, achieving these goals is still associated with several challenges that must be overcome in the future.

Despite all the breakthroughs achieved in the last few years, there is still a long way to go until the effective use of these nanodelivery systems in current clinical practices. So far, the advances in this field essentially remain in engineering a variety of potential NPs and in performing some preclinical tests, mostly in in vitro studies. Thus, more preclinical and clinical studies are awaited to ultimately support the use of nanodelivery systems for the treatment of inflammatory conditions.

Meanwhile, improving the safety of NSAID therapy remains a growing necessity, not only to improve the quality of life of chronic users, but also to enable the use of these pharmaceuticals in the treatment of a broader spectrum of diseases. Indeed, nanocarriers have also been developed to deliver NSAIDs to cancer cells [137–140], since these drugs appear to exert antiproliferative effects. Therefore, if the toxicity of NSAIDs could be circumvented by nanomedicine, nanodelivery systems for NSAIDs may be a valuable option for the prevention and/or treatment of cancer. Either way, nanomedicine seems to be a promising tool for the future of NSAID therapy.

Acknowledgments

José Lopes-de-Araújo acknowledges Fundação para a Ciência e a Tecnologia (FCT) through the FCT PhD program and Programa Operacional Capital Humano (POCH), specifically the BiotechHealth Programme (Doctoral Programme on Cellular and Molecular Biotechnology Applied to Health Sciences), reference PD/00016/2012, for his PhD grant (PD/BD/114012/2015). Catarina Pereira-Leite thanks FCT, POCH, and the European Union, as well as Conselho Nacional de Desenvolvimento Científico e Tecnológico (CNPq) for her PhD grants (SFRH/BD/109621/2015 and 160446/2013-9, respectively). Cláudia Nunes thanks FCT, POCH, and EU for her investigator grant (IF/00293/2015). Iolanda M. Cuccovia thanks FAPESP (Projeto Temático 2013/08166-5), INCT-FCx, and CNPq (Proc. 301250/2013-8) for financial support. This work also received financial support from the European Union (FEDER funds POCI/01/0145/FEDER/007728) and National Funds (Fundação para a Ciência e a Tecnologia and Ministério da Educação e Ciência [FCT/MEC]) under the partnership agreement PT2020 UID/MULTI/04378/2013.

References

1. Zhou, Y., et al. Trends in the use of aspirin and nonsteroidal anti-inflammatory drugs in the general U.S. population. *Pharmacoepidemiol Drug Saf*, 2014, **23**(1):43–50.
2. Patrignani, P., et al. Cyclooxygenase inhibitors: from pharmacology to clinical read-outs. *Biochim Biophys Acta*, 2015, **1851**(4):422–432.
3. Sweetman, S. *Martindale: The Complete Drug Reference*, 37th ed. Pharmaceutical Press, London, 2011.
4. Sostres, C., et al. Nonsteroidal anti-inflammatory drugs and upper and lower gastrointestinal mucosal damage. *Arthritis Res Ther*, 2013, **15**(Suppl 3):S3.
5. Brune, K., et al. New insights into the use of currently available non-steroidal anti-inflammatory drugs. *J Pain Res*, 2015, **8**:105–118.
6. Harirforoosh, S., et al. Adverse effects of nonsteroidal antiinflammatory drugs: an update of gastrointestinal, cardiovascular and renal complications. *J Pharm Pharm Sci*, 2013, **16**(5):821–847.
7. Heath, J.R. Nanotechnologies for biomedical science and translational medicine. *Proc Natl Acad Sci USA*, 2015, **112**(47):14436–14443.
8. Khan, I., et al. Nanobiotechnology and its applications in drug delivery system: a review. *IET Nanobiotechnol*, 2015, **9**(6):396–400.
9. Yazdanian, M., et al. The "high solubility" definition of the current FDA Guidance on Biopharmaceutical Classification System may be too strict for acidic drugs. *Pharm Res*, 2004, **21**(2):293–299.
10. Payyappilly, S.S., et al. The heat-chill method for preparation of self-assembled amphiphilic poly(ε-caprolactone)-poly(ethylene glycol) block copolymer based micellar nanoparticles for drug delivery. *Soft Matter*, 2014, **10**(13):2150–2159.
11. Mennini, N., et al. Development of a chitosan-derivative micellar formulation to improve celecoxib solubility and bioavailability. *Drug Dev Ind Pharm*, 2014, **40**(11):1494–1502.
12. Abdelbary, G., et al. Adoption of polymeric micelles to enhance the oral bioavailability of dexibuprofen: formulation, in-vitro evaluation and in-vivo pharmacokinetic study in healthy human volunteers. *Pharm Dev Technol*, 2014, **19**(6):717–727.
13. Kuskov, A.N., et al. Preparation and characterization of amphiphilic poly-N-vinylpyrrolidone nanoparticles containing indomethacin. *J Mater Sci Mater Med*, 2010, **21**(5):1521–1530.

14. Kuskov, A.N., et al. Amphiphilic poly-N-vinylpyrrolidone nanoparticles as carriers for non-steroidal anti-inflammatory drugs: characterization and in vitro controlled release of indomethacin. *Int J Mol Med*, 2010, **26**(1):85–94.

15. Deniz, A., et al. Celecoxib-loaded liposomes: effect of cholesterol on encapsulation and in vitro release characteristics. *Biosci Rep*, 2010, **30**(5):365–373.

16. Sugihara, H., et al. Effectiveness of submicronized chitosan-coated liposomes in oral absorption of indomethacin. *J Liposome Res*, 2012, **22**(1):72–79.

17. Zhang, L., et al. Drug-in-cyclodextrin-in-liposomes: a novel drug delivery system for flurbiprofen. *Int J Pharm*, 2015, **492**(1–2):40–45.

18. Raffin, R.P., et al. Natural lipid nanoparticles containing nimesulide: synthesis, characterization and in vivo antiedematogenic and antinociceptive activities. *J Biomed Nanotechnol*, 2012, **8**(2):309–315.

19. Baviskar, D.T., et al. Modulation of drug release from nanocarriers loaded with a poorly water soluble drug (flurbiprofen) comprising natural waxes. *Pharmazie*, 2012, **67**(8):701–705.

20. Kheradmandnia, S., et al. Preparation and characterization of ketoprofen-loaded solid lipid nanoparticles made from beeswax and carnauba wax. *Nanomedicine*, 2010, **6**(6):753–759.

21. Potta, S.G., et al. Preparation and characterization of ibuprofen solid lipid nanoparticles with enhanced solubility. *J Microencapsul*, 2011, **28**(1):74–81.

22. Din, F.U., et al. Novel dual-reverse thermosensitive solid lipid nanoparticle-loaded hydrogel for rectal administration of flurbiprofen with improved bioavailability and reduced initial burst effect. *Eur J Pharm Biopharm*, 2015, **94**:64–72.

23. Ozturk, K., et al. Cytotoxicity and in vitro characterization studies of synthesized Jeffamine-cored PAMAM dendrimers. *J Microencapsul*, 2014, **31**(2):127–136.

24. Koc, F.E., et al. Solubility enhancement of non-steroidal anti-inflammatory drugs (NSAIDs) using polypolypropylene oxide core PAMAM dendrimers. *Int J Pharm*, 2013, **451**(1–2):18–22.

25. Jain, S., et al. Poly propyl ether imine (PETIM) dendrimer: a novel non-toxic dendrimer for sustained drug delivery. *Eur J Med Chem*, 2010, **45**(11):4997–5005.

26. Ahmadi, E., et al. Synthesis and surface modification of mesoporous silica nanoparticles and its application as carriers for sustained drug delivery. *Drug Deliv*, 2014, **21**(3):164–172.

27. Kamarudin, N.H., et al. Variation of the crystal growth of mesoporous silica nanoparticles and the evaluation to ibuprofen loading and release. *J Colloid Interface Sci*, 2014, **421**:6–13.

28. Chen, Z., et al. Mesoporous silica nanoparticles with manipulated microstructures for drug delivery. *Colloids Surf B*, 2012, **95**:274–278.

29. Carriazo, D., et al. Inclusion and release of fenbufen in mesoporous silica. *J Pharm Sci*, 2010, **99**(8):3372–3380.

30. Yang, Y.J., et al. Fluorescent mesoporous silica nanotubes incorporating CdS quantum dots for controlled release of ibuprofen. *Acta Biomater*, 2009, **5**(9):3488–3496.

31. Cooper, D.L., et al. Effect of formulation variables on preparation of celecoxib loaded polylactide-co-glycolide nanoparticles. *PLoS One*, 2014, **9**(12):e113558.

32. Roullin, V.G., et al. Optimised NSAIDs-loaded biocompatible nanoparticles. *Nano-Micro Lett*, 2010, **2**(4):247–255.

33. Javadzadeh, Y., et al. Preparation and physicochemical characterization of naproxen-PLGA nanoparticles. *Colloids Surf B*, 2010, **81**(2):498–502.

34. Turk, C.T., et al. Formulation and optimization of nonionic surfactants emulsified nimesulide-loaded PLGA-based nanoparticles by design of experiments. *AAPS PharmSciTech*, 2014, **15**(1):161–176.

35. Khachane, P., et al. Eudragit EPO nanoparticles: application in improving therapeutic efficacy and reducing ulcerogenicity of meloxicam on oral administration. *J Biomed Nanotechnol*, 2011, **7**(4):590–597.

36. Adibkia, K., et al. Naproxen-eudragit RS100 nanoparticles: preparation and physicochemical characterization. *Colloids Surf B*, 2011, **83**(1):155–159.

37. Nita, L.E., et al. Indomethacin-loaded polymer nanocarriers based on poly(2-hydroxyethyl methacrylate-co-3,9-divinyl-2,4,8,10-tetraoxaspiro (5.5) undecane): preparation, in vitro and in vivo evaluation. *J Biomed Mater Res B Appl Biomater*, 2012, **100**(4):1121–1133.

38. Nita, L.E., et al. An in vitro release study of indomethacin from nanoparticles based on methyl methacrylate/glycidyl methacrylate copolymers. *J Mater Sci Mater Med*, 2010, **21**(12):3129–3140.

39. Michailova, V., et al. Nanoparticles formed from PNIPAM-g-PEO copolymers in the presence of indomethacin. *Int J Pharm*, 2010, **384**(1–2):154–164.

40. Wahab, A., et al. Development of poly(glycerol adipate) nanoparticles loaded with non-steroidal anti-inflammatory drugs. *J Microencapsul*, 2012, **29**(5):497–504.

41. Rodrigues, M.R., et al. Preparation, in vitro characterization and in vivo release of naproxen loaded in poly-caprolactone nanoparticles. *Pharm Dev Technol*, 2011, **16**(1):12–21.

42. Khachane, P., et al. Positively charged polymeric nanoparticles: application in improving therapeutic efficacy of meloxicam after oral administration. *Pharmazie*, 2011, **66**(5):334–338.

43. Racles, C. Polydimethylsiloxane-indomethacin blends and nanoparticles. *AAPS PharmSciTech*, 2013, **14**(3):968–976.

44. Radwan, M.A., et al. Pharmacokinetics of ketorolac loaded to polyethylcyanoacrylate nanoparticles using UPLC MS/MS for its determination in rats. *Int J Pharm*, 2010, **397**(1–2):173–178.

45. Wang, B., et al. Novel PEG-graft-PLA nanoparticles with the potential for encapsulation and controlled release of hydrophobic and hydrophilic medications in aqueous medium. *Int J Nanomed*, 2011, **6**:1443–1451.

46. Haroun, A.A., et al. Synthesis and in vitro release study of ibuprofen-loaded gelatin graft copolymer nanoparticles. *Drug Dev Ind Pharm*, 2014, **40**(1):61–65.

47. Homayouni, A., et al. Comparing various techniques to produce micro/nanoparticles for enhancing the dissolution of celecoxib containing PVP. *Eur J Pharm Biopharm*, 2014, **88**(1):261–274.

48. Oh, D.H., et al. Flurbiprofen-loaded nanoparticles prepared with polyvinylpyrrolidone using Shirasu porous glass membranes and a spray-drying technique: nano-sized formation and improved bioavailability. *J Microencapsul*, 2013, **30**(7):674–680.

49. Suksiriworapong, J., et al. Functionalized (poly(varepsilon-caprolactone))(2)-poly(ethylene glycol) nanoparticles with grafting nicotinic acid as drug carriers. *Int J Pharm*, 2012, **423**(2):562–570.

50. Kumar, R., et al. In-vitro and in-vivo study of indomethacin loaded gelatin nanoparticles. *J Biomed Nanotechnol*, 2011, **7**(3):325–333.

51. Varga, N., et al. BSA/polyelectrolyte core-shell nanoparticles for controlled release of encapsulated ibuprofen. *Colloids Surf B*, 2014, **123**:616–622.

52. Tran, P.H., et al. Enhanced solubility and modified release of poorly water-soluble drugs via self-assembled gelatin-oleic acid nanoparticles. *Int J Pharm*, 2013, **455**(1–2):235–240.

53. Narayanan, D., et al. Poly-(ethylene glycol) modified gelatin nanoparticles for sustained delivery of the anti-inflammatory drug ibuprofen-sodium: an in vitro and in vivo analysis. *Nanomedicine*, 2013, **9**(6):818–828.

54. Shi, Y., et al. Experimental and mathematical studies on the drug release properties of aspirin loaded chitosan nanoparticles. *Biomed Res Int*, 2014, **2014**:613619.

55. Abioye, A.O., et al. Controlled electrostatic self-assembly of ibuprofen-cationic dextran nanoconjugates prepared by low energy green process - a novel delivery tool for poorly soluble drugs. *Pharm Res*, 2015, **32**(6):2110–2131.

56. El-Habashy, S.E., et al. Ethyl cellulose nanoparticles as a platform to decrease ulcerogenic potential of piroxicam: formulation and in vitro/in vivo evaluation. *Int J Nanomed*, 2016, **11**:2369–2380.

57. Morgen, M., et al. Polymeric nanoparticles for increased oral bioavailability and rapid absorption using celecoxib as a model of a low-solubility, high-permeability drug. *Pharm Res*, 2012, **29**(2):427–440.

58. Hassani Najafabadi, A., et al. Synthesis and evaluation of PEG-O-chitosan nanoparticles for delivery of poor water soluble drugs: ibuprofen. *Mater Sci Eng C Mater Biol Appl*, 2014, **41**:91–99.

59. Daus, S., et al. Xylan-based nanoparticles: prodrugs for ibuprofen release. *Macromol Biosci*, 2010, **10**(2):211–220.

60. Zhao, X., et al. Enhanced bioavailability of orally administered flurbiprofen by combined use of hydroxypropyl-cyclodextrin and poly(alkyl-cyanoacrylate) nanoparticles. *Eur J Drug Metab Pharmacokinet*, 2014, **39**(1):61–67.

61. Mura, P., et al. Development of a new delivery system consisting in 'drug-in cyclodextrin-in PLGA nanoparticles'. *J Microencapsul*, 2010, **27**(6):479–486.

62. Kommineni, S., et al. Sugar coated ceramic nanocarriers for the oral delivery of hydrophobic drugs: formulation, optimization and evaluation. *Drug Dev Ind Pharm*, 2012, **38**(5):577–586.

63. Dian, L., et al. Cubic phase nanoparticles for sustained release of ibuprofen: formulation, characterization, and enhanced bioavailability study. *Int J Nanomed*, 2013, **8**:845–854.

64. Sha, S., et al. Suppression of NSAID-induced small intestinal inflammation by orally administered redox nanoparticles. *Biomaterials*, 2013, **34**(33):8393–8400.

65. Yoshitomi, T., et al. Indomethacin-loaded redox nanoparticles improve oral bioavailability of indomethacin and suppress its small intestinal inflammation. *Ther Deliv*, 2014, **5**(1):29–38.

66. Bilthariya, U., et al. Folate-conjugated albumin nanoparticles for rheumatoid arthritis-targeted delivery of etoricoxib. *Drug Dev Ind Pharm*, 2015, **41**(1):95–104.

67. Chandrasekar, D., et al. The development of folate-PAMAM dendrimer conjugates for targeted delivery of anti-arthritic drugs and their pharmacokinetics and biodistribution in arthritic rats. *Biomaterials*, 2007, **28**(3):504–512.

68. Chandrasekar, D., et al. Folate coupled poly(ethyleneglycol) conjugates of anionic poly(amidoamine) dendrimer for inflammatory tissue specific drug delivery. *J Biomed Mater Res A*, 2007, **82**(1):92–103.

69. Kandadi, P., et al. Albumin coupled lipid nanoemulsions of diclofenac for targeted delivery to inflammation. *Nanomedicine*, 2012, **8**(7):1162–1171.

70. Kang, D.I., et al. Preparation and in vitro evaluation of anti-VCAM-1-Fab'-conjugated liposomes for the targeted delivery of the poorly water-soluble drug celecoxib. *J Microencapsul*, 2011, **28**(3):220–227.

71. Jukanti, R., et al. Drug targeting to inflammation: studies on antioxidant surface loaded diclofenac liposomes. *Int J Pharm*, 2011, **414**(1–2):179–185.

72. Sousa, C.T., et al. pH sensitive silica nanotubes as rationally designed vehicles for NSAIDs delivery. *Colloids Surf B*, 2012, **94**:288–295.

73. Guo, M., et al. Multilayer nanoparticles with a magnetite core and a polycation inner shell as pH-responsive carriers for drug delivery. *Nanoscale*, 2010, **2**(3):434–441.

74. Agotegaray, M., et al. Novel chitosan coated magnetic nanocarriers for the targeted Diclofenac delivery. *J Nanosci Nanotechnol*, 2014, **14**(5):3343–3347.

75. Arias, J.L., et al. Development of iron/ethylcellulose (core/shell) nanoparticles loaded with diclofenac sodium for arthritis treatment. *Int J Pharm*, 2009, **382**(1–2):270–276.

76. Debrassi, A., et al. Synthesis, characterization and in vitro drug release of magnetic N-benzyl-O-carboxymethylchitosan nanoparticles loaded with indomethacin. *Acta Biomater*, 2011, **7**(8):3078–3085.

77. Ribeiro, L.N., et al. Bionanocomposites containing magnetic graphite as potential systems for drug delivery. *Int J Pharm*, 2014, **477**(1–2):553–563.

78. Mahdavinia, G.R., et al. Magnetic/pH-responsive beads based on caboxymethyl chitosan and kappa-carrageenan and controlled drug release. *Carbohydr Polym*, 2015, **128**:112–121.

79. Surnar, B., et al. Stimuli-responsive poly(caprolactone) vesicles for dual drug delivery under the gastrointestinal tract. *Biomacromolecules*, 2013, **14**(12):4377–4387.

80. Cetin, M., et al. Formulation and in vitro characterization of Eudragit(R) L100 and Eudragit(R) L100-PLGA nanoparticles containing diclofenac sodium. *AAPS PharmSciTech*, 2010, **11**(3):1250–1256.

81. Dupeyron, D., et al. Design of indomethacin-loaded nanoparticles: effect of polymer matrix and surfactant. *Int J Nanomed*, 2013, **8**:3467–3477.

82. Shen, X., et al. Electrospun diclofenac sodium loaded Eudragit(R) L 100-55 nanofibers for colon-targeted drug delivery. *Int J Pharm*, 2011, **408**(1–2):200–207.

83. Zhu, Y., et al. Highly efficient nanomedicines assembled via polymer-drug multiple interactions: tissue-selective delivery carriers. *J Controlled Release*, 2011, **152**(2):317–324.

84. Yuan, Z., et al. Chitosan-graft-beta-cyclodextrin nanoparticles as a carrier for controlled drug release. *Int J Pharm*, 2013, **446**(1–2):191–198.

85. Lopes-de-Araújo, J., et al. Oxaprozin-loaded lipid nanoparticles towards overcoming NSAIDs side-effects. *Pharm Res*, 2016, **33**(2):301–314.

86. Gavini, E., et al. Development of solid nanoparticles based on hydroxypropyl-beta-cyclodextrin aimed for the colonic transmucosal delivery of diclofenac sodium. *J Pharm Pharmacol*, 2011, **63**(4):472–482.

87. Huanbutta, K., et al. Application of multiple stepwise spinning disk processing for the synthesis of poly(methyl acrylates) coated chitosan-diclofenac sodium nanoparticles for colonic drug delivery. *Eur J Pharm Sci*, 2013, **50**(3–4):303–311.

88. Spada, G., et al. Solid lipid nanoparticles with and without hydroxypropyl-beta-cyclodextrin: a comparative study of nanoparticles designed for colonic drug delivery. *Nanotechnology*, 2012, **23**(9):095101.

89. Abdel-Mottaleb, M.M., et al. Lipid nanocapsules for dermal application: a comparative study of lipid-based versus polymer-based nanocarriers. *Eur J Pharm Biopharm*, 2011, **79**(1):36–42.

90. Gonüllü, U., et al. Formulation and characterization of solid lipid nanoparticles, nanostructured lipid carriers and nanoemulsion of lornoxicam for transdermal delivery. *Acta Pharm*, 2015, **65**(1):1–13.

91. Gaur, P.K., et al. Ceramide-2 nanovesicles for effective transdermal delivery: development, characterization and pharmacokinetic evaluation. *Drug Dev Ind Pharm*, 2014, **40**(4):568–576.

92. Gaur, P.K., et al. Preparation, characterization and permeation studies of a nanovesicular system containing diclofenac for transdermal delivery. *Pharm Dev Technol*, 2014, **19**(1):48–54.

93. Caddeo, C., et al. Inhibition of skin inflammation in mice by diclofenac in vesicular carriers: liposomes, ethosomes and PEVs. *Int J Pharm*, 2013, **443**(1–2):128–136.

94. El Zaafarany, G.M., et al. Role of edge activators and surface charge in developing ultradeformable vesicles with enhanced skin delivery. *Int J Pharm*, 2010, **397**(1–2):164–172.

95. Ammar, H.O., et al. Proniosomes as a carrier system for transdermal delivery of tenoxicam. *Int J Pharm*, 2011, **405**(1–2):142–152.

96. Tavano, L., et al. Drug compartmentalization as strategy to improve the physico-chemical properties of diclofenac sodium loaded niosomes for topical applications. *Biomed Microdevices*, 2014, **16**(6):851–858.

97. Ahad, A., et al. Enhanced anti-inflammatory activity of carbopol loaded meloxicam nanoethosomes gel. *Int J Biol Macromol*, 2014, **67**:99–104.

98. Al-Mahallawi, A.M., et al. Investigating the potential of employing bilosomes as a novel vesicular carrier for transdermal delivery of tenoxicam. *Int J Pharm*, 2015, **485**(1–2):329–340.

99. Bragagni, M., et al. Comparative study of liposomes, transfersomes and ethosomes as carriers for improving topical delivery of celecoxib. *Drug Deliv*, 2012, **19**(7):354–361.

100. Duangjit, S., et al. Comparative study of novel ultradeformable liposomes: menthosomes, transfersomes and liposomes for enhancing skin permeation of meloxicam. *Biol Pharm Bull*, 2014, **37**(2):239–247.

101. Duangjit, S., et al. Evaluation of meloxicam-loaded cationic transfersomes as transdermal drug delivery carriers. *AAPS PharmSciTech*, 2013, **14**(1):133–140.

102. Duangjit, S., et al. Role of the charge, carbon chain length, and content of surfactant on the skin penetration of meloxicam-loaded liposomes. *Int J Nanomed*, 2014, **9**:2005–2017.

103. El-Menshawe, S.F., et al. Formulation and evaluation of meloxicam niosomes as vesicular carriers for enhanced skin delivery. *Pharm Dev Technol*, 2013, **18**(4):779–786.

104. Ghanbarzadeh, S., et al. Enhanced transdermal delivery of diclofenac sodium via conventional liposomes, ethosomes, and transfersomes. *Biomed Res Int*, 2013, **2013**:616810.

105. Nagarsenker, M.S., et al. Potential of cyclodextrin complexation and liposomes in topical delivery of ketorolac: in vitro and in vivo evaluation. *AAPS PharmSciTech*, 2008, **9**(4):1165–1170.

106. Szura, D., et al. The impact of liposomes on transdermal permeation of naproxen--in vitro studies. *Acta Pol Pharm*, 2014, **71**(1):145–151.

107. Uchino, T., et al. Characterization and skin permeation of ketoprofen-loaded vesicular systems. *Eur J Pharm Biopharm*, 2014, **86**(2):156–166.

108. Wen, M.M., et al. Nano-proniosomes enhancing the transdermal delivery of mefenamic acid. *J Liposome Res*, 2014, **24**(4):280–289.

109. Dasgupta, S., et al. In vitro & in vivo studies on lornoxicam loaded nanoemulsion gels for topical application. *Curr Drug Deliv*, 2014, **11**(1):132–138.

110. Rezaee, M., et al. Formulation development and optimization of palm kernel oil esters-based nanoemulsions containing sodium diclofenac. *Int J Nanomed*, 2014, **9**:539–548.

111. Ngawhirunpat, T., et al. Cremophor RH40-PEG 400 microemulsions as transdermal drug delivery carrier for ketoprofen. *Pharm Dev Technol*, 2013, **18**(4):798–803.

112. Fouad, S.A., et al. Microemulsion and poloxamer microemulsion-based gel for sustained transdermal delivery of diclofenac epolamine using in-skin drug depot: in vitro/in vivo evaluation. *Int J Pharm*, 2013, **453**(2):569–578.

113. Gonzalez-Mira, E., et al. Potential use of nanostructured lipid carriers for topical delivery of flurbiprofen. *J Pharm Sci*, 2011, **100**(1):242–251.

114. Gaur, P.K., et al. Solid lipid nanoparticles of guggul lipid as drug carrier for transdermal drug delivery. *Biomed Res Int*, 2013, **2013**:750690.

115. Gaur, P.K., et al. Formulation and evaluation of guggul lipid nanovesicles for transdermal delivery of aceclofenac. *Sci World J*, 2014, **2014**:534210.

116. Han, F., et al. Nanostructured lipid carriers (NLC) based topical gel of flurbiprofen: design, characterization and in vivo evaluation. *Int J Pharm*, 2012, **439**(1–2):349–357.

117. Khalil, R.M., et al. Nanostructured lipid carriers (NLCs) versus solid lipid nanoparticles (SLNs) for topical delivery of meloxicam. *Pharm Dev Technol*, 2014, **19**(3):304–314.

118. Liu, D., et al. Solid lipid nanoparticles for transdermal delivery of diclofenac sodium: preparation, characterization and in vitro studies. *J Microencapsul*, 2010, **27**(8):726–734.

119. Vega, E., et al. Flurbiprofen PLGA-PEG nanospheres: role of hydroxy-beta-cyclodextrin on ex vivo human skin permeation and in vivo topical anti-inflammatory efficacy. *Colloids Surf B*, 2013, **110**:339–346.

120. Vucen, S.R., et al. Improved percutaneous delivery of ketoprofen using combined application of nanocarriers and silicon microneedles. *J Pharm Pharmacol*, 2013, **65**(10):1451–1462.

121. Degim, I.T., et al. Cheap, suitable, predictable and manageable nanoparticles for drug delivery: quantum dots. *Curr Drug Deliv*, 2013, **10**(1):32–38.

122. Uchino, T., et al. Glyceryl monooleyl ether-based liquid crystalline nanoparticles as a transdermal delivery system of flurbiprofen: characterization and in vitro transport. *Chem Pharm Bull (Tokyo)*, 2015, **63**(5):334–340.

123. Cohen-Avrahami, M., et al. Sodium diclofenac and cell-penetrating peptides embedded in H(II) mesophases: physical characterization and delivery. *J Phys Chem B*, 2011, **115**(34):10189–10197.

124. Cohen-Avrahami, M., et al. Penetratin-induced transdermal delivery from H(II) mesophases of sodium diclofenac. *J Controlled Release*, 2012, **159**(3):419–428.

125. Cohen-Avrahami, M., et al. On the correlation between the structure of lyotropic carriers and the delivery profiles of two common NSAIDs. *Colloids Surf B*, 2014, **122**:231–240.

126. Schwengber, A., et al. Carbon nanotubes buckypapers for potential transdermal drug delivery. *Mater Sci Eng C Mater Biol Appl*, 2015, **57**:7–13.

127. Giri, A., et al. Polymer hydrogel from carboxymethyl guar gum and carbon nanotube for sustained trans-dermal release of diclofenac sodium. *Int J Biol Macromol*, 2011, **49**(5):885–893.

128. Fetih, G., et al. Liposomal gels for site-specific, sustained delivery of celecoxib: in vitro and in vivo evaluation. *Drug Dev Res*, 2014, **75**(4):257–266.

129. Kumbhar, D., et al. Niosomal gel of lornoxicam for topical delivery: in vitro assessment and pharmacodynamic activity. *AAPS PharmSciTech*, 2013, **14**(3):1072–1082.

130. Khurana, S., et al. Preparation and evaluation of solid lipid nanoparticles based nanogel for dermal delivery of meloxicam. *Chem Phys Lipids*, 2013, **175–176**:65–72.

131. Jana, S., et al. Carbopol gel containing chitosan-egg albumin nanoparticles for transdermal aceclofenac delivery. *Colloids Surf B*, 2014, **114**:36–44.

132. Spizzirri, U.G., et al. Spherical gelatin/CNTs hybrid microgels as electro-responsive drug delivery systems. *Int J Pharm*, 2013, **448**(1):115–122.

133. Huang, B., et al. Dendrimer-coupled sonophoresis-mediated transdermal drug-delivery system for diclofenac. *Drug Des Devel Ther*, 2015, **9**:3867–3876.

134. Tomoda, K., et al. Enhanced transdermal delivery of indomethacin-loaded PLGA nanoparticles by iontophoresis. *Colloids Surf B*, 2011, **88**(2):706–710.

135. Tomoda, K., et al. Enhanced transdermal delivery of indomethacin using combination of PLGA nanoparticles and iontophoresis in vivo. *Colloids Surf B*, 2012, **92**:50–54.

136. Shi, Y., et al. Design and in vitro evaluation of transdermal patches based on ibuprofen-loaded electrospun fiber mats. *J Mater Sci Mater Med*, 2013, **24**(2):333–341.

137. Sengel-Turk, C.T., et al. Comparative evaluation of nimesulide-loaded nanoparticles for anticancer activity against breast cancer cells. *AAPS PharmSciTech*, 2017, **18**(2):393–403.

138. Sengel-Turk, C.T., et al. Preparation and in vitro evaluation of meloxicam-loaded PLGA nanoparticles on HT-29 human colon adenocarcinoma cells. *Drug Dev Ind Pharm*, 2012, **38**(9):1107–1116.

139. Erdog, A., et al. In vitro characterization of a liposomal formulation of celecoxib containing 1,2-distearoyl-sn-glycero-3-phosphocholine, cholesterol, and polyethylene glycol and its functional effects against colorectal cancer cell lines. *J Pharm Sci*, 2013, **102**(10):3666–3677.

140. Bonelli, P., et al. Ibuprofen delivered by poly(lactic-co-glycolic acid) (PLGA) nanoparticles to human gastric cancer cells exerts antiproliferative activity at very low concentrations. *Int J Nanomed*, 2012, **7**:5683–5691.

Chapter 13

Innovative Target-to-Treat Nanostrategies for Rheumatoid Arthritis

Virgínia Moura Gouveia, Cláudia Nunes, and Salette Reis
UCIBIO, REQUIMTE, Department of Chemical Sciences, Faculty of Pharmacy, University of Porto, Portugal
virginia.mgouveia@gmail.com

RA is a chronic inflammatory and autoimmune disease that manifests in the synovium of joints, hence causing severe functional limitations. Currently, available treatment options include glucocorticoids (GCs) and disease-modifying antirheumatic drugs (DMARDs), used either in monotherapy or in combination therapy. Nevertheless, treatment options only suppress inflammatory symptoms and not the irreversible joint damage. Thus, the disease's chronic nature and off-target therapy often lead to serious adverse side effects. To overcome the drawbacks of conventional therapies, there is increasing interest in nanotherapy. The increasing interest in the development of therapeutic delivery systems at the nanoscale level promises to revolutionize RA treatment management. Nanocarriers' huge potential is based on targeting strategies to selectively aim for and treat inflamed synovial tissues, allowing in situ therapeutic

Nanoparticles in Life Sciences and Biomedicine
Edited by Ana Rute Neves and Salette Reis

ISBN 978-981-4745-98-7 (Hardcover), 978-1-351-20735-5 (eBook)
www.panstanford.com

efficacy and hence remission of disease activity. The therapeutic potential of target-to-treat nanostrategies, aiming to defeat disease progression over the last 10 years of research, is reviewed in this section.

13.1 Rheumatoid Arthritis

Rheumatoid arthritis (RA) is a chronic systemic inflammatory autoimmune disease that afflicts approximately 1% of the population worldwide [1]. Particularly, the average prevalence rates is higher in North America and in northern European countries, compared with southern Europe and developing countries [2].

RA is mainly characterized by progressive inflammation of the synovial tissue of multiple body joints and hence structural joint damage [1, 2]. At the early stage, disease manifestations are the swelling and pain of joints, especially small joints with frequent movement, such as wrists, neck, hands, and feet. Symptoms in weight-bearing joints, such as hips, knees, spine, and ankles, develop later. RA may also affect tissues and organs, which makes it a systemic disease. In later stages, the chronic inflammatory nature of this disease leads to cartilage degradation, bone erosion, and joint stiffness as the disease progresses [1, 3]. In fact, the joint's inflammation is the hallmark of RA to measure disease severity [1, 3].

In that way, although not regarded as a lethal disease, depending on the severity, RA causes severe functional joint limitations [1]. Plus, long-term joint disability and, ultimately, implications for patients′ quality of life shorten their mean life expectancy by approximately 10 years [4].

13.2 Treat with What?

There is no cure for RA. During the first half of the 20th century the disease's progression from symptom onset to major joint deformity was often inevitable [5]. The European League Against Rheumatism (EULAR) and American College of Rheumatology (ACR) guidelines for RA treatment crucially acclaim an early intervention, immediately after disease diagnosis, to delay the beginning of joint damage [6, 7].

So, joint replacement surgery doesn't have to be the final outcome, and patients with RA can maintain a comfortable and active lifestyle with the more suitable treatment [8]. Clinically, the requirements for the management of RA are relied on primarily to provide the patient symptomatic relief, then slow down the disease's inflammatory activity, and finally amend the course of disease remission, hence preventing irreversible joint destruction [9, 10]. Nowadays, EULAR and ACR recommendations for early RA treatment include the use of glucocorticoids (GCs) in combination therapy, preferably with synthetic [6, 7]. Then, further introduction of biological disease-modifying antirheumatic drugs (DMARDs) in the treatment regime should be restricted to patients with advanced prognostic RA signs that did not responded adequately to the previous treatments [6].

GCs, such as prednisolone, methylprednisolone, dexamethasone, and budesonide, represent highly effective drugs in the control of the inflammatory response in the synovium and additionally promote rapid pain relief and reduce the swelling and stiffness manifested in arthritic joints [3, 11, 12]. These steroidal drugs inhibit the inflammatory reaction by controlling synovial neovascularization, thus reducing the release of endothelial growth factors and consequently the migration of leukocytes [3, 11, 13]. Moreover, GCs further prevent inflammation through downregulation of specific pro-inflammatory cytokines expression [3, 11, 13]. Although GCs are the most potent anti-inflammatory drugs in suppressing earlier disease symptoms, the long-term use and high doses can cause severe side effects [10, 12], namely impaired wound healing, skin atrophy, hypertension, weight gain, increased risks of cardiovascular (CV) diseases, osteoporosis, muscle atrophy, glaucoma, gastric ulcer, and manifestation of latent diabetes, leading ultimately to premature mortality [10, 12]. Additionally, their low bioavailability and off-targeted biodistribution profile often limit their therapeutic efficiency in controlling RA symptoms. Thereby, GCs are usually used in low doses in combination therapies meant to control inflammatory symptoms during periods when the disease activity is at its peak [10]. Contrary to GCs, which are mainly used for the symptomatic relief rather than remission of disease activity, synthetic DMARDs such as methotrexate (MTX), sulfasalazine, and also gold salts have specific antirheumatic activity to prevent disease progression [10]. The exact mechanism of action of synthetic

DMARDs is still unclear. Nonetheless evidence indicates that these drugs modulate inflammatory and immune responses [10, 14]. Specifically, MTX is the most used synthetic DMARD in the treatment of both early and established RA, as it has a rapid onset of action, as well as high, prolonged anti-inflammatory efficacy [10]. However, despite proven therapeutic efficacy, the long-term use of MTX is hampered by serious systemic effects, including infection, hepatitis, cirrhosis, and renal dysfunction, often resulting in the cessation of therapy [13, 15]. Furthermore, synthetic DMARDs aren't known for having a direct activity on the relief of inflammatory symptoms, as the suppression of immunologic response is not apparent until months after treatment begins [5, 10]. Therefore, EULAR and ACR recommendations for the treatment of early RA stages support the use of synthetic DMARDs in combination therapy with GCs in low doses and only for short time periods [6, 15].

In the last decade, significant advances in the treatment of RA have been emerging with the introduction of biologic DMARDs, especially for patients who do not respond to treatment with synthetic DMARDs.

Biologic DMARDs are genetically engineered molecules, including antibodies and nucleic acids (deoxyribonucleic acid [DNA] or small interfering ribonucleic acid [siRNA]), able to control inflammatory and autoimmune responses and further modulate disease progression, leading to a more effective prevention of joint disability [16]. The biologic agents' mechanism of action enables the inhibition of the activity of specific inflammatory mediators, like cytokines and chemokines, or their receptors and block the interaction of immune cells with VECs, thereby reducing the migration of immune cells into the synovium [16]. Therapies using biologic agents for the treatment of RA are being developed continuously, including tumor necrosis factor-α (TNF-α) antagonists (such as etanercept, infliximab, and adalimumab), interleukin (IL)-1 receptor antagonist (IL-1RA) anakinra, and IL-6RA tocilizumab [3]. However, despite the promising progress in enhancing RA treatment, the widespread distribution of these biologic molecules in off-target tissues results in severe systemic side effects, requiring repeated and expensive long-term treatment to achieve therapeutic efficiency, which often leads to patient noncompliance [3, 16]. Additionally, treatment-related immune system depressions enhance the possibility of occurrence and re-emergence of viral and bacterial infections (e.g., tuberculosis), as well as increased risk of multiple sclerosis and congestive heart failure [3, 16].

Owing to the above-mentioned limitations, the ideal therapy for RA remains a challenge. Despite current available therapies being highly effective in the inhibition of the inflammatory and immune responses of RA, there are important issues that should be regarded about their therapeutic efficacy and safety, especially upon long-term administration.

13.3 Target What?

Although the exact origin of RA onset remains unknown, it is believed that arthritic inflammation is triggered by an autoimmune system response attacking the joint's synovium [1]. The synovial membrane of an arthritic joint becomes thicker, and the synovium undergoes a sustained inflammatory reaction, ultimately resulting in cartilage tissue damage. The innate immune system response is tangled in synovial pathophysiology, which mainly involves the recruitment of inflammatory cells to the synovium and intracellular signaling pathways for cellular activation and hence enduring inflammation [17].

Inflammatory diseases are commonly characterized by the enhanced permeability of the vasculature tissue, allowing the migration of immune cells, especially T and B cells, to the inflamed tissue [17]. Conversely, B cells mediate the immunologic process by promoting synovial vascularization and angiogenesis [18–20]. T cells are responsible for initiating the disease progression, followed by inflammation [21]. In early synovitis, the inflammatory response is mediated by synovial cells with the production of inflammatory mediators like cytokines and chemokines, which leads to the early recruitment of inflammatory cells to the synovium [17–19]. The synovial inflammation endures through activation of inflammatory cells. In turn, the process of joint destruction is mediated by the activation of intracellular signaling pathways, involving transcription factors, cytokines, chemokines, growth factors, cellular ligands, and adhesion molecules [17]. In the synovium, activated macrophages play a crucial role in perpetuating chronic inflammatory reaction through overproduction of pro-inflammatory cytokines and chemokines, such as TNF-α, IL-1β, and IL-6, which stimulate the proliferation of synovial fibroblasts, also known as synoviocytes [17–19].

Both TNF-α and IL-1β induce activated synoviocytes differentiation and proliferation of osteoclasts, leading to synovial pannus formation [17]. Additionally, IL-6 acts with IL-1β, stimulating synoviocytes to produce tissue degrading matrix metalloproteases (MMPs), which are responsible for cartilage tissue destruction and further bone erosion [18, 19, 22]. Remarkably, synovial fibroblasts are considered to be responsible for the progressing inflammation from one arthritic joint to other unaffected joints, the role bearing resemblance to that of metastatic tumor cells in cancer [23].

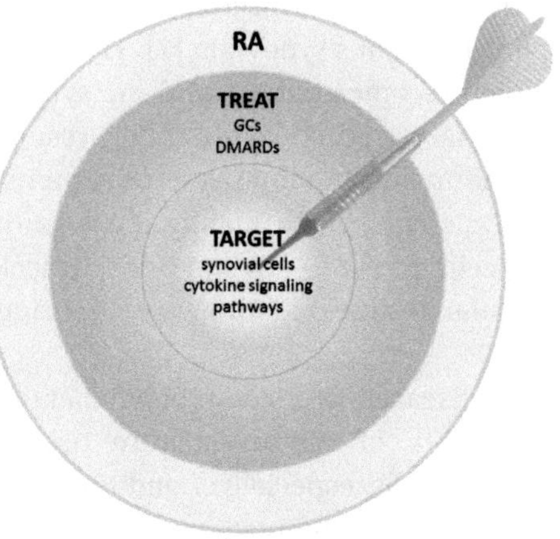

Figure 13.1 Target-to-treat strategies for RA.

Likewise, synoviocytes play a disease-promoting role by the activation of vascular endothelial growth factor (VEGF), inducing synovial vascularization and prominent angiogenesis, comparable to the one that occurs during tumor growth [20, 24]. Cells found within the synovium, mainly macrophages and synoviocytes, are major players in maintaining disease activity mediated by cytokine-dependent pathways. Through cellular interactions between these cells and the innate immune system, plus the production of inflammatory mediators, as well as their involvement in angiogenesis, synovial cells are responsible for the perpetuation of joint inflammation and ultimately the joint's destruction [17]. In this regard, synovial cells and cytokines (Fig. 13.1) have moved into the spotlight as interesting targets for the specific and selective delivery of therapeutic agents for RA treatment.

13.4 Target-to-Treat Nanostrategies

The main drawback of current therapies is the drug off-target activity that increases the risk of off-target toxic effects. Progress in the field of nanotherapy promises interesting selective strategies, based on nanocarriers that deliver therapeutic agents to disease-affected tissues and cells, promoting in situ drug activity and overcoming potential adverse effects on normal tissues. Nanocarrier selectivity can be achieved through a process known as targeting, which basically depends on the properties of the system allowing the delivery of the drug into the site of action [12, 25]. One of the attractive properties of drug delivery systems is their nanometric size, which increases the surface area relatively to the volume, allowing a higher biological interaction and enhanced cell uptake [26]. Moreover, molecules' encapsulation within nanocarriers is useful either to provide protection from physiological degradation or for off-target activation in blood circulation [26]. Therefore, unlike conventional RA therapy, target-to-treat nanostrategies promise to increase stability, specificity, and bioavailability of therapeutic agents in the inflamed tissue and hence enhance their efficiency, while reducing the risk of well-known systemic side effects.

Nanocarriers are useful for increasing the stability of the drug during the long circulation time in blood and resistance to degradation until they reach the disease-affected tissues. In fact, intravenous NP administration often results in rapid plasma protein recognition and hence clearance by cells of the reticuloendothelial system (RES), mainly the cells of the liver and spleen [12, 25, 27]. To avoid NP clearance by RES cells and hence increase their half-life in blood circulation, the nanocarriers should be sterically stable in physiologic fluids [25]. Polymer conjugation is an efficient and well-known approach to modifying the pharmacokinetic properties of NPs. The most successful polymer conjugation is with PEG chains, or PEGylation [28]. NP PEGylation increases the hydrophilicity, as it generates a protective layer over the surface [25, 28]. Thus, the presence of highly hydrated PEG chains results in repulsive interactions with plasma proteins and cellular biological components and, consequently, provides serum stability [29]. Additionally, conjugated PEG chains confer to the NPs the ability of passive tissue targeting, as a result of leaky vasculature and inadequate lymphatic drainage, an effect known as enhanced permeation and retention (EPR) [25, 30]. Whereas lymphatic drainage in inflamed synovial tissues appears to be still functioning, the emphasis is the prominent angiogenesis, similar to that in tumor tissues [21, 25, 30]. Indeed, synovial tissue is characterized by significantly increased vascular permeability, which allows PEGylated NPs to passively accumulate within the inflamed disease-tissue by the EPR effect, in an attempt to further enhance their cellular binding and uptake [21, 25].

The potential of PEGylated nanocarriers, including liposomes, micelles, and polymeric NPs, to passively target the site of inflammation and deliver drugs into it has been reported in several in vitro and in vivo studies (Table 13.1). For example, GC-loaded PEGylated liposomes were described by several authors to exhibit an enhanced passive accumulation within the synovium of arthritic joints compared to healthy joints upon intravenous administration [31–41]. Despite GCs playing a prominent role in the therapeutic management of RA, the occurrence of significant severe effects is a serious limitation with respect to the clinical application of GCs. In this way, Anderson and coworkers investigated the efficacy of intravenous injection of dexamethasone phosphate (DXP) PEGylated liposomes in comparison to the free drug in rats with established adjuvant-induced arthritis (AIA) [38]. Results show that inflammation in the AIA model was reduced, plus histological signs and expression of IL-1β and IL-6 by peritoneal macrophages were verified [38]. In fact, the suppression was long-lasting, as DXP-loaded PEG-conjugated liposomes considerably reduced the dose dependence by a factor of 3–10 compared to the injected free DXP [38]. Thereby, the therapeutic efficiency of DXP-loaded PEGylated liposomes was enhanced, limiting the unwanted toxicity of GC therapy in RA [38]. Similarly, Ishihara and colleagues obtained the same prolonged anti-inflammatory

Table 13.1 PEGylated target-to-treat nanocarriers for RA

Nanocarrier	Therapeutic agent	Preclinical tests	Ref.
Liposomes	PLP	In vitro fibroblast and macrophages cell lines In vivo AbIA murine model	[31] [32] [33] [34] [35]
Liposomes	DXP	In vivo CIA mouse model AIA rat model	[36] [37] [38]
Liposomes	BSP	In vivo AA mouse model	[39]
Liposomes	MPS BMS	In vivo AIA and AA rat models	[40] [41]
Micelles	DXP	In vivo AbIA mouse model	[42]
Polymeric nanoparticles	BSP	In vivo AA rat model AbIA mouse model	[43] [44]
Liposomes	MTX	In vivo AA rat model	[45]
Metallic nanoparticles	Gold	In vivo CIA rat model	[46]

AA, adjuvant arthritis; AbIA, antibody-induced arthritis; AIA, adjuvant-induced arthritis; BMS, betamethasone hemisuccinate; BSP, betamethasone sodium phosphate; CIA, collagen-induced arthritis; DXP, dexamethasone phosphate; MPS, methylprednisolone hemisuccinate; MTX, methotrexate; PLP, prednisolone phosphate.

effect with NPs of PLGA and poly(lactic acid) (PLA) conjugated with PEG for betamethasone sodium phosphate (BSP) delivery within the synovium [43, 44]. These polymeric NPs were then intravenously administered to rats with adjuvant arthritis (AA) and mice with anti-type II collagen antibody-induced arthritis (AbIA). In AA rats, a single injection of BSP-loaded PEGylated NPs resulted in 35% joint inflammation reduction within 1 day, which was maintained for 9 days [43]. On the other hand, in AbIA mice, a single injection resulted in complete remission of the inflammatory response after

1 week [43]. Moreover, the same authors have observed that due to PEG by the EPR effect BSP-loaded PEGylated NPs preferentially accumulated in the inflammatory lesion in AIA mice models [44]. Then, the loss of PEG and subsequent uptake by inflammatory macrophages allow intracellular drug release [44]. In another study, Hofkens and coworkers investigated the effect on the inhibition of protease expression after intravenous injection of prednisolone phosphate (PLP)-loaded liposomes in the AbIA murine model [33]. In vivo results revealed that a single injection promotes a noticeable suppression of synovial immune cell accumulation compared to the control [33]. Further in vitro experiments showed that there was efficient uptake of PLP-loaded PEGylated liposomes by macrophages and, consequently, an inhibition of the expression of IL-1β and MMPs in the synovium was observed after induced inflammation [33]. Thus, PEG-conjugated liposomes encapsulated with PLP could target the synovium, to internalize within macrophages and inhibit destruction of the cartilage matrix in AbIA model [33]. PEGylated nanocarriers increased drug therapeutic effect in arthritis models, possibly due to prolonged blood circulation, thus preventing interactions with RES cells. Moreover, the passive targeting of the inflamed synovial tissues increases the drug´s bioavailability in situ.

13.4.1 Synovial Cell Targeting

Over the years, targeting strategies (Fig. 13.2) have been explored to maximize a nanocarrier's selectivity to further allow specific NP-cell binding and uptake [12, 27]. In this context, the targeting potential is based on the unique pathophysiological features of RA, namely the accumulation of inflammatory cells within the synovium, the production of inflammatory mediators, and angiogenesis.

Targeting strategies take advantage of the roles of the cells found within the synovium. In contrast to other inflammatory diseases, arthritic inflammation involves the collaboration of numerous different cells, including mesenchymal cells (such as synovial fibroblast), macrophages, endothelial cells, dendritic cells, and other cells of the immune system [17, 19]. Thus, in addition to signaling pathophysiologic pathways, the cellular interactions in the synovium are also the key to perpetuating cellular proliferation and chronic synovitis [17].

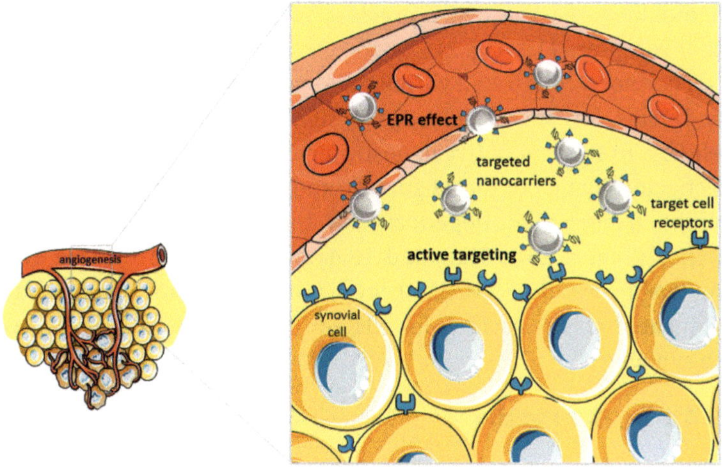

Figure 13.2 Targeting strategies for RA.

The activation of synovial mesenchymal cells, especially synoviocytes, is believed to be crucial in the pathogenesis involved in joint destruction through the synthesis of tissue-degrading molecules. Synoviocytes, which in healthy joints are involved in synovial homeostasis, also contribute directly to synovial chronic inflammation by excreting pro-inflammatory cytokines and chemokines for the signaling and proliferation of other mesenchymal and immune cells [17].

For example, T cell proliferation is enhanced by synovial fibroblasts through the release of IL-16. In addition, the activation of T cells in the synovium is believed to be strongly B cell dependent [17]. Remarkably, synoviocytes promote B cell survival and differentiation into plasma cells and produce B cell activation factors [17]. Conversely, macrophages that line the synovium play a key role in the inflammatory process, mainly by producing potent pro-inflammatory cytokines and chemokines that perpetuate inflammation [19]. Particularly, IL-1β and TNF-α stimulate synoviocyte activation and hence lead to irreversible joint destruction [19]. Furthermore, the inflamed synovium line is characterized by an extensive accumulation of T cells, B cells, plasma cells, natural killer cells, and dendritic cells [17]. The interaction between these cells and macrophages results in the production of chemokines and the overexpression of chemokine receptors and other receptors [17, 19].

Therefore, cells found within the synovium are interesting target candidates for successful specific and selective intracellular NP delivery. To this end, active nanostrategies focus on the identification of effector ligands to target specific and selective cellular receptors by the nanocarrier. Depending on the cell of interest, several studies (Tables 13.2, 13.3, and 13.4) have reported the use of an FA, hyaluronic acid (HA), or cyclic Arg-Gly-Asp (RGD) peptide conjugated in nanocarriers for effector ligand-mediated targeting of the synovium.

13.4.1.1 FA targeting

Folic acid (FA) is a water-soluble vitamin that can induce receptor-mediated endocytosis, providing cytosolic drug delivery [47]. Folate receptors (FRs) are cysteine-rich cell surface glycoproteins that can exist in two isoforms: FRα and FRβ. Usually, FRα is addressed as a target for therapy and imaging in oncology, whereas FRβ has been reported in some studies as the cell effector receptor in inflammatory diseases. [48] This has unlocked the possibility to explore the potential of FA to target therapeutic agents to treat inflammatory RA. The feasibility of FRβ to mediate the active targeting delivery of nanocarriers is based on its specific overexpression on activated macrophages found in the synovial fluid, unlike the slight expression of this receptor on quiescent resident macrophages and other blood cells [49–52]. Activated macrophages are the key cells in RA pathogenesis that secrete multiple pro-inflammatory mediators with a central role in intracellular cytokine-mediated pathways involved in synovial inflammation and tissue-matrix degeneration [17].

Recent in vitro studies (Table 13.2) reported FRβ's potential to mediate specific delivery to activated macrophages in the synovial tissue of RA patients [47, 49, 51–54]. Likewise, in vivo studies on animal models of arthritis have shown that FA nanocarriers can selectively deliver therapeutic agents to joints' synovial tissues, showing reduced collateral toxicity comparing to normal tissues

[47, 51, 52, 55, 56]. Thomas and coworkers investigated the anti-inflammatory potential of FA-conjugated polyamidoamine (PAMAM) dendrimers to target macrophages and deliver loaded MTX in a collagen-induced arthritis (CIA) rat model [47]. The in vitro results showed that FA conjugation significantly increased PAMAM dendrimers' uptake rate via FR-mediated endocytosis expressed in activated synovial macrophages [47]. Moreover, in vivo studies revealed that FA-PAMAM dendrimers loaded with MTX increased the therapeutic anti-inflammatory index over free MTX. As arthritis-induced parameters of inflammation were reduced, an enhancement of the maximum tolerated dose in the used arthritic model was also observed [47]. Thus, the use of FA-targeted dendrimers to specifically delivery MTX into macrophages within the synovium can provide an efficient active targeting approach for RA treatment [47].

Table 13.2 FA nanocarriers targeting synovial cells

Nanocarrier	Therapeutic agent	Preclinical tests	Ref.
Dendrimers	MTX	In vitro macrophage cell line In vivo CIA rat model	[47]
Dendrimers	IND	In vivo AIA rat model	[57] [58]
Polymeric nanoparticles	IL-1 receptor antagonist gene	In vivo AIA rat model	[56] [55]
Lipid nanoparticles	NF-κB	In vitro macrophage cell line	[59]

AIA, adjuvant-induced arthritis; CIA, collagen-induced arthritis; FA, folic acid; IL, interleukin; IND, indomethacin; MTX, methotrexate; NF-κB, nuclear factor kappa B.

FA has been also used to achieve active targeting of macrophages in gene therapy. Fernandes and, later, Shi designed FA-conjugated chitosan NPs to deliver IL-1RA gene encoding plasmid DNA [55, 56]. The formulation administered intravenously in AIA rats showed to inhibit paw inflammation. Indeed, an improved transfection efficiency of FA-chitosan NPs loading IL-1RA resulted in a significant downregulation of TNF-α and IL-1β [55]. Further in vitro analysis of macrophages isolated from mouse arthritic joints suggested that IL-1RA plasmid DNA was efficiently and specifically delivered through FR-mediated targeting of macrophages [56]. In another study, Hattori and colleagues investigated the intracellular delivery of a nuclear factor kappa B (NF-κB) decoy using FA-conjugated lipid-based NPs into macrophages [59]. NF-κB is a key regulator of gene transcription highly active in the synovial membrane. NF-κB-dependent signaling pathway in activated macrophages directly induce osteoclast for cartilage and bone destruction [17]. In vitro experiments suggested that FA-lipid NPs effectively delivered the NF-κB decoy into the macrophage's cytoplasm, resulting in an inhibitory effect on the translocation of NF-κB into the nucleus, providing protection against bone resorption by osteoclasts [59].

13.4.1.2 HA targeting

HA is a glycosaminoglycan responsible for cellular growth and tissue integrity maintenance, thus being an essential component of the extracellular matrix [60, 61]. HA has the potential to selectively bind to CD44 antigen spliced variant isoforms CD44v4 and v6, which are known to be involved in cell migration and in the regulation of inflammation through lymphocyte activation [60, 62, 63]. Several in vitro and in vivo studies have shown that inflamed synovial tissue contains high expression levels of the CD44 receptor on both synoviocytes and macrophages compared to healthy normal tissue [25, 61–65]. Therefore, CD44's importance in the chemotaxis phenomenon, as well as in mediating the specific binding of HA during inflammation, makes it a suitable candidate for the nanocarrier's specific delivery within the inflamed synovium.

Several studies (Table 13.3) have investigated the potential of HA for active targeting of synovial cells plus for the safe delivery of therapeutic agents in arthritis animal models. Shin and colleagues investigated the therapeutic effect of MTX-loaded HA-conjugated NPs in CIA mouse models [66]. In vivo studies showed that HA NPs preferentially accumulated in the inflamed joints and were substantially more potent than the free drug in controlling the clinical score of joint swelling [66]. These results suggest that the enhanced therapeutic efficiency of MTX might owe to the conjugate's ability to enable the release of MTX under acidic conditions found in both extracellular inflamed synovium and intracellular compartments of macrophages [66]. Further in vitro experiments revealed an efficient significant uptake of conjugated HA-MTX by the activated macrophage cell line [66]. Similarly, Kamat and coworkers developed iron oxide–based magnetic NPs conjugated to HA to study their specific recognition by the CD44 receptor present on activated macrophages [67]. In vitro cell uptake studies also demonstrated significant uptake of HA NPs by an activated macrophage cell line [67]. Moreover, fluorescein contained on the NPs was found to be delivered to the cell nucleus [67]. This means that HA-conjugated nanocarriers are suitable for both molecular imaging and drug delivery and targeting of inflamed disease-tissue. In another study, not only is HA used as a targeting strategy, but Mero et al. and Ryan et al. have been using the polymeric properties of HA to develop NPs able to accumulate within the synovium and reduce the EPR-driven leakage from the synovial joint upon intra-articular injection. In vivo studies by Mero and colleagues revealed that salmon calcitonin (sCT)-conjugated HA NPs avoided EPR-driven leakage, and hence systemic off-target effects, known to occur upon intra-articular injection of sCT formulation in a CIA mouse model [68]. Conversely, Ryan and coworkers' investigation showed that sCT conjugation with HA NPs strongly inhibited the production of pro-inflammatory cytokine and matrix-degrading enzymes, such as MMPs [69]. Indeed, the NF-κB mouse model of arthritis and the rabbit model of early osteoarthritis used in the studies presented reduced signs of inflammation after treatment [69].

Synovial vascularization and prominent angiogenesis are crucial for enduring synovitis and progressive joint destruction [20]. Likewise, the VEGF and its receptor found in the synovium are essential elements for neovascularization [20]. In this regard, in addition to HA's synovial-targeting ability, Lee and coworkers developed a HA-conjugated tocilizumab (TCZ) gold NP, used simultaneously for a targeting and biologic therapeutic approach. TCZ is a monoclonal therapeutic IL-6 antibody against IL-6 signaling by binding to the interleukin-6 receptor (IL-6R) [73]. IL-6 present in the synovium actively contributes to the RA pathogenesis by stimulating synoviocyte cell proliferation and osteoclast activation, plus the production of MMPs, hence promoting cartilage destruction. Additionally, this pro-inflammatory cytokine influences the VEGF expression for the vascularization of inflamed synovial tissue. Therefore, IL-6 is an interesting cytokine target used for RA treatment [17]. Remarkably, the HA-gold NP TCZ complex administered intravenously in a CIA mouse model, aside from being able to target synovial tissues and accumulate within the synovium, also reveals TCZ ability to target angiogenic blood vessels [72, 73]. The observed reduction of VEGF and IL-6 expression in the synovium suggests that treatment with the HA-gold NP TCZ complex had an anti-inflammatory and antiangiogenic effect in arthritic mouse models [72].

Table 13.3 HA nanocarriers targeting synovial cells

Nanocarrier	Therapeutic agent	Preclinical tests	Ref.
Polymeric nanoparticles	MTX	In vitro macrophage cell line In vivo CIA mouse model	[66]
Polymeric nanoparticles	γ-secretase	In vivo CIA mouse model	[70]
Polymeric nanoparticles	sCT	In vivo arthritis rabbit and NF-κB mouse models	[68] [69]
PEG-polymeric nanoparticles	TRAIL	In vivo CIA mouse model	[71]
Metallic nanoparticles	TCZ gold	In vivo CIA mouse model	[72]
Magnetic nanoparticles	Fluorescent imaging molecule	In vitro macrophage cell line	[67]
Polymeric nanoparticles	Photosensitizers	In vitro macrophage cell line	[62]

CIA, collagen-induced arthritis; HA, hyaluronic acid; MTX, methotrexate; NF-κB, nuclear factor kappa B; PEG, polyethylene glycol; sCT, salmon calcitonin; TCZ, tocilizumab; TRAIL, TNF-related apoptosis-inducing ligand.

An alternative therapeutic targeting and biologic approach was also investigated by Kim and colleagues [71]. Instead of IL-6, in this study proapoptotic members of the TNF family, such as tumor necrosis factor–related apoptosis-inducing ligand (TRAIL), were efficiently delivered in CIA mouse models using the HA targeting strategy. In fact, in vivo experiments suggested that HA conjugation to PEGylated NP loading TRAIL resulted in prolonged and sustained delivery of TRAIL with increased therapeutic effect over the unconjugated PEG-TRAIL formulation [71].

13.4.1.3 RGD targeting

Synovial vascularization represents a potential therapeutic target area, possibly by being the first barrier that blood leukocytes face before their migration into the target synovial tissue [20]. In fact, the interaction between cell adhesion molecules expressed on the endothelial cells and migrating blood leukocytes may direct specific leukocyte subsets to the target tissue [20]. In this regard, some studies (Table 13.4) have been reporting cell adhesion receptors, such as integrins, expressed on endothelial cells as therapeutic targets for RA. Particularly, the RGD peptide displays high affinity to the $\alpha v\beta 3$ integrin, which is overexpressed in VECs and macrophages in inflamed disease-tissues [74].

Table 13.4 RGD nanocarriers targeting synovial cells

Nanocarrier	Therapeutic agent	Preclinical tests	Ref.
PEG-liposomes	DXP	In vivo AIA rat model	[75]
PEG-liposomes	PLP core peptide	In vivo AIA rat model	[76]
Polymeric nanoparticles	MTX gold	In vivo CIA mouse model	[77]

AIA, adjuvant-induced arthritis; CIA, collagen-induced arthritis; DXP, dexamethasone phosphate; MTX, methotrexate; PEG, polyethylene glycol; PLP, prednisolone phosphate; RGD, Arg-Gly-Asp.

Koning and coworkers investigated the therapeutic efficacy of DXP-loaded RGD conjugated to PEGylated liposomes using an AIA rat model [75]. The in vivo results revealed that RGD-PEG liposomes' retention time at the inflammation site was greater compared with PEGylated ones. Besides, in vitro studies showed that RGD enhances the uptake in proliferating human VECs [75], thereby suggesting that the specific RGD targeting is more effective than the EPR effect in terms of targeting inflamed disease-tissue in early stages of arthritis [75]. Moreover, in vivo studies showed that a single intravenous injection of DXP-loaded RGD-PEG liposomes prolonged the anti-inflammatory effect in arthritic joints, thus limiting off-target tissue toxicity [75]. Similarly, Vanniasinghe and colleagues developed RGD-PEG liposomes to deliver either PLP or an immunosuppressive core peptide within the synovium [76]. In this study, liposome formulations were tested in vitro regarding their ability to specifically internalize on both synovial fibroblast and endothelial cells [76]. Additionally, in vivo studies were performed using a rat model of AIA, revealing that the accumulation of RGD-PEG liposomes increased 7- to 10-fold in inflamed joints compared to unaffected joints [76]. Thereby, the specific delivery of loaded therapeutic agents to inflamed joints led to efficient and prolonged anti-inflammatory efficacy [76].

Lately, a strategy has been developed to improve the ability to mediate intracellular controlled delivery of nanocarriers, along with triggering the release of loaded therapeutic agents, by hyperthermia. To this end, Lee and coworkers developed MTX-loaded PLGA NPs coated with gold half-shell and conjugated to the RGD peptide [77]. In vivo results using CIA mice showed that RGD allowed tissue-specific accumulation of NPs and cell targeting through $\alpha V\beta 3$ integrin [77]. Then, upon application of local near-infrared photothermic stimulation on arthritic joints, whereas heat was locally generated due to gold half-shells, a MTX intracellular release increase was observed [77]. Hence, prolonged therapeutic efficacy was observed when compared with that of the free drug, which simultaneously minimizes MTX-dose-related side effects in the treatment of RA [77]. Similarly, in another study, MTX-loaded PLGA NPs containing gold NPs were investigated for near-infrared photothermic application [78]. In vitro experiments suggested that the incorporation of gold NPs resulted in a temperature-dependent and sustained drug release profile [78]. Further results revealed that the successful uptake of MTX-loaded PLGA/gold NPs by macrophages led to a significant reduction of pro-inflammatory cytokine expression after induced inflammation [78], suggesting a hyperthermia therapy as a promising strategy to improve the drug´s anti-inflammatory efficiency [78]. Remarkably, gold has been also described to exhibit therapeutic potential in RA treatment. Indeed, Tsai and coworkers used gold NPs to target VEGF in human RA synovial fluid, showing that these NPs inhibit VEC proliferation and migration [46]. Additionally, in vivo studies in CIA mice revealed that intra-articular injection of gold NPs suppresses synovial cells' inflammatory response, presenting reduced clinical, radiographic, and histologic arthritic features [46].

13.4.2 Cytokine Targeting

The imbalance of pro-inflammatory cytokines and anti-inflammatory cytokines plays a key role in the pathogenesis of RA. In the inflamed synovium, pro-inflammatory cytokines, such as TNF-α, IL-1, IL-6, IL-12, and chemokines, are elevated, whereas the levels of anti-inflammatory cytokines, such as IL-10 and IL-4, are reduced [17]. Aside from stimulate inflammatory cells' migration to the inflamed synovium, cytokines are also responsible for perpetuating both autoimmune and inflammatory responses [3]. Moreover, cytokine-dependent signaling pathways mediate cellular interactions within the synovium, leading to persistent disease activity and hence irreversible joint destruction [17].

Table 13.5 Target-to-treat nanocarriers toward cytokines

Nanocarriers	Therapeutic agent	Preclinical tests	Ref.
Polymeric nanoparticles	Etanercept	In vivo CIA rat model	[79]
Polymeric nanoparticles	TNF-α siRNA	In vivo AbIA mouse model	[80]
Polymeric nanoparticles	TNF-α siRNA	In vivo CIA mouse model	[81] [82]
Liposomes	TNF-α siRNA IL-1/IL-6/IL-18 siRNA	In vivo CIA mouse model	[83] [84]
Liposomes	TNF-α siRNA	In vivo CIA mouse model	[85] [86]
Liposomes	IL-10 plasmid DNA	In vivo CIA mouse model	[87] [88]
Polymeric nanoparticles	IL-2/IL-15Rβ siRNA	In vivo AIA rat model	[89] [90]
Polymeric nanoparticles	TNF-α siRNA	In vitro macrophage cell line	[91]

AbIA, antibody-induced arthritis; AIA, adjuvant-induced arthritis; CIA, collagen-induced arthritis; DNA, deoxyribonucleic acid; IL, interleukin; siRNA, small interfering ribonucleic acid; TNF-α, tumor necrosis factor-α.

The selective downregulation of inflammatory mediators either by preventing their expression or by antagonizing their receptors using antibodies is simultaneously a promising therapeutic approach as well as a targeting approach used in the treatment of RA. Thus, mainly due to their involvement within the pathogenesis of RA, cytokines have been used as targets for monoclonal antibody therapy in RA. Formulations of monoclonal antibodies effectively inhibit inflammation and synovial tissue damage. Still after systemic administration many patients continue to experience severe off-target side effects and less symptomatic relief [21]. Furthermore, cytokines can be targeted therapeutically by gene transfer and gene-silencing approaches. Gene therapy is based on the delivery of nucleic acids within the cell either for overexpression or for silencing of a target cytokine [92]. More recently, nanotechnology has been combined with gene therapy, offering an enhanced strategy for targeting and treating RA after it can potentially lead to a safer, more selective, and more efficient delivery of therapeutic gene at the inflamed disease-tissue, hence avoiding adverse off-target systemic toxicity.

In the last decade, several studies (Table 13.5) have been developed for treating synovial inflammation by silencing the expression of target pro-inflammatory cytokines with targeted gene delivery systems. Regarding gene delivery in vivo, cationic nanocarriers represent a safe and selective strategy once positively charged moieties enable electrostatic complexation with negatively charged nucleic acids, intracellularly promoting the nanocarriers' ability to mediate endosomal escape and improve cytosolic delivery. Cationic nanocarriers can be developed using cationic polymers, such as poly(ethylene imine) (PEI), poly-L-lysine, and chitosan, or cationic lipids, such as dioleoylphosphatidylethanolamine (DOPE) and dioleoyltrimethylammoniumpropane (DOTAP).

Pro-inflammatory cytokines, such as TNF-α, IL-1β, and IL-6, have a crucial role in inflammation. It is well known that TNF-α is mainly produced by macrophages that line the synovium [17]. TNF-α upregulates the production of IL-1β in loop response, hence enduring chronic inflammation. As both cytokines induce the expression of cell adhesion molecules on endothelial cells, they promote inflammatory and immune cells' migration into the synovium. In addition, these cytokines stimulate the production of chemokines important for intracellular signaling events. At the same time, they increase synoviocytes' and chondrocytes' destructive potential by inducing the production of MMPs and other matrix-degrading enzymes responsible for cartilage erosion [17]. Among all pro-inflammatory cytokines, TNF-α siRNA–loaded nanotherapeutic approaches have been successful in the inhibition of TNF-α pro-inflammatory activity in animal models of arthritis. Komano and coworkers investigated the therapeutic potential of cationic liposomes loading TNF-α siRNA in a CIA mouse model [85]. Mice were intravenously injected with Cy5-labeled TNF-α siRNA–loaded liposomes, wherefore the Cy5 fluorescence was used to assess the tissue distribution. Results showed that the Cy5 intensity of fluorescence was higher in arthritic inflamed joints than in off-target tissues. In fact, the fluorescence remained higher up to 48 h after treatment. Further experiments suggested that Cy5 fluorescence intensity was higher in synovial macrophages cells than in splenocytes, bone marrow cells, and blood leukocytes. Thus, treatment with TNF-α siRNA–loaded liposomes resulted in significant decrease in arthritis severity, allowing an efficient and targeted gene delivery to the inflamed synovium [85]. Similarly, Khoury and colleagues developed DOPE cationic liposome for systemic delivery of TNF-α siRNA in a CIA mouse model [83]. Results revealed that blocking TNF-α receptors effectively decreases local and systemic inflammation, as well as prevents cartilage damage in arthritic mice [83]. The same authors, in another study, investigated gene silencing via siRNA of other well-known pro-inflammatory cytokines (IL-1β, IL-6, and IL-18) instead of TNF-α [84]. Treatment in established arthritic mice revealed that a combination of the three cytokine-specific siRNAs reduced all arthritic pathological features [83]. In fact, compared to TNF-α siRNA treatment, IL-1β/IL-6/IL-18 siRNA–loaded liposomes improved overall anti-inflammatory efficiency, preventing cartilage and bone destruction [83]. Thus, multi-IL-specific siRNA can be a promising biologic nanostrategy for improving RA treatment, possibly shifting the supremacy of anti-TNF agents for standard care in patients not responding to synthetic DMARDs.

Apart from liposomes for gene delivery, te Boekhorst and colleagues developed PLGA/DOTAP NPs for intra-articular injection of loaded TNF-α siRNA [80]. The intra-articular injection of TNF-α siRNA–loaded NPs in the AbIA mouse model resulted in a significant reduction in the paw swelling and joint inflammation [80].

Synoviocytes have been reported to play a central role in the chronic nature of RA, being key cells to target as they participate in inflammatory events of joint damage through the production of many pro-inflammatory cytokines and MMPs [17]. Particularly, IL-15 has been shown to have a role in these events, as it induces the production of other cytokines, such as TNF-α and IL-17, by T cells through a cell contact–dependent mechanism [17]. In turn, T cells induce the expression of IL-15 and IL-6 by synoviocytes, creating a feedback loop that favors persistent synovial inflammation [17]. In this regard, PEI-nucleic acid complexes have been developed by Zhang et al. and Duan et al. to deliver siRNA targeting the IL-2 and IL-15 receptor β (IL-2/15Rβ) on synovial fibroblasts for RA therapy [89, 90]. In both studies, following intravenous administration in rats with AIA, IL-2/15Rβ siRNA–loaded PEI nanocomplexes effectively accumulated in arthritic paws and subsequently were internalized by inflamed synovial cells [89, 90]. Thereby, results revealed that PEI-mediated siRNA delivery has the potential to treat RA, as anti-IL-2/15Rβ siRNA treatment decreases disease progression in arthritic mice [89, 90].

In another study, Xiao and colleagues developed mannosylated PEI NPs loaded with anti-TNF-α siRNA to specifically target the mannose receptor overexpressed on macrophages found within the synovium line [91]. Both in vitro and ex vivo studies showed that mannosylated NPs were efficiently internalized by macrophages, hence mediating the local delivery of TNF-α siRNA [91]. More results showed an increased silencing of TNF-α expression, which improved the anti-inflammatory efficiency of RA treatment [91].

Chitosan-based nanocarriers have also been used for gene delivery and target synovial cells. Howard and coworkers formulated chitosan NPs encapsulating TNF-α siRNA, aiming to target resident peritoneal macrophages upon intraperitoneal administration in CIA mouse models [81]. In vivo results showed effective anti-inflammatory responses that involved the formulation silencing the TNF-α gene on macrophages [81]. With a similar intent, Lee and colleagues used thiolated glycol chitosan NPs, showing enhanced accumulation at the arthritic joint of CIA mouse model and also efficient in vitro TNF-α gene silencing [82].

The effectiveness of downregulation of pro-inflammatory cytokines in reducing inflammatory activity has been proven. Nonetheless, an alternative biologic strategy to pro-inflammatory-cytokine-targeting gene-silencing approach is the gene transfer of anti-inflammatory cytokines, such as IL-10. Despite IL-10's potent anti-inflammatory activity, its clinical use is hampered by serious cytokine-related side effects [87]. To overcome this problem, cationic liposomes were developed by Fellowes and colleagues for plasmid DNA coding of IL-10 gene delivery [87]. In vivo experiments revealed that a single intraperitoneal injection of cationic liposomes containing human IL-10 plasmid DNA in CIA mice resulted in significant anti-inflammatory efficacy [87]. In fact, the anti-inflammatory activity in arthritic joints was prolonged by up to 30 days after injection. Further experiments showed that high accumulation and overexpression of IL-10 plasmid DNA in synovial inflamed joints allowed efficient gene transfer by synovial macrophages [87].

Indeed, a growing number of nanocarriers have been developed either for selective downregulation of pro-inflammatory cytokines or for overexpression of anti-inflammatory cytokines, amending the course of RA biologic treatment and improving its cost-effectiveness.

13.5 Final Remarks

RA is a serious health problem, the actual treatment of which has high rates of patient noncompliance, as available treatment options either are associated with risk of infection and potentially deleterious side effects or entail high medical costs.

Owing to these limitations, the ideal therapy for RA remains a challenge. In the past decade, nanotherapy has emerged as an enhanced clinical approach for the treatment of RA. Innovative target-to-treat nanostrategies focused on the inflamed disease-tissue have proven to be safe and efficient in the long term. Unlike conventional pharmacologic therapy, target-to-treat nanotherapies embody the concept of a magic bullet, conceived by Paul Ehrlich, enabling antirheumatic therapeutic agents to be selectively delivered to their designated synovial targets, enhancing tissue specificity and hence their therapeutic efficacy, while diminishing off-target toxicities that compromise the effectiveness of RA treatment.

The uncovering of novel cellular and molecular synovial targets opens the gate for a new era of therapeutic intervention with tailored target-to-treat nanotherapies. The therapeutic potential of these targets lies in RA pathogenesis amending nanocarriers' synovial tissue-targeting ability. To this end, exploiting the leaky vasculature of inflamed tissues and synovial-targeting strategies might then lead to treatments with improved clinically relevant therapeutic potency and pharmacologic safety profiles.

Acknowledgments

The authors thank the European Union (FEDER funds POCI/01/0145/FEDER/007728) and National Funds (Fundação para a Ciência e a Tecnologia and Ministério da Educação e Ciência [FCT/MEC]) under the partnership agreement PT2020 UID/MULTI/04378/2013 for the financial support received. Virgínia Moura Gouveia and Cláudia Nunes also thank FCT for the PhD grant (PD/BD/128388/2017) and the investigator grant (IF/00293/2015), respectively.

References

1. Gouveia, V.M., et al. Non-biologic nanodelivery therapies for rheumatoid arthritis. *J Biomed Nanotechnol*, 2015, **11**(10):1701–1721.
2. Lundkvist, J., et al. The burden of rheumatoid arthritis and access to treatment: health burden and costs. *Eur J Health Econ*, 2008, **8**:S49–S60.
3. Quan, L.D., et al. The development of novel therapies for rheumatoid arthritis. *Expert Opin Ther Pat*, 2008, **18**(7):723–738.
4. Kvien, T.K. Epidemiology and burden of illness of rheumatoid arthritis. *Pharmacoeconomics*, 2004, **22**(2 Suppl 1):1–12.
5. Van Vollenhoven, R.F. Treatment of rheumatoid arthritis: state of the art 2009. *Nat Rev Rheumatol*, 2009, **5**:531–541.
6. Smolen, J.S., et al. EULAR recommendations for the management of rheumatoid arthritis with synthetic and biological disease-modifying antirheumatic drugs: 2013 update. *Ann Rheum Dis*, 2014, **73**(3):492–509.
7. Singh, J.A., et al. 2015 American College of Rheumatology guideline for the treatment of rheumatoid arthritis. *Arthritis Rheumatol*, 2016, **68**(1):1–26.

8. Firestein, G.S. Evolving concepts of rheumatoid arthritis. *Nature*, 2003, **423**:356–361.
9. Upchurch, K.S., et al. Evolution of treatment for rheumatoid arthritis. *Rheumatology (Oxford)*, 2012, **51**:vi28–36.
10. Lee, S.J., et al. Pharmacological treatment of established rheumatoid arthritis. *Best Pract Res Clin Rheumatol*, 2003, **17**(5):811–829.
11. Patel, J., et al. Novel drug delivery technologies for the treatment of rheumatoid arthritis. *Internet J Med Technol*, 2009, **5**.
12. Pham, C.T. Nanotherapeutic approaches for the treatment of rheumatoid arthritis. *Wiley Interdiscip Rev Nanomed Nanobiotechnol*, 2011, **3**(6):607–619.
13. Gaffo, A., et al. Treatment of rheumatoid arthritis. *Am J Health Syst Pharm*, 2006, **63**(24):2451–2465.
14. Montesinos, M.C., et al. The anti-inflammatory mechanism of methotrexate depends on extracellular conversion of adenine nucleotides to adenosine by ecto-5'-nucleotidase: findings in a study of ecto-5'-nucleotidase gene-deficient mice. *Arthritis Rheum*, 2007, **56**:1440–1445.
15. Susan Jung-Ah Lee, A.K. Pharmacological treatment of established rheumatoid arthritis. *Best Pract Res Clin Rheumatol*, 2003, **17**(5):811–829.
16. Keyser, F.D. Choice of biologic therapy for patients with rheumatoid arthritis: the infection perspective. *Curr Rheumatol Rev*, 2011, **7**(1):77–87.
17. Muller-Ladner, U., et al. Mechanisms of disease: the molecular and cellular basis of joint destruction in rheumatoid arthritis. *Nat Clin Pract Rheumatol*, 2005, **1**(2):102–110.
18. Karouzakis, E., et al. Molecular and cellular basis of rheumatoid joint destruction. *Immunol Lett*, 2006, **106**(1):8–13.
19. Kinne, R.W., et al. Cells of the synovium in rheumatoid arthritis. Macrophages. *Arthritis Res Ther*, 2007, **9**(6):224.
20. Szekanecz, Z., et al. Angiogenesis and vasculogenesis in rheumatoid arthritis. *Curr Opin Rheumatol*, 2010, **22**(3):299–306.
21. Tran, T.H., et al. Targeted delivery systems for biological therapies of inflammatory diseases. *Expert Opin Drug Deliv*, 2015, **12**(3):393–414.
22. Takayanagi, H., et al. Involvement of receptor activator of nuclear factor kappaB ligand/osteoclast differentiation factor in osteoclastogenesis from synoviocytes in rheumatoid arthritis. *Arthritis Rheum*, 2000, **43**(2):259–269.

23. Lefevre, S., et al. Synovial fibroblasts spread rheumatoid arthritis to unaffected joints. *Nat Med*, 2009, **15**(12):1414–1420.
24. Szekanecz, Z., et al. Angiogenesis and its targeting in rheumatoid arthritis. *Vascul Pharmacol*, 2009, **51**(1):1–7.
25. Bader, R.A. The development of targeted drug delivery systems for rheumatoid arthritis treatment, in *Rheumatoid Arthritis: Treatment*, Lemmey, A.B., ed. InTech, 2012.
26. Gill, S., et al. Nanoparticles: characteristics, mechanisms of action, and toxicity in pulmonary drug delivery: a review. *J Biomed Nanotechnol*, 2007, **3**:107–119.
27. Garg, A., et al. pH-Sensitive PEGylated liposomes functionalized with a fibronectin-mimetic peptide show enhanced intracellular delivery to colon cancer cell. *Curr Pharm Biotechnol*, 2011, **12**(8):1135–1143.
28. Ferrari, M., et al. Trojan horses and guided missiles: targeted therapies in the war on arthritis. *Nat Rev Rheumatol*, 2015, **11**(6):328–337.
29. Vanniasinghe, A.S., et al. The potential of liposomal drug delivery for the treatment of inflammatory arthritis. *Semin Arthritis Rheum*, 2009, **39**(3):182–196.
30. Elbayoumi, T.A., et al. Current trends in liposome research. *Methods Mol Biol*, 2010, **605**:1–27.
31. Hofkens, W., et al. Intravenously delivered glucocorticoid liposomes inhibit osteoclast activity and bone erosion in murine antigen-induced arthritis. *J Controlled Release*, 2011, **152**(3):363–369.
32. Hofkens, W., et al. Liposomal targeting of prednisolone phosphate to synovial lining macrophages during experimental arthritis inhibits M1 activation but does not favor M2 differentiation. *PLoS One*, 2013, **8**(2):e54016.
33. Hofkens, W., et al. Liposomal targeting of glucocorticoids to the inflamed synovium inhibits cartilage matrix destruction during murine antigen-induced arthritis. *Int J Pharm*, 2011, **416**(2):486–492.
34. Hofkens, W., et al. Safety of glucocorticoids can be improved by lower yet still effective dosages of liposomal steroid formulations in murine antigen-induced arthritis: comparison of prednisolone with budesonide. *Int J Pharm*, 2011, **416**(2):493–498.
35. Harigai, T., et al. Prednisolone phosphate-containing TRX-20 liposomes inhibit cytokine and chemokine production in human fibroblast-like synovial cells: a novel approach to rheumatoid arthritis therapy. *J Pharm Pharmacol*, 2007, **59**(1):137–143.

36. Rauchhaus, U., et al. Targeted delivery of liposomal dexamethasone phosphate to the spleen provides a persistent therapeutic effect in rat antigen-induced arthritis. *Ann Rheum Dis*, 2009, **68**(12):1933–1934.

37. Rauchhaus, U., et al. Separating therapeutic efficacy from glucocorticoid side-effects in rodent arthritis using novel, liposomal delivery of dexamethasone phosphate: long-term suppression of arthritis facilitates interval treatment. *Arthritis Res Ther*, 2009, **11**(6):R190.

38. Anderson, R., et al. Liposomal encapsulation enhances and prolongs the anti-inflammatory effects of water-soluble dexamethasone phosphate in experimental adjuvant arthritis. *Arthritis Res Ther*, 2010, **12**(4):R147.

39. van den Hoven, J.M., et al. Optimizing the therapeutic index of liposomal glucocorticoids in experimental arthritis. *Int J Pharm*, 2011, **416**(2):471–477.

40. Avnir, Y., et al. Amphipathic weak acid glucocorticoid prodrugs remote-loaded into sterically stabilized nanoliposomes evaluated in arthritic rats and in a Beagle dog: a novel approach to treating autoimmune arthritis. *Arthritis Rheum*, 2008, **58**(1):119–129.

41. Ulmansky, R., et al. Glucocorticoids in nano-liposomes administered intravenously and subcutaneously to adjuvant arthritis rats are superior to the free drugs in suppressing arthritis and inflammatory cytokines. *J Controlled Release*, 2012, **160**(2):299–305.

42. Crielaard, B.J., et al. Glucocorticoid-loaded core-cross-linked polymeric micelles with tailorable release kinetics for targeted therapy of rheumatoid arthritis. *Angew Chem Int Ed Engl*, 2012, **51**(29):7254–7258.

43. Ishihara, T., et al. Treatment of experimental arthritis with stealth-type polymeric nanoparticles encapsulating betamethasone phosphate. *J Pharmacol Exp Ther*, 2009, **329**(2):412–417.

44. Ishihara, T., et al. Preparation and characterization of a nanoparticulate formulation composed of PEG-PLA and PLA as anti-inflammatory agents. *Int J Pharm*, 2010, **385**(1–2):170–175.

45. Prabhu, P., et al. Investigation of nano lipid vesicles of methotrexate for anti-rheumatoid activity. *Int J Nanomed*, 2012, **7**:177–186.

46. Tsai, C.Y., et al. Amelioration of collagen-induced arthritis in rats by nanogold. *Arthritis Rheum*, 2007, **56**(2):544–554.

47. Thomas, T.P., et al. Folate-targeted nanoparticles show efficacy in the treatment of inflammatory arthritis. *Arthritis Rheum*, 2011, **63**(9):2671–2680.

48. Low, P.S., et al. Discovery and development of folic-acid-based receptor targeting for imaging and therapy of cancer and inflammatory diseases. *Acc Chem Res*, 2008, **41**(1):120–129.

49. van der Heijden, J.W., et al. Folate receptor beta as a potential delivery route for novel folate antagonists to macrophages in the synovial tissue of rheumatoid arthritis patients. *Arthritis Rheum*, 2009, **60**(1):12–21.

50. Xia, W., et al. A functional folate receptor is induced during macrophage activation and can be used to target drugs to activated macrophages. *Blood*, 2009, **113**(2):438–446.

51. Paulos, C.M., et al. Folate receptor-mediated targeting of therapeutic and imaging agents to activated macrophages in rheumatoid arthritis. *Adv Drug Deliv Rev*, 2004, **56**(8):1205–1217.

52. Paulos, C.M., et al. Folate-targeted immunotherapy effectively treats established adjuvant and collagen-induced arthritis. *Arthritis Res Ther*, 2006, **8**(3):R77.

53. Nagayoshi, R., et al. Effectiveness of anti-folate receptor beta antibody conjugated with truncated Pseudomonas exotoxin in the targeting of rheumatoid arthritis synovial macrophages. *Arthritis Rheum*, 2005, **52**(9):2666–2675.

54. Puig-Kroger, A., et al. Folate receptor beta is expressed by tumor-associated macrophages and constitutes a marker for M2 anti-inflammatory/regulatory macrophages. *Cancer Res*, 2009, **69**(24):9395–9403.

55. Shi, Q., et al. Hydrodynamic delivery of chitosan-folate-DNA nanoparticles in rats with adjuvant-induced arthritis. *J Biomed Biotechnol*, 2011, **2011**:148763.

56. Fernandes, J.C., et al. Bone-protective effects of nonviral gene therapy with folate-chitosan DNA nanoparticle containing interleukin-1 receptor antagonist gene in rats with adjuvant-induced arthritis. *Mol Ther*, 2008, **16**(7):1243–1251.

57. Chandrasekar, D., et al. The development of folate-PAMAM dendrimer conjugates for targeted delivery of anti-arthritic drugs and their pharmacokinetics and biodistribution in arthritic rats. *Biomaterials*, 2007, **28**(3):504–512.

58. Chandrasekar, D., et al. Folate coupled poly(ethyleneglycol) conjugates of anionic poly(amidoamine) dendrimer for inflammatory tissue specific drug delivery. *J Biomed Mater Res A*, 2007, **82**(1):92–103.

59. Hattori, Y., et al. Folate-linked lipid-based nanoparticles deliver a NFkappaB decoy into activated murine macrophage-like RAW264.7 cells. *Biol Pharm Bull*, 2006, **29**(7):1516–1520.

60. Sladek, Z., et al. Expression of macrophage CD44 receptor in the course of experimental inflammatory response of bovine mammary gland induced by lipopolysaccharide and muramyl dipeptide. *Res Vet Sci*, 2009, **86**(2):235–240.

61. Naor, D., et al. CD44 in rheumatoid arthritis. *Arthritis Res Ther*, 2003, **5**(3):105–115.

62. Schmitt, F., et al. Chitosan-based nanogels for selective delivery of photosensitizers to macrophages and improved retention in and therapy of articular joints. *J Controlled Release*, 2010, **144**(2):242–250.

63. Vachon, E., et al. CD44 is a phagocytic receptor. *Blood*, 2006, **107**(10):4149–4158.

64. Vachon, E., et al. CD44-mediated phagocytosis induces inside-out activation of complement receptor-3 in murine macrophages. *Blood*, 2007, **110**(13):4492–4502.

65. Golan, I., et al. Expression of extra trinucleotide in CD44 variant of rheumatoid arthritis patients allows generation of disease-specific monoclonal antibody. *J Autoimmun*, 2007, **28**(2–3):99–113.

66. Shin, J.M., et al. A hyaluronic acid-methotrexate conjugate for targeted therapy of rheumatoid arthritis. *Chem Commun (Camb)*, 2014, **50**(57):7632–7635.

67. Kamat, M., et al. Hyaluronic acid immobilized magnetic nanoparticles for active targeting and imaging of macrophages. *Bioconjug Chem*, 2010, **21**(11):2128–2135.

68. Mero, A., et al. A hyaluronic acid-salmon calcitonin conjugate for the local treatment of osteoarthritis: chondro-protective effect in a rabbit model of early OA. *J Controlled Release*, 2014, **187**:30–38.

69. Ryan, S.M., et al. An intra-articular salmon calcitonin-based nanocomplex reduces experimental inflammatory arthritis. *J Controlled Release*, 2013, **167**(2):120–129.

70. Heo, R., et al. Hyaluronan nanoparticles bearing gamma-secretase inhibitor: in vivo therapeutic effects on rheumatoid arthritis. *J Controlled Release*, 2014, **192**:295–300.

71. Kim, Y.J., et al. Ionic complex systems based on hyaluronic acid and PEGylated TNF-related apoptosis-inducing ligand for treatment of rheumatoid arthritis. *Biomaterials*, 2010, **31**(34):9057–9064.

72. Lee, H., et al. Hyaluronate-gold nanoparticle/tocilizumab complex for the treatment of rheumatoid arthritis. *ACS Nano*, 2014, **8**(5):4790–4798.

73. Nishimoto, N., et al. Study of active controlled tocilizumab monotherapy for rheumatoid arthritis patients with an inadequate response to methotrexate (SATORI): significant reduction in disease activity and serum vascular endothelial growth factor by IL-6 receptor inhibition therapy. *Mod Rheumatol*, 2009, **19**(1):12–19.

74. Cox, D., et al. Integrins as therapeutic targets: lessons and opportunities. *Nat Rev Drug Discov*, 2010, **9**(10):804–820.

75. Koning, G.A., et al. Targeting of angiogenic endothelial cells at sites of inflammation by dexamethasone phosphate-containing RGD peptide liposomes inhibits experimental arthritis. *Arthritis Rheum*, 2006, **54**(4):1198–1208.

76. Vanniasinghe, A.S., et al. Targeting fibroblast-like synovial cells at sites of inflammation with peptide targeted liposomes results in inhibition of experimental arthritis. *Clin Immunol*, 2014, **151**(1):43–54.

77. Lee, S.M., et al. Targeted chemo-photothermal treatments of rheumatoid arthritis using gold half-shell multifunctional nanoparticles. *ACS Nano*, 2013, **7**(1):50–57.

78. Costa Lima, S.A., et al. Temperature-responsive polymeric nanospheres containing methotrexate and gold nanoparticles: A multi-drug system for theranostic in rheumatoid arthritis. *Colloids Surf B*, 2015, **133**:378–387.

79. Jung, Y.S., et al. Temperature-modulated noncovalent interaction controllable complex for the long-term delivery of etanercept to treat rheumatoid arthritis. *J Controlled Release*, 2013, **171**(2):143–151.

80. te Boekhorst, B.C., et al. MRI-assessed therapeutic effects of locally administered PLGA nanoparticles loaded with anti-inflammatory siRNA in a murine arthritis model. *J Controlled Release*, 2012, **161**(3):772–780.

81. Howard, K.A., et al. Chitosan/siRNA nanoparticle-mediated TNF-alpha knockdown in peritoneal macrophages for anti-inflammatory treatment in a murine arthritis model. *Mol Ther*, 2009, **17**(1):162–168.

82. Lee, S.J., et al. TNF-alpha gene silencing using polymerized siRNA/thiolated glycol chitosan nanoparticles for rheumatoid arthritis. *Mol Ther*, 2014, **22**(2):397–408.

83. Khoury, M., et al. Efficient new cationic liposome formulation for systemic delivery of small interfering RNA silencing tumor necrosis factor alpha in experimental arthritis. *Arthritis Rheum*, 2006, **54**(6):1867–1877.

84. Khoury, M., et al. Efficient suppression of murine arthritis by combined anticytokine small interfering RNA lipoplexes. *Arthritis Rheum*, 2008, **58**(8):2356–2367.

85. Komano, Y., et al. Arthritic joint-targeting small interfering RNA-encapsulated liposome: implication for treatment strategy for rheumatoid arthritis. *J Pharmacol Exp Ther*, 2012, **340**(1):109–113.

86. Presumey, J., et al. Cationic liposome formulations for RNAi-based validation of therapeutic targets in rheumatoid arthritis. *Curr Opin Mol Ther*, 2010, **12**(3):325–330.

87. Fellowes, R., et al. Amelioration of established collagen induced arthritis by systemic IL-10 gene delivery. *Gene Ther*, 2000, **7**(11):967–977.

88. Kageyama, Y., et al. Plasmid encoding interleukin-4 in the amelioration of murine collagen-induced arthritis. *Arthritis Rheum*, 2004, **50**(3):968–975.

89. Zhang, T., et al. Systemic delivery of small interfering RNA targeting the interleukin-2/15 receptor beta chain prevents disease progression in experimental arthritis. *PLoS One*, 2013, **8**(11):e78619.

90. Duan, J., et al. Polyethyleneimine-functionalized iron oxide nanoparticles for systemic siRNA delivery in experimental arthritis. *Nanomedicine (Lond)*, 2014, **9**(6):789–801.

91. Xiao, B., et al. Mannosylated bioreducible nanoparticle-mediated macrophage-specific TNF-alpha RNA interference for IBD therapy. *Biomaterials*, 2013, **34**(30):7471–7482.

92. Robbins, P.D., et al. Gene therapy for arthritis. *Gene Ther*, 2003, **10**(10):902–911.

Part VIII

Gene Delivery Approaches

Chapter 14

Nonviral Therapeutic Approaches for Modulation of Gene Expression: Nanotechnological Strategies to Overcome Biological Challenges

Ana M. Cardoso,[a] Ana L. Cardoso,[a] Maria C. Pedroso de Lima,[a] and Amália S. Jurado[a,b]
[a]*Center for Neuroscience and Cell Biology (CNC), University of Coimbra, Coimbra, Portugal*
[b]*Department of Life Sciences, University of Coimbra, Coimbra, Portugal*
asjurado@bioq.uc.pt

The treatment of a disease at its origin rather than ameliorating its symptoms has been a longstanding aim of medicine. A wide range of disorders can be the object of a gene therapy–based approach. The potential of nucleic acids as therapeutic molecules has been extensively studied, both in monogenic diseases, such as X-linked severe combined immunodeficiency (X-SCID), cystic fibrosis, and hemophilia, and in multifactorial diseases, including cancer and cardiovascular diseases. Depending on the therapeutic objective, modulation of gene expression can be achieved using different types

Nanoparticles in Life Sciences and Biomedicine
Edited by Ana Rute Neves and Salette Reis

ISBN 978-981-4745-98-7 (Hardcover), 978-1-351-20735-5 (eBook)
www.panstanford.com

of nucleic acids, such as plasmid DNAs (pDNAs) to introduce genes, small interfering RNAs (siRNAs) to silence genes, splice-switching oligonucleotides (ssONs) to correct splicing errors, and micro-RNA (miRNA) modulators (mimics or inhibitors) to tune gene expression. Different sequential arrangements of synthetic nucleotides, easily produced in the laboratory and displaying high versatility, allow for the design of sequences for specific targets, granting them a place in targeted therapies that can hardly be fulfilled by conventional drugs. However, the extraordinary complexity of living organisms presents a plethora of challenges to the delivery of nucleic acids to target cells, including the challenges of (i) transposing systemic barriers, such as immune system activation, (ii) achieving specific cell targeting, and (iii) overcoming cellular barriers, such as the cytoplasmic membrane and intracellular degradation pathways. This chapter will address the design of new nanotechnological platforms and the physicochemical properties that make them effective in mediating nucleic acid–based therapies, by overcoming biological impediments. Overall, our aim will be to provide an integrated view of how nonviral delivery systems, differing in their chemical nature and properties, can be modulated in order to improve their ability to carry and deliver nucleic acids to target cells, for example, by exploring the strategies displayed by viruses, which have naturally evolved the capacity to elude the obstacles presented by their host organisms.

14.1 Gene Therapy Overview

Following the discovery of antibiotics, which allowed the eradication of most kinds of infections, gene-related diseases gained a high relevance in the context of medical research—the neurodegenerative disorders sporadic Alzheimer's disease (AD) and Parkinson's disease (PD) being the most prevalent, whose etiology remains elusive. Both of these disorders are characterized by increased expression of proteins (amyloid-β peptide [Aβ] and α-synuclein [α-syn], respectively), whose cytoplasmic and extracellular aggregates have been assigned as responsible for the disease progression [1–3]. Therefore, strategies that modulate dysregulated expression of genes, such as those coding for proteins involved in Aβ and α-syn production and degradation or coding for regulatory elements of these pathways, are expected to efficiently counteract the development of these diseases. Although cancer has been described to have occurred in the primordial days of humanity [4], in recent years it has become one of the leading health concerns, the annual cancer cases being predicted to afflict 22 million people in 2032 [5]. Genomic instability, which characterizes cancer cells, facilitates gene mutation and, thus, aberrant gene expression [6]. Despite the differences in the genetic profiles displayed by various cancer types, some pathways are transversally affected, sharing a set of characteristics designated as "cancer hallmarks." These include sustained proliferation ability, resistance to cell death, increased angiogenesis, invasion ability, and metastasis formation [7], which have been reported to be facilitated by a pro-inflammatory status [7] and energy metabolism dysregulation, the latter being regarded as a stemness promoter [8].

The treatment of all these diseases, as well as others displaying a genetic component, namely monogenic disorders, characterized by a mutation in a single gene, can be addressed using gene therapy, which aims to treat the disorder at its origin rather than to merely alleviate its symptoms. Research in this field started in the 1970s with pseudovirus-mediated DNA delivery into mice embryo cells [9]. In this first approach, pseudoviruses, consisting of fragments of DNA from the host cells in which viruses have been produced, encapsidated in viral protein coats, were shown to release DNA upon contact with other cells, suggesting a high potential of those particles to transduce animal cells [9]. The possibility of introducing foreign DNA into cells inaugurated the field of gene therapy and was explored in the context of monogenic diseases, namely X-linked severe combined immunodeficiency (X-SCID) which was successfully treated using retroviruses as gene delivery systems. However, insertional oncogenesis, arising from the random genome insertion of the transgene, led to the development of leukemia in four of nine treated patients [10] and propelled the field into finding safer strategies to overcome the secondary effects resulting from retroviral infection.

The discovery of other nucleic acids also contributed to the development of new approaches to gene therapy application. Importantly, a brave new world emerged from the discovery of RNA interference (RNAi) in 1998 in the nematode *Caenorhabditis elegans* [11]. In fact, the first evidences of RNAi were observed in petunia flowers when a chimeric chalcone synthase (CHS) gene was introduced in this organism in an attempt to overexpress the CHS gene to produce more intensely pigmented flowers. Unexpectedly, this genetic intervention resulted in a population of petunias totally white or with white and pigmented flowers [12]. However, it was only in 1998 that the RNAi mechanism started to be unraveled [11]. RNAi technology is based on the fact that oligonucleotides (ONs) complementary to messenger RNA (mRNA) sequences have the ability to prevent the translation of the corresponding genes into proteins (reviewed in "RNA Interference: Biology, Mechanism, and Applications" [13]). RNAi is based on total complementarity between siRNAs and their target mRNAs, which leads to target degradation. However, due to their small size (20- to 25-nucleotide long), RNAi molecules can display off-target complementarity. In fact, synthetic siRNAs can interfere with the efficiency of mRNA translation by mimicking the miRNA mechanism of action and pairing with untargeted mRNAs through imperfect complementarity, which results in safety concerns regarding their therapeutic application. This issue gains relevance because imperfect complementarity is extremely probable, and it is this particular feature that allows the same miRNA molecule to regulate hundreds of mRNAs, contributing to the interconnection of several cellular pathways. In fact, the miRNA preferential targets depend on the availability of the miRNA and complementary mRNA in the cell at a given moment. A corollary of this premise is that miRNA preferential targets can be specific to tissue type and cell type. In this context, an interesting example regards miRNA 122 (miR-122), which is expressed specifically and in large amounts in the liver. This miRNA was shown to be necessary for the replication of hepatitis C viral genome, probably by facilitating viral RNA folding or its interaction with the protein complexes involved in its replication [14]. Viral dependency on a host miRNA expressed in a specific tissue determines its infection ability and explains the tissue selectivity of hepatitis C virus.

The safety concerns raised by the pervasive nature of miRNAs and the siRNA's ability to trigger immune responses, as well as the attempt to originate ONs more stable in biological fluids, less prone to endonuclease degradation, and with a higher ability to be internalized by cells propelled the development of second-generation ONs with chemical modifications (reviewed in "Chemical Modifications to Improve the Cellular Uptake of Oligonucleotides" and "RNA Therapeutics: Beyond RNA Interference and Antisense

Oligonucleotides" [15, 16]). In this regard, nucleic acid delivery systems were also manipulated in order to help overcome the drawbacks associated with viral gene therapy. After the tragedy of the X-SCID clinical trial, nonviral vectors took an important step forward and a plethora of new systems, based on lipids [17, 18], peptides [19, 20], polymers [21, 22], and surfactants [23], were developed to safely surpass biological barriers and efficiently achieve nucleic acid delivery into target cells.

14.2 Biological Challenges on the Way to a Successful Targeted Gene Therapy

From the beginning, it became obvious that cellular uptake of exogenous nucleic acids has to be assisted, since the negatively charged nucleic acid backbone prevents them from crossing cellular membranes [24]. Proper uptake of these molecules has been achieved despite the physical challenge of the cell membrane integrity with an electric field by bombardment with nucleic acid–coated metallic particles or, importantly, through less invasive processes, based on the use of viral- and nonviral nucleic acid delivery systems. This chapter, rather than exhaustively describing the nucleic acid delivery systems developed over the years, will explore the various strategies used to overcome physiological barriers to gene therapy, presenting examples of nonviral delivery systems that have been designed in order to accomplish this objective.

Strategies to successfully overcome each biological barrier toward improvement of the efficacy of the carrier system to mediate nucleic acid delivery and therapeutic activity depend on the target tissue characteristics and have to be defined according to disease idiosyncrasies. Thus, a "universal recipe" to produce an ideal nucleic acid delivery system does not exist. This chapter aims, therefore, to provide an overview of the challenges posed to gene therapy and of the strategies used to surpass them in the context of specific diseases. In this regard, how some obstacles can be turned into advantages and a clear picture of the multiple possibilities already available to efficiently deliver nucleic acids into cells will be presented. Table 14.1 summarizes nanotechnological strategies to approach the biological barriers addressed in this chapter.

Table 14.1 Nanotechnological strategies designed to overcome challenges posed to gene therapy

	Challenges	Strategies
Systemic challenges	Immune system activation	• Chemical modification of ONs to decrease TLR recognition (2'-O-methylation) • PEGylation of nanoparticles to decrease interactions with immune system components
	Stability in the extracellular space	• Chemical modifications of ONs to increase resistance to nucleases (2'-O-methoxyethyl- or 2'-O-methyl-modification) • Formulation of ONs in lipid-based vehicles (containing cholesterol, HSA, folate, transferrin, R3V6 peptide, and PEG) for protection against nuclease degradation
	Capillary retention	• Formulation of ONs into small, stealthy particles unable to self-aggregate or aggregate with blood components (PEGylation)
	Biodistribution and cell selectivity	• Use of cell-specific ligands (endogenous: transferrin, folate, and LRP1; exogenous: CTX)
Cellular challenges	Plasma membrane	• Use of internalization pathway–specific ligands (CME: transferrin; raft-mediated endocytosis: folate, GM1, and cholera toxin subunit B) • Use of elements favoring direct membrane crossing (cell-penetrating peptides, for example, $S4_{13}PV$)
	Intracellular degradation	• Use of elements that promote the proton sponge effect in the endosome (pKa < 7.0: PEI, histidine, and lysine) • Use of nonlamellar phase forming elements (H_{II} phase: DOPE; H_I phase: surfactants containing sugar-based spacers)
	Nuclear membrane	• Use of nuclear-targeting molecules (signaling to the nucleus: NLS from the SV40 large T-antigen and NLS from the heterogeneous nuclear ribonucleoprotein-A1; binding to TFs: SV40 DNA nuclear–targeting sequence)
	Mitochondrial membrane system	• Use of compounds that selectively interact and mix with cardiolipin-containing lipid structures • Use of mitochondrial-targeting molecules (MTS)
	Genome correction	• Use of site-specific genome editors (ZFNs, TALENs, and CRISPR/Cas9)

ONs, oligonucleotides; TLR, toll-like receptor; HAS, human serum albumin; PEG, polyethylene glycol; LRP1, low-density-lipoprotein receptor-related protein 1; CTX, chlorotoxin; CME, clathrin-mediated endocytosis; GM1, monosialotetrahexosyl ganglioside; PEI, poly(ethylene imine); H_{II}, inverted hexagonal; H_I, normal hexagonal; DOPE, dioleoylphosphatidyl ethanolamine; TF, transcription factor; NLS, nuclear localization sequence; MTS, mitochondria-targeting sequence; ZFN, zinc finger nuclease; TALEN, transcription activator–like effector nuclease; CRISPR, clustered regularly interspaced short palindromic repeats; Cas9, CRISPR-associated proteins 9.

14.2.1 Systemic Barriers

Although some diseases can be approached through in situ application of the therapeutic agent, such as retina genetic disorders (reviewed in "A Comprehensive Review of Retinal Gene Therapy" [25]), other diseases, including neurodegenerative diseases and some types of cancers, require systemic administration. In the latter case, in order to efficiently reach the target organ, strategies have to be applied to overcome the systemic biological barriers imposed against nucleic acid delivery, which include the first-line immune response to exogenous agents, the nuclease degradation of circulating nucleic acids, and the retention in capillaries in specific tissues.

14.2.1.1 Immune system activation

The activation of toll-like receptors (TLRs) has been described to be promoted by short ONs, such as siRNAs (reviewed in "Interfering with Disease: A Progress Report on siRNA-Based Therapeutics" [26]), whose ability to elicit immune responses has put a hurdle against their therapeutic use, thus propelling the development of new strategies to reduce this undesired side effect. TLR-mediated innate immune activation results in the production of pro-inflammatory cytokines and antiviral interferons. This effect has evolved as a pathogen-fighting mechanism, which recognizes and targets unmethylated CpG dinucleotides, or even specific nucleic acid sequences, which are common in bacterial genomic DNA but are rare in vertebrates and thus act as effective foreign agent-detection patterns that are recognized by TLR9 [27]. Viral double-stranded RNAs are also able to activate TLR3, through stimulation of IFN-α and IFN-β, which trigger mechanisms of viral RNA degradation [28, 29]. Furthermore, TLR activation favors the development of adaptive immune responses by stimulating antigen-presenting cells and the complementary system [30].

Immune system activation can be a drawback, in particular when the siRNA-based therapy is designed to tackle diseases that already present a pro-inflammatory component and/or when such activation emerges in very sensitive tissues, such as the brain. In this regard, astrocytes were shown to produce a nonspecific innate immune response when challenged with siRNAs formulated in lipid-based complexes (oligofectamine/siRNAs), which was reported not to be caused by the lipidic delivery system but rather depended on the siRNA sequence [31]. In this study, performed in primary mouse astroglial cells, microglia were discarded as the one responsible for the observed increase of Stat1 expression and cytokine and chemokine release. Furthermore, siRNA chemical modification consisting in the replacement of either the uridine or the guanosine nucleotides by 2'-O-methylated nucleotides on the sense strand strongly reduced the immune-associated effects, although not affecting the siRNA-specific silencing ability [31].

Because siRNAs have to be recognized by the ribonucleic protein, termed the RNA-induced silencing complex (RISC), for sequence-specific cleavage of mRNA, only a limited amount of modifications can be introduced in these molecules, in order to preserve their function. In this context, contrary to the siRNA sense (passenger) strand, which can carry heavy modifications without compromising siRNA loading onto RISC, the guide strand can only accommodate one or two 2'-O-methylated nucleotide substitutions and phosphorothioate internucleotide linkages at the 3′ end (reviewed in "RNA Therapeutics: Beyond RNA Interference and Antisense Oligonucleotides" [16]).

Another strategy to overcome ON-mediated immune system activation involves the use of nucleic acid delivery systems with the ability to protect the carried nucleic acids from interaction with immune system components. This can be achieved using lipid-based delivery systems decorated with polyethyleneglycol (PEG) moieties, whose hydrophilic character leads to the formation of a hydration layer at the delivery system surface, which contributes to a reduced interaction and coating of the lipid systems by circulating proteins and to a decreased immune system recognition (reviewed in "PEGylation as a Strategy for Improving Nanoparticle-Based Drug and Gene Delivery" [32]). In fact, a study comparing liposomes of egg phosphatidylcholine (PC) and cholesterol (1:1 molar ratio), decorated or not with PEG, showed that the PEGylated nanoparticles presented improved stability in phosphate-buffered saline (PBS) containing 90% of human serum, as compared with the unmodified nanoparticles [33]. In addition, the density of PEG moieties at the surface of polystyrene nanoparticles was shown to impact the extent of interaction between the nanoparticles and cells (in vitro), as well as their lifetime in circulation, avoiding recognition by the mononuclear phagocytic system (in vivo) [34].

Although immune system activation has mostly been regarded as a safety concern in the context of some diseases, it was also recognized as a therapeutic opportunity in other types of diseases, namely in cancer. Indeed, cancer cells are widely described to suppress immunity, which contributes to preventing them from being detected and eliminated by the organism. Therefore, a therapeutic approach that elicits immunity can come as advantageous. An elegant strategy using immunostimulation to fight cancer was reported in mice bearing melanoma and colon carcinoma [35]. In this study, antitumor activity was achieved by TLR9 activation through both peritumoral and systemic administration of CpG ON sequences, with well-described immunostimulatory activity, fused to siRNAs targeting the immune suppressor Stat3, which resulted in enhanced immunomodulation [35].

14.2.1.2 Stability in the extracellular space

Another constraint to ON-based therapies is the instability of these molecules in biological fluids. Nucleases are abundant in blood serum and extracellular matrix, being responsible for the short circulation lifetime of ONs. To overcome this drawback two strategies, similar to those described above, were devised: (i) introduction of chemical modifications in the nucleic acid structure to increase stability and (ii) formulation of the nucleic acids into delivery systems able to protect their cargo from interaction with endogenous enzymes and thus preventing its degradation.

The first approach was used to improve the stability of ONs whose function does not require their recognition and loading onto intracellular enzyme complexes. This particularity allows for extensive chemical modifications and can be applied to antisense oligonucleotides (ASOs) (reviewed in "RNA Therapeutics: Beyond RNA Interference and Antisense Oligonucleotides" [16]). ASOs are designed to bind to a specific mRNA sequence and are usually based on a phosphorothioate backbone, whose resistance to nuclease degradation is increased by nucleotide replacement by 2'-O-methoxyethyl (2'-MOE)- or 2'-O-methyl (2'-OMe)-modified residues on each end. These heavily modified portions flank an unmodified "gap," which is recognized by ribonuclease H (RNase

H), an endonuclease that specifically degrades the RNA strand of an RNA–DNA hybrid. However, the introduction of flanking 2'-OMe residues can also prevent RNase H from binding to the ASO-target mRNA duplex [36]. Alternatively, the entire ASO molecule can be 2'-MOE or 2'-OMe modified, its strong binding to the target mRNA being enough to prevent mRNA translation. Due to the increased stability of the ASO-mRNA hybrid, these molecules can act through a steric block mechanism (reviewed in "RNA Therapeutics: Beyond RNA Interference and Antisense Oligonucleotides" [16]), resulting in reduced off-target effects [36].

The design of stable nanoparticles with the ability to carry nucleic acids and protect them from nuclease degradation, thus ensuring their integrity during routing to the target cells, was devised as a strategy for achieving efficient delivery of RNAi ONs and pDNA. Complexation of nucleic acids with lipid molecules showed to be a nonimmunogenic alternative to the use of viral particles, while promoting an extended nucleic acid half-life in biological fluids. In particular, the presence of cholesterol has been found to improve stability of nucleic acid–carrying complexes, allowing them to reach their target without being degraded (reviewed in "Gene Delivery by Lipoplexes and Polyplexes" [37]). In fact, liposomes of 1,2-dioleoyl-3-(trimethylammonium)propane or 1-palmitoyl-2-oleoyl-sn-glycero-3-ethylphosphocholine and cholesterol (DOTAP:Chol or EPOPC:Chol) were used to produce lipid/DNA complexes, which, at optimal charge ratios, were shown to maintain most of the carried DNA inaccessible to the DNA-intercalating probe ethidium bromide, in the presence of serum, and confer efficient DNA protection from DNase I degradation [38]. Importantly, association of human serum albumin (HSA) to EPOPC:Chol/DNA complexes, prepared at an optimal (+/–) charge ratio, did not affect DNA protection and, in contrast to plain complexes (lacking HSA), their enhanced biological activity was not inhibited in the presence of 60% serum, a condition that mimics the physiological environment [38]. Incubation of complexes of DOTAP:Chol/DNA or EPOPC:Chol/DNA with human plasma showed that those bearing a heavy positive charge were more prone to interact with albumin and with the C3 human complement protein, which might limit their use in vivo [38, 39]. The association of specific ligands, such as folate, with DOTAP:Chol/DNA complexes resulted in a significant enhancement of their biological activity in the presence of serum [40], while association of transferrin to complexes with the same lipid composition carrying siRNA showed an increase in their ability to protect their cargo when incubated with serum, with respect to their counterparts lacking the ligand [41]. Another example is the use of amphiphilic peptides, such as R3V6 (composed of three arginine and six valine residues), which was found to protect anti-miR-21 ASOs from degradation in the presence of serum with higher efficiency than 25 kDa poly(ethylene imine) (PEI25k), the golden standard of polymer-based transfection reagents. Importantly, in two glioblastoma cell lines, R3V6-mediated ASO delivery resulted in decreased expression of miR-21 and enhanced apoptosis [42].

Furthermore, the above-mentioned PEG-conferred stealth properties contribute to increasing the stability of lipid-based particles carrying nucleic acids. Indeed, by avoiding immediate clearance from the bloodstream, PEG contributes to increasing the particle circulation time, and, consequently, to enhancing the bioavailability of the nucleic acids at the target site. In this regard, Tagalakis and coworkers [43] tested the potential of targeted PEGylated liposomes of dioleoylphosphatidylglycerol (DOPG) and dioleoylphosphatidylethanolamine (DOPE) for in vivo delivery of anti-β-secretase siRNA using a rat model of AD. The authors demonstrated that administration of PEGylated siRNA nanoparticles into the rat brain by convection-enhanced delivery resulted in the specific silencing of the enzyme, which was found to be highly efficient compared to that achieved using their non-PEGylated counterparts [43].

14.2.1.3 Capillary retention

Systemic intravenous (IV) administration of gene therapy agents has to account for the capillary beds the agents encounter once in circulation. The most significant filtration systems are the pulmonary and hepatic capillary beds, but skin, muscle, and intestines can also contribute to this filtration effect [44]. To prevent filter removal of the therapeutic molecules, these are often formulated into delivery systems able to condense the nucleic acids into small particles (<100 nm). However, this may not suffice to prevent capillary retention. In fact, small particles bearing an overall positive surface charge can interact with blood components and aggregate into larger structures, which are then filtered out or eliminated through activation of the mononuclear phagocytic system [44]. In this regard, capillary bed retention can be avoided by the use of inert small nucleic acid delivery particles, whose "stealthiness" properties can be conferred by PEG moieties, similarly to what was previously described for avoiding clearance by the mononuclear phagocytic system. In fact, PEGylated polymeric particles of poly(lactic-co-glycolic) acid (PLGA) were shown to have an extended lifetime in circulation as compared to non-PEGylated PLGA nanoparticles (ca. 10 times higher dose 3 h after IV injection in mice) [45].

14.2.1.4 Biodistribution and cell selectivity

Stealthiness and stability of nucleic acid delivery systems in biological fluids are required but not enough for efficient transfection of the target cells. In fact, to achieve this goal, the therapeutic agent must reach the target cells in sufficient concentration. This could be guaranteed by increasing the administered dose, which may also enhance nonspecific side effects and promote toxicity due to alteration of gene expression in nontarget cells. Therefore, a balance should be established between the amount of administered nucleic acids, formulated into nanoparticles, and the undesirable effects associated with their action in nontarget cells and tissues. Strategies that optimize the amount of nanoparticles that reach and accumulate in the target cells also improve their therapeutic effect, while decreasing the toxic effect by reducing their concentration in the healthy tissues.

Ligands that bind to specific cell types or diseased cell receptors can be used to improve targeting selectivity. In this regard, cancer is an interesting example, since tumor cells overexpress specific receptors on their surface, as compared to their healthy counterparts, thus allowing for selective targeting by association of nucleic acid complexes with tumor-receptor-specific ligands. In fact, transferrin [46], folate [47, 48], and angiopep receptors (low-density lipoprotein receptor-related protein-1, LRP1) [49, 50] were described to be upregulated in malignant tumors and have been widely exploited for targeted therapeutic delivery to cancer. One example of such approach is the association of folate to EPOPC:Chol liposomes, which was employed to mediate antitumoral activity upon application of herpes simplex virus-thimidine kinase (HSV-tk) suicide gene therapy, in an in vivo animal model of oral cancer. This strategy resulted in a considerable reduction of tumor growth, as compared to that observed with plain lipoplexes (lacking folate) or folate lipoplexes containing a reporter gene [51]. The ligand transferrin has also proven to be useful in the targeting of neuronal cells. The transferrin receptor is highly expressed both on endothelial cells, composing the blood–brain barrier (BBB), and in neurons themselves. The use of antibodies against this receptor [52] or the association of the ligand itself to lipid-based vectors has allowed nonviral delivery systems to cross the BBB through transcytosis and reach neurons throughout the brain or to be directly delivered, using stereotactic injection, to specific brain regions [53]. In this context, Cardoso et al. have shown that the transferrin-associated DOTAP:Chol liposomes containing anti-c-Jun siRNAs, and directly administered into the hippocampus, were able to reduce excitotoxic damage in this region in a model of acute brain injury [53]. Silencing of the transcription factor (TF) c-Jun was able to decrease neuronal loss and seizure intensity, following administration of kainic acid in the lateral ventricle, while simultaneously reducing microglia activation and astrocyte proliferation in the hippocampus of the injected animals [53].

In addition to endogenous ligands, other natural molecules are able to target cancer cells. An interesting example is chlorotoxin (CTX), a peptide first isolated from scorpion venom and identified as an inhibitor of Cl^- channels, which binds to a protein complex anchored in lipid rafts [54]. CTX was found to selectively bind gliomas [55] and, thus, can be used to target nanoparticles at brain tumors. In fact, CTX-coupled stable nucleic acid lipid particles (SNALPs) were developed and successfully applied for targeted delivery of encapsulated nucleic acids, including anti-miR-21, both to glioblastoma cultured cells and to an orthotopic glioblastoma mouse model. Of note, IV administration of the targeted SNALPs-formulated anti-miR-21 resulted in their preferential accumulation in the brain tumor and efficient downregulation of miR-21 in tumor cells, without affecting peripheral tissues or healthy brain [56], which supports the targeting ability of CTX peptide.

14.2.2 Cellular Barriers

Once the extracellular obstacles have been overcome and the carrier system has reached the target cells, additional barriers are encountered that need to be surpassed for efficient intracellular delivery of the nucleic acid molecules so that they can exert their therapeutic function. Depending on the machinery necessary for the processing and action of each specific nucleic acid, the site of delivery should be the cytoplasm, the nucleus, or the mitochondria. In fact, siRNAs, mature miRNA (mimics or inhibitors), and pre-miRNAs (short hairpin molecules that are further processed by the endonuclease Dicer to originate mature miRNA molecules) perform their action in the cytoplasm, after incorporation into the RISC [57]. To efficiently deliver these small nucleic acids, the delivery systems should be able to cross the cytoplasmic membrane and release their cargo into the cytosol. Depending on the mechanism favored for the entry of nucleic acid–carrying systems into the cells, endosomal release can itself pose a challenge. Regarding nucleic acids whose function depends on their transcription into mRNA, such as pDNA encoding a gene for a protein or short hairpin RNAs that ultimately originate siRNAs or miRNAs, their delivery requires vectors that, besides overcoming the cellular membrane, are competent to mediate nuclear internalization, releasing their cargo into the nucleus. Mitochondria, as other organelles that contain the machinery necessary for transcription, are also interesting targets for nucleic acid delivery.

14.2.2.1 Plasma membrane

The cytoplasmic membrane, basically consisting of a lipid bilayer with embedded proteins, some lipids and proteins being glycosylated, provides a selective barrier between the extracellular and the intracellular environments. Cellular membranes also act as a communication platform between neighboring cells and between distant cells, thus allowing controlled release and uptake of small molecules through specific transporters and producing vesicular structures to engulf or expel larger molecules. The uptake of large particles from the extracellular environment can occur through distinct endocytic pathways, which differ in vesicle size and structure, nature of cargo, and the mechanism involved in vesicle formation (including clathrin-mediated endocytosis [CME], lipid-raft-dependent endocytosis, clathrin- independent and lipid-raft-independent endocytosis, micropinocytosis, and phagocytosis) [58]. While phagocytosis mostly occurs in specialized cells, such as immune cells that clear pathogens, diseased cells, and debris from the organism, and to a low extent, also in fibroblasts and endothelial and epithelial cells (reviewed in "Nanocarriers' Entry into the Cell: Relevance to Drug Delivery" [59]), all other endocytosis processes occur in any kind of cells. CME, which is essential for cell communication and nutrient uptake, can occur in a receptor-independent manner (a slow internalization process) or in a receptor-dependent manner (a fast rate process) in nonlipid raft regions of the cytoplasmic membrane rich in clathrin [60]. Receptors involved in CME can be targeted by nanoparticles displaying ligands, such as transferrin [59], on their surfaces. In this regard, studies conducted in vitro by Cardoso et al. showed that liposomes of DOTAP:Chol associated with transferrin and complexed with anti-c-Jun siRNA were efficiently internalized by murine hippocampal cells, mediating 50% silencing of c-Jun expression, which resulted in neuronal protection from glutamate-induced cytotoxicity, as compared to control complexes lacking transferrin, which presented much lower efficiency [61].

Unlike CME, a number of endocytic pathways are created in lipid rafts, cholesterol-rich domains of the lipid membrane, playing an important role in the removal of cell membrane receptors, such as epidermal growth factor (EGF) and transforming growth factor-β (TGF-β) receptors, which through CME would be recycled to the membrane. Thus, attenuation of signaling mediated by these receptors is more efficiently achieved through their internalization by a non-CME mechanism [60]. Several proteins are involved in endocytic pathways being created in lipid rafts, including caveolin, flotillin, GTPase regulator associated with focal adhesion kinase-1 (GRAF1), adenosine diphosphate-ribosylation factor 6 (Arf6), and Ras homolog gene family member A (RhoA) (reviewed in "Endocytosis of Gene Delivery Vectors: From Clathrin-Dependent to Lipid Raft-Mediated Endocytosis" [60]), although the most well described is caveolin-mediated endocytosis. Internalization of nanoparticles through caveolin-mediated endocytosis can be triggered by the presence of the ligand folate on the nanoparticle surface. In fact, folate-decorated poly(ethylene imine) (PEI)/DNA complexes were observed to co-localize with caveolin-1, and their internalization was inhibited by sequestration of cholesterol by methyl-β-cyclodextrin, as well as by loss of caveolin-1 function, mediated by genistein [62].

In another study, GM1 ganglioside was described to trigger endocytosis of simian virus 40 peptide through lipid rafts [63], and thus, this glycolipid can be explored for targeting nucleic acid delivery systems at lipid rafts so that they can be internalized through these membrane domains. In particular, GM1 is extremely abundant in neuronal cells and binds the cholera toxin subunit B, thus making it an attractive ligand to target nucleic acid delivery systems at neurons. Complexes of cholera toxin–conjugated polylysine and fluorescently labeled pDNA were shown to be internalized by neuronal PC12 cells with higher efficiency than unconjugated complexes [64].

In addition to the above-mentioned pathways, macropinocytosis is a nonspecific mechanism of endocytosis involving the formation of large vacuoles (macropinosomes) through actin-dependent ruffling of the cell membrane [60]. In a recent work, the mechanisms of internalization of complexes consisting of bis-quaternary gemini surfactants of the alkanediyl-α,ω-bis(alkyldimethylammonium bromide) family and a reporter gene were evaluated in HeLa cells using chemical inhibitors of CME (chlorpromazine), endocytosis taking place in lipid rafts (filipin III), and macropinocytosis (amiloride hydrochloride). In this study, 20%–40% transfection was found to depend on internalization through macropinocytosis, but no contribution from CME or lipid raft–mediated endocytosis was apparent [65]. An alternative internalization pathway to those previously described is the direct translocation through the cytoplasmic membrane, a mechanism that can take place when components of the nucleic acid delivery system interact with the cell surface, inserting into the lipid palisade and creating irregularities in the bilayer structure. In fact, the cell-penetrating peptide $S4_{13}PV$, derived from the Dermaseptin S4 antimicrobial peptide, to which the nuclear localization sequence (NLS) from the simian virus SV40 large T-antigen was added, was shown to efficiently deliver nucleic acids into HeLa cells and, on its own, to be able to promote lateral phase separation in model anionic lipid membranes and to induce the formation of quasi-hexagonal nonlamellar phases in the membrane of those cells [66]. However, when used as a nucleic acid delivery system, a complex prepared with cell-penetrating peptides may not be able to cross the cell membrane directly, since the electrostatic interactions established with the nucleic acids can alter its conformation and, thus, its ability to interact with and penetrate the lipid bilayer. In fact, the trans-activating transcriptional (TAT) activator peptide from the human HIV-1 virus, which, in a way similar to $S4_{13}PV$, was described to directly translocate across the cell membranes [67], was found to mediate endocytosis when formulated into lipid-based nucleic acid delivery systems [68].

14.2.2.2 Intracellular degradation pathways

After reaching the intracellular environment, complexes internalized by endocytic mechanisms become entrapped in endocytic vesicles, which leads to lysosomal degradation unless the components of the complexes are able to mediate endosomal release. To accomplish this "escape," the pH drop occurring along the endocytic pathway can be exploited by preparing complexes containing molecules that promote endosomal swelling and rupture ("proton sponge" effect) or that, upon undergoing conformational changes in the acidic milieu, interact with the inner leaflet of the endosomal membrane and promote lipid bilayer destabilization, thus allowing cytoplasmic release of the cargo. The proton sponge effect occurs when the components of complexes are able to undergo protonation in the pH range of endosome/lysosome pathway. This effect was first described for PEI polymers, whose secondary amines are protonated at pH 5.5. An increase in the amount of protons pumped to the endosome/lysosome compartment is necessary in order to accommodate the protonation of PEI and to promote pH decrease. As a consequence, import of Cl^- ions and water into the endosome/lysosome is also promoted to compensate the ionic and osmotic imbalance, respectively, causing endosome/lysosome swelling and rupture. Therefore, the "buffering" ability of polycationic molecules induces the release of the entrapped contents [69]. In a way similar to PEI, histidine residues can also undergo protonation in acidic environments. Interestingly, the pKa of the histidine pyrrole NH is 6.04 and that of its carboxylic acid moiety is 1.82. On the basis of this knowledge, the HIV-1 Tat cell-penetrating peptide was modified by the addition of histidine residues to facilitate endosomal release of the peptide/nucleic acid complexes, which resulted in improved transfection efficiency [70]. To achieve the same goal, five histidine residues have been added to the N-terminal of the cell-penetrating peptide $S4_{13}PV$ [71]. However, a comparison between the "wild type" $S4_{13}PV$ and the histidine-$S4_{13}PV$ regarding the ability to mediate gene silencing showed that the superiority of the latter peptide results from its higher capacity to promote the uptake of small ONs, rather than to promote a more efficient endosomal escape [71]. In another study, DNA was complexed with the gemini surfactant 1,9-bis(dodecyl)-1,1,9,9-tetramethyl-5-imino-1,9-nonanediammonium dibromide (12-7NH-12) or with its derivative with a glycyl-lysine dipeptide-substituted spacer, both complexes showing to be internalized by CME. The complexes prepared with the 12-7NH-12 surfactant were rapidly degraded in the lysosome, whereas those prepared with the surfactant containing the glycyl-lysine spacer were able to mediate the proton sponge effect, which was attributed to the buffering ability of lysine (pKa of 2.18) [72].

The destabilization of the endosomal membrane mediated by the complexes, besides promoting endosomal release, can result in the disassembling of the complexes necessary for nucleic acid activity. In this regard, the zwitterionic lipid DOPE is often used as a coadjuvant lipid in the design of nucleic acid delivery systems due to its propensity to form inverted hexagonal phases (H_{II}) [73, 74]. Lipids or lipid-like molecules may promote membrane rearrangements by inducing other nonlamellar structures, such as type I hexagonal phases (H_I), which also facilitate endosomal escape. This phenomenon was observed with complexes of pH-sensitive gemini surfactants containing sugar-based spacers of glucose or mannose, complexes that efficiently delivered pDNA to Chinese hamster ovary (CHO) cells and facilitated its expression, by mediating the formation of H_I structures in the endosomal membrane, thus allowing nucleic acid release from the endosomal compartment [75, 76].

In the cytosol, the nucleic acids should still be protected from degradation mediated by intracellular endonucleases so that their function will be fully accomplished. In this regard, the kinetics of nucleic acid release from the complexes play an important role in their effectiveness. A rapid destabilization of the complex structure after internalization, such as that described for 12-7NH-12 gemini surfactant-based complexes, which interact extensively with the cellular membrane, resulted in early release of the carried nucleic acids with subsequent degradation and low transfection efficiency [72].

14.2.2.3 Nuclear membrane

Contrary to siRNA or miRNA, pDNA needs to reach the nuclei of cells in order to be transcribed into the desired RNA molecule, either a protein-coding mRNA or a small RNA precursor. Thus, the nuclear membrane appears as an additional challenge to pDNA delivery. Like the cytoplasmic membrane, the nuclear membrane does not pose an insurmountable barrier, but, in fact, has nuclear pores that are responsible for the controlled trafficking of mRNA molecules from the nucleus to the cytosol for further processing and of nuclear proteins that are synthesized in the cytoplasm and are addressed to the nucleus. For fast dividing cells, it is usually assumed that nuclear membrane fission during the mitotic process facilitates the translocation of nanocarriers from the cytoplasm to the nucleus. Nonetheless, even in actively dividing cells, only 1%–10% of the pDNA copies internalized by the cells reach the nucleus (reviewed in "Progress and Prospects: Nuclear Import of Nonviral Vectors" [77]). However, the efficiency of this process can be enhanced using different strategies to modify gene delivery systems. In particular, advantage can be taken of cellular signaling sequences that target proteins at the nucleus. Peptide NLSs are used by cells to address proteins to the nuclear compartment, through binding to karyopherins (importins), which mediate their translocation across the nuclear pore complex (NPC) (reviewed in "Progress and Prospects: Nuclear Import of Nonviral Vectors" [77]). Typical examples of NLSs are the SV40 large T-antigen (PKKKRKV) and the 38-amino acid sequence M9 of the heterogeneous nuclear ribonucleoprotein-A1, a mRNA-binding protein that mediates nuclear uptake of otherwise cytoplasmic proteins [78]. This peptide, covalently linked to a scrambled version of the SV40 large T-antigen, has been preincubated with pDNA and subsequently with lipofectamine. Fluorescence microscopy images taken of murine 3T3 fibroblasts transfected with the resulting ternary complexes carrying a rhodamine-labeled pDNA showed co-localization of the pDNA with cell nuclei, whereas the control system lacking the peptide was found in the cytoplasm. The transfection efficiency of these systems correlated with their ability to mediate nuclear entry of the pDNA [79].

In addition to the peptide-based NLS, specific nucleic acid sequences can promote their own translocation to the nucleus by binding to TFs. The first identified sequence to mediate this kind of transport was a 72-base pair (bp) sequence in the SV40 enhancer region, called SV40 DNA nuclear–targeting sequence (DTS). In fact, this sequence was found to have multiple binding sites for TFs, which are translocated to the nuclear compartment after translation. Therefore, pDNA containing TF-binding sequences may bind to TFs in the cytoplasm and then be carried to the nucleus along with the TF (reviewed in "Progress and Prospects: Nuclear Import of Nonviral Vectors" [77]). Cells microinjected with a pDNA containing the 72-bp sequence or with a similar pDNA lacking this sequence have been compared in terms of nuclear translocation. The presence of the 72-bp sequence on the pDNA resulted in efficient nuclear localization, whereas in its absence, the pDNA remained in the cytoplasm [80]. The nuclear factor kappa-light-chain-enhancer of activated B cells (NF-κB) is among the TFs described to facilitate nuclear translocation of pDNA. Fluorescent polymeric particles of PEI carrying a fluorescently labeled pDNA containing or lacking a NF-κB-binding motif were imaged by fluorescence microscopy over time to follow the intracellular trafficking of the pDNA. In this study, a 6-fold higher extent of nuclear internalization was observed for the NF-κB-binding-motif-containing pDNA as compared to its counterpart lacking the TF-binding sequence [81].

14.2.2.4 Mitochondrial membrane system

Besides the nucleus, chloroplasts and mitochondria, whose ancestors have been proposed as being primitive prokaryotic organisms, are endowed with all the machinery necessary to transcribe their own DNA and express the encoded proteins. Although over millions of years of evolution, mitochondria relocated most of their genes to the nuclear genome, 13 subunits of the mitochondrial respiratory complexes are still encoded in the mitochondrial genome, as well as 22 transfer RNAs (tRNA) and 2 ribosomal RNAs (rRNA) from small and large ribosome subunits. Mutations in mitochondrial DNA have been associated with a number of diseases, most of which display symptoms in organs with high energy requirements, including the heart and the brain [82]. Such mutations occur at a higher rate than in genomic DNA due to the extremely high production of reactive oxygen species in mitochondria, rather than due to the lack or less efficient protection/reparation mechanisms (reviewed in "The Maintenance of Mitochondrial DNA Integrity: Critical Analysis and Update" [83]). Contrary to the nucleus, whose DNA contains two alleles of each gene, mitochondria have multiple copies of their genome. In addition, for each disease induced by mitochondrial DNA mutation, a threshold of mutated DNA molecules exists, above which the mitochondria become dysfunctional. Promoting gene expression in mitochondria, by delivering normal copies of otherwise defective genes, could, therefore, maintain the level of the mutated genes below the threshold, thus avoiding phenotypic manifestations of the disease [82]. To efficiently promote mitochondrial gene expression, the specific mitochondrial DNA codon-amino acid code must be taken into account. Thus, the difference between nuclear and mitochondrial genetic code in four codons can be used for specific mitochondrial expression of the delivered pDNA. With this purpose, nucleic acids should be engineered to contain the mitochondria initiation codon AUA (which codes for Ile in the nucleus), a Try coded by UGA (STOP codon in the nucleus), or the mitochondria-specific STOP codons AGA or AGG (which code for Arg in the nucleus) [84]. On the basis of this knowledge, a green fluorescent protein (GFP)-coding mitochondrial pDNA (mpDNA) was designed, containing the UGA codon [85]. HeLa cells transfected with this mpDNA complexed with conventional and serine-derived bis-quaternary ammonium surfactants displayed a significant GFP expression in mitochondria [86]. In the same study, interaction of the mpDNA/surfactant complexes with model lipid membranes with compositions mimicking different cellular organelles was assessed. Complexes able to transfect and mediate gene expression in mitochondria, as compared to those that did not share this property, showed to interact to a greater extent with lipid vesicles enriched in cardiolipin, mimicking the mitochondrial inner membrane, thus promoting lipid mixing and release of their aqueous contents [86]. Since cardiolipin exists exclusively in mitochondria membranes in eukaryotic systems, a selective interaction of certain components of nucleic acid delivery systems with this lipid could constitute a key event for the success of mitochondria gene regulation.

Mitochondria-resident proteins that are transcribed in the nucleus and translated in the cytoplasm have a mitochondria-targeting sequence (MTS) for addressing them to the appropriate compartment. This MTS is recognized by the translocase of the outer membrane (TOM) and by the translocase of the inner membrane 23 (TIM23), which mediate the entry into the mitochondria [87]. Therefore, similarly to the NLS, the MTS could be used to modify delivery system components to improve mitochondrial accumulation of the desired pDNA. Although to our knowledge, this approach has not been used yet, it deserves to be explored in terms of its potential therapeutic application in mitochondrial genetic diseases.

14.2.3 Molecular Targeting: Genome Editors

One of the most challenging barriers to foreign gene expression is the difficulty of correcting defective genes in their genomic location. This has been surpassed by the silencing or overexpression of a gene of interest, but correction of the target locus has been a hurdle that can now be overcome by the use of genomic editors. Zinc-finger nucleases (ZFNs) and transcription activator-like effector nucleases (TALENs) have been around since the year 2000. Both of these systems comprise a protein domain that recognizes specific nucleotides in the DNA and address the protein complexes to a specific genome location for cleavage and repair. The major disadvantage of these editing technologies is the need to design proteins for each gene correction. Although libraries of ZFNs and TALENs are currently available, the discovery of the clustered regularly interspaced short palindromic repeat (CRISPR)-associated proteins (Cas) system resulted in an extraordinary leap in the field (reviewed in "Delivery and Therapeutic Applications of Gene Editing Technologies ZFNs, TALENs, and CRISPR/Cas9" and "Current and Future Delivery Systems for Engineered Nucleases: ZFN, TALEN and RGEN" [88, 89]). The major advantage of this system, with respect to ZFNs and TALENs, is the nature of the DNA recognition domain, which, in the case of CRISPR, is nucleic acid based and guides the nuclease Cas to the cleavage location. Thus, designing CRISPRs to target a specific genomic sequence is a simpler process than engineering functional and targeted custom ZFNs and TALENs for each target DNA sequence.

The CRISPR/Cas system was first identified as part of the bacterial adaptive immune response against viral infection (reviewed in "The CRISPR-Cas Immune System: Biology, Mechanisms and Applications" [90]). The mechanism underlying CRISPR/Cas-based immunity has three steps: (i) adaptation, which consists of the insertion of a sequence of the viral genome in the CRISPR locus of the host (spacer acquisition), (ii) expression of the CRISPR locus and processing of the CRISPR RNA, and (iii) interference, which is the detection of the viral transposons by CRISPR RNA and their degradation by Cas proteins (reviewed in "The CRISPR-Cas Immune System: Biology, Mechanisms and Applications" [90]).

In a therapeutic intervention with the CRISPR/Cas system, a pDNA is designed to code for the Cas protein, as well as to contain the guide CRISPR RNA (crRNA) and, for Cas9, the trans-activating RNA (tracrRNA), which base-pairs with the crRNA for RNA duplex recognition by this nuclease. This pDNA can then be delivered to cells using viral or nonviral gene delivery systems.

The major disadvantage of these genome editing techniques remains their imperfect specificity, which can pose a risk to the patient as severe as that observed in X-SCID patients treated with retroviral-based gene therapy. In fact, in a study aiming to correct the human hemoglobin β (HBB) and the C-C chemokine receptor type 5 (CCR5) mutated genes, several guide RNA strands were designed to assess the double strand break (DSB) efficiency and specificity of the CRISPR/Cas9 system. After cell transfection with each CRISPR construct, genomic DNA was harvested and sequenced to evaluate the on- and off-target effects of each guide strand. This study demonstrated that genes with high homology, such as those presented by the human hemoglobin δ (HBD) and the C-C chemokine receptor type 2 (CCR2), with respect to HBB and CCR5, respectively, suffered an increased risk of off-target effects, which can be as small as a point mutation or as large as the inversion of a chromosome portion [91]. Thus, cautious evaluation of the off-target effects is necessary to guarantee the safety of the CRISPR/Cas system for a given target.

Nevertheless, the CRISPR/Cas system has been used widely since its first description and has been constantly updated and improved, both through the discovery of new versions of bacterial proteins that fulfill the Cas role and through the engineering of Cas proteins to make them simpler and more target-accurate (reviewed in "The CRISPR-Cas Immune System: Biology, Mechanisms and Applications" [90]). In addition, due to the extraordinary potential of genome editing strategies for the treatment of genetic diseases recognized by gene therapy researchers, new tools to predict and determine the off-target effects of CRISPR/Cas are being developed at a fast rate (reviewed in "Methods for Optimizing CRISPR-Cas9 Genome Editing Specificity" [92]). One example is the development of a reporter system to determine the efficiency of the DSBs mediated by Cas9 in the presence of complementary guide RNAs. A reporter construct expressing a protospacer RNA sequence under the same promoter as the GFP gene was designed, and different combinations of guide RNA sequences and Cas nucleases differing in the N-terminus sequence were compared in terms of efficiency to promote DSB in the target protospacer. Due to the location of the protospacer, an efficient cleavage resulted in a decrease of the GFP expression [93].

14.3 Concluding Remarks

Living organisms are remarkably well adapted to their environments and able to withstand most of the potentially dangerous insults that come from the exterior. A series of barriers protect humans from diseases caused by external agents, but they also prevent therapeutic molecules from efficiently reaching their targets. Therefore, the treatment of genetic diseases requires the development of nucleic acid delivery systems with the ability to overcome a number of nontrivial obstacles, imposing different hurdles and requiring different strategies to be surpassed. A challenge for the future is to construct such a system whose components are organized in layers that would be degraded sequentially, after fulfilling their respective functions. The perfect system does not exist, but efforts are being made to gain insight into the processes that prevent successful nucleic acid delivery so that nanoparticles could be modified in order to elude human defense mechanisms or to take advantage of the intrinsic properties of natural barriers.

Acknowledgments

This work was financed by the European Regional Development Fund (ERDF), through the Centro 2020 Regional Operational Programme under the project CENTRO-01-0145-FEDER-000008: BrainHealth 2020 and through the COMPETE 2020 - Operational Programme for Competitiveness and Internationalisation and Portuguese national funds via Fundação para a Ciência e a Tecnologia (FCT), I.P., under projects POCI-01-0145-FEDER-016390: CANCEL STEM and POCI-01-0145-FEDER-007440. Ana M. Cardoso and Ana L. Cardoso are recipients of fellowships from FCT with references SFRH/BPD/99613/2014 and SFRH/BPD/108312/2015, respectively.

References

1. Oddo, S., et al. Triple-transgenic model of Alzheimer's disease with plaques and tangles. *Neuron*, 2003, **39**(3):409–421.
2. Pacheco, C., et al. An extracellular mechanism that can explain the neurotoxic effects of alpha-synuclein aggregates in the brain. *Front Physiol*, 2012, **3**:297.
3. Ruiperez, V., et al. Alpha-synuclein, lipids and Parkinson's disease. *Prog Lipid Res*, 2010, **49**(4):420–428.
4. Odes, E.J., et al. Earliest hominin cancer: 1.7-million-year-old osteosarcoma from Swartkrans Cave, South Africa. *S Afr J Sci*, 2016, **112**(7/8):1–5.
5. World Cancer Report 2014, B.W.S.a.C.P. Wild, Editor 2014, International Agency for Research on Cancer: Lyon, France.
6. Ferguson, L.R., et al. Genomic instability in human cancer: molecular insights and opportunities for therapeutic attack and prevention through diet and nutrition. *Semin Cancer Biol*, 2015, **35**(Suppl):S5–S24.
7. Hanahan, D., et al. Hallmarks of cancer: the next generation. *Cell*, 2011, **144**(5):646–674.
8. Menendez, J.A., et al. Metabostemness: a new cancer hallmark. *Front Oncol*, 2014, **4**:262.
9. Osterman, J.V., et al. DNA and gene therapy: uncoating of polyoma pseudovirus in mouse embryo cells. *Proc Natl Acad Sci USA*, 1970, **67**(1):37–40.

10. Hacein-Bey-Abina, S., et al. Insertional oncogenesis in 4 patients after retrovirus-mediated gene therapy of SCID-X1. *J Clin Invest*, 2008, **118**(9):3132–3142.

11. Fire, A., et al. Potent and specific genetic interference by double-stranded RNA in Caenorhabditis elegans. *Lett Nat*, 1998, **391**:808–811.

12. Napoli, C., et al. Introduction of a chimeric chalcone synthase gene into petunia results in reversible co-suppression of homologous genes in trans. *Plant Cell*, 1990, **2**:279–289.

13. Agrawal, N., et al. RNA interference: biology, mechanism, and applications. *Microbiol Mol Biol Rev*, 2003, **67**(4):657–685.

14. Jopling, C.L., et al. Modulation of hepatitis C virus RNA abundance by a liver-specific MicroRNA. *Science*, 2005, **309**(5740):1577–1581.

15. Debart, F., et al. Chemical modifications to improve the cellular uptake of oligonucleotides. *Curr Top Med Chem*, 2007, **7**:727–737.

16. Kole, R., et al. RNA therapeutics: beyond RNA interference and antisense oligonucleotides. *Nat Rev Drug Discov*, 2012, **11**(2):125–140.

17. Simões, S., et al. Cationic liposomes for gene delivery. *Expert Opin Drug Deliv*, 2005, **2**(2):237–254.

18. Pensado, A., et al. Current strategies for DNA therapy based on lipid nanocarriers. *Expert Opin Drug Deliv*, 2014, **11**(11):1721–1731.

19. Heitz, F., et al. Twenty years of cell-penetrating peptides: from molecular mechanisms to therapeutics. *Br J Pharmacol*, 2009, **157**:195–206.

20. Trabulo, S., et al. Cell-penetrating peptide-based systems for nucleic acid delivery: a biological and biophysical approach. *Methods Enzymol*, 2012, **509**:277–300.

21. Wagner, E. Polymers for nucleic acid transfer-an overview. *Adv Genet*, 2014, **88**:231–261.

22. Kim, J., et al. Targeted polymeric nanoparticles for cancer gene therapy. *J Drug Target*, 2015, **23**(7–8):627–641.

23. Cardoso, A.M.S., et al. Gene delivery mediated by gemini surfactants, in *Engineering of Nanobiomaterials*, 1st ed., Alexandru Grumezescu, ed., Elsevier, 2016, pp. 227–256.

24. Felgner, P.L., et al. Lipofection: a highly efficient, lipid-mediated DNA-transfection procedure. *Proc Natl Acad Sci USA*, 1987, **84**:7413–7417.

25. Boye, S.E., et al. A comprehensive review of retinal gene therapy. *Mol Ther*, 2013, **21**(3):509–519.

26. de Fougerolles, A., et al. Interfering with disease: a progress report on siRNA-based therapeutics. *Nat Rev*, 2007, **6**:443–453.

27. Krieg, A.M. CpG motifs in bacterial DNA and their immune effects. *Annu Rev Immunol*, 2002, **20**:709–760.

28. Alexopoulou, L., et al. Recognition of double-stranded RNA and activation of NF-kappaB by Toll-like receptor 3. *Nature*, 2001, **413**:732–738.

29. Davidson, B.L., et al. Current prospects for RNA interference-based therapies. *Nat Rev Genet*, 2011, **12**(5):329–340.

30. Song, W.C. Crosstalk between complement and toll-like receptors. *Toxicol Pathol*, 2012, **40**(2):174–182.

31. Gorina, R., et al. Astrocytes are very sensitive to develop innate immune responses to lipid-carried short interfering RNA. *Glia*, 2009, **57**(1):93–107.

32. Suk, J.S., et al. PEGylation as a strategy for improving nanoparticle-based drug and gene delivery. *Adv Drug Deliv Rev*, 2016, **99**(Pt A):28–51.

33. Klibanov, A.L., et al. Amphipathic polyethyleneglycols effectively prolong the circulation time of liposomes. *FEBS Lett*, 1990, **268**(1):235–237.

34. Dunn, S.E., et al. Polystyrene-poly(ethylene glycol) (PS-PEG2000) particles as model systems for site specific drug delivery. 2. The effect of PEG surface density on the in vitro cell interaction and in vivo biodistribution. *Pharm Res*, 1994, **11**(7):1016–1022.

35. Kortylewski, M., et al. In vivo delivery of siRNA to immune cells by conjugation to a TLR9 agonist enhances antitumor immune responses. *Nat Biotechnol*, 2009, **27**(10):925–932.

36. Yoo, B.H., et al. 2'-O-methyl-modified phosphorothioate antisense oligonucleotides have reduced non-specific effects in vitro. *Nucleic Acids Res*, 2004, **32**(6):2009–2016.

37. Tros de Ilarduya, C., et al. Gene delivery by lipoplexes and polyplexes. *Eur J Pharm Sci*, 2010, **40**(3):159–170.

38. Faneca, H., et al. Association of albumin or protamine to lipoplexes: enhancement of transfection and resistance to serum. *J Gene Med*, 2004, **6**(6):681–692.

39. Faneca, H., et al. Evaluation of lipid-based reagents to mediate intracellular gene delivery. *Biochim Biophys Acta*, 2002, **1567**:23–33.

40. Duarte, S., et al. Non-covalent association of folate to lipoplexes: a promising strategy to improve gene delivery in the presence of serum. *J Controlled Release*, 2011, **149**(3):264–272.

41. Cardoso, A.L., et al. siRNA delivery by a transferrin-associated lipid-based vector: a non-viral strategy to mediate gene silencing. *J Gene Med*, 2007, **9**(3):170–183.

42. Song, H., et al. Delivery of anti-microRNA-21 antisense-oligodeoxynucleotide using amphiphilic peptides for glioblastoma gene therapy. *J Drug Target*, 2015, **23**(4):360–370.

43. Tagalakis, A.D., et al. Multifunctional, self-assembling anionic peptide-lipid nanocomplexes for targeted siRNA delivery. *Biomaterials*, 2014, **35**(29):8406–8415.

44. Dash, P.R., et al. Factors affecting blood clearance and in vivo distribution of polyelectrolyte complexes for gene delivery. *Gene Ther*, 1999, **6**:643–650.

45. Avgoustakis, K. Effect of copolymer composition on the physicochemical characteristics, in vitro stability, and biodistribution of PLGA–mPEG nanoparticles. *Int J Pharm*, 2003, **259**(1–2):115–127.

46. Gao, J.Q., et al. Glioma targeting and blood-brain barrier penetration by dual-targeting doxorubincin liposomes. *Biomaterials*, 2013, **34**(22):5628–5639.

47. Cagle, P.T., et al. Folate Receptor in Adenocarcinoma and Squamous Cell Carcinoma of the Lung. *Arch Pathol Lab Med*, 2013, **137**:241–244.

48. Parker, N., et al. Folate receptor expression in carcinomas and normal tissues determined by a quantitative radioligand binding assay. *Anal Biochem*, 2005, **338**(2):284–293.

49. Catasus, L., et al. Low-density lipoprotein receptor-related protein 1 (LRP-1) is associated with high-grade, advanced stage and p53 and p16 alterations in endometrial carcinomas. *Histopathology*, 2011, **59**(3):567–571.

50. Ruan, S., et al. Fluorescent carbonaceous nanodots for noninvasive glioma imaging after angiopep-2 decoration. *Bioconjug Chem*, 2014, **25**(12):2252–2259.

51. Duarte, S., et al. Folate-associated lipoplexes mediate efficient gene delivery and potent antitumoral activity in vitro and in vivo. *Int J Pharm*, 2012, **423**(2):365–377.

52. Boado, R.J., et al. The Trojan horse liposome technology for nonviral gene transfer across the blood-brain barrier. *J Drug Deliv*, 2011, **2011**:296151.

53. Cardoso, A.L., et al. Tf-lipoplex-mediated c-Jun silencing improves neuronal survival following excitotoxic damage in vivo. *J Controlled Release*, 2010, **142**(3):392–403.

54. DeBin, J.A., et al. Chloride channel inhibition by the venon of the scorpion Leirus Quinquestriatus. *Toxicology*, 1991, **29**(11):1403–1408.

55. Lyons, S.A., et al. Chlorotoxin, a scorpion-derived peptide, specifically binds to gliomas and tumors of neuroectodermal origin. *Glia*, 2002, **39**(2):162–173.

56. Costa, P.M., et al. MiRNA-21 silencing mediated by tumor-targeted nanoparticles combined with sunitinib: a new multimodal gene therapy approach for glioblastoma. *J Controlled Release*, 2015, **207**:31–39.

57. Krol, J., et al. The widespread regulation of microRNA biogenesis, function and decay. *Nat Rev Genet*, 2010, **11**(9):597–610.

58. Conner, S.D., et al. Regulated portals of entry into the cell. *Nature*, 2003, **422**:37–44.

59. Hillaireau, H., et al. Nanocarriers' entry into the cell: relevance to drug delivery. *Cell Mol Life Sci*, 2009, **66**(17):2873–2896.

60. El-Sayed, A., et al. Endocytosis of gene delivery vectors: from clathrin-dependent to lipid raft-mediated endocytosis. *Mol Ther*, 2013, **21**(6):1118–1130.

61. Cardoso, A.L., et al. Tf-lipoplexes for neuronal siRNA delivery: a promising system to mediate gene silencing in the CNS. *J Controlled Release*, 2008, **132**(2):113–123.

62. Gabrielson, N.P., et al. Efficient polyethylenimine-mediated gene delivery proceeds via a caveolar pathway in HeLa cells. *J Controlled Release*, 2009, **136**(1):54–61.

63. Ewers, H., et al. GM1 structure determines SV40-induced membrane invagination and infection. *Nat Cell Biol*, 2010, **12**(1):11–18; sup pp. 1–12.

64. Barrett, L.B., et al. Targeted transfection of neuronal cells using a poly(D-lysine)-cholera-toxin b chain conjugate. *Biochem Soc Trans*, 1999, **27**(6):851–857.

65. Cardoso, A.M., et al. Bis-quaternary gemini surfactants as components of nonviral gene delivery systems: a comprehensive study from physicochemical properties to membrane interactions. *Int J Pharm*, 2014, **474**(1–2):57–69.

66. Cardoso, A.M., et al. S4(13)-PV cell-penetrating peptide induces physical and morphological changes in membrane-mimetic lipid systems and cell membranes: implications for cell internalization. *Biochim Biophys Acta*, 2012, **1818**(3):877–888.

67. Vivès, E., et al. A Truncated HIV-1 tat protein basic domain rapidly translocates through the plasma membrane and accumulates in the cell nucleus. *J Biol Chem*, 1997, **272**(25):16010–16017.

68. Li, G.H., et al. Molecular mechanisms in the dramatic enhancement of HIV-1 Tat transduction by cationic liposomes. *FASEB J*, 2012, **26**(7):2824–2834.

69. Benjaminsen, R.V., et al. The possible "proton sponge" effect of polyethylenimine (PEI) does not include change in lysosomal pH. *Mol Ther*, 2013, **21**(1):149–157.

70. Lo, S.L., et al. An endosomolytic Tat peptide produced by incorporation of histidine and cysteine residues as a nonviral vector for DNA transfection. *Biomaterials*, 2008, **29**(15):2408–2414.

71. Cardoso, A.M., et al. Comparison of the efficiency of complexes based on S4(13)-PV cell-penetrating peptides in plasmid DNA and siRNA delivery. *Mol Pharm*, 2013, **10**(7):2653–2666.

72. Singh, J., et al. Evaluation of cellular uptake and intracellular trafficking as determining factors of gene expression for amino acid-substituted gemini surfactant-based DNA nanoparticles. *J Nanobiotechnol*, 2012, **10**:7.

73. Duzgunes, N., et al. Intracellular delivery of therapeutic oligonucleotides in pH-sensitive and cationic liposomes, in *Liposome Technology: Interactions of Liposomes with the Biological Milieu*, Gregoriadis, G., ed. Informa Healthcare, New York, 2007, pp. 253–275.

74. Hoekstra, D., et al. Gene delivery by cationic lipids: in and out of an endosome. *Biochem Soc Trans*, 2007, **35**:68–71.

75. Wasungu, L., et al. Transfection mediated by pH-sensitive sugar-based gemini surfactants; potential for in vivo gene therapy applications. *J Mol Med*, 2006, **84**(9):774–784.

76. Wasungu, L., et al. Lipoplexes formed from sugar-based gemini surfactants undergo a lamellar-to-micellar phase transition at acidic pH. Evidence for a non-inverted membrane-destabilizing hexagonal phase of lipoplexes. *Biochim Biophys Acta*, 2006, **1758**(10):1677–1684.

77. Lam, A.P., et al. Progress and prospects: nuclear import of nonviral vectors. *Gene Ther*, 2010, **17**(4):439–447.

78. Siomi, H., et al. A nuclear localization domain in the hnRNP A1 protein. *J Cell Biol*, 1995, **129**(3):551–560.

79. Byrnes, C.K., et al. A nuclear targeting peptide, M9, improves transfection efficiency in fibroblasts. *J Surg Res*, 2002, **108**:85–90.

80. Dean, D.A., et al. Sequence requirements for plasmid nuclear import. *Exp Cell Res*, 1999, **253**(2):713–722.

81. Breuzard, G., et al. Nuclear delivery of NFkappaB-assisted DNA/polymer complexes: plasmid DNA quantitation by confocal laser scanning microscopy and evidence of nuclear polyplexes by FRET imaging. *Nucleic Acids Res*, 2008, **36**(12):e71.

82. DiMauro, S., et al. Mitochondrial DNA mutations in human disease. *Am J Med Genet*, 2001, **106**:18–26.

83. Alexeyev, M., et al. The maintenance of mitochondrial DNA integrity--critical analysis and update. *Cold Spring Harb Perspect Biol*, 2013, **5**(5):a012641.

84. Yoon, Y.G., et al. Re-engineering the mitochondrial genomes in mammalian cells. *Anat Cell Biol*, 2010, **43**(2):97–109.

85. Lyrawati, D., et al. Expression of GFP in the mitochondrial compartment using DQAsome-mediated delivery of an artificial mini-mitochondrial genome. *Pharm Res*, 2011, **28**(11):2848–2862.

86. Cardoso, A.M., et al. Gemini surfactants mediate efficient mitochondrial gene delivery and expression. *Mol Pharm*, 2015, **12**(3):716–730.

87. de la Cruz, L., et al. The intermembrane space domain of Tim23 is intrinsically disordered with a distinct binding region for presequences. *Protein Sci*, 2010, **19**(11):2045–2054.

88. LaFountaine, J.S., et al. Delivery and therapeutic applications of gene editing technologies ZFNs, TALENs, and CRISPR/Cas9. *Int J Pharm*, 2015, **494**(1):180–194.

89. Ul Ain, Q., et al. Current and future delivery systems for engineered nucleases: ZFN, TALEN and RGEN. *J Controlled Release*, 2015, **205**:120–127.

90. Rath, D., et al. The CRISPR-Cas immune system: biology, mechanisms and applications. *Biochimie*, 2015, **117**:119–128.

91. Cradick, T.J., et al. CRISPR/Cas9 systems targeting beta-globin and CCR5 genes have substantial off-target activity. *Nucleic Acids Res*, 2013, **41**(20):9584–9592.

92. Tycko, J., et al. Methods for optimizing CRISPR-Cas9 genome editing specificity. *Mol Cell*, 2016, **63**(3):355–370.

93. Zhang, J.H., et al. Improving the specificity and efficacy of CRISPR/CAS9 and gRNA through target specific DNA reporter. *J Biotechnol*, 2014, **189**:1–8.

Part IX
Theranostic Approaches

Chapter 15

Theranostics: Simultaneous Treatment and Diagnosis Made Possible by Nanotechnology

João Albuquerque, Ana Rute Neves, and Salette Reis
UCIBIO, REQUIMTE, Department of Chemical Sciences, Faculty of Pharmacy, University of Porto, Portugal
shreis@ff.up.pt

Recently in the field of nanotechnology, the understanding and manipulation of materials with dimensions within the nanoscale have been on the rise and have led to the emergence of several more specific fields. The application of nanotechnology in medicine, for example, has opened up a wide variety of possibilities for new treatments and diagnosis methods. Nanomedicine takes advantage of nanotechnology to propose alternatives to current or even past treatments. For instance, it allows the encapsulation of therapeutic agents into nanoparticles (NPs) that can be tailored and even targeted at a specific application, organ, or tissue. This nanomedicine approach increases the bioavailability of the drugs and lowers their

Nanoparticles in Life Sciences and Biomedicine
Edited by Ana Rute Neves and Salette Reis

ISBN 978-981-4745-98-7 (Hardcover), 978-1-351-20735-5 (eBook)
www.panstanford.com

toxicity by decreasing the overall amount that has to be used while also increasing the percentage that reaches the desired site, thus increasing the efficacy of the treatment. The same approach can be applied to imaging or contrasting agents, increasing the specificity and, therefore, the detail and reliability of the obtained imaging results.

Nowadays, nanomedicine can go even further when considering nanocarrier approaches. In fact, instead of encapsulating either therapeutic or imaging agents, nanotechnology is able to encapsulate both of them within the same particle. This coencapsulation allows for simultaneous treatment and imaging, this being a new concept known as theranostics: the combination of therapy with diagnosis. This concept has recently caught the attention of many researchers across the world, and much effort has been put into attempting to use it in the clinic, from the encapsulation of drugs for systemic diseases with iron as a magnetic resonance imaging (MRI) contrast agent, to the use of magnetic NPs that act as both therapeutic agent and imaging contrast agent.

This chapter will focus on the nanotechnology applications for theranostic-based approaches in medicine, reviewing what is already being researched in the labs and their possible applications in the clinics, as well as other possibilities in the future of medicine and health care.

15.1 Introduction

The term "theranostic" was first used in 2002 and was described as the capability of a compound or methodology to function simultaneously as a diagnostic and a therapeutic agent. This emerging novel field arose with the acceptance that a universal elixir that can treat all diseases and conditions simply does not exist and that most diseases, even for different patients with the same disease, sometimes require individualized treatment [1–3]. The main goal of theranostics is to develop methodologies that allow treatment and simultaneous follow-up of every patient, with the hope that in the long term the treatment can be tuned for each individual, increasing its efficacy and improving the prognosis [1–4].

15.2 Nanotechnology-Based Approaches

Nanotechnology is a scientific field that encompasses the understanding and manipulation of materials with dimensions in the nanoscale [5–10]. It comprises knowledge from fundamental physics, chemistry, biology, and engineering, with applications spreading to a wide variety of scientific areas, from computer engineering and electronics to medicine and food products [3, 8, 11–13].

Within the field of nanotechnology, several more specific subfields exist, such as the application of nanotechnology to the medicine and health sciences (nanomedicine). In the past few years, nanomedicine has taken advantage of different strategies and materials in an attempt to improve diagnosis, imaging, and treatment of several diseases and conditions [1, 4, 5, 14, 15]. These attempts include drug delivery to bypass toxicity and improve bioavailability [16], the use of iron NPs that can be guided to a specific location and ablate tumors via application of magnetic field [17], and the double encapsulation of a therapeutic and an imaging agent within surface-modified NPs for targeted theranostics [18].

Prior to the concept of theranostics, nanomedicine had already used NPs to encapsulate drugs and imaging agents, although in separate carriers [6, 19, 20]. This earlier and independent concept was achieved through drug delivery, and its main objectives [6, 19, 20] were:

- *Increased overall bioavailability*, by encapsulating the agent within nanocarriers that are able to avoid recognition in the blood and improve the stability of these agents in the body
- *Reduced toxicity to healthy organs/tissues*, by using selective or targeted nanocarriers that only accumulate in the intended organ or tissue, allowing reduction in the total amount of drugs

Nowadays, nanomedicine has gone a step further in the field of drug delivery by altering the NPs to improve their applications and the efficacy of the proposed formulation. The surface of the particles can be functionalized with different molecules to achieve different goals, such as immune system avoidance, specific targeting, and increased cell uptake [21–23].

Figure 15.1 is a schematic representation of a nanoparticle to be used in drug delivery, including some examples of surface functionalization and a summary of the possibilities that nanomedicine provides.

Figure 15.1 Schematic representation of multifunctional NPs for drug delivery applications.

15.2.1 Coencapsulation/Association of Therapeutic and Imaging Agents

Theranostics brought a new light to the drug delivery concept. In fact, if we can encapsulate a therapeutic agent or an imaging agent inside a NP, why can't we encapsulate both of them at the same time and inside the same NP? In this context, theranostic approaches based on the coencapsulation of a therapeutic and an imaging agent arose.

There are numerous agents that have already been successfully coencapsulated inside a wide variety of NPs. In this section some of these proposed formulations will be presented, grouped by NP type.

15.2.1.1 Lipid nanoparticles

Lipid NPs can be divided into three major types. Solid lipid nanoparticles (SLNs) are NPs composed of lipids that are solid at body temperature (37°C) and that arrange themselves in spherical solid particles with a crystal-like matrix [24, 25]; nanostructured lipid carriers (NLCs) are NPs composed of a mixture of lipids that are solid and liquid at body temperature (37°C) and arrange themselves in spherical solid particles with an amorphous matrix [24, 25]; and liposomes are NPs composed of amphiphilic lipids that arrange themselves in a thin, spherical bilayer of lipids, containing an aqueous phase inside, similar to a cellular membrane [13, 24, 26, 27].

All these NPs have proven to be biocompatible and are capable of therapeutic and imaging agent coencapsulation within their matrix or membrane and core [24, 27, 28]. It is also possible to functionalize the surface of these NPs either by taking advantage of existing functional groups or by altering the surface to present additional available groups [18, 29, 30]. These properties, combined with the possibility to scale up lipid NP production, make these NPs promising candidates for theranostic applications. Table 15.1 presents some examples of theranostic applications based on lipid NPs.

Table 15.1 List of some theranostic applications of lipid NPs

Type of NPs	Therapeutic agent	Imaging agent	Application	Ref.
SLN	Paclitaxel and small interfering RNA	Quantum dots	Cancer theranostics	[31]
SLN	Methotrexate	Superparamagnetic iron oxide NPs	Rheumatoid arthritis treatment	[18]
NLC	DIM-C-pPhC6H5	XenolightDiR	Lung cancer treatment	[32]
Liposome	Small interfering RNA	Quantum dots	Improved and monitored gene silencing	[33]
Liposome	Doxorubicin	Quantum dots	Cancer treatment	[34]
Liposome	Plasmid DNA	Gold NPs	DNA transfection	[35]

15.2.1.2 Polymeric nanoparticles

Polymeric NPs can be composed of a wide variety of polymers, both synthetic—poly(lactic-co-glycolic) acid (PLGA), poly(ε-caprolactone) (PCL), etc.—and natural (chitosan, alginate, etc.) and produced by several different methods, depending on the needs of the encapsulated agent [1, 36]. The polymer PLGA has inclusively been approved by the Food and Drug Administration (FDA). But regardless of the method used, the production of these NPs relies on condensation reaction between monomers, which can all be the same molecule or different molecules. In the latter case, the ratio between different monomers can vary and this variation influences molecular weight and other properties [36]. The research performed with polymeric NPs for drug delivery and theranostic purposes is extensive and widespread. Some examples of the suggested applications are listed in Table 15.2.

Table 15.2 List of some theranostic applications of polymeric NPs

Polymer	Therapeutic agent	Imaging agent	Application	Ref.
PLGA	Meso-tetraphenyl-porpholactol	Meso-tetraphenyl-porpholactol	Dual imaging/photodynamic therapy of tumors	[37]
PLGA	Methotrexate	Superparamagnetic iron oxide NPs	Rheumatoid arthritis treatment	[38]
PLA or PCL	Doxorubicin	Perfluoropentane	Ultrasound-assisted drug delivery and imaging in cancer	[39]
HPMA	Doxorubicin	(123I or 125I) iodine	Treatment and imaging in melanomas and mammary carcinomas	[40]
Chitosan	Paclitaxel	Cy5.5	Cancer treatment and imaging	[41]

PLGA, poly(lactic-co-glycolic) acid; PLA, poly(L-lactide); PCL, poly(ε-caprolactone); HPMA, N-(2-hydroxypropyl)methacrylamide.

15.2.1.3 Dendrimers

A dendrimer is synthetic macromolecule composed of a central core and surrounded by several repeated branches of dendrons. This composition and organization results in a nearly perfect three-dimensional geometric shape, and its size and molecular weight depend on the generation of the dendrimer [42]. Dendrimers are produced by stepwise chemical methods, which allow for different levels of branches to be produced. Every level or generation consists of a "layer" of dendrons linked to the previous layer, the first level being linked to the central core [13, 42]. Due to their production mechanism, dendrimers have a uniform size, shape, and molecular weight. Since the generations are produced in a stepwise manner, it is also relatively easy to incorporate functional groups on the surface of dendrimer NPs and ensure they are well distributed [13].

The most commonly studied dendrimers for theranostic applications are composed of polyamidoamine (PAMAM), coated or not with poly(ethylene glycol) (PEG), but other molecules have also been suggested. These dendrimer-based approaches have been proposed for several theranostic applications, as shown in Table 15.3.

Table 15.3 List of some theranostic applications of dendrimers

Dendrimer	Therapeutic agent	Imaging agent	Application	Ref.
G5 PAMAM	Doxorubicin	FITC	Glioblastoma treatment	[43]
G5 PAMAM	α-Tocopheryl succinate	Gold NPs	Cancer treatment and imaging	[44]
G3.5 PAMAM	Doxorubicin	Superparamagnetic iron oxide NPs	Tumor treatment	[45]

For instance, G5 PAMAM dendrimers modified with fluorescein isothiocyanate (FITC) and arginine-glycine-aspartic acid (RGD) with a terminal PEG have been developed to encapsulate doxorubicin. FITC functions as a fluorescent imaging agent, RGD-PEG was used as a targeting molecule, and doxorubicin is a well-known anticancer drug. These agents conferred theranostic and targeting capabilities on the dendrimers that have been studied for applications in glioblastomas [43]. Besides that, G5 PAMAM dendrimers loaded with gold NPs and surface-modified with folic acid and α-tocopheryl succinate, both with PEG spacers, have also been established. The gold NPs were used in computed tomography imaging, with folic acid as a targeting agent, and α-tocopheryl succinate is known as an anticancer drug. The different agents combined resulted in theranostic capabilities for the dendrimers and were suggested for cancer treatment and imaging [44].

Finally, G3.5 PAMAM dendrimers modified with PEG and folic acid, with a core consisting of superparamagnetic iron oxide nanoparticles (SPIONs) and conjugated with doxorubicin have also been produced. SPIONs were used as contrast agents in MRI, folic acid enabled the active targeting of folic receptors, and doxorubicin is a well-known anticancer drug. The combination of these agents enabled targeted theranostics when using these dendrimers, proposed for use in tumors [45].

15.2.1.4 Silica nanoparticles

Silica is considered a safe material and has inclusively been used in surgical implants in the past. Silica NPs can be synthetized by both chemical and physical methods, and their size and morphology have been reported to be accurately controlled by the production parameters [1, 3, 46]. It is also possible to include functional groups on the surface by adding coprecursors during NP synthesis, such as aminopropyltrimethoxysilane and mercaptopropylmethoxysilane, which supply amine and thiol groups, respectively [46]. Biomolecules or smaller NPs can then be integrated either by previous coupling to the precursor molecules or by direct encapsulation within the silica matrix [1, 3, 46]. Silica NPs with highly porous matrices have also been reported, with the advantage of allowing a postencapsulation plugging through the pores, slowing agent release [46].

Molecules can be easily introduced into silica NPs during their production. For example, the coencapsulation of 2-devinyl-2-(1-hexyloxyethyl)pyropheophorbide (HPPH) and 9,10-bis(4'-[4"-aminostyryl]styryl)anthracene (BDSA) within silica NPs using such methods has been proposed for cancer treatment [47]. HPPH is a fluorescent compound that has been shown to have increased fluorescent activity when encapsulated in silica NPs. BDSA is a two-photon absorbing dye that can absorb near-infrared (NIR) rays and transfer the energy to HPPH, which, in turn, enables its photodynamic therapeutic properties [47]. Another approach consisted in using luminescent silica NPs to deliver doxorubicin to tumors. The luminescence was generated by the defects located at the interfaces of Si-SiO_2 and by quantum confinement effects [48].

15.2.1.5 Iron oxide nanoparticles

Iron oxide nanoparticles (IONPs) are nanocrystals of magnetite or hematite usually coated with other molecules that depend on the production method and alter the colloidal properties of the NPs [17, 46, 49]. IONPs smaller than 20 nm have been shown to be superparamagnetic, a state where particles show zero magnetism unless subjected to a magnetic field [49].

Traditionally, IONPs have been synthetized in aqueous solution by the coprecipitation of iron ions (II and III), where additives have been added to confer colloidal properties to the formulations. More recently, other methods, such as high-temperature decomposition, have been proposed and proven to have a more accurate control of the NPs' size and increased crystallinity, which translates into better magnetic properties. However, the latter method presents the disadvantage of NPs being coated with a thick alkyl layer, which may have to be substituted by ligand exchange or ligand addition for some applications [46, 49].

Regardless of the production method IONPs have been shown to be biocompatible and relatively cheap. They also possess superior magnetic properties and have been surface-functionalized to allow the coupling of biomolecules to their surface [17, 46, 49]. Considering this latter case, IONPs can be used in theranostics with two main approaches for conjugation: (i) the IONPs can be coencapsulated within a larger NP together with a therapeutic agent and (ii) a therapeutic agent can be coupled to the surface of IONPs. In both cases, the IONPs are the imaging agents commonly used as contrast agents in MRI [17, 46]. Examples of the proposed theranostic approaches already implemented are shown in Table 15.4.

Table 15.4 List of some theranostic applications of IONPs

Type of NPs	Therapeutic agent	Imaging agent	Application	Ref.
IONPs	Cisplatin	IONPs	Breast cancer treatment	[50]
IONPs	Small interfering RNA	IONPs	Gastric cancer treatment	[51]
Lipid	Small interfering RNA	IONPs	Proof of concept	[52]
Polymeric	Doxorubicin	IONPs	Lung carcinoma treatment	[53]

For instance, porous IONPs have been produced by controlled oxidation at 250°C followed by acid etching. The NPs have been loaded with cisplatin and coupled with herceptin for targeted delivery and controlled release in breast cancer. Cisplatin is an anticancer drug and was adsorbed in the pores of the IONPs, while herceptin provided targeting capability to the NPs [50]. IONPs coated with cationic lipids and coupled with small interfering RNA have also been developed, which outperformed the existing commercial particles in gastric cancer models. The IONPs were guided using external magnetic fields and accumulation and increased efficacy were observed at the targeted site [51]. Moreover, small interfering RNA and PEGylated cyclic RGD peptide have been also conjugated to IONPs as a promising strategy. The RGD functioned as the targeting agent, and the small interfering RNA was successfully transfected in a proof-of-concept study [52]. Furthermore, doxorubicin has been loaded into nonfouling polymer-coated IONPs for applications in lung carcinomas. In this case, the polymer was poly([3-(trimethoxysily) propyl]-r-[poly(ethylene glycol) methyl ether methacrylate]) (poly(TMSMA-r-PEGMA), the IONPs were used as MRI contrast agents, and doxorubicin was the therapeutic agent [53].

15.2.1.6 Gold nanoparticles

Gold nanoparticles (AuNPs) are metallic NPs that have been shown to possess unique properties in imaging applications [46, 54, 55]. They can be produced in many sizes and shapes, and both of these parameters will result in NPs with different optical properties.

AuNPs can be produced by several methods, both chemical and physical (such as laser ablation), and the production parameters can be fine-tuned to accurately control the NP size and morphology, both of which influence the absorption spectrum of the NPs [54, 55]. Additionally, Au surfaces have a high affinity for thiol groups, enabling simple and effective surface functionalization of these NPs. The affinity arises by use of molecules with a thiol group on one end or even by simply adding a thiol tag to a selected molecule [21, 54].

This tunable absorption spectrum and ease of surface functionalization, coupled with a good biodistribution and a low toxicity, have made AuNPs good candidates for several proposed bioapplications with theranostic approaches [13, 46, 54, 55]. These theranostic approaches have strategies similar to those for IONPs. The AuNP can be either coencapsulated with a therapeutic agent inside a bigger NP or can be directly coupled to a therapeutic agent, with the AuNP also functioning as the imaging agent [54]. Some examples can be found in Table 15.5.

Table 15.5 List of some theranostic applications using AuNPs

Shape of NPs	Therapeutic agent	Imaging agent	Application	Refs.
Nanospheres	Hairpin DNA labeled with a fluorophore	AuNPs	Colorectal carcinoma treatment	[56]
Nanospheres	Plasmonic AuNPs covered by a thin layer of medium	AuNPs	Prostate cancer treatment	[57, 58]
Nanospheres	Insulin	AuNPs	Diabetes management	[59]
Nanocages	Yb-2,4-dimethoxy-hemato-porphyrin	Yb-2,4-dimethoxy-hemato-porphyrin	Photothermic and photodynamic therapy simultaneously with imaging of cervical tumors	[60]
Nanorods	Doxorubicin	AuNPs	Cancer treatment	[61]
Nanostars	Methylene blue	Methylene blue	Breast cancer treatment	[62]

15.2.1.7 Quantum dots

Quantum dots (QDs) consist of nanocrystals formed by semiconductor materials that have been shown to possess unique optical properties [21, 46, 63]. QDs are brighter, have a narrower emission spectrum, and are more photostable when compared to organic dyes and fluorescent proteins [46, 63]. The first QDs developed showed emissions around the visible spectrum, but recently QDs have been created that emit light mainly in the infrared and NIR region, making them more promising for bioimaging considering the low tissue penetration of visible light. Moreover, the optical properties of QDs can be accurately adjusted by tuning their size and composition [46, 63].

The production method of QDs is very similar to the previously described method of pyrolysis for IONP production [46]. The selected metallic precursors are heated in organic solvents to achieve particle formation, and adequate surfactants are added to control particle growth. These particles are also usually coated with a ZnS layer, which has been proven to enhance optical properties and also provides a shell that prevents leakage of the toxic core constituents. As a result of the production method, QDs are coated with a thick alkyl layer, rendering them hydrophobic [46, 63]. This layer can be replaced using ligand exchange or ligand addition, again similar to IONPs [46]. Recently, the production of magnetic QDs has also been reported. These magnetic QDs consist of an Fe_3O_4 core and a ZnO shell and may provide QDs with alternative targeting and functional capabilities [64].

QDs' unique and superior optical properties as well as their proven considerable renal clearance have made them a promising candidate for in vivo imaging approaches. QDs can also be used in theranostic applications if we apply strategies similar to those used for IONPs and AuNPs, either by coencapsulation with a therapeutic agent or by direct coupling of a therapeutic agent [21, 46, 63]. Several theranostic approaches, comprising both previously mentioned strategies, using QDs have been proposed and studied (Table 15.6).

Table 15.6 List of some theranostic applications using QDs

Type of NP	Therapeutic agent	Imaging agent	Application	Refs.
QDs	Doxorubicin or mitoxantrone	QDs	Tumor treatment	[65, 66]
Liposomes	Apomorphine	QDs	Brain tumor treatment	[67, 68]
Polymeric	Small interfering RNA	QDs	Cancer treatment	[69]

For instance, QDs functionalized with A10 RNA aptamer loaded with doxorubicin, by intercalation of the drug with the aptamer sequence, have been studied. The A10 RNA aptamer recognizes the prostate cell membrane and together with doxorubicin acts as a fluorescence-quencher, resulting in the attenuation of the fluorescence activity of the formulation. However, when the QDs enter cells, doxorubicin begins to be gradually released and the fluorescent properties of the QDs are recovered, resulting in both therapy and imaging of the targeted tumor cells [65]. A similar attempt with positive results has been proposed involving direct binding of mitoxantrone to QDs [66]. PEGylated QDs can also be loaded into the aqueous core of cationic liposomes. The liposomes increase the uptake by tumor cells, resulting in superior staining of the tumor cells. It is also possible to surface-functionalize the liposome and load therapeutic agents in order to achieve theranostic effects [67]. A similar approach has also shown improved treatment and imaging in mouse brain tumors; the approach involves loading QDs inside the bilayer of liposomes while dissolving apomorphine in the aqueous core of the liposomes [68]. Hydrophobic triblock biodegradable copolymers, composed of PEG, PCL, and poly(ethylene imine), have been coloaded with QDs and fluorescently labeled "small interfering RNA." The polymeric NPs increased cellular uptake of the agents, and the coencapsulated agents acted as a prototype pair for fluorescence resonance energy transfer (also known as Förster resonance energy transfer [FRET]) with considerable efficiency [69].

15.2.1.8 Nanocarbons

Carbon nanomaterials, or nanocarbons, are structures composed solely of carbon and can be zero-dimensional (fullerene and carbon dots [CDs]), one-dimensional (carbon nanotubes [CNTs]), two-dimensional (graphene), or three-dimensional (nanodiamonds) [70]. Many of these nanocarbons possess inherent optical properties, making them useful in imaging, and some of them, like CNTs and graphene, also present exceptional electrical properties that could make them very useful in sensing applications of biomedicine. Moreover, nanocarbons, especially single-walled nanocarbons, have a very high surface area available for functionalization or therapy/imaging agent coupling. This functionalization and coupling can occur by both covalent and noncovalent methods [13, 46, 70, 71].

Table 15.7 List of some theranostic applications using nanocarbons

Type of NP	Therapeutic agent	Imaging agent	Application	Refs.
Carbon nanotubes	Small interfering RNA	Carbon nanotubes	Human T cells and primary cells applications	[72, 73]
Nanodiamonds	Doxorubicin	Radiation-damaged nanodiamonds	Colorectal carcinoma treatment	[74, 75]
Fullerene	Fullerene	Gd^{3+}	Tumor treatment	[76, 77]
Carbon dots	Oxaliplatin(IV) –COOH derivative	Carbon dots	Cancer treatment	[78]
Graphene	Graphene	Cu	Endothelial tumor treatment	[79]

However, some nanocarbons, like CNTs, have been found to be toxic to the organism as well as nonbiodegradable, which could limit their use in bioapplications [46, 70]. New studies have shown that nanocarbon's toxicity is greatly influenced by their surface and that surface functionalized nanocarbons have much lower toxicity than nonfunctionalized ones [71]. Still, more studies are required to assess long-term complications of nanocarbons before their safety can be correctly evaluated.

Regardless, the inherent properties and the ease of their functionalization combined with their ability to be taken up by cells make them promising candidates in future theranostic applications [46, 71]. In fact, some theranostic approaches based on nanocarbons have already been proposed and studied, as can be seen in Table 15.7.

15.2.2 Theranostic Agents and Nanoparticles

In the previous section, theranostic approaches that relied on the association of a therapeutic agent and an imaging agent were presented. This section will focus on NPs and agents that can simultaneously function as therapeutic and imaging agents themselves.

15.2.2.1 Theranostic nanoparticles

Some of the NPs described in the previous chapter have the potential to be theranostic agents on their own. For example, besides the imaging potential described in the previous section, IONPs can be used for hyperthermia [17, 46]. A magnetic field can be used to guide the IONPs to the targeted area and once there, if an alternate magnetic field is applied, the IONPs will convert the electromagnetic energy into heat, resulting in the thermal ablation of the surrounding tissue [2, 17]. Considering gold nanoshells, these NPs are gold shells, each NP having a silica core (60 nm) and a gold outer shell (10–12 nm) [1, 55]. They have been shown to both absorb and scatter incident light, resulting in the rise of the temperature around the NPs. Therefore, these NPs are capable of functioning as both imaging contrasting agents and photothermic actuators [55]. QDs have also great potential for imaging applications coupled with photodynamic therapy (PDT) [46, 63]. It has been found that QDs can be activated by light and then transfer their triplet state energy to oxygen molecules in the proximity, resulting in reactive oxygen radicals that damage the surrounding tissues [46]. Finally, nanocarbons also present optical properties that can be used for theranostic applications. CNTs and graphene derivatives with strong absorbance in the NIR region can be useful for photothermic ablation [80], and fullerenes can be used for photodynamic and photothermic therapies and as contrasting agents in MRI [77].

The discovery of these theranostic NPs brings new possibilities to the future of medicine, such as a potential simultaneous diagnosis, treatment, and imaging without the use of toxic drugs with adverse side effects for the patients.

15.2.2.2 Theranostic compounds

Similar to NPs that can have potential theranostic applications on their own some bioactive compounds and molecules also have this capability. For example, photosensitizers are capable of being both imaging and therapeutic agents. Photosensitizers are molecules that are able to produce changes in other molecules when subjected to light [81, 82]. They are essential in PDT since when excited by light (the wavelength depends on the compound) the photosensitizers produce highly reactive singlet oxygen, which then reacts with and destroys nearby molecules, cells, and tissues. Most photosensitizers are also bright fluorophores by nature, and almost all of them tend to emit light in the NIR spectra, making them useful in imaging applications [81, 82]. There are also recent reports about the existence of activatable molecules that only gain photosensitizer properties when activated, by either temperature, pH, or another chemical agent [81]. These latter photosensitizers are proposed to be used as smart drugs in the future [81].

Examples of such compounds are porphyrins, molecules consisting of four pyrrole subunits linked together by four methine bridges [83]. These molecules have well-described photosensitizing properties and have also been shown to be capable of passively accumulating in tumor tissues and residing there for a considerable period of time. Considering these properties, porphyrins have been considered for several imaging approaches, such as fluorescence and acoustic imaging and MRI, as well as for different therapies, such as photo- and sonodynamic therapy [83]. These approaches can be combined to create theranostic strategies using only one agent, which is simultaneously an imaging and therapeutic one.

However, porphyrins, and many other photosensitizers, such as bacteriochlorophyll-α, methylene blue, silicon naphthalocyanine, rose Bengal, and Chlorin-e_6, all of which are commonly used [81], have been found to have poor tumor selectivity and accumulate at high rates in the skin [83]. This limitation can be overcome by using nanotechnology approaches. Photosensitizers can be encapsulated in or coupled to one of the NPs described in the previous sections to increase their bioavailability and to promote a targeted and more specific delivery to the intended site [83]. Overall, the combination of a theranostic agent with nanotechnology can greatly improve the efficacy of the formulation.

15.3 Concluding Remarks

Nowadays, nanotechnology has been largely applied to medicine and biomedical applications. Nanomedicine has brought the possibility to achieve better treatment and imaging capabilities to the clinic and even to achieve both of them at the same time. As was shown, the studies regarding the application of NPs toward theranostic are very widespread and comprise a considerable variety of NPs, therapeutic agents, and imaging agents. Much work and understanding are still required, but these days, when a need for individualized medicine is on the rise, nanotechnology may potentiate existing treatments and imaging methods and result in an overall superior life span and quality.

Acknowledgments

This work received financial support from the European Union (FEDER funds) and National Funds (*Fundação para a Ciência e Tecnologia* and *Ministério da Educação e Ciência* [FCT/MEC]) under the partnership agreement PT2020 UID/MULTI/04378/2013 - POCI/01/0145/FEDER/ 007728. João Albuquerque thanks FCT, SANFEED Doctoral Programme, and PREMIX® for his PhD grant reference PD/BDE/114426/2016. Ana Rute Neves also thanks ICETA for her postdoctoral grant (FOOD_RL3_PHD_GABAI_02) under the project NORTE-01-0145-FEDER-00001.

References

1. Kelkar, S.S., et al. Theranostics: combining imaging and therapy. *Bioconjug Chem*, 2011, **22**(10):1879–1903.

2. Xie, J., et al. Magnetic nanoparticle-based theranostics. *Theranostics*, 2012, **2**:122–124.
3. Choi, K.Y., et al. Theranostic nanoplatforms for simultaneous cancer imaging and therapy: current approaches and future perspectives. *Nanoscale*, 2012, **4**(2):330–342.
4. Prabhu, P., et al. The upcoming field of theranostic nanomedicine: an overview. *J Biomed Nanotechnol*, 2012, **8**(6):859–882.
5. Bamrungsap, S., et al. Nanotechnology in therapeutics: a focus on nanoparticles as a drug delivery system. *Nanomedicine (Lond)*, 2012, **7**(8):1253–1271.
6. Sehgal, R. Nanotechnology and its applications in drug delivery: a review. *Int J Med Mol Med*, 2013, **3**(1):WMC002867.
7. Sanna, V., et al. Targeted therapy using nanotechnology: focus on cancer. *Int J Nanomed*, 2014, **9**:467–483.
8. Westesen, K., et al. Physicochemical characterization of lipid nanoparticles and evaluation of their drug loading capacity and sustained release potential. *J Controlled Release*, 1997, **48**(2–3):223–236.
9. Svenson, S., et al. *Multifunctional Nanoparticles for Drug Delivery Applications: Imaging, Targeting, and Delivery*. Springer, New York, 2012.
10. Shang, L., et al. Engineered nanoparticles interacting with cells: size matters. *J Nanobiotechnol*, 2014, **12**:5.
11. Panyam, J., et al. Biodegradable nanoparticles for drug and gene delivery to cells and tissue. *Adv Drug Deliv Rev*, 2003, **55**(3):329–347.
12. Scott, N.R. Nanotechnology and animal health. *Rev Sci Tech Off Int Epiz*, 2005, **24**(1):425–432.
13. Chan, W.C.W. *Bio-Applications of Nanoparticles*. Springer Science + Business Media, New York; Landes Bioscience, Austin, Tex., 2007.
14. Moghimi, S.M., et al. Nanomedicine: current status and future prospects. *FASEB J*, 2005, **19**(3):311–330.
15. Sumer, B., et al. Theranostic nanomedicine for cancer. *Nanomedicine*, 2008, **3**(2):137–140.
16. Zhang, Z.H., et al. Solid lipid nanoparticles modified with stearic acid-octaarginine for oral administration of insulin. *Int J Nanomed*, 2012, **7**(1):3333–3339.
17. Kandasamy, G., et al. Recent advances in superparamagnetic iron oxide nanoparticles (SPIONs) for in vitro and in vivo cancer nanotheranostics. *Int J Pharm*, 2015, **496**(2):191–218.

18. Albuquerque, J., et al. Solid lipid nanoparticles: a potential multifunctional approach towards rheumatoid arthritis theranostics. *Molecules*, 2015, **20**(6):11103–11118.

19. Couvreur, P. Nanoparticles in drug delivery: past, present and future. *Adv Drug Deliv Rev*, 2013, **65**(1):21–23.

20. Sun, T.M., et al. Engineered nanoparticles for drug delivery in cancer therapy. *Angew Chem Int Ed*, 2014, **53**(46):12320–12364.

21. Sperling, R.A., et al. Surface modification, functionalization and bioconjugation of colloidal inorganic nanoparticles. *Philos Trans A Math Phys Eng Sci*, 2010, **368**(1915):1333–1383.

22. Sapsford, K.E., et al. Functionalizing nanoparticles with biological molecules: developing chemistries that facilitate nanotechnology. *Chem Rev*, 2013, **113**(3):1904–2074.

23. Pan, H., et al. Programmable nanoparticle functionalization for in vivo targeting. *FASEB J*, 2013, **27**(1):255–264.

24. Mashaghi, S., et al. Lipid nanotechnology. *Int J Mol Sci*, 2013, **14**(2):4242–4282.

25. Weber, S., et al. Solid lipid nanoparticles (SLN) and nanostructured lipid carriers (NLC) for pulmonary application: a review of the state of the art. *Eur J Pharm Biopharm*, 2013, **86**(1):7–22.

26. Charron, D.M., et al. Theranostic lipid nanoparticles for cancer medicine. *Cancer Treat Res*, 2015, **166**:103–127.

27. Al-Jamal, W.T., et al. Liposomes: from a clinically established drug delivery system to a nanoparticle platform for theranostic nanomedicine. *Acc Chem Res*, 2011, **44**(10):1094–1104.

28. Ekambaram, P., et al. Solid lipid nanoparticles: review. *Sci Rev Chem Commun*, 2012, **2**(1):80–102.

29. Mehnert, W., et al. Solid lipid nanoparticles: production, characterization and applications. *Adv Drug Deliv Rev*, 2001, **47**(2–3):165–196.

30. Shastri, V., et al. Functionalized solid lipid nanoparticles and methods of making and using same, 2006, Google Patents: US.

31. Bae, K.H., et al. Optically traceable solid lipid nanoparticles loaded with sirna and paclitaxel for synergistic chemotherapy with in situ imaging. *Adv Healthcare Mater*, 2013, **2**(4):576–584.

32. Patel, A.R., et al. Theranostic tumor homing nanocarriers for the treatment of lung cancer. *Nanomedicine*, 2014, **10**(5):1053–1063.

33. Chen, A.A., et al. Quantum dots to monitor RNAi delivery and improve gene silencing. *Nucleic Acids Res*, 2005, **33**(22):e190.

34. Weng, K.C., et al. Targeted tumor cell internalization and imaging of multifunctional quantum dot-conjugated immunoliposomes in vitro and in vivo. *Nano Lett*, 2008, **8**(9):2851–2857.

35. Li, D., et al. The enhancement of transfection efficiency of cationic liposomes by didodecyldimethylammonium bromide coated gold nanoparticles. *Biomaterials*, 2010, **31**(7):1850–1857.

36. Wang, Z., et al. Polymeric materials for theranostic applications. *Pharm Res*, 2014, **31**(6):1358–1376.

37. McCarthy, J.R., et al. Polymeric nanoparticle preparation that eradicates tumors. *Nano Lett*, 2005, **5**(12):2552–2556.

38. Moura, C.C., et al. Co-association of methotrexate and SPIONs into anti-CD64 antibody-conjugated PLGA nanoparticles for theranostic application. *Int J Nanomed*, 2014, **9**(1):4911–4922.

39. Rapoport, N., et al. Multifunctional nanoparticles for combining ultrasonic tumor imaging and targeted chemotherapy. *J Natl Cancer Inst*, 2007, **99**(14):1095–1106.

40. Pimm, M.V., et al. Gamma scintigraphy of the biodistribution of 123I-labelled N-(2-hydroxypropyl)methacrylamide copolymer-doxorubicin conjugates in mice with transplanted melanoma and mammary carcinoma. *J Drug Target*, 1996, **3**(5):375–383.

41. Kim, K., et al. Tumor-homing multifunctional nanoparticles for cancer theragnosis: Simultaneous diagnosis, drug delivery, and therapeutic monitoring. *J Controlled Release*, 2010, **146**(2):219–227.

42. Sk, U.H., et al. Dendrimers for theranostic applications. *Biomol Concepts*, 2015, **6**(3):205–217.

43. He, X., et al. RGD peptide-modified multifunctional dendrimer platform for drug encapsulation and targeted inhibition of cancer cells. *Colloids Surf B*, 2015, **125**:82–89.

44. Zhu, J., et al. Targeted cancer theranostics using alpha-tocopheryl succinate-conjugated multifunctional dendrimer-entrapped gold nanoparticles. *Biomaterials*, 2014, **35**(26):7635–7646.

45. Chang, Y., et al. Synthesis and characterization of DOX-conjugated dendrimer-modified magnetic iron oxide conjugates for magnetic resonance imaging, targeting, and drug delivery. *J Mater Chem*, 2012, **22**(19):9594–9601.

46. Xie, J., et al. Nanoparticle-based theranostic agents. *Adv Drug Deliv Rev*, 2010, **62**(11):1064–1079.

47. Roy, I., et al. Ceramic-based nanoparticles entrapping water-insoluble photosensitizing anticancer drugs: a novel drug-carrier system for photodynamic therapy. *J Am Chem Soc*, 2003, **125**(26):7860–7865.
48. Park, K., et al. New generation of multifunctional nanoparticles for cancer imaging and therapy. *Adv Funct Mater*, 2009, **19**(10):1553–1566.
49. Lin, M.M., et al. Development of superparamagnetic iron oxide nanoparticles (SPIONS) for translation to clinical applications. *IEEE Trans Nanobiosci*, 2008, **7**(4):298–305.
50. Cheng, K., et al. Porous hollow Fe(3)O(4) nanoparticles for targeted delivery and controlled release of cisplatin. *J Am Chem Soc*, 2009, **131**(30):10637–10644.
51. Namiki, Y., et al. A novel magnetic crystal-lipid nanostructure for magnetically guided in vivo gene delivery. *Nat Nanotechnol*, 2009, **4**(9):598–606.
52. Lee, J.H., et al. All-in-one target-cell-specific magnetic nanoparticles for simultaneous molecular imaging and siRNA delivery. *Angew Chem Int Ed Engl*, 2009, **48**(23):4174–4179.
53. Yu, M.K., et al. Drug-loaded superparamagnetic iron oxide nanoparticles for combined cancer imaging and therapy in vivo. *Angew Chem Int Ed Engl*, 2008, **47**(29):5362–5365.
54. Ashraf, S., et al. Gold-Based nanomaterials for applications in nanomedicine. *Top Curr Chem*, 2016, **370**:169–202.
55. Bardhan, R., et al. Theranostic nanoshells: from probe design to imaging and treatment of cancer. *Acc Chem Res*, 2011, **44**(10):936–946.
56. Conde, J., et al. Gold-nanobeacons for simultaneous gene specific silencing and intracellular tracking of the silencing events. *Biomaterials*, 2013, **34**(10):2516–2523.
57. Wagner, D.S., et al. The in vivo performance of plasmonic nanobubbles as cell theranostic agents in zebrafish hosting prostate cancer xenografts. *Biomaterials*, 2010, **31**(29):7567–7574.
58. Lukianova-Hleb, E.Y., et al. Cell-specific multifunctional processing of heterogeneous cell systems in a single laser pulse treatment. *ACS Nano*, 2012, **6**(12):10973–10981.
59. Bhumkar, D.R., et al. Chitosan reduced gold nanoparticles as novel carriers for transmucosal delivery of insulin. *Pharm Res*, 2007, **24**(8):1415–1426.

60. Khlebtsov, B., et al. Nanocomposites containing silica-coated gold-silver nanocages and Yb-2,4-dimethoxyhematoporphyrin: multifunctional capability of IR-luminescence detection, photosensitization, and photothermolysis. *ACS Nano*, 2011, **5**(9):7077–7089.

61. Venkatesan, R., et al. Doxorubicin conjugated gold nanorods: A sustained drug delivery carrier for improved anticancer therapy. *J Mater Chem B*, 2013, **1**(7):1010–1018.

62. Fales, A.M., et al. Silica-coated gold nanostars for combined surface-enhanced Raman scattering (SERS) detection and singlet-oxygen generation: a potential nanoplatform for theranostics. *Langmuir*, 2011, **27**(19):12186–12190.

63. Ho, Y.P., et al. Quantum dot-based theranostics. *Nanoscale*, 2010, **2**(1):60–68.

64. Singh, S.P. Multifunctional magnetic quantum dots for cancer theranostics. *J Biomed Nanotechnol*, 2011, **7**(1):95–97.

65. Bagalkot, V., et al. Quantum dot-aptamer conjugates for synchronous cancer imaging, therapy, and sensing of drug delivery based on bi-fluorescence resonance energy transfer. *Nano Lett*, 2007, **7**(10):3065–3070.

66. Yuan, J., et al. Anticancer drug-DNA interactions measured using a photoinduced electron-transfer mechanism based on luminescent quantum dots. *Anal Chem*, 2009, **81**(1):362–368.

67. Al-Jamal, W.T., et al. Functionalized-quantum-dot-liposome hybrids as multimodal nanoparticles for cancer. *Small*, 2008, **4**(9):1406–1415.

68. Wen, C.J., et al. Theranostic liposomes loaded with quantum dots and apomorphine for brain targeting and bioimaging. *Int J Nanomed*, 2012, **7**:1599–1611.

69. Endres, T., et al. Amphiphilic biodegradable PEG-PCL-PEI triblock copolymers for FRET-capable in vitro and in vivo delivery of siRNA and quantum dots. *Mol Pharm*, 2014, **11**(4):1273–1281.

70. Liu, Z., et al. Nano-carbons as theranostics. *Theranostics*, 2012, **2**(3):235–237.

71. Yang, S.T., et al. Pharmacokinetics, metabolism and toxicity of carbon nanotubes for biomedical purposes. *Theranostics*, 2012, **2**(3):271–282.

72. Kam, N.W., et al. Functionalization of carbon nanotubes via cleavable disulfide bonds for efficient intracellular delivery of siRNA and potent gene silencing. *J Am Chem Soc*, 2005, **127**(36):12492–12493.

73. Liu, Z., et al. siRNA delivery into human T cells and primary cells with carbon-nanotube transporters. *Angew Chem Int Ed Engl*, 2007, **46**(12):2023–2027.

74. Huang, H., et al. Active nanodiamond hydrogels for chemotherapeutic delivery. *Nano Lett*, 2007, **7**(11):3305–3314.

75. Vaijayanthimala, V., et al. Functionalized fluorescent nanodiamonds for biomedical applications. *Nanomedicine (Lond)*, 2009, **4**(1):47–55.

76. Liu, J., et al. Preparation of PEG-conjugated fullerene containing Gd3+ ions for photodynamic therapy. *J Controlled Release*, 2007, **117**(1):104–110.

77. Chen, Z., et al. Applications of functionalized fullerenes in tumor theranostics. *Theranostics*, 2012, **2**(3):238–250.

78. Zheng, M., et al. Integrating oxaliplatin with highly luminescent carbon dots: An unprecedented theranostic agent for personalized medicine. *Adv Mater*, 2014, **26**(21):3554–3560.

79. Shi, S., et al. Tumor vasculature targeting and imaging in living mice with reduced graphene oxide. *Biomaterials*, 2013, **34**(12):3002–3009.

80. Yang, K., et al. The influence of surface chemistry and size of nanoscale graphene oxide on photothermal therapy of cancer using ultra-low laser power. *Biomaterials*, 2012, **33**(7):2206–2214.

81. Lovell, J.F., et al. Activatable photosensitizers for imaging and therapy. *Chem Rev*, 2010, **110**(5):2839–2857.

82. Cengel, K.A., et al. PDT: what's past is prologue. *Cancer Res*, 2016, **76**(9):2497–2499.

83. Zhou, Y., et al. Porphyrin-loaded nanoparticles for cancer theranostics. *Nanoscale*, 2016, **8**(25):12394–12405.

Chapter 16

Quantum Dots: Light Emitters for Diagnostics and Therapeutics

João L. M. Santos, José X. Soares, S. Sofia M. Rodrigues, and David S. M. Ribeiro
LAQV, REQUIMTE, Department of Chemical Sciences, Faculty of Pharmacy, University of Porto, Portugal
joaolms@ff.up.pt

In the past two decades, colloidal nanocrystals, or quantum dots, have captured the attention of the scientific community as an important class of nanomaterials with singular properties and great potential of application. In the biomedical field, quantum dots have emerged as versatile modular fluorescent tools providing a range of useful features that combine high brightness and stability with size, composition, surface, and morphological tunability, which could be exploited for tagging biological molecules, cellular labeling, tissue imaging, event signaling, biosensing or bioimaging, drug delivery, photodynamic therapy, etc.

In this section, relevant issues regarding the biological application of semiconductor quantum dots, either as diagnostic or as therapeutic tools, will be discussed. Emphasis will be on the

Nanoparticles in Life Sciences and Biomedicine
Edited by Ana Rute Neves and Salette Reis

ISBN 978-981-4745-98-7 (Hardcover), 978-1-351-20735-5 (eBook)
www.panstanford.com

properties, preparation, and surface modification of the nanocrystals and their usage as fluorescence probes for in vitro and in vivo imaging, in FRET bioanalysis, DNA sensing, and photodynamic therapy. Finally, the toxicity concerns associated with their use will be also examined.

16.1 Quantum Dots: Properties, Synthesis, and Bioconjugation

"Quantum dots" (QDs), a term coined by Mark Reed in 1988 [1], were firstly introduced by Brus and coworkers during the 1980s, when they demonstrated the occurrence of carrier confinement in semiconductor crystallites [2]. Initially applied to semiconductor nanocrystals the expression has in recent years been extended to other materials exhibiting similar chemical, physical, and optical properties, such as carbon-based graphene quantum dots (GQDs) and carbon dots (CDs).

Acclaimed for their unique optical and physical properties QDs have found relevant applications in fields as different as light-emitting devices, photovoltaic cells, optical and electronic devices, memory elements, photodetectors, chemical sensors, absorption filters, etc. In diagnostics and therapeutics, QDs have been investigated as advantageous substitutes for organic fluorophore or fluorescent protein in vivo or in vitro imaging and targeting, as active scaffolds in drug delivery, and in photodynamic therapy (PDT).

Fluorescence probes and labels have become crucial tools for clinical diagnostics, high-throughput screening, and several other biomedical applications. The fluorescence tagging of biological molecules for biomedical purposes has been mostly based on fluorescent proteins, such as green fluorescent protein (GFP); on conventional fluorophores, like xanthene-derived or cyanines [3]; or on lanthanide chelates, which are still regularly used, due mostly to their small size, simplicity of utilization, and the ease with which they undergo bioconjugation. These fluorescent dyes exhibit, however, several limitations, including low chemical stability and susceptibility to photodegradation, narrow excitation, and broad asymmetric emission spectra with red tailing, which significantly restrain their applicability.

In recent years, inorganic luminescent QDs have been capable of providing noteworthy advantages regarding their organic counterparts, emerging as alternative fluorescent tools for distinct bioanalytical, biological, or imaging applications.

QDs are monodisperse colloidal semiconductor nanocrystals comprising a few hundred to a few thousand atoms, with a size between 1 and 10 nm, lower than the exciton Bohr radius of their constituent bulk materials. Due to the low size, which is of the same order of magnitude as the de Broglie wavelength of electrons and holes, the nanomaterials exhibit a quantum confinement regime. Accordingly, the energy states of free charge carriers are no longer continuous, unlike with bulk materials, but quantized, as in atoms and molecules, and, similar to a particle-in-a-box model, assume discrete levels whose spacing depends on the nanocrystal size: the smaller a nanocrystal, the greater is the separation between the energy levels. In view of that, it is possible to modulate the nanocrystal's energy levels by adjusting its diameter, for instance, in terms of crystal growth, during the synthesis process. Considering that the energy gap between the excited and the fundamental state determines the wavelength of the fluorescence emission, in essence it is possible to prepare QDs emitting from the entire visible region to the NIR region of the optical spectrum simply by size-tuning (Fig. 16.1).

Moreover, QDs absorb light throughout a wide range of wavelengths, meaning that it is possible to use a single excitation source for an entire population of distinctly sized QDs. This characteristic not only greatly simplifies the optical set-up but also allows simultaneously the implementation of multiplexed detection applications.

The versatility of QD materials relies not only on their size but also on their composition, which could as well affect the bandgap energy. A multiplicity of QDs have been prepared involving the manipulation of the NP's core composition, especially with elements of the II-VI, III-V, and IV-VI groups, the most relevant being CdSe [4], ZnSe [5], CdS [4], PbS [6], CdTe [4], InP [7], and CdSe/ZnS [8]. They were prepared by employing a variety of synthetic approaches that enabled the tailoring of QD size, morphology, composition, and surface functionality. The latter is critical not only in terms of reactivity but also with respect to solution stability.

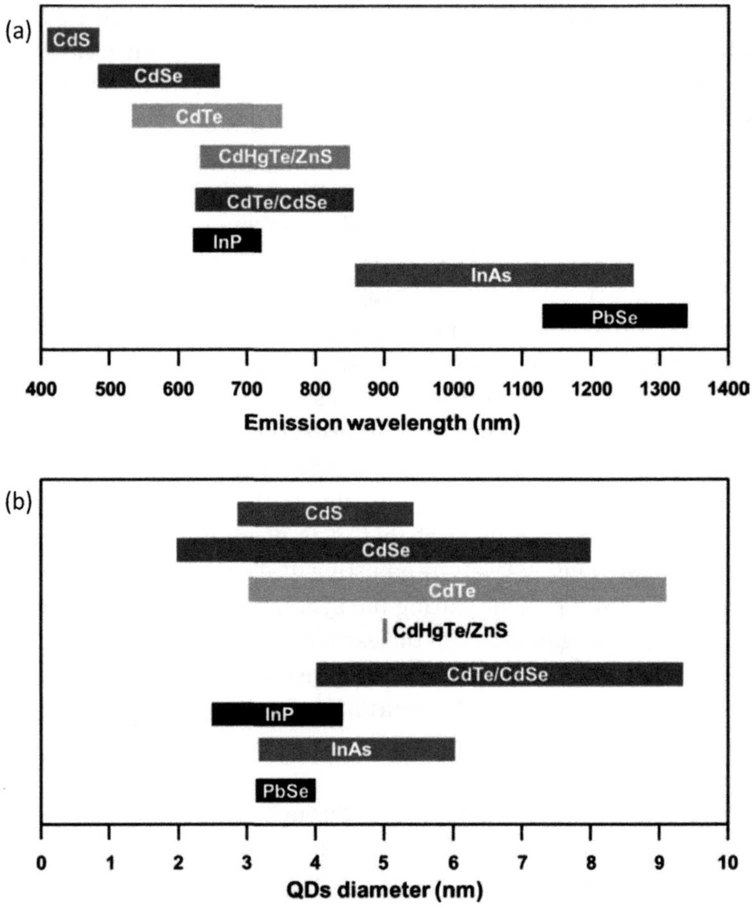

Figure 16.1 (a) Emission wavelength ranges of QDs prepared with elements from different groups (II-VI: CdS, CdSe, and CdTe; III-V: InP and InAs; IV-VI: PbSe). (b) Typical size ranges of the referred QDs.

Besides an incapacity to interact with biological systems, bare nanocrystals show a tendency to aggregate and precipitate. The introduction of surface ligands ensures adequate nanocrystal solubility, preventing their agglomeration. The occurrence of ligand effects could also affect nanocrystal size, size distribution, and shape. Further coating with polymeric materials or the encapsulation into a silica shell could amend the photochemical properties, while decorating with appropriate molecules imparts biofunctionality and allows implementing biorecognition mechanisms.

16.1.1 Properties

In spite of their small size QDs behave like typical semiconductors. The overlap of the metal and nonmetal atomic orbitals leads to the formation of two bands separated by an energy gap EG (bandgap): a low-energy valence band (VB) grouping the bonding molecular orbitals and a high-energy conductance band (CB) grouping the antibonding orbitals. Between the VB and CB lies a region of forbidden energy states. Upon photoexcitation by a photon with energy higher than EG, an electron is promoted from a filled orbital in the VB up to an empty orbital in the CB, leaving a positively charged "hole" in the VB. The electron–hole pair is attracted to each other by electrostatic Coulomb forces behaving like a quasiparticle, an exciton, with slightly lower energy than the unbound electron and hole. Upon deactivation, the electron returns from the LUMO (lowest unoccupied molecular orbital) energy state on the CB to the HOMO (highest occupied molecular orbital) state on the VB and recombines with the hole, emitting radiation at a wavelength that corresponds to the LUMO/HOMO energy gap.

QDs are zero-dimensional NPs and therefore subject to strong spatial confinement in all three directions. Quantum confinement effects give rise to unique optical and electronic properties that provide significant advantages as imaging and biosensing tools. These include the following:

- Size-controlled fluorescence, which enables tailoring of the nanocrystals during the synthesis to emit at the required wavelength (Fig. 16.2). Indeed, since the spacing of the LUMO and HOMO is determined by size, the energy levels can be tuned by adjustment of nanocrystal diameter, guaranteeing specifically designed optical properties [9].
- Wide absorption and narrow (full width at half maximum [FWHM], 30–50 nm) and symmetric emission bands, which allow the excitation of mixed-sized QDs, emitting at a multiplicity of wavelengths, by using a single wavelength while simplifying the discrimination of the resulting emission since the bands are less prone to overlap; this enables the simultaneous detection of multiple target molecules, preventing overheating of cells or tissues during conventional multicolor imaging [10].

- High photostability that affords the carrying out of measurements for extended periods, given that the nanomaterials, resistant to photobleaching, provide stable readouts; this attribute concurred with longer photoluminescence lifetimes that minimize interference from background autofluorescence during the imaging of living cells.
- Large molar attenuation coefficients providing large effective Stokes shifts that also facilitate multiplexing and high-fluorescence quantum yields that ensure enhanced sensitivity (Fig. 16.2).

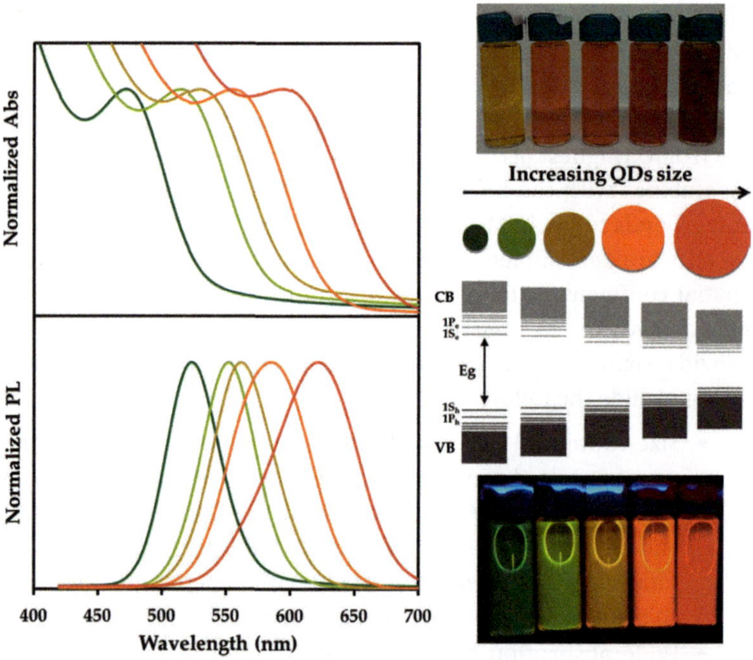

Figure 16.2 Absorption and emission spectra of distinctly sized QDs along with representative color (top right) and photoluminescence (bottom right). The maximum absorption and emission wavelengths shifted toward the red region as the QD size increased and bandgap decreased.

Some specific features of QDs regarding bulk materials rely on two main differences: a high surface-to-volume ratio and a great chemical reactivity. Indeed, due to the nanomaterial's small size a large fraction of its component atoms reside on the surface [11]. For this reason, they are more prone to participating in reactional schemes, being efficient catalysts. On the other hand, surface atoms are devoid of neighboring counterparts exhibiting unoccupied molecular orbitals. These orbitals, commonly referred as dangling bonds or surface traps, could lead to nonradiative recombination of charge carriers, affecting negatively the fluorescence quantum yield. Finally, surface atoms are imperfectly passivated, being highly reactive, which could affect nanocrystals' stability in biological media.

16.1.2 Synthesis

The synthetic route used for QD preparation is crucial because it determines not only the QD's solubility, luminescence efficiency, and size distribution but also the surface functionality influencing the QD's ability to interact with the target species. Despite the different synthetic approaches that could be used to obtain QDs of appropriate size and with appropriate crystalline properties, which further determine their optical response, the singularities associated with their biological application, mostly related with the need for a flagrant aqueous solubility, justify the success of bottom-up colloidal synthesis methods. Nevertheless, a panoply of alternative top-down process has been deployed, ranging from X-ray lithography [11] to molecular beam epitaxy [13] and ion implantation [14].

One of the greatest breakthroughs in the colloidal synthesis of QDs was achieved by the Bawendi group in 1993 [4], which was able to prepare high-quality nanocrystals of CdSe with a very narrow size distribution and fairly high quantum yields by using a synthetic route resorting to organometallic precursors, that is, molecules that deliver the monomers of the NPs, and a high reaction temperature [15]. The reaction scheme involved the decomposition of the precursors dimethylcadmium (CdMe2) and trioctylphosphine selenide (TOP:Se), which were swiftly injected into a hot (280°C–300°C) nonpolar coordinating solvent (trioctylphosphine oxide [TOPO]). The TOPO layer stabilizes the nanocrystals in organic solvents like chloroform or toluene, preventing agglomeration and precipitation. This led to high oversaturation of monomers, which favored nanocrystal growth [16]. The experiments also demonstrated that temperature and composition of the solvent had a remarkable influence on crystal growth kinetics and nanocrystal shape. Upon adjustment of the precursors' concentration it was possible to modulate the QD size [7]. Peng et al. [7] improved this reaction pathway using less unstable, less toxic, and less pyrophoric precursors, such as Cd acetate and CdO, and were able to prepare other types of colloidal QDs. Moreover, it was found that the epitaxial growth of a layer of wider bandgap semiconductor material such as ZnS, acting as a passivating shell around the QD core, could noticeably enhance the quantum yield by up to 80% [17, 18] and increase resistance to photodegradation. The resulting core/shell NPs were more fluorescent. In addition, the passivating shell was capable of preventing Cd leakage, minimizing QD toxicity.

One of the obvious shortcomings of the organometallic synthetic route is that the as-prepared nanocrystals are capped with organic TOP/TOPO ligands with high hydrophobicity [19]. Consequently, they demand the carrying out of distinct surface modification stratagems to adjust their water solubility in order to make them compatible with biological applications, while preserving their stability and optical properties. These could be generally achieved either by exchanging the QD capping layer or by encapsulating them within shells of block copolymer or into phospholipidic micelles or by silica coating. The exchanging of the TOPO surfactant layer with more polar species, such as bifunctional ligands like mercaptoacids (mercaptoacetic, mercaptopropionic, or mercaptodecanoic acids), thiolated amines, or even cysteine-terminated peptides, is the simplest way to obtain a highly hydrophilic surface: the thiol group binds to the inorganic nanocrystal surface and the carboxyl group, which is deprotonated at neutral pH, promotes not only affinity to the aqueous environment but also prevents QD agglomeration due to electrostatic repulsion. Thiol ligands modified with poly(ethylene glycol) (PEG) are also used to ensure colloidal stability [20]. Moreover, the bifunctional molecules, such as cysteamine or cysteine, that are used to replace the hydrophobic ligands provide reactive groups for additional bioconjugation.

Encapsulation with amphiphilic polymers, such as poly(acrylic acid) or block copolymer micelles consisting of PEG and a phospholipid, maintains the original nanocrystal quantum yield but increases significantly their size. A silica coating suffers from a similar drawback, although ensuring robust, stable, and easily functionalized NPs.

Alternative synthetic routes to prepare biocompatible QDs involve their direct preparation in an aqueous solution by employing hydrophilic capping ligands. Rogach et al. [21] used 2-mercaptoethanol and 1-thioglycerol as passivating agents to synthesize very small (1.3–2.4 nm) aqueous CdTe QDs with narrow size distribution. Similar synthetic approaches allowed the obtaining of nanomaterials with a small size and high quantum yield, including glutathione-capped CdTe QDs [22] and nontoxic core/shell cysteine-capped CdTe/ZnTe QDs.

16.1.3 Bioconjugation

Whether prepared from organometallic precursors or prepared by using the hydrothermal synthesis, QDs have a surface chemistry that is not particularly apt to target any specific molecules. This presumption is even more relevant in the case of the QDs fabricated by the organometallic route as they have no intrinsic aqueous solubility, exhibit low biocompatibility, and show no reactive surface groups for conjugation with the appropriate biomolecules [10]. In addition, QDs have a tendency for nonspecific binding, in particular to cellular membranes and proteins, which causes the adsorption of a protein layer on the nanocrystal surface, which further aggravates the problem [23].

Regardless of the ultimate objective, the application of QDs in almost all biological contexts will indubitably depend on their ability to be bioconjugated in a careful and controlled manner [24, 25]. Bioconjugation could be defined as the pairing of a NP with a biomolecule, via covalent or noncovalent bonding, creating a biofunctional entity. The biomolecule, such as antibody, enzyme, oligonucleotide, peptide, or small molecule-binding protein, should be able to mediate target-specific interactions. The QD bioconjugate should preserve the initial photophysical properties of the nanocrystals, exhibit appropriate stability and solubility, and guarantee the biological functions of the ligands.

Different strategies have been applied to attach biomolecules to the QD surface [26], such as passive adsorption [27] and the establishment of electrostatic interactions [28], although the most commonly used approach relies on direct covalent conjugation of reactive functional groups present on the QD's organic capping to relevant functionalities in the biomolecules by means of cross-linking agents (Fig. 16.3). sulfo-NHS (*N*-hydroxysulfosuccinimide) and 1-ethyl-3-(3-dimethylaminopropyl) carbodiimide (EDC) and are used to couple carboxylic acid terminal groups on the nanocrystal to the amine group (e.g., lysine residues of proteins) on the affinity molecule, forming the corresponding amide bond. Active ester maleimide-mediated covalent coupling could be used to attach thiol and amine groups [29].

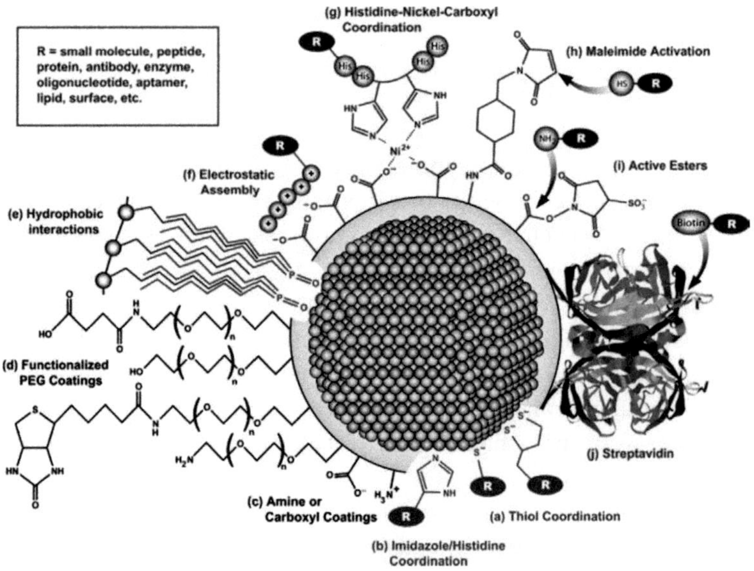

Figure 16.3 Illustrative scheme of selected surface chemistries and conjugation strategies that are commonly applied to QDs. The gray zone around the QD represents a general coating, which can be associated with the surface of the QD via hydrophobic interactions or ligand coordination. The exterior of the coating mediates distinct strategies for improved aqueous solubility and for bioconjugation. Reprinted from Ref. [25], Copyright (2010), with permission from Elsevier

Affinity-based systems relying on noncovalent receptor-ligand interactions are also subject to widespread utilization. Biotin and avidin (or other variants, such as streptavidin) could be covalently conjugated to QDs and selected biomolecules, commonly by EDC/NHS, and the extremely high affinity and specificity of their bond could be subsequently exploited for labeling or targeting purposes. The availability of a large variety of biotinylation reagents and of biotin or avidin-modified biomolecules further reinforces the convenience of this method.

The strong affinity between polyhistidine and charged metal ions, such as Ni(II) and Zn(II), has been also employed to link, via direct coordination, distinct biomolecules (tagged with polyhistidine on N- or C-terminus) to the QD surface, usually core/shell CdSe/ZnS nanocrystals [30].

More recently, Han et al. [31] developed a bio-orthogonal and modular conjugation method using a norbornene-tetrazine cycloaddition. They first prepare norbornene-coated QDs, which were coupled to a dye modified with a tetrazine derivative via a Cu-catalyst-free cycloaddition reaction. The approach was tested in the targeting of live cancer cells labeled with tetrazine-modified proteins by norbornene-coated QDs.

16.2 Diagnostics

Sensing of biorelevant molecules as a means of ascertaining the occurrence of biological agents or toxic products, for developing immunoassays, for monitoring biomarkers associated with specific pathological conditions, and for controlling intra- or extracellular processes, is a core issue for biomedical diagnosis. Since the pioneering works of Alivisatos [32] and Nie [33], the number of papers reporting the utilization of QDs in biomedical research has increased steeply. Either for in vitro or for in vivo assays, QDs have been used as fluorescence labels and probes; as donors or acceptors in Förster resonance energy transfer (FRET), coherent resonance energy transfer (CRET), or photoinduced electron transfer processes; and as active species in photoelectrochemical (PEC) bioanalysis, in solution or immobilized on a solid support. In this regard, due to the potential toxicity issues the in vitro application of QDs for diagnostic purpose is undoubtedly more relevant than their utilization in vivo.

Whether for the analysis of liquid samples, cells, or tissues, the remarkable optical properties of QDs have provided a stable, highly photoluminescent and multiplexed tool to improve the sensitivity of biodiagnostic assays, such as flow cytometry, fluorescence in situ hybridization (FISH), and immunohistochemistry (IHC), or for the implementation of lateral flow immunochromatographic tests, lab-on-a-chip devices, or microfluidic devices.

The great applicability of QDs arises from a multiplicity of attributes, including easily tailored surface chemistry and similarity of size with biomolecules, such as polynucleic acids, proteins, antibodies, and antigens, which offer an excellent platform for the establishment of selected interactions with the target species. Moreover, they are not subject to noteworthy kinetic or steric hindrance problems [34].

QDs' high reactivity and tendency to nonspecific binding limit their direct application and demand the implementation of suitable strategies to improve selectivity. The required biorecognition mechanisms are usually established upon conjugation of QDs with selected molecules. These could involve, for instance, enzymes recognizing specific substrates or biomolecules capable of establishing strong binding affinity interactions, such as antibody-antigen (Ab-Ag) immunocomplexes or DNA hybridization [35]. For instance, by coupling biotin-modified glucose oxidase with FITC–avidin-modified CdSe/ZnS QDs Gill et al. [36] developed a ratiometric fluorescence assay for monitoring glucose levels based upon the quenching effect of the biocatalytically generated H_2O_2. The method has been also applied to the analysis of acetylcholine esterase inhibitors by monitoring their effect on the H_2O_2 levels generated by the acetylcholine esterase–choline oxidase cascade. In 2009, the same authors developed an optical sensing approach for biocatalytic processes involving NAD+-dependent enzymes (where NAD is nicotinamide adenine dinucleotide) and for the screening of anticancer drugs by monitoring intracellular metabolism with NAD(P)H-sensitive Nile-blue-functionalized CdSe/ZnS QDs [37].

QDs could be also functionalized with molecular imprinted polymers (MIPs). The MIPs provide template (target) specificity, while the fluorescence modulation (either quenching or enhancing) undergone by the QDs in the presence of the template signalizes the recognition event. As an example, Zhang et al. [38] designed a QD-based MIP-coated composite for selective recognition of cytochrome c (Cyt) with increased affinity. In this case, the QD's fluorescence was quenched by Cyt. Results confirmed that the developed approach could be used as a selective sensing system for protein recognition.

16.2.1 Imaging

Fluorescence imaging is one of the most valuable imaging techniques, becoming a powerful tool in biological research and medical diagnostic.

The remarkable optical properties of QDs, in particular the brightness of the photoluminescence and the possibility of multicolor labeling for extended visualization, are some of the features that raised greater expectations in terms of their applications in fluorescence imaging. However, other physical properties, such as size, surface groups, and charge, could modulate the QDs' delivery into the distinct subcellular compartments and their accumulation, affecting the overall brightness of the fluorescent bioconjugate. On the other hand, imaging agents should be also photochemically stable and principally nontoxic and biocompatible. One of the major drawbacks of semiconductor QDs, restraining, for instance, their utilization in vivo, is the occurrence of heavy metals, such as cadmium, which exhibit strong toxicity even at low concentrations. The acute toxicity problem can be circumscribed by adding a protective shell made of insulating or semiconductor material, like ZnS. Moreover, passivation of the native core with a layer of a material with a wider bandgap, in order to provide electronic insulation and ensure charge carriers' confinement in the core, can also improve significantly the quantum yield and therefore the fluorescence emission of the resulting core/shell QDs [39]. Although shells made of one to two monolayers ensure enhanced quantum yield, thicker shells composed of four to six monolayers provide superior protection against photooxidation and degradation [40]. Encapsulation with amine-modified polymers or coating with a silica layer could also be used to attain adequate water solubility. Coating with a neutral hydrophilic polymer, such as PEG, could be adopted to reduce nonspecific binding.

Larson et al. [41] investigated the photophysical properties of water-soluble CdSe/ZnS nanocrystals encapsulated within an amphiphilic polymer, confirming that these fluorescent probes have the largest two-photon cross sections of any label used in multiphoton microscopy. This is an important feature because two-photon absorption enables QDs' excitation in the NIR region, providing imaging upon deep tissue penetration.

QDs' advanced fluorescence imaging applications range from in vitro cellular imaging, multiplexed quantitative analysis of cellular phenotypes, tissue staining, and biomolecular tracking for real-time monitoring of cellular processes [41] to in vivo molecular imaging, tumor and vascular imaging, and in vivo QD biodistribution and tracking assays.

Several in vitro studies have targeted the cellular membrane or intracellular structures. Highly selective labeling and tracking of sialic acid was accomplished with QDs conjugated to phenylboronic acid [42]. CdSe/ZnS QDs coupled to IgC and streptavidin [43] were used for labeling distinct targets at a subcellular level, including breast cancer marker human epidermal growth factor receptor 2 (HER2), nuclear antigens, and actin and microtubule fibers, with great specificity and with more brightness and photostability than the ones obtained with comparable organic dyes. Multiplex detection of two cellular targets was also achieved by using QDs with different emission spectra and identical excitation wavelength.

One of the most used methods for targeting QDs at cells is antibody-based labeling. However, it produces unstable large-sized conjugates. To overcome these shortcomings Howarth et al. [44] used *Escherichia coli* biotin ligase to site-specifically biotinylate an acceptor peptide (AP) sequence encoded at the epidermal growth factor (EGF) in HeLa cells and at α-amino-3-hydroxy-5-methyl-4-isoxazolepropionic acid (AMPA) receptors in neurons. The biotin group was subsequently used for targeting streptavidin-conjugated QDs. Seeking to obtain QDs with a reduced size the same authors [45] synthesized 605 nm emitting monovalent CdSe-ZnCdS core/shell QDs passivated with carboxy-terminated PEG ligand (DHLA-PEG8-CO2H). The NPs were conjugated with monovalent streptavidin or antibody to carcinoembryonic antigen for labeling cell-surface proteins. Results confirmed that the small size facilitated access of glutamate receptors labeled with the QDs to neuronal synapses, while monovalency prevented EphA3 tyrosine kinase activation.

The multicolor and multiplexing potential of QDs was explored in the detection and characterization of Hodgkin and Reed-Sternberg (HRS) cells, a class of low-abundant malignant cells in Hodgkin's lymphoma, by the simultaneous detection of four protein biomarkers directly on human tissue biopsies [46]. Multiplexed CdSe/CdTe core/shell QDs (emitting at 655 and 605 nm) combined with four protein biomarkers (E-cadherin, high-molecular-weight cytokeratin, p63, and alpha-methylacyl CoA racemase) were also used by the same authors to map tumor heterogeneity on human prostate cancer tissue specimens [47]. These and similar results demonstrated that the prompt assessment of multiple biomarkers by using multiplexed QDs imaging could be a valuable strategy not only for cancer-targeting but also for evaluating tumor type and malignancy.

The enhanced photostability of QDs facilitate long-term live-cell imaging for studying intracellular behavior of adeno-associated virus serotype 2 (AAV2), labeled with the QDs, in living target cells, including mechanisms of virus entry and viral motility [48]. In a study to evaluate QDs' IHC in the detection of HER2, which is important for breast cancer treatment and diagnostic, Chen et al. [49] demonstrated that QD photoluminescence could remain stable for as long as 75 days, guaranteeing sensitivity and accuracy superior to conventional immunohistochemistry techniques.

For in vivo applications, which hitherto have been restrained to animals, the biological optical window that delineates the light wavelengths capable of tissue penetration is relatively narrow and therefore usage is usually restricted to QDs with excitation/emission on the NIR region (650–900 nm) [50]. These QDs (type II) [51] have usually a core/shell structure, and the selection of the core and shell materials favors the relaxation of the CB electrons into the VB of the shell [52]. Since core and shell band alignment is staggered QDs' effective bandgap is smaller than the one observed for core and shell constituents.

In a pioneering work, Kim et al. [53] (Fig. 16.4) prepared CdTe/CdSe core/shell type II QDs emitting at 840–860 nm, with a polydentate phosphine coating to ensure serum solubility, for carrying out sentinel lymph node mapping in large animals. Integrins are transmembrane receptors highly expressed in tumor cells, which could be targeted by probes such as arginine-glycine-aspartic acid (RGD) peptide for diagnostic purposes. In 2006, Cai et al. [54] reported the in vivo targeting and imaging of tumor vasculature in mice bearing subcutaneous U87MG human glioblastoma tumors using CdTe/ZnS core/shell QDs (λ_{em} = 705 nm) labeled with RGD peptide.

In an effort to reduce toxicity associated with the occurrence of heavy metals, such as Cd, on the QDs structure, a new type of stable and biocompatible dendron-coated InP/ZnS core/shell NIR emitting (710 nm) QD was reported [55].

The NP (QD710-Dendron) conjugated to RGD peptide dimers was used for in vivo targeting of integrin avb3-positive tumor cells, showing both high stability, selectivity, and tumor uptake and long retention at the tumor sites.

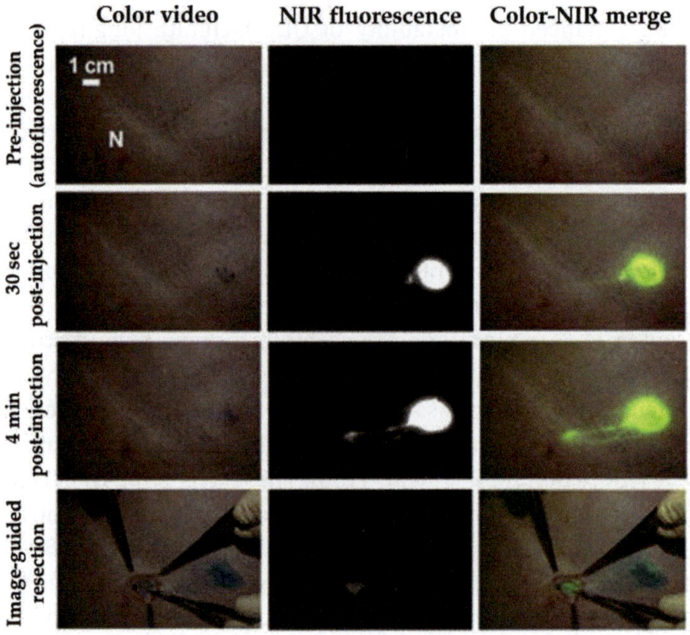

Figure 16.4 Sentinel lymph node mapping using CdTe/CdSe core/shell type II QDs emitting at 840–860 nm. Images of the surgical field in a pig injected intradermally with 400 pmol of NIR QDs in the right groin. For each time point, color video (left), NIR fluorescence (middle), and color-NIR merge (right) images are shown. Fluorescence images have identical exposure times and normalization. To create the merged image, the NIR fluorescence image was pseudocolored lime green and superimposed on the color video image. The position of a nipple (N) is indicated. Reprinted by permission from Macmillan Publishers Ltd: [*Nature Biotechnology*] (Ref. [53]), copyright (2003).

Besides application in direct fluorescence imaging, more recently research has been also focused on the development of new QD-based probes that could be used as contrast agents in MRI, computed tomography (CT), or positron emission tomography

(PET). In one of the first developments, and aiming at application not only in fluorescence imaging but also with PET, Duconge et al. [56] coupled Fluorine-18 to phospholipid QD micelles to design a new nanoprobe exhibiting long circulation half-time and slow uptake by the reticuloendothelial system. In addition, it provides highly versatile surface chemistry to conjugate multiple chemicals and biomolecules.

16.2.2 Förster Resonance Energy Transfer

FRET is a nonradiative energy transfer from an excited donor to an acceptor over distances of up to 10 nm [57]. It could be used to transduce into a photoluminescence response a biological event or a molecular recognition that resulted in a concentration-related modulation of the photochemical properties of the species participating in the FRET process. In addition, it could be extremely useful to measure variations in the distance between the donor and the acceptor, resulting from associative, dissociative, or distance modulated processes [58], making it particularly attractive for monitoring the establishment of biomolecules' interactions, conformational changes on the target species, enzyme activity assay [59] (Fig. 16.5), etc.

QDs could be used either as acceptors or donors although their unique optical properties are indeed fully exploited when used as donors: the wide absorption spectra, the large molar attenuation coefficient, and the versatility provided in terms of tuning the emission for wavelengths that show an adequate spectral overlap with the acceptor dye justify the widespread use of QDs as effective and reliable FRET donors. AuNPs, known for their capacity to induce strong QD fluorescence quenching, are frequently used as paired acceptors.

The sensing mechanism relies on the establishment of interactions between the target and the QD-AuNP system, which disturbs the FRET process. Conjugation of QDs and AuNPs with selected biomolecules imparts selectivity. A hybrid FRET nanobiosensor combining QDs-ConA-beta-CDs-AuNPs was developed for the determination of glucose in serum. Sensing was based on FRET between concanavalin A-conjugated CdTe quantum QDs (donor) and beta-cyclodextrin-modified AuNPs (acceptor).

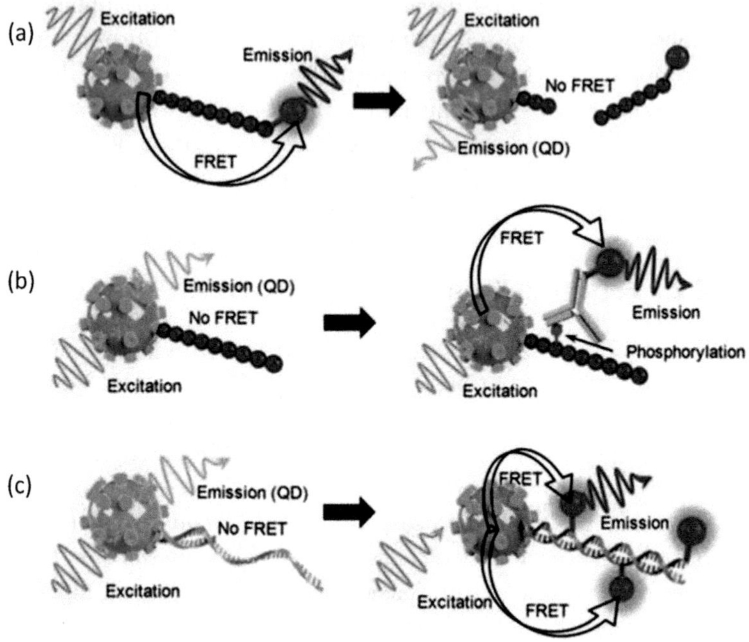

Figure 16.5 QD-FRET schemes for the evaluation of distinct enzyme activity: (a) protease, (b) protein kinase, and (c) DNA polymerase. Reprinted from Ref. [59] with permission from Theranostics.

In the presence of glucose, the AuNP-beta-CD component of the nanobiosensor was displaced by glucose, which competed with beta-CDs on the binding sites of ConA, resulting in the fluorescence recovery of the initially quenched QDs [60].

QDs could be also combined with other fluorophore acceptors, such as organic dyes or fluorescent proteins. For example, two ratiometric pH sensors combining CdSe/ZnS core/shell QDs with pH-sensitive squaraine dye and mOrange, a pH-sensitive monomeric fluorescent protein, were developed by Snee et al. [61] and Dennis et al. [62], respectively, for intracellular pH monitoring.

QDs coated with PEG and conjugated with single-domain antibodies were used as acceptors in a time-gated detection Tb-to-QD FRET-based immunoassay aiming at subnanomolar detection of soluble epidermal growth factor receptor (EGFR) in serum samples [63].

QDs have also been used for implementing FRET processes for DNA sensing, which will be discussed in more detail in the next section. Zhang et al. [64] reported an ultrasensitive FRET nanosensor capable of detecting DNA at low concentrations. The developed approach used CdSe/ZnS core/shell QDs (605 nm) linked to DNA probes to capture DNA targets. The target strand bound to a reporter strand labeled with Cyanine 5, thus forming a FRET donor-acceptor ensemble (Fig. 16.6).

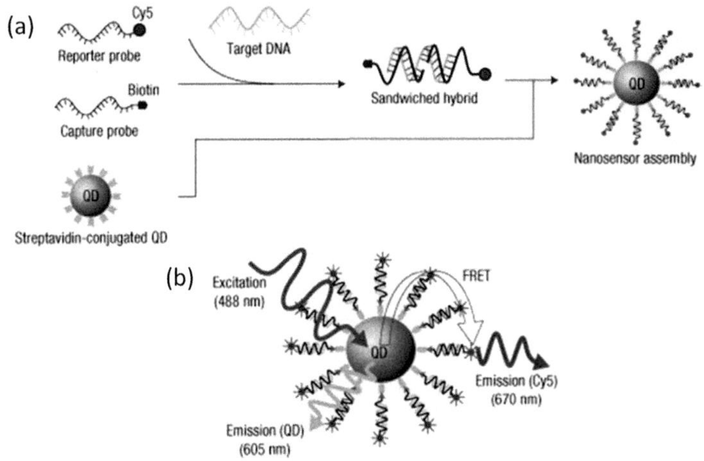

Figure 16.6 Schematic illustration of single-QD-based DNA nanosensors showing (a) formation of a nanosensor assembly in the presence of targets and (b) the FRET process between a QD donor and cyanine 5 (Cy5) acceptor upon nanosensor assembly. Reprinted by permission from Macmillan Publishers Ltd: [*Nature Materials*] (Ref. [64]), copyright (2005).

A multiplexed FRET combining two CdSe/ZnS QDs (emitting at 524 and 606 nm) conjugated with selected oligonucleotide probes, acting as energy donors, with two fluorophores Cy3 and Alexa Fluor 647 (Alexa647), as acceptors, was implemented by Algar et al. [65]. The acceptor dyes were labeled with the target oligonucleotides and upon hybridization with the donor probes established the two-color FRET scheme, allowing picomolar detection by using a single UV excitation source (Fig. 16.7).

Seeking to implement prompt point-of-care bioassays combined with smartphone readout, Petryayev et al. [66] conjugated CdSe/CdS/ZnS QDs emitting at 630 nm with Alexa647 labeled with peptide substrates for thrombin. The nanoprobes were immobilized on paper test strips to develop a point-of-care FRET-based bioassay to detect enzyme activity in serum and whole blood, using thrombin as a model analyte. The developed FRET system ensured fast quantitative results with a good limit of detection and low sample consumption.

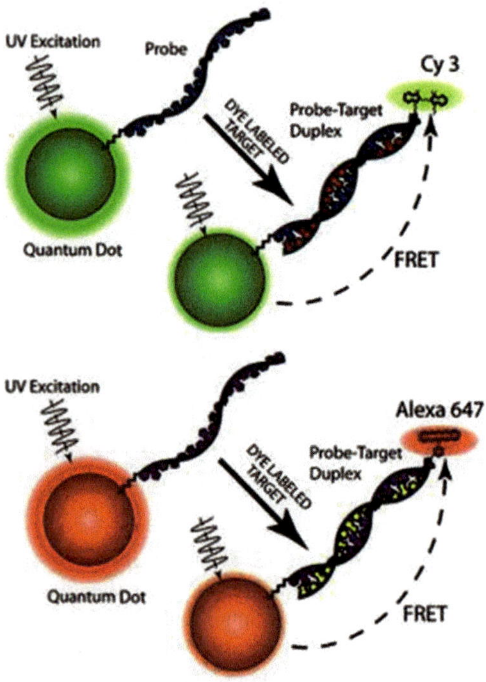

Figure 16.7 QD-FRET-based strategy for two-color nucleic acid detection. Simultaneous excitation of green and red quantum dots in the ultraviolet-region prevented significant excitation of Cy3 or Alexa647 dyes. When probe oligonucleotides were conjugated to QDs, hybridization with the Cy3- or Alexa647-labeled target oligonucleotide yielded FRET-sensitized emission from the dyes, which was used as the analytical signal. Adapted from Ref. [65], Copyright (2007), with permission from Elsevier.

16.2.3 Nucleic Acid Sensing

The development of DNA biosensors has gained noteworthy importance in clinical diagnosis since DNA could act as a sort of functional biomarkers for many diseases, including distinct types of cancers. In this regard, DNA-related detections, such as gene sequencing or the monitoring of the amount of gene occurrence, could play a role of extreme relevance in the early detection of a given pathology, even before the appearance of the first symptoms.

One of the most common applications of QDs in DNA sensing is as fluorescent labels in hybridization assays between a target DNA and its complementary probe, in which they are used to label the DNA probe. Used as alternative labels for conventional organic dyes QDs allow overcoming some of the problems faced with the latter: a susceptibility to photobleaching that induces cleavage of the DNA and a tendency to impair DNA-protein interactions [19]. Crut et al. [67] exploited this concept by resorting to biotin- or digoxigenin-modified DNA fragments that were covalently linked to both extremities of a DNA molecule. These fragments were subsequently signaled by multicolor fluorescence microscopy using two-color QDs conjugated to streptavidin or antibody.

Alternative DNA sensing schemes contemplate the usage of molecular beacon probes, sandwich-structured assays, and competitive systems (either homogeneous or heterogeneous). Cui et al. [68] reported an ultrasensitive DNA or antigen detection method based on the self-assembly of multiwalled CNTs and CdSe QDs. To achieve this goal, both nanomaterials were previously functionalized with oligonucleotide DNA or antibody. In the presence of the target complementary oligonucleotide or the antigen, CNTs and QDs assembled into a nanohybrid structure via oligonucleotide hybridization, which promoted a pronounced decrease in the QDs' fluorescence.

As was previously exemplified QD-based FRET detection plays a major role in nucleic acid sensing (likewise it happens with immunoassay setting up), because of FRET versatility in terms of possible mechanisms of recognition/signaling/transduction and a panoply of available donor/acceptor pairs. In a recent work, Tian et al. [69] used single-strand DNA-modified CdTe QDs paired with oxidized carbon nanotubes (oxCNTs) for H5N1 DNA detection. oxCNTs acted as the QDs' initial quenchers, but, upon target recognition, the strong affinity of the latter toward QDs-ssDNA reversed the quenching interactions between QDs-ssDNA and oxCNTs, leading to a recovery of the QDs' fluorescence.

DNA-functionalized CdSe/ZnS QDs of different sizes have also been used as fluorophore-modified signaling probes in a variety of DNA microarray-based applications. In one of these applications, Gerion et al. [70] carried out the detection of single-nucleotide polymorphism mutation in the human p53 tumor suppressor gene as well as multiallele detections by identifying simultaneously hepatitis B virus (HBV) and hepatitis C virus (HCV).

The high suitability of QDs for DNA sensing is clearly manifest in their ability for multiplex DNA detection. Multicolor optical coding upon excitation of three distinctly sized CdSe/ZnS QDs, embedded into polymeric microbeads, with a single light source, was explored to design a DNA hybridization system using oligonucleotide probes [71]. The coding and target signals provided identification of the DNA sequence and quantified the abundance of that sequence. A two-color FRET based on a similar approach was previously referred [64]. The same authors succeeded in developing a single QD-based nanosensor for multiple DNA detection, applied for the screening of HIV-1 and HIV-2 at the single-molecule level [72].

QD barcodes have also been successfully explored for multiplexed DNA sensing. Giri et al. [73] developed a QD-barcode-based assay for fast multiplex analysis of genetic biomarkers of the bloodborne pathogens HIV, malaria, hepatitis B and C, and syphilis, at the femtomol range.

Like DNA, RNA is among the most important biological molecules playing crucial roles in encoding, regulating, transmitting, and expressing genetic information. For this reason, the detection of RNA is of utmost importance in early diagnosis of cancer. By functionalizing the surface of CdSe/ZnS QDs (λ_{em} = 490 nm) with A10 RNA aptamer, Bagalkot et al. [74] implemented a QD-aptamer(Apt)-Dox (doxorubicin) conjugate sensing system for cancer targeting, imaging, and therapy. A10 RNA aptamer, which is able to recognize the extracellular domain of the prostate-specific membrane antigen, acted as the targeting element, while doxorubicin, a widely used antineoplastic anthracycline drug, was the therapeutic agent.

16.2.4 Immunoassays

Immunoassays relying on specific molecular recognition of an antigen by its antibody have been extensively implemented for quantitative clinical analysis of selected biomolecules, mostly protein biomarkers, in biological samples. This is a valuable tool for detecting low analyte concentrations in early disease diagnosis with the inherent advantages in terms of therapeutic efficiency [75].

Goldman et al. [76] developed a multiplexed fluoroimmunoassay combining distinctly sized CdSe/Zn core/shell QDs, emitting different colors, with pertinent antibodies for the simultaneous detection of cholera toxin, ricin, shiga-like toxin 1, and staphylococcal enterotoxin B. A microfluidic device was proposed by Klostranec et al. [77] for the multiplexed, high-throughput monitoring of serum biomarkers of hepatitis B, hepatitis C, and HIV. The detection system consisted of three QD-barcodes combining antigen-coated CdSe/ZnS QDs embedded in polystyrene beads and provided sensitivity at near picomolar level.

Several immunoassay approaches have been designed relying on the utilization of selected quenchers of QDs' fluorescence, either by photoinduced electron transfer or by energy transfer mechanisms. This strategy functions as "turn on" sensors as the QDs' fluorescence is "switched on" in the presence of the target specie. For instance, Chen et al. [78] described an immunoassay based on streptavidin-conjugated QDs, emitting at 525 and 605 nm, and biotinylated antibodies for the simultaneous determination of Human Enterovirus 71 (EV71) and Coxsackievirus B3 (CVB3). Graphene oxide (GO) was used as a QD quencher. Upon target recognition the QD-GO interaction was cleaved and fluorescence was restored.

The rapid, specific, and sensitive detection of pathogenic bacteria is an area of extreme relevance. Yang and Li [79] developed a fluoroimmunoassay for the simultaneous detection of two species of foodborne pathogenic bacteria, *E. coli* O157: H7 and *Salmonella typhimurium*, by using two streptavidin-conjugated CdSe/ZnS QDs, emitting at 525 and 705 nm, and magnetic beads coated with anti-*Salmonella* and anti-*E. coli* biotinylated antibodies. The magnetic beads allowed separation and isolation of target bacteria from samples while the QDs ensured two-color detection.

By combining QDs and a lateral flow test strip Li et al. [80] developed a portable fluorescence immunosensor for the quantitative detection of ceruloplasmin, a biomarker for cardiovascular disease, lung cancer, and stress response to smoking, in human plasma. The obtained results confirmed the potential of QD-based lateral flow tests for biomarkers screening in point-of-care analysis (Fig. 16.8).

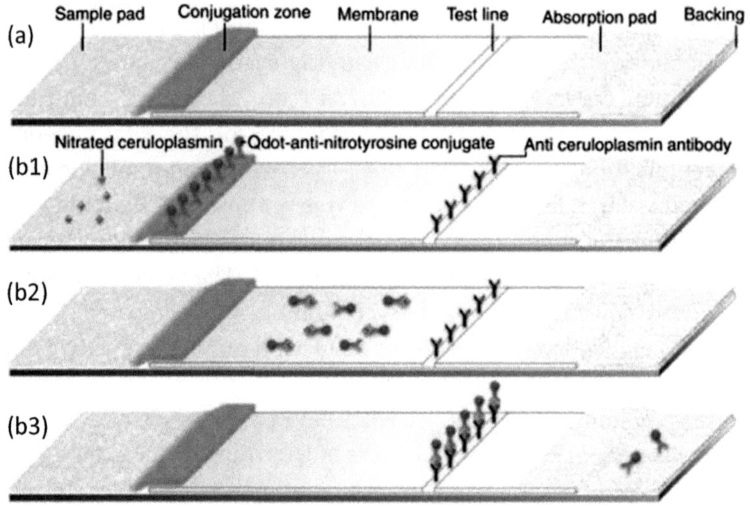

Figure 16.8 Lateral flow test strip (LFTS) assay for the detection of nitrated ceruloplasmin (NC). (a) An aqueous sample containing NC is applied to the sample pad (b1) and transported to the conjugation zone, where the preabsorbed and dried QD-antinitrotyrosine antibodies (QD–AB1) are rehydrated and washed off and QD-AB1-NC biocomplexes are formed. The sample further migrates to the test line (b2) where anticeruloplasmin antibodies (AB2) are immobilized, leading to the formation of AB2-NC-AB1-QD immunosandwich conjugates on the test line, whereas the free QD-AB1 without NC further migrates to the absorption pad (b3). The NC concentration is measured by QD fluorescence intensity on the test line using a commercial test strip reader. Reprinted with permission from Ref. [80]. Copyright (2010) American Chemical Society.

16.3 Therapeutics

Drug delivery by multifunctional QD-based nanocomposites capable of drug loading, tissue targeting, and controlled release represents a significant breakthrough in nanomedicine seeking to overcome the downfalls of conventional drug delivery. In the more common configuration, the nanocomposite includes a drug carrier, usually a liposome or a micelle, and the QDs are either attached to their surface or integrated within the carrier. Acting as fluorescent labels,

QDs allow carrier tracking, monitoring of drug release, distribution, pharmacokinetics, etc. [40]. A significant limitation of this approach is that the QDs' loading, either internal or external, is low in order to facilitate cell internalization, which results in poor detection sensitivity. Alternative delivery vehicles, such as PLGA polymeric NPs, show a higher loading capacity, but their larger size restrains biodistribution and tissue penetration.

Weng et al. [34] reported a targeted drug delivery system for cancer diagnosis and treatment comprising QD-conjugated immunoliposome-based NPs (QD-ILs). The prepared QD-ILs exhibited efficient receptor-mediated endocytosis in HER2-overexpressing SK-BR-3 and MCF-7/HER2 cells. Loaded with doxorubicin the NPs demonstrated effective anticancer activity. Lai et al. [81] described the synthesis of a mesoporous silica-based controlled-release delivery system, which allowed the encapsulation of several pharmaceutical drugs and neurotransmitters. CdS QDs were used as caps for closing the openings of the mesoporous channels and physically block the drugs/neurotransmitters from leaching out. These QDs were responsive to chemical stimuli (disulfide bond–reducing molecules) used as controlled-release triggers.

Multimodal QDs able to simultaneously perform multiple tasks, such as separation, sensing, and imaging, have attracted great attention. Li et al. [82] prepared magnetic and fluorescent nanocomposites made of Fe(3)O(4) hollow spheres coated with multilayers of CdTe QDs via layer-by-layer assembly, which combined MRI, magnetic targeting, fluorescent imaging, and drug delivery multifunctionalities.

Photodynamic therapy (PDT) is a viable therapeutic option in the treatment of a wide variety of medical conditions, including distinct forms of cancer. It is based on the generation of cytotoxic oxygen–based molecular species, in particular singlet oxygen, by the photoactivation of a photosensitizer. QDs could be used either as passive or as active PDT NPs (Fig. 16.9). The former act simply as energy carriers for the photosensitizer (usually an organic dye), while the latter are able to themselves produce the reactive species. Samia et al. [83] evaluated the potential of semiconductor QDs in PDT by studying the interaction between CdSe QDs and an organic dye (phthalocyanine 4) used as a photosensitizer. The study demonstrated that the QDs could not only activate the dye by energy transfer but also generate directly singlet oxygen from molecular oxygen.

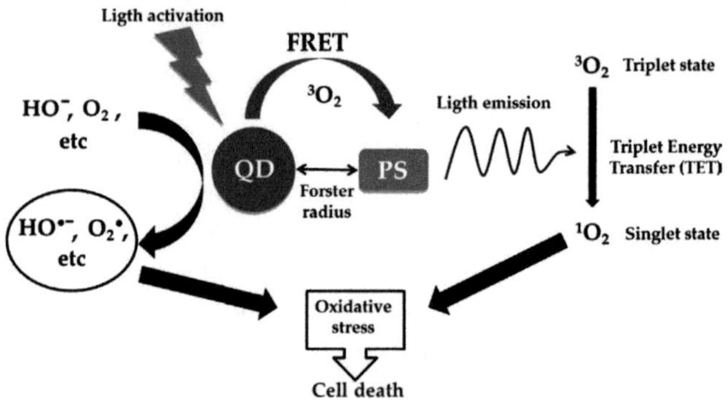

Figure 16.9 Scheme of PDT using QDs. The QD-PS FRET pair confined to the site of cancer is activated by light of a specific wavelength in the presence of molecular oxygen ($^3O_2$) to generate singlet oxygen, which kills cancer cells. Activation of the QD by light can also produce free radicals that play a similar role.

The ability of photoexcited QDs to generate radical species was exploited to kill a wide range of multidrug-resistant bacterial clinical isolates, including methicillin-resistant *Staphylococcus aureus*, carbapenem-resistant *E. coli*, and extended-spectrum beta-lactamase-producing *Klebsiella pneumoniae* and *S. typhimurium*. [84]. Results confirmed that the killing effect was independent of material and controlled by the redox potentials of the photogenerated charge carriers. Since the mammalian cells remained intact the authors suggest that QDs could be used in clinical phototherapy for the treatment of infections

16.4 Toxicity

Semiconductor QDs' intrinsic toxicity is an issue of great concern and debate, which has seriously moderated the huge initial expectations regarding the application of QDs in the biomedical field. Mostly related with the occurrence of heavy metals in the nanocrystal core but also dependent on factors such as size, morphology, and surface chemistry, the cytotoxicity of QDs has been observed in a number of in vitro and in vivo studies, affecting both cell growth and viability. Overcoating with a semiconductor shell, such as ZnS, or the addition of a protective hydrophilic PEG coating, has been demonstrated to prevent the heavy metal's leakage although long-term in vivo utilization still raises concerns in terms of bioaccumulation and chronic toxicity. Nonetheless, the direct correlation of these results with the potential for negative effects of QDs on humans remains unclear [85] and more trustworthy and comprehensive studies, along with critical assessment of risk versus benefit, are needed before QD utilization moves into the clinical research phase.

These facts probably explain some part of the wide interest that other nanomaterials, such as silicon QDs (SiQDs) [86], and carbon NPs, like CDs or graphene quantum dots (GQDs), have recently attracted as less toxic alternatives to conventional semiconductor QDs [87]. However, even in these latter cases, reports on toxicity do not guarantee the desirable innocuousness [88].

16.5 Summary and Outlook

The huge potential of QDs for health care and medical research applications, in particular for in vitro and in vivo imaging and diagnostic, and the tremendous advantages they offer, from bright emissions to the simplicity of implementation of distinct detection modes, clearly justify the continuation of the research on these nanomaterials, even considering that there are several issues regarding their clinical acceptance that should be addressed.

Challenges remaining ahead will necessarily involve improvement of analytical performance, in particular in what concerns assay selectivity. This will involve not only the search for new recognition elements, new hybrid materials, or strategies to integrate the recognition and signaling mechanisms, which could facilitate targeting and detection while preventing nonspecific adsorption of interfering species, but also the design of more efficient and specifically goal-oriented conjugation strategies. On the other hand, the solution stability of these nanomaterials, either in aqueous buffers or the biological environment, still suffers from noteworthy hindrances, demanding more effective passivating coatings. An interesting perspective to limit these drawbacks could be the immobilization of QDs onto solid supports. In this regard, the development of new sensing platforms for carrying out QD-based bioassays, like those relying on lateral flow immunochromatographic and paper-based tests or microfluidic devices, could be valuable assets not only to guarantee stability but also to implement point-of-care expeditious diagnostic devices.

FRET-based analysis and imaging is one of the areas that has experienced some of the most relevant advances as a consequence of research on QDs, but at the same time it is one of those with greatest potential for future progress. In a similar manner, PEC bioanalysis has been subject to a huge impulse in recent years. The advantages provided by combining QDs with the PEC process and electrochemical bioanalysis are expected to gain a noteworthy interest in the near future in PEC DNA analysis, PEC immunoassay, and PEC enzymatic sensing [35].

Although not discussed in this chapter, silicon- and carbon-based QDs have recently emerged as interesting alternatives to semiconductor fluorescent probes in bioimaging applications, due to the higher biocompatibility, lower toxicity, and chemical inertia, and are therefore expected to play a preponderant role in research and development related to QDs, in the near future.

Acknowledgments

David S. M. Ribeiro thanks Fundação para a Ciência e Tecnologia (FCT) and Programa Operacional Potencial Humano (POPH) for the postdoctoral grant ref. SFRH/BPD/104638/2014. S. Sofia M. Rodrigues thanks the financial support from Operação NORTE-01-0145-FEDER-000011 - Qualidade e Segurança Alimentar—uma abordagem (nano) tecnológica. José X. Soares thanks FCT and POPH for his PhD grant ref. SFRH/98105/2013 and also the Biotech Health Programme (Doctoral Programme on Cellular and Molecular Biotechnology Applied to Health Sciences), Reference PD/00016/2012. This work received financial support from the European Union (FEDER funds POCI/01/0145/ FEDER/007265) and National Funds (FCT/MEC) under the partnership agreement PT2020UID/QUI/50006/2013.

References

1. Reed, M.A., et al. Observation of discrete electronic states in a zero-dimensional semiconductor nanostructure. *Phys Rev Lett*, 1988, **60**:535–537.
2. Rossetti, R., Nakahara, S. and Brus, L.E. Quantum size effects in the redox potentials, resonance Raman spectra and electronic spectra of CdS crystallites in aqueous solution. *J Chem Phys*, 1983, **79**:1086–1088.
3. Patsenker, L., et al. Fluorescent probes and labels for biomedical applications. *Ann NY Acad Sci*, 2008, **1130**:179–187.
4. Murray, C.B., Norris, D.J. and Bawendi, M.G. Synthesis and characterization of neraly monodisperse CdE (E=S, Se, Te) semiconductor nanocrystallites. *J Am Chem Soc*, 1993, **115**:8706–8715.
5. Pradhan, N., et al. An alternative of CdSe nanocrystal emitters: pure and tunable impurity emissions in ZnSe nanocrystals. *J Am Chem Soc*, 2005, **127**:17586–17587.
6. Ellingson, R.J., et al. Highly efficient multiple exciton generation in colloidal PbSe and PbS quantum dots. *Nano Lett*, 2005, **5**:865–871.
7. Peng, X.G., Norris, D.J. and Alivisatos, A.P. Kinetics of II-VI and III-V colloidal semiconductor nanocrystal growth - focusing of size distributions. *J Am Chem Soc*, 1998, **120**:5343–5344.
8. Dabbousi, B.O., et al. (CdSe)ZnS core-shell quantum dots: synthesis and characterization of a size series of highly luminescent nanocrystallites. *J Phys Chem C*, 1997, **101**:9463–9475.
9. Frigerio, C., et al. Application of quantum dots as analytical tools in automated chemical analysis: a review. *Anal Chim Acta*, 2012, **735**:9–22.
10. Martín-Palma, R.J., Manso, M. and Torres-Costa, V. Optical biosensors based on semiconductor nanostructures. *Sensors*, 2009, **9**:5149–5172.
11. Kuno, M. *Introductory Nanoscience: Physical and Chemical Concepts*. Garland Science, New York, 2012.
12. Bertino, M.F., et al. Quantum dots by ultraviolet and x-ray lithography. *Nanotechnology*, 2007, **18**:315603.
13. Nakata, Y., et al. Molecular beam epitaxial growth of InAs self-assembled quantum dots with light emission at 1.3 microm. *J Cryst Growth*, 2000, **208**:93–99.
14. Birudavolu, S., et al. Selective area growth of InAs quantum dots formed on a patterned GaAs substrate. *Appl Phys Lett*, 2004, **85**:2337.

15. Murray, C.B., Kagan, C.R. and Bawendi, M.G. Synthesis and characterization of monodisperse nanocrystals and close-packed nanocrystal assemblies. *Ann Rev Mater Sci*, 2000, **30**:545–610.
16. Kudera, S., et al. Growth mechanism, shape and composition control of semiconductor nanocrystals, in *Semiconductor Nanocrystal Quantum Dots*, Rogach, A.L., ed. Springer-Verlag/Wien, New York, 2008, pp. 1–34.
17. Hines, M.A. and Guyot-Sionnest, P. Synthesis and characterization of strongly luminescent ZnS-capped CdSe nanocrystals. *J Phys Chem*, 1996, **100**:468–471.
18. Peng, X.G., et al. Epitaxial growth of highly luminescent CdSe/CdS core/shell nancrystals with photostability and electronic acessability. *J Am Chem Soc*, 1997, **119**:7019–7029.
19. Azzazy, H.M.E., Mansour, M.M.H. and Kazmierczak, S.C. From diagnostics to therapy: prospects of quantum dots. *Clin Chem*, 2007, **40**:917–927.
20. Uyeda, H. T., et al. Synthesis of compact multidentate ligands to prepare stable hydrophilic quantum dot fluorophores. *J Am Chem Soc*, 2005, **127**:3870–3878.
21. Rogach, A.L., et al. Synthesis and characterization of thiol-stabilized CdTe nanocrystals. *Ber Bunsenges Phys Chem*, 1996, **100**:1772–1778.
22. Zheng, Y.G., Gao, S.J. and Ying, J.Y. Synthesis and cell-imaging applications of glutathione-capped CdTe quantum dots. *Adv Mater*, 2007, **19**:376–380.
23. Kairdolf, B.A., et al. Semiconductor quantum dots for bioimaging and biodiagnostic applications. *Annu Rev Anal Chem*, 2013, **6**:143–162.
24. Blanco-Canosa, J.B., et al. Recent progress in the bioconjugation of quantum dots. *Coord Chem Rev*, 2014, **263–264**:101–137.
25. Algar, W.R., Tavares, A.J. and Krull, U.J. Beyond labels: a review of the application of quantum dots as integrated components of assays, bioprobes, and biosensors utilizing optical transduction. *Anal Chim Acta*, 2010, **673**:1–25.
26. Sperling, R.A and Parak, W.J. Surface modification, functionalization and bioconjugation of colloidal inorganic nanoparticles. *Phil Trans R Soc A*, 2010, **368**:1333–1383.
27. Carvalho, K.H.G., et al. Fluorescence plate reader for quantum dot-protein bioconjugation analysis. *J Nanosci Nanotechnol*, 2014, **14**:3320–3327.

28. Ji, X.J., et al. (CdSe)ZnS quantum dots and organophosphorus hydrolase bioconjugate as biosensors for detection of paraoxon. *J Phys Chem C*, 2005, **109**:3793–3799.

29. Parak, W.J., et al. Conjugation of DNA to silanized colloidal semiconductor nanocrystalline quantum dots. *Chem Mater*, 2002, **14**:2113–2119.

30. Sapsfor, K.E., et al. Kinetics of metal-affinity driven self-assembly between proteins or petides and CdSe-ZnS quantum dots. *J Phys Chem C*, 2007, **111**:11528–11538.

31. Han, H.S., et al. Development of a bioorthogonal and highly efficient conjugation method for quantum dots using tetrazine-norbornene cycloaddition. *J Am Chem Soc*, 2010, **132**:7838–7839.

32. Bruchez, M., et al. Semiconductor nanocrystals as fluorescent biological labels. *Science*, 1998, **281**:2013–2016.

33. Chan, W.C. and Nie, S. Quantum dot bioconjugates for ultrasensitive nonisotopic detection. *Science*, 1998, **281**:2016–2018.

34. Weng, K.C., et al. Targeted tumor cell internalization and imaging of multifunctional quantum dot-conjugated immunoliposomes in vitro and in vivo. *Nano Lett*, 2008, **8**(9):2851–2857.

35. Zhao, W.W., Xu, J.J. and Chen, H.Y. Photoelectrochemical bioanalysis: the state of the art. *Chem Soc Rev*, 2015, **44**:729–741.

36. Gill, R., et al. Optical detection of glucose and acetylcholine esterase inhibitors by H2O2-sensitive CdSe/ZnS quantum dots. *Angew Chem Int Ed*, 2008, **47**:1676–1679.

37. Freeman, R., et al. Biosensing and probing of intracellular metabolic pathways by NADH-sensitive quantum dots. *Angew Chem Int Ed*, 2009, **48**:309–313.

38. Zhang, W., et al. Composite of CdTe quantum dots and molecularly imprinted polymer as a sensing material for cytochrome c. *Biosens Bioelectron*, 2011, **26**:2553–2558.

39. Matoussi, H., Palui, G. and Na, H.B. Luminescent quantum dots as platforms for probing in vitro and in vivo biological processes. *Adv Drug Deliev Rev*, 2012, **64**:138–166.

40. Zrazhevsky, P., Sena, M. and Gao, X. Designing multifunctional quantum dots for bioimaging, detection and drug delivery. *Chem Soc Rev*, 2010, **39**:4326–4354.

41. Larson, D.R., et al. Water-soluble quantum dots for multiphoton fluorescence imaging in vivo. *Science*, 2003, **300**:1434–1436.

42. Liu, A.P., et al. Quantum dots with phenylboronic acid tags for specific labeling of sialic acids on living cells. *Anal Chem*, 2011, **83**:1124–1130.

43. Wu, X.Y., et al. Immunofluorescent labeling of cancer marker Her2 and other cellular targets with semiconductor quantum dots. *Nat Biotech*, 2003, **23**:41–46.

44. Howarth, M., et al. Targeting quantum dots to surface proteins in living cells with biotin ligase. *Proc Natl Acad Sci USA*, 2005, **102**:7583–7588.

45. Howarth, M., et al. Monovalent, reduced-size quantum dots for imaging receptors on living cells. *Nat Methods*, 2008, **5**:397–399.

46. Liu, J.A., et al. Multiplexed detection and characterization of rare tumor cells in Hodgkin's lymphoma with multicolor quantum dots. *Anal Chem*, 2010, **82**:6237–6243.

47. Liu, J., et al. Molecular mapping of tumor heterogeneity on clinical tissue specimens with multiplexed quantum dots. *ACS Nano*, 2010, **4**:2755–2765.

48. Joo, K.I., et al. Enhanced real-time monitoring of adeno-associated virus trafficking by virus-quantum dot conjugates. *ACS Nano*, 2011, **5**:3523–3535.

49. Chen, C., et al. Quantum dots-based immunofluorescence technology for the quantitative determination of HER2 expression in breast cancer. *Biomaterials*, 2009, **30**:2912–2918.

50. Doane, T. L. and Burda, C. The unique role of nanoparticles in nanomedicine: imaging, drug delivery and therapy. *Chem Soc Rev*, 2012, **41**:2885–2911.

51. Ivanov, S.A., et al. Type-II core/shell Cds/ZnSe nanocrystals: synthesis, elctronic structures, and spectroscopic properties. *J Am Chem Soc*, 2007, **129**:708–711.

52. Lacroix, L.M., et al. New generation of magnetic and luminescent nanoparticles for in vivo real-time imaging. *Interface Focus*, 2013, **3**:1–20.

53. Kim, S., et al. Near-infrared fluorescent type II quantum dots for sentinel lymph node mapping. *Nat Biotechnol*, 2004, **22**:93–97.

54. Cai, W.B., et al. Peptide-labeled near-infrared quantum dots for imaging tumor vasculature in living subjects. *Nano Lett*, 2006, **6**:669–676.

55. Gao, J.H., et al. A novel clinically translatable fluorescent nanoparticle for targeted molecular imaging of tumors in living subjects. *Nano Lett*, 2012, **12**:281–286.

56. Duconge, F., et al. Fluorine-18-labeled phospholipid quantum dot micelles for in vivo multimodal imaging from whole body to cellular scales. *Bioconjug Chem*, 2008, **19**:1921–1926.

57. Riegler, J. and Nann, T. Application of luminescent nanocrystals as labels for biological molecules. *Anal Bioanal Chem*, 2004, **379**:913–919.

58. Delehanty, J.B., et al. Active cellular sensing with quantum dots: transitioning from research tool to reality: a review. *Anal Chim Acta*, 2012, **750**:63–81.

59. Zhang, Y. and Wang, T.-H. Quantum dot enabled molecular sensing and diagnostics. *Theranostics*, 2012, **2**:631–654.

60. Tang, B., et al. A new nanobiosensor for glucose with high sensitivity and selectivity in serum based on fluorescence resonance energy transfer (FRET) between CdTe quantum dots and an nanoparticles. *Chem Eur J*, 2008, **14**:3637–3644.

61. Snee, P.T., et al. A ratiometric CdSe/ZnS nanocrystal pH sensor. *J Am Chem Soc*, 2006, **128**:13320–13321.

62. Dennis, A.M., et al. Quantum dot-fluorescent protein FRET probes for sensing intracellular pH. *ACS Nano*, 2012, **6**:2917–2924.

63. Wegner, K.D., et al. Nanobodies and nanocrystals: highly sensitive quantum dot-based homogeneous FRET immunoassay for serum-based EGFR detection. *Small*, 2014, **10**:734–740.

64. Zhang, C.Y., et al. Single-quantum-dot-based DNA nanosensor. *Nat Mater*, 2005, **4**:826–831.

65. Algar, W.R. and Krull, U.J. Towards multi-colour strategies for the detection of oligonucleotide hybridization using quantum dots as energy donors in fluorescence resonance energy transfer (FRET). *Anal Chim Acta*, 2007, **581**:193–201.

66. Petryayeva, E. and Algar, W.R. Single-step bioassays in serum and whole blood with a smartphone, quantum dots and paper-in-PDMS chips. *Analyst*, 2015, **140**:4037–4045.

67. Crut, A., et al. Detection of single DNA molecules by multicolor quantum-dot end-labeling. *Nucleis Acids Res*, 2005, **33**:1–9.

68. Cui, D.X., et al. Self-assembly of quantum dots and carbon nanotubes for ultrasensitive DNA and antigen detection. *Anal Chem*, 2008, **80**:7996–8001.

69. Tian, J.P., et al. Detection of influenza A virus based on fluorescence resonance energy transfer from quantum dots to carbon nanotubes. *Anal Chim Acta*, 2012, **723**:83–87.

70. Gerion, D., et al. Room-temperature single-nucleotide polymorphism and multiallele DNA detection using fluorescent nanocrystals and microarrays. *Anal Chem*, 2003, **75**:4766–4772.

71. Han, M.Y., et al. Quantum-dot-tagged microbeads for multiplexed optical coding of biomolecules. *Nat Biotech*, 2001, **19**:631–635.

72. Zhang, C.Y. and Hu, J. Single quantum dot-based nanosensor for multiple DNA detection. *Anal Chem*, 2010, **82**:1921–1927.

73. Giri, S., et al. Rapid screening of genetic biomarkers of infectious agents using quantum dot barcodes. *ACS Nano*, 2011, **5**:1580–1587.

74. Bagalkot, V., et al. Quantum dot - aptamer conjugates for synchronous cancer imaging, therapy, and sensing of drug delivery based on Bi-fluorescence resonance energy transfer. *Nano Lett*, 2007, **7**:3065–3070.

75. Qian, J., et al. Versatile immunosensor using a quantum dot coated silica nanosphere as a label for signal amplification. *Anal Chem*, 2010, **82**:6422–6429.

76. Goldman, E.R., et al. Multiplexed toxin analysis using four colors of quantum dot fluororeagents. *Anal Chem*, 2004, **76**:684–688.

77. Klostranec, J.M., et al. Convergence of quantum dot barcodes with microfluidics and signal processing for multiplexed high-throughput infectious disease diagnostics. *Nano Lett*, 2007, **7**:2812–2818.

78. Chen, L., et al. Simultaneous determination of human enterovirus 71 and coxsackievirus B3 by dual-color quantum dots and homogeneous immunoassay. *Anal Chem*, 2012, **84**:3200–3207.

79. Yang, L.J. and Li, Y.B. Simultaneous detection of Escherichia coli O157 : H7 and *Salmonella typhimurium* using quantum dots as fluorescence labels. *Analyst*, 2006, **131**:394–401.

80. Li, Z.H., et al. Rapid and sensitive detection of protein biomarker using a portable fluorescence biosensor based on quantum dots and a lateral flow test strip. *Anal Chem*, 2010, **82**:7008–7014.

81. Lai, C.Y., et al. A mesoporous silica nanosphere-based carrier system with chemically removable CdS nanoparticle caps for stimuli-responsive controlled release of neurotransmitters and drug molecules. *J Am Chem Soc*, 2003, **125**:4451–4459.

82. Li, L.L., et al. Preparation and characterization of quantum dots coated magnetic hollow spheres for magnetic fluorescent multimodal imaging and drug delivery. *J Nanosci Nanotech*, 2009, **9**:2540–2545.

83. Samia, A.C.S., Chen, X.B. and Burda, C. Semiconductor quantum dots for photodynamic therapy. *J Am Chem Soc*, 2003, **125**:15736–15737.

84. Courtney, C.M., et al. Photoexcited quantum dots for killing multidrug-resistant bacteria. *Nat Mater*, 2016, **15**:1–7.

85. Jamieson, T., et al. Biological applications of quantum dots. *Biomaterials*, 2007, **28**:4717–4732.

86. Cheng, X.Y., et al. Colloidal silicon quantum dots: from preparation to the modification of self-assembled monolayers (SAMs) for bio-applications. *Chem Soc Rev*, 2014, **43**:2680–2700.

87. Lim, S.Y., Shen, W. and Gao, Z.Q. Carbon quantum dots and their applications. *Chem Soc Rev*, 2015, **44**:362–381.

88. Havrdova, M., et al. Toxicity of carbon dots: effect of surface functionalization on the cell viability, reactive oxygen species generation and cell cycle. *Carbon*, 2016, **99**:238–248.

Part X
Cytotoxicity

Chapter 17

Pro-Inflammatory and Toxic Effects of Silver Nanoparticles

Marisa Freitas,[a] Daniela Ribeiro,[a] Paula Silva,[b] José L. F. C. Lima,[c] Félix Carvalho,[d] and Eduarda Fernandes[a]
[a]*UCIBIO, REQUIMTE, Laboratory of Applied Chemistry, Department of Chemical Sciences, Faculty of Pharmacy, University of Porto, Portugal*
[b]*UCIBIO, REQUIMTE, Laboratory of Histology and Embryology, Institute of Biomedical Sciences Abel Salazar (ICBAS), University of Porto, Portugal*
[c]*LAQV, REQUIMTE, Laboratory of Applied Chemistry, Department of Chemical Sciences, Faculty of Pharmacy, University of Porto, Portugal*
[d]*UCIBIO, REQUIMTE, Laboratory of Toxicology, Department of Biological Sciences, Faculty of Pharmacy, University of Porto, Portugal*
marisa.freitas@ff.up.pt

Nanoparticles (NPs) have been the subject of intense research due to their extensive potential applications, namely in the biotechnological, industrial, and military areas. Silver NPs (AgNPs) are components of most commercialized nanomaterials worldwide,

Nanoparticles in Life Sciences and Biomedicine
Edited by Ana Rute Neves and Salette Reis

ISBN 978-981-4745-98-7 (Hardcover), 978-1-351-20735-5 (eBook)
www.panstanford.com

taking advantage of the use of NPs in human health products, among other applications. With the growing use of AgNPs, the general population is at a greater risk of occupational and environmental exposure to significant levels of these NPs in daily life. Still, and despite all the possible applications, the widespread use of AgNPs is not exempt from safety concerns related to their size, chemical composition, and coating; their ability to bind and affect biological sites; as well as their putative pro-inflammatory effects. This chapter provides an overview of the main factors that are involved in AgNP toxicity, focusing on the mechanisms by which AgNPs exert pro-inflammatory effects.

17.1 Introduction

Silver is a soft, white, lustrous transition metal with high electrical and thermal conductivity. Its medical and therapeutic benefits have been known since antiquity. In ancient Greece and Rome, silver was widely used to control infections and prevent food spoilage. Since the 20th century silver can be found in many forms—coins, vessels, solutions, foils, sutures, lotions, colloids, and ointments, to name a few. It is the foremost used therapeutic agent in medicine to prevent infection-related diseases [1]. Silver nanoparticles (AgNPs) have received attention due to their unique physical, chemical, and biological properties compared to their macroscale counterparts [1]. Their distinctive physicochemical properties, including a high electrical and thermal conductivity, surface-enhanced Raman scattering, chemical stability, catalytic activity, and nonlinear optical behavior, make these nanoparticles (NPs) attractive in the production of medical imaging, microelectronics, and ink products [2]. AgNPs are currently used in diagnosis, imaging, therapeutics, drug delivery, the treatment of vascular diseases, wound healing, and the development of materials and medical devices with antimicrobial properties [3, 4]. The use of AgNPs in cancer therapy has also been a hot topic. In fact, their toxicity against tumor cells in mammalian models revealed them as potential agents that can be used to modulate some cancer types. Promising results have been obtained in the modulation of leukemia, hepatoma, breast, lung, skin, and oral cancers [5]. They can be also used in the textile industry by incorporating them into the fiber (spun) or employed in filtration membranes of water purification systems [3, 4].

According to the Project on Emerging Nanotechnologies (PEN, (http://www.nanotechproject.org), over 1300 manufacturer-identified, nanotechnology-enabled products have entered the commercial market around the world. Among them, there are 313 products employing AgNPs (24% of the listed products). This has made AgNPs the largest- and fastest- growing class of NPs in anthropogenic product applications [6].

The synthesis of AgNPs is mainly based on the reduction of Ag^+ by plant extracts, polysaccharides, amine, aldehydes, ascorbic acid, D-glucose, and citrate, in aqueous solution. The synthesis processes have been improved in order to avoid particles' aggregation and to obtain final particles free from impurities. More recently, green synthesis has been gaining importance as it uses benign reactants, solvents, and byproducts, producing particles with interesting features for medical applications. Microorganisms can also be used to produce AgNPs (the biogenic route), originating particles with smaller sizes than the ones obtained by the chemical route. The proteins produced by microbial metabolism are used to stabilize the particles. The developments made in the characterization techniques have undoubtedly been contributing to the advances in the production and applications of NPs. These techniques, namely X-ray diffraction, scanning electron microscopy, energy dispersive X-ray analysis, and UV–visible spectroscopy, allow the structural, morphological, compositional, and optical behavior analysis of NPs [4].

Despite all the possible above-mentioned applications, the widespread use of AgNPs is not exempt from safety concerns, with their pro-inflammatory effects being possibly responsible for the adverse effects attributed to them. In the following sections, we compile and explore the main factors that are involved in the toxicity of AgNPs, focusing on their pro-inflammatory effects.

17.2 Factors Influencing Silver Nanoparticles' Toxicity

Due to the above-mentioned applications of AgNPs and the numerous related products that are available in the market, humans can be exposed to AgNPs by different routes, namely dermal, oral, inhalation, and intravenous. This direct contact with AgNPs may be responsible for adverse health effects that have been discussed in several reviews [4, 7, 8].

It is important to note that the toxicological effects of AgNPs are variable and depend on the source of the AgNPs tested (generated in the laboratory or from commercially available materials), their size, and the presence or absence of capping agents, as well as other factors, such as dose, exposure time, and temperature [1, 5]. Experimentally, the type of cells used, the animal features (species, strain, sex, age, etc.), and the overall experimental design (dose, exposure time, endpoints for sampling, etc.) can influence the results obtained with AgNPs. The toxicity of AgNPs is also dependent on their transformation in biological and environmental media, including surface oxidation, release of Ag^+ ions, and interaction with biological macromolecules [9]. For example, it is currently accepted that both AgNPs and Ag^+ can interact with proteins and amino acids [9]. The studies using cell culture imply the use of media containing fetal bovine serum (FBS). It is important to note that FBS is a complex mixture of proteins that may interfere with the results. In accordance, Liu et al. [10] reported that Ag^+ levels in cells cultured in medium without FBS increased when compared to those cultured in FBS-containing medium. This result can be related to the ability of AgNPs to bind to the proteins present in FBS, decreasing the amount of AgNPs that can enter cells. The topics ahead describe some of the main factors that may interfere and affect AgNP toxicity.

17.2.1 Interaction with Cells

In general, the toxicity of NPs depends on the type of cellular interaction established. It is generally accepted that NPs' uptake by mammalian cells occurs by mechanisms such as phagocytosis, pinocytosis, and endocytosis mediated by clathrin, caveolae, and lipid raft. In a review by Bartłomiejczyk et al. [11] these types of uptake are described in various cellular models. It is clear that the uptake kinetic is conditioned by the cell type. For example, uptake of NPs by red blood cells, due to their lack of phagocytic abilities, is directly dependent on the NP size. In this case, the NP charge or the material type has little importance. On the other hand, when NPs come in contact with platelets the particle charge plays an essential role in their uptake and their influence on blood clot formation [12]. AgNP uptake is also likely to depend on cell type, although current research does not provide sufficient information for drawing conclusions on this subject. For instance, Greulich et al. [13] showed that nonfunctionalized AgNPs are taken up better by peripheral monocytes but not by T cell populations, but their results did not clarify whether that difference was related to the uptake mechanism. More studies are needed to explain the cell-type-specific uptake of AgNPs and to take the results as a consideration in AgNP design and dosage. The uptake, intracellular localization, and exocytosis also depend on the AgNP's size, surface characteristics, and ability to form aggregates. What is noteworthy is that it seems that the physicochemical properties are more important than the particle size [11]. The review in "Silver Nanoparticles - Allies or Adversaries?" [11] presents a handy diagram of the mechanisms of AgNPs' uptake, transport pathways inside the cell, and intracellular targets. The intracellular localization of AgNPs could vary according to the cell type used and also the applied method. For example, a study with human mesenchymal stem cells, using fluorescent probes, showed that AgNPs (50 ± 20 nm) were absent in the cell nucleus, the endoplasmic reticulum (ER), or the Golgi complex but were present mainly within endolysosomal structures, forming agglomerates in the perinuclear region. On the contrary, a study with transmission electron microscopic (TEM) analysis indicated the presence of AgNPs inside the nucleus and mitochondria. In other studies, using human lung adenocarcinoma epithelial cells (A549) and human hepatocellular carcinoma cells (HepG2), AgNPs (20 nm) were found in both cytoplasm and mitochondria [11].

17.2.2 Size

It is recognized that the size of AgNPs is correlated with their toxicological effects. Soares et al. [14] studied the effect of poly(*N*-vinyl pyrrolidone) (PVP)-coated AgNPs (10 and 50 nm) on human neutrophils' toxicity. The observed toxicity was size and time dependent, the tested 10 nm AgNPs being more toxic than the 50 nm AgNPs. Interestingly, and corroborating the above-mentioned importance of the cell interaction with the NPs, this work showed, through TEM images, that 10 nm AgNPs were internalized in human neutrophils, localized inside the cell throughout the cytosol and also more intrinsically in phagosomes, in contrast with 50 nm AgNPs, which were not able to enter the cells, remaining outside, close to the cellular membrane (Fig. 17.1) [14]. In endothelial cells from rat brain, the size of citrate-coated AgNPs (10, 50, and 100 nm) determined their cytotoxicity, which was higher when the smallest AgNPs were used [15]. Similar results were obtained with the cell culture of human macrophages treated with PVP AgNPs. For example, 3.12 μg/ml of 4 nm AgNPs decreased the cell viability to 40%, while, with the same time of exposure, 50 μg/ml of 70 nm AgNPs did not affect the cell viability [16]. Also reported was a clear size-dependent toxicity for AgNPs in human lung cells, where only the 10 nm citrate- or PVP-coated AgNPs were cytotoxic for the human bronchial epithelial cells (BEAS-2B), when compared to the other tested sizes, 40 and 75 nm [17]. Liu et al. [10] found that 5 nm PVP AgNPs were more toxic than 20 and 50 nm ones in four cell lines: A549, human stomach cancer cells (SGC-7901), HepG2, and human breast adenocarcinoma cells (MCF-7). Despite this known size-dependent toxicity, most studies have been performed using only one particle size, neglecting the contribution of particle size to the toxicity of AgNPs.

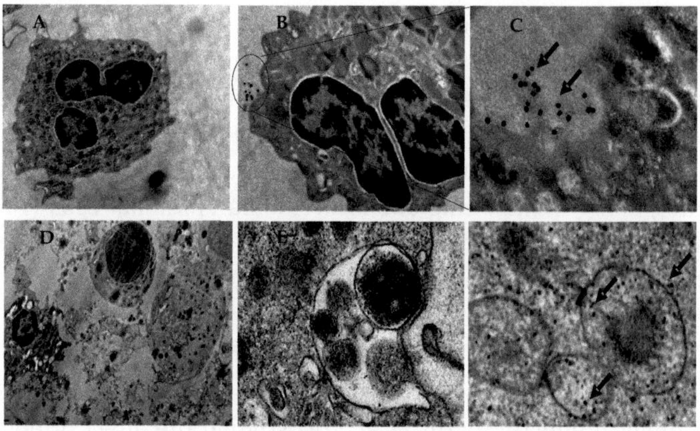

Figure 17.1 TEM images of ultrathin sections of neutrophils incubated for 16 h: (A) without AgNPs (control), (B) with 50 nm AgNPs (100 μg/ml), (C) with 50 nm AgNPs surrounding the neutrophil membrane (arrows), (D) disintegrated neutrophils after incubation with 10 nm AgNPs, (E) phagosome with 10 nm AgNPs, and (F) 10 nm AgNPs randomly distributed throughout the phagosome (arrows). Reprinted from Ref. [14], Copyright (2016), with permission from Elsevier.

17.2.3 Surface Coatings

Due to their tendency to aggregate, AgNPs are typically synthesized with surface coatings to stabilize them in suspension. Citrate is the most commonly used reducing agent and stabilizer [18]. Polymers have also been used to coat AgNPs, such as PVP and poly(ethylene glycol) (PEG). Citrate provides both a negative surface charge of NPs and colloidal stability through electrostatic repulsions. In turn, low-molecular-weight PEG neutralizes surface charge and stabilizes NPs through steric hindrance [18]. It has been argued that the surface coating can affect the affinity of AgNPs for a cell surface. It is suspected that different capping agents interfere with AgNP toxicity in different forms. However, despite this, little attention has been given so far to the coating-dependent toxicity of AgNPs. A study in keratinocytes that investigated the cellular effects of well-characterized AgNPs with citrate and PEG coatings led to the conclusion that citrate AgNPs were more cytotoxic than PEG AgNPs. Among other factors, the cellular uptake of AgNPs can justify the obtained results, since PEG AgNPs were taken up by cells to a lower extent than citrate AgNPs [18]. In a study using J774A.1 macrophage and human colonic epithelial cells (HT29), the uncoated AgNPs were found to be more cytotoxic than their counterparts (PVP and citrate AgNPs). Comparing the PVP- and citrate-coated AgNPs, the first ones were more toxic than the second ones [19]. The results of these two studies seem to indicate that a citrate surface coating increases the AgNP's toxicity. Suresh et al. [20] compared the toxic effects of poly(diallyldimethylammonium) chloride, bionic, oleate coated, and uncoated colloidal AgNPs in RAW 264.7 macrophages and lung epithelial C-10 cell lines and found that uncoated AgNPs were the least toxic compared to coated counterparts [20]. Moreover, the toxicity varies among the coating agents used, since poly(diallyldimethylammonium) chloride was the most toxic, followed by bionic and oleate coated. In turn, the surface coating alone did not have any impact on the cell viability. Interestingly, dynamic light scattering data showed that poly(diallyldimethylammonium) chloride AgNPs tend to agglomerate more than oleate AgNPs, justifying the differences in the toxicity. Nevertheless, this issue is controversial in literature, since Gliga et al. [17] compared the cytotoxicity levels of uncoated, PVP-coated, and citrate-coated

AgNPs in bronchial BEAS-2B cells and reported no coating-dependent differences in cytotoxicity. It is known that aggregation of NPs not only enhances but speeds up the cytotoxic response [20, 21]. It is important to remember that the culture media used can influence the susceptibility of the NPs to aggregate. For example, in a study by Bastos et al. [18] the characterization results revealed that citrate and PEG AgNPs exhibited different time-course behavior in a culture medium (Dulbecco's modified Eagle's medium) depending on the coating surface. The hydrodynamic diameter of citrate AgNPs increased immediately after suspension in the culture medium, and it seems to be due to some aggregation induced by the high ionic strength of the culture medium or due to the formation of a protein corona, as it was reported before for metal NPs. Contrariwise, the hydrodynamic diameter of PEG AgNPs did not show significant variation in culture medium, indicating high colloidal stability, which is likely to derive from steric repulsions [18].

17.2.4 Silver Release

Nowadays, there is a scientific challenge to precisely distinguish whether the toxicity described for AgNPs is due to AgNPs per se or due to the release of Ag^+ ions. It is currently accepted that AgNPs might also act as a "Trojan horse," in which the NPs entering into cells, by bypassing their barriers, release high amounts of Ag^+, which will damage cell machinery [22]. Nevertheless, the exact explanation of how AgNPs might work in a living system has been problematic and controversial [22]. It has been described that Ag^+ ions are involved in the cytotoxicity caused by AgNPs in several cell types [23]. Park et al. [24] studied the cytotoxic effect of AgNPs using RAW 264.7 and reported the ionization of AgNPs in cells, with the subsequent release of free Ag^+ [24]. The cytotoxic effects observed in A549 cells after exposure to AgNPs were also attributed to the considerable release of Ag^+ ions [7]. Corroborating the above-mentioned results, Hsiao et al. [25] found that, despite the cell type (microglia and astrocytes), AgNPs were taken up similarly to, or more rapidly than, Ag (I). The ratio of AgNPs to total Ag—$AgNPs^+$:Ag (I)—in both cells was lower than that in outside media, suggesting that the Trojan horse mechanism is one true possibility to AgNP-induced cytotoxicity.

It is important to keep in mind that AgNPs' dissolution could occur, or even be enhanced, in cell culture media and ions detected in cells might also come from extracellular media and not only from AgNPs [25]. However, recently, Bastos et al. [18] contradicted these results, since the authors reported that citrate AgNPs and PEG AgNPs showed different cytotoxicity levels, while both AgNPs had similar dissolution behavior in culture medium. Thus, in this case, extracellular Ag^+ could not justify the difference in the toxicity levels of the two AgNPs, which can be attributed to the different coating agents used.

17.3 Pro-Inflammatory Effects of Silver Nanoparticles

AgNPs induce cytotoxicity by several mechanisms, and most of them are directly and/or indirectly related to their ability to induce the production of pro-inflammatory mediators. There are multiple mechanisms by which AgNPs exert their pro-inflammatory effects. It is described that AgNPs are able to induce the production of reactive species (RS), activate transcription factors, alter the mitochondrial membrane permeability, produce cytokines, etc. The inflammatory process is defined as a complex cascade of a series of intracellular and extracellular events that are interrelated by amplification loops. In the following sections we outline a simplified description of the main pro-inflammatory effects of AgNPs, through which they may exert their toxicity.

17.3.1 Reactive Species

It is an undeniable fact that NPs can act on the immune system. When NPs enter the human body, they are immediately confronted with the innate immune system, of which monocytes, macrophages, and neutrophils represent an important part, by eliminating threats through phagocytosis, oxidative burst, and degrading enzymes [5]. During inflammatory processes two concurrent events occur: one oxygen independent, involving the release of enzymatic or antimicrobial protein content, and the other, oxygen dependent, known as oxidative burst, during which occurs the formation of reactive oxygen species (ROS) and reactive nitrogen species (RNS). RS are generally present in all cells and are part of the redox equilibrium. During the inflammatory process, however, there is an imbalance between the endogenous antioxidant defenses and the RS production, and the latter accumulates in cells, potentially leading to damage [26].

The toxicity induced by AgNPs may be linked to the production of RS by immune cells. Indeed, the accumulation of these RS inevitably leads to oxidative damage in proteins and DNA and is considered one of the key factor in the biological effects of AgNPs in vivo and in vitro [11]. Soares et al. [14] reported that, in contrast to 50 nm PVP AgNPs, the 10 nm PVP AgNPs induced human neutrophils' nicotinamide adenine dinucleotide phosphate (NADPH) oxidase–dependent oxidative burst. NADPH oxidase is a complex enzymatic system that is responsible for the production of the first ROS, superoxide anion radical ($O_2^{\bullet-}$), by phagocytes. These results were corroborated by Paino et al., who showed an increased production of RS by human neutrophils stimulated with polyvinyl-alcohol AgNPs [27]. In macrophages, AgNPs also induced the production of RS as measured by dichloro-dihydro-fluorescein diacetate [28]. Besides NADPH oxidase, mitochondria are also responsible for the production of $O_2^{\bullet-}$. PVP AgNPs caused leakage of cathepsins from lysosomes and an efflux of intracellular K^+ in monocytes. These two events seem to be involved in the AgNP-induced $O_2^{\bullet-}$ production by mitochondria. $O_2^{\bullet-}$ can be rapidly dismutated into hydrogen peroxide (H_2O_2), either spontaneously, especially at low pH, or through enzyme catalysis, by superoxide dismutase (SOD) contained in phagocyted microorganisms [29]. Reactions between H_2O_2 and AgNPs seem to be one of the factors leading to Ag^+ release in vivo. The possible chemical reaction involves silver oxidation by H_2O_2 [5]. PVP AgNPs can cause an increased production of H_2O_2 in monocytes, in which smaller AgNPs, of 5 nm, are more effective than the larger ones (28 nm) [30].

An in vivo study using bronchoalveolar lavage fluid (BALF) reported that AgNPs caused reduction in SOD activity, depletion of glutathione content, and increase in the malondialdehyde (MDA) levels [31]. During ROS formation, peroxyl radicals ($ROO^\bullet$), among which the most common is hydroperoxyl radical ($HOO^\bullet$), are responsible for the induction of lipid peroxidation, resulting in the production of MDA [29].

The production of RNS starts with the expression of inducible nitric oxide synthase, with subsequent production of near-micromolar amounts of nitric oxide (•NO), produced for long periods [29]. AgNPs induced the production of •NO in two immune cell types, human promyelocytic leukemia cells (HL-60) and human monocytic cells (U-937) [32]. Also in macrophages, RAW 264.7, AgNPs induced the production of •NO [24, 28]. Similar results were obtained by Liu et al. in an in vivo study, through the analysis of BALF from rats exposed to AgNPs by intratracheal instillation [31]. Taken together, these results show that RS production is one of the possible mechanisms by which AgNPs are cytotoxic.

17.3.2 Transcription Factors

The production of inflammatory mediators is regulated by transcription factors. Their expression is also modulated by these same mediators. In this way a counter-regulatory cycle is created and the interference in this loop is crucial for regulating the inflammatory response [26]. AgNPs are described to activate signaling pathways that involve various protein kinases. The main groups of kinases involved are p38 and c-Jun N-terminal kinases (JNKs); but the redox-sensitive transcription factors, nuclear factor erythroid 2–related factor 2 (Nrf-2) and nuclear factor kappa-light-chain-enhancer of activated B cells (NF-κB), are also involved. The outcome of this activation may be the stimulation of proliferation, inflammation, and/or apoptotic cell death, depending, once more, on the NP size, concentration, and cell type [11].

NF-κB is involved in almost all of the inflammatory phenomena, and it is crucial in the orchestration of the inflammatory response. NF-κB has been reported to mediate the synthesis of several cytokines, such as tumor necrosis factor (TNF), interleukin (IL)-1β, IL-6, and IL-8, as well as cyclooxygenase (COX)-2 [26]. In a study that assessed early transcription factor involvement in primary human dermal fibroblasts (HDFs) and in NF-κB knockdown cells (these cells inhibit NF-κB nuclear translocation, preventing subsequent DNA binding and genetic upregulation) exposed to citrate-stabilized AgNPs, an extracellular signal–regulated kinase (ERK)1/2 upregulation was observed that either precedes or is nearly concurrent with phosphorylation of the NF-κB inhibitor. Of note, AgNPs were also able to induce p38 phosphorylation. In the NF-κB knockdown cells this phosphorylation was less severe, corroborating the idea that NF-κB inhibition can lead to anti-inflammatory effects. These findings suggest that NF-κB is responsible for controlling the major part of the inflammatory response in the dermal cells tested. p38 phosphorylation may be stimulated by IL-1 upregulation, which supports the activation of IL-6 by AgNPs. As already mentioned, this transcriptional cascade leads to ROS production and upregulation of cytokines [33]. In another cell model, human umbilical vein endothelial cells (HUVECs), AgNPs were shown to activate IκB kinase (IKK)/NF-kB, which is associated with oxidative stress, and consequently lead to HUVEC dysfunction. Interestingly, in this study the authors proposed an AgNP-mediated signaling pathway [34]:

1. Oxidative stress
2. IKKα/β phosphorylation
3. IkBα (inhibitor of kappa B) phosphorylation
4. IkBα release; then NF-kB activation
5. NF-kB nuclear translocation
6. DNA binding activity of NF-kB
7. Abnormal genes' and proteins' expression—cell dysfunction

Indeed, in RAW 264.7 mouse macrophages exposed to AgNPs an induced nuclear translocation of NF-κB p65 and p50 proteins was evident and this effect was dependent on the NP size and time of exposure. This effect was only reverted when an antioxidant was preincubated with the cells before addition of the AgNPs, indicating the crucial interference of RS in the activation of NF-κB by NPs [34]. HepG2 cells treated with AgNPs revealed activation of genes related to the functionality of the NF-κB signaling pathway. In this same study, the A549 cell line was also tested, but no activation was found, indicating that cellular response to AgNPs is cell type specific [35]. In another study using high-throughput reporter genes in HepG2 cells, Prasad et al. [36] found that exposure to AgNPs activated Nrf-2 to a greater magnitude than either NF-κB or activator protein 1 (involved in inflammation) [36]. The Nrf2-antioxidant response element signaling pathway controls the expression of genes involved in the detoxification and elimination of reactive oxidants and electrophilic agents by enhancing cellular antioxidant capacity [37]. AgNP-induced activation of Nrf2 was also observed in human colorectal carcinoma cell line (Caco-2). AgNPs activated the translocation of Nrf2 from the cytoplasm to the nucleus. Moreover, they also induced an increase in the expression of heme oxygenase-1 (HO-1), and the obtained results enabled the conclusions that Nrf2 has a role in the control of HO-1 expression and implicated AgNPs in oxidative stress induction [38]. In Jurkat T cells (immortalized line of human T lymphocyte cells), AgNPs also activated p38 mitogen-activated protein kinase (MAPK) through Nrf2 and NF-κB signaling pathways, later inducing DNA damage, cell cycle arrest, and apoptosis. p38 MAPK activation, as other transcription factors already described, is associated with human conditions of inflammation and autoimmunity and is implicated in cell apoptosis. Remarkably, these findings were only seen for AgNPs, not for Ag^+ ions. This may indicate that p38 MAPK activation is possibly involved in the selective toxicity of AgNPs over that of Ag^+ ions in Jurkat T cells [39]. JNK and ERK are other MAPK, and they generally regulate several important biological processes, such as cell mitosis, metabolism, survival, apoptosis, proliferation, and differentiation. The apoptosis induction via mitochondrial pathway was associated with AgNP-induced JNK activation in NIH3T3 fibroblast cells [40]. The phosphorylation of JNK and ERK was induced by AgNPs in human epithelial embryonic cell (EUE) line, corroborating the effects already reported for AgNPs in MAPK expression. But, and contrary to the results of Eom and Choi [39], in this study the activation of p38 was not observed. This may be attributed to the variation in the time rate response to NPs, depending on the difference in the susceptibility of cells, which, in turn, depends on the NP type [41].

In another in vivo study with ovalbumin-induced bronchial inflammation and hyperresponsiveness in C57BL/6 mice, AgNP administration significantly reduced the NF-κB increased levels (cytosolic NF-κB p65). In general, the authors concluded that AgNPs attenuated airway inflammation and hyperresponsiveness through the modulation of ROS generation, which then attenuated the inflammatory cytokine (IL-4, IL-5, and IL-13) expression and NF-κB activation [42]. This in vivo study somewhat contradicts the above-described works; however, it highlights the importance of distinguish between an activity per se or an activity studied after an earlier induced inflammatory condition.

17.3.3 Cytokines/Chemokines

An inflammatory response is accompanied by the secretion of signaling molecules as cytokines and chemokines, which provide communication between immune cells and coordinate molecular events [43]. These small proteins are often described as having a redundant effect as various cytokines can target the same receptor and a single cytokine can have multiple, even contradictory, effects [26].

It has been shown that many immunostimulatory reactions initiated by AgNPs are mediated by the production of inflammatory cytokines. As can be seen in the text below, immunotoxicity assays with AgNPs have been focused on a wide spectrum of cytokines produced by several cell types. Immune cells, namely macrophages, isolated from exposed animals or primary cells, and cell lines, are the most studied.

In immune cells, as THP-1 monocytes and primary blood monocytes, AgNPs induced the production of IL-1, IL-6, and TNF, suggesting a rapid response to these NPs in circulation [44]. In addition, Yang et al. reported that in primary human monocytes, AgNPs elicited the production of IL-1β inversely dependent on the size [30]. In macrophages, AgNPs induced the production of several cytokines, such as IL-8 [45], IL-6 [28, 46], IL-1β [47, 48], IL-4 [48], IL-12(p70) [48], IL-10 [46], eotaxin [48], TNF [28, 46–49], macrophage inhibitory protein (MIP-2) [47], and the chemokine (C-X-C motif) ligand 1 (CXCL1) [48].

The ability of AgNPs to induce the production of cytokines was also reported in skin cells. For example, human keratinocytes were exposed to AgNPs coated either with citrate or with PEG, and significant differences were noted only for monocyte chemoattractant protein-1, while there was no effect on the other cytokines studied (IL-1β, IL-6, IL-10, and TNF-α) [18]. In contrast, in the same cell model, Samberg et al. [50] reported an increase in IL-1β, IL-6, IL-8, and TNF levels after treatment with AgNPs. These differences could be explained by the use of different coatings, concentrations, and incubation times in both studies.

Astrocytes, which play a pivotal role in neuroinflammatory responses, were also used to study the effect of AgNPs, and it was seen that, in contrast to Ag^+ ions, AgNPs were able to cause a significant increase in the secretion of multiple cytokines, including several ILs, interferon-gamma-induced protein 10 (IP-10), L-selectin, and thymus chemokine, suggesting that AgNPs could be involved in neuroinflammation [51].

The treatment of human mesenchymal stem cells with AgNPs resulted in the production of IL-6, IL-8, and vascular endothelial growth factor (VEGF) [52]. In what concerns IL-8, these results were corroborated by Greulich et al. [53]. The levels of IL-6 and VEGF, however, were concomitantly decreased after AgNP treatment [53]. Once again, the variability in the type and/or size of AgNPs used can explain the discrepant results.

AgNPs also increased the levels of IL-8 in several intestinal cell lines, such as HT29 [48], intestinal epithelium cell line (LoVo cell) [54], Caco-2 [55], and human epithelial colorectal adenocarcinoma cell line (SW480) [55].

In human lung epithelial cells, AgNPs induced the production of IL-1β and IL-6 [56]. Similar results were obtained in an in vivo study where AgNPs and $AgNO_3$ were injected into the mouse lung. The pro-inflammatory cytokine, IL-1β, and neutrophils number counting increased in BALF after both treatments. Interestingly, BALF IL-1β in $AgNO_3$-treated mice was lower than that in AgNP-treated mice 4 h after instillation, having similar values after 24 h of instillation [57]. Those results were corroborated by Haberl et al. [58], who, besides IL-1β, reported an increase of IL-6, IL-12(p709), MIP-1α and MIP-2, cytokine-induced neutrophil chemoattractant, and macrophage-/colony-stimulating factor in BALF after AgNP treatment.

In an in vivo study, rats received a standard diet enriched with AgNPs for 81 days, which resulted in an increase in the levels of inflammatory cytokines, such as TNF and IL-6, in the liver [59]. Park et al. [60] in an oral 28-day study in mice have also found increased levels of several cytokines, such as IL-1, IL-6, IL-4, IL-10, IL-12, and transforming growth factor-β, in a dose-dependent manner [60]. These results suggest that repeated oral administration of AgNPs may cause an inflammatory response in mice.

17.3.4 Eicosanoids

Eicosanoids are lipid-derived molecules that are not stored but synthesized de novo as a result of cell activation by several factors, such as mechanical trauma, growth factors, and cytokines. This important family of regulatory molecules is derived from arachidonic acid, which is released from the lipid bilayer of the nuclear envelope, the ER, and the Golgi apparatus. There are multiple subfamilies of eicosanoids that play a crucial role in immunity, namely in the inflammatory process, including prostaglandins (PGs), thromboxanes, and leukotrienes, as well as lipoxins [26]. Despite the relevance of eicosanoids during the inflammatory process there are just a few studies about whether AgNPs influence the expression of these mediators.

COX-2 is known to mediate the production of PGs. In macrophages, a study using several NPs, including silver (Ag), aluminum (Al), carbon black (CB), carbon-coated silver (CAg), and gold (Au), showed that among all of the tested NPs, AgNPs induced the higher expression of COX-2. Once again, the effect was inversely dependent on the size, since the smallest NPs, 15 nm, induced the higher effect when compared with 40 nm AgNPs [34]. In contrast, in skin cells, normal human dermal fibroblasts (NHDFs) and normal human epidermal keratinocytes (NHEKs), AgNPs decreased the expression of COX-2 after 24 h of incubation [61]. These discrepant results can be related with several variables among the studies, the use of different type of cells being the one with the highest impact. To the best of our knowledge, these two studies are the only ones that discuss the effect of AgNPs on eicosanoid production and on COX-2 expression. There is an obvious lack of studies in literature about this issue. As can be seen throughout this chapter, AgNPs have a pro-inflammatory activity and due to the important role of eicosanoids during the inflammatory process, this subject should be further studied.

17.3.5 Cell Death

AgNPs can induce cell death through a variety of mechanisms that are directly and/or indirectly related with the production of RS, including (i) lipid peroxidation and increased membrane permeability, (ii) abnormalities of the autophagy flux, (iii) DNA damage and cell growth arrest, and (iv) activation of signaling cascades involving the mitochondrial pathway [62]. The last one is undoubtedly the most studied mechanism for AgNPs, possibly due to the fact that mitochondria are a key regulator of cell viability. The mitochondrial membrane depolarization is considered the first step in the activation of the intrinsic cell death pathway. Moreover, RS, due to their damaging potential, are responsible for the formation of mitochondrial permeability transition pore (MPTP), which will lead to activation of mitochondria-dependent cell death pathways [63]. Ma et al. [64] studied the toxic effects of AgNPs, 30 nm, in the murine hippocampal HT22 cell line. The authors found AgNPs to induce a moderate mitochondrial membrane depolarization, suggesting the formation of MPTP. Interestingly, they also found AgNPs to increase oxygen consumption, which may reflect not only the mitochondrial response to stress in the early phase but also an increased electron flow associated with increased RS formation. This last aspect was confirmed by the increased ROS formation in cells treated with AgNPs. Moreover, the number of caspase-3-positive neurons also increased, reflecting the activation of a mitochondria-caspase-dependent cell death pathway by AgNPs [64]. The apoptotic process induced by AgNPs in human Chang liver cells seemed to be also governed by mitochondrial release of cytochrome c, followed by caspase-9 and caspase-3 activation and loss of mitochondrial membrane potential. This loss of membrane potential was due to the downregulation of B-cell lymphoma 2 (Bcl-2) and upregulation of apoptosis regulator BAX. Bcl-2 inactivation may result from their phosphorylation by JNK, a transcription factor that participates in apoptotic signaling. Thus, it is proposed that AgNPs led to apoptosis via mitochondria- and caspase-dependent pathways mediated by JNK [65].

AgNPs caused significant dissipation of mitochondrial membrane potential in RAW 264.7 macrophages cell line, which was associated with a significant increase in the number of apoptotic cells. The authors concluded that free radicals, however, do not play a major role in AgNP cytotoxicity, as they observed higher activity of weak antioxidants in contrast to relative ineffectiveness of potent antioxidants in preventing AgNP-induced toxicity. The authors also concluded that the most effective antioxidants contained higher molar content of thiol groups [28]. It is known that particle internalization location depends on NP size and smaller NPs (<100 nm) localize in organelles, such as mitochondria, leading to mitochondrial architecture disruption [12, 66]. Yang et al. [30] observed the collapse of the mitochondrial membrane in human blood monocytes after treatment with 28 nm and 5 nm AgNPs. The results revealed mitochondrial swelling, probably due to the presence of 28 nm AgNPs within the mitochondria.

Mitochondrial function can also be affected by the ER, as the depletion of ER calcium stores causes ER stress and leads to the increase of mitochondrial calcium levels. AgNPs, ≤100 nm, significantly increased mitochondrial calcium overloading in human Chang liver cells and Chinese hamster lung fibroblasts (V79-4). This indicates that alterations in calcium homeostasis may be implicated in AgNP-induced apoptosis. Moreover, AgNPs also induced ER stress and it was concluded that AgNP-induced apoptosis is mediated by the ER stress–signaling pathway [67].

Autophagy is a survival mechanism used by cells to deliver cytoplasmic constituents to lysosomes and eliminate long-lived proteins and damaged organelles and might be one of the defense mechanisms of the cell against oxidative stress. It has already been described that AgNP treatment induces excessive ROS production, leading to the disruption of the autophagy flux, resulting in either apoptosis or autophagic cell death [62]. In addition, a study on human neutrophils showed that AgNPs impaired lysosomal activity, culminating in neutrophils' necrosis. That effect was size dependent, 10 nm AgNPs being more effective than the 50 nm ones [14].

It is known that RS constitutes one of the major sources of spontaneous damage to DNA. Among RS, a hydroxyl radical is one of the most harmful, as it can cause DNA damage to generate 8-hydroxyguanine, leading to a decrease in the stability of repetitive sequences and single- and double-strand breaks. Subsequently DNA damage will culminate in cell cycle arrest, and once the damage is too extensive, cells may irreversibly undergo into apoptosis [62]. The genotoxic effects of AgNPs have already been discussed by Kim and Ryu [68]. The authors reported that many biochemical and molecular changes related to genotoxicity are promoted by AgNPs in cultured cells, as AgNPs induce DNA breakage. For example, Eom and Choi [39] reported that in Jurkat T cells, AgNPs induced ROS production, subsequently resulting in DNA strand breaks, cell cycle arrest at the

G2/M phase, and cell viability decline in AgNP-treated cells, via p38 MAPK and Nrf-2 signaling pathways [39]. Moreover, it is known that AgNPs possess a strong tendency to interact with the thiol groups of enzymes, and the phosphate group of DNA bases, impairing their activity. It was also described that AgNPs increased the expression of a DNA damage repair protein, Rad51, precluding safe DNA repair damage [68]. The formation of bulky DNA adducts induced by AgNPs were reverted by the use of antioxidants, proving the important role of RS in the induction of DNA damage [68]. Furthermore, as has been mentioned above, the interaction of AgNPs with mitochondria results in the release of caspase-3, an effector protein, which has also the ability to induce the cleavage of DNA [69].

17.4 Conclusions

AgNPs have a wide spectrum of applications that reach far beyond therapeutics, being easily found in the everyday life of humans. The increased use of AgNPs has raised several concerns about their side effects on biological systems and cellular compartments. In fact, there are numerous reports about the toxicity of AgNPs, and in this chapter we focused on their pro-inflammatory effect as one of the main toxicity mechanisms. Figure 17.2 summarizes the present state of knowledge about the effect of AgNPs on the crucial mediators that participate actively in the inflammatory process cascade. To date, several studies have shown that AgNPs can interact with different cell types of the immune system, resulting in the production of RS and activation of transcription factors with the subsequent production of cytokines/chemokines and consequently the induction of cell death. These interactions, however, are diverse, complex, and not yet well understood. Several points of concern should be taken into account when these kinds of studies are analyzed and compared, since the shape, size, coating agent, and composition of AgNPs can have a significant impact on their function and possible risks to human health. Centuries ago Paracellsus said that "everything is a poison and nothing is a poison, it is only a matter of dose." In the case of AgNPs it is clear that their effects are dependent on both dose and particle size. The smaller AgNPs tend to have higher pro-inflammatory effects. Furthermore, the type of cellular model, the time of exposure, and the cell culture medium used can directly influence the results. Additional attention should be given to the selection of appropriate analytical methods, since AgNPs can interfere with common procedures. Moreover, extensive research is needed to fully understand the contribution from the ionic form versus the contribution from the nanoform of silver in their possible toxic effects. In this context, there is the need to elaborate standard protocols that include detailed AgNP characterization, including the analysis of their effects under different conditions, to decrease the possible discrepancies in the available data.

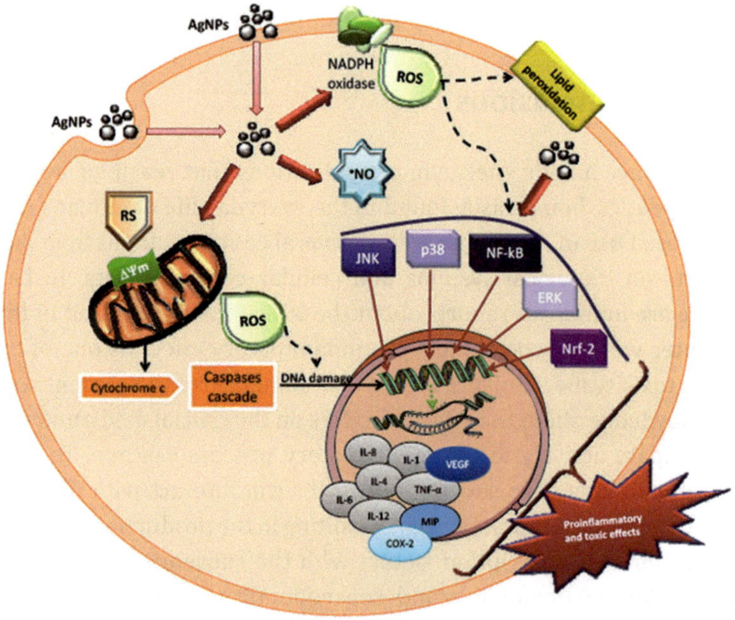

Figure 17.2 Schematic and summarized effects of AgNPs on the crucial mediators that participate actively in the inflammatory process cascade. COX-2, cyclooxygenase-2; ERKs, extracellular signal–regulated kinases; IL, interleukin; JNKs, c-Jun N-terminal kinases; MIP, macrophage inhibitory protein; NF-κB, nuclear factor kappa-light-chain-enhancer of activated B cells; •NO, nitric oxide; Nrf-2, nuclear factor erythroid 2–related factor 2; ROS, reactive oxygen species; RS, reactive species; TNF, tumor necrosis factor; Δψm, variations of the mitochondrial transmembrane potential; VEGF, vascular endothelial growth factor.

Despite the great challenge of controlling several variables, and application of standardized methods, new findings about the interactions of AgNPs with the immune system, and the knowledge about their effects on several points of the inflammatory process, will allow safer use of these NPs.

Acknowledgments

The authors acknowledge the financial support from National Funds (Fundação para a Ciência e Tecnologia and Ministério da Educação e Ciência [FCT/MEC]) and European Union funds (Fundo Europeu de Desenvolvimento Regional [FEDER]) under the program PT2020 (PT2020 UID/MULTI/04378/2013 - POCI/01/0145/FEDER/007728), the framework of QREN (NORTE-01-0145-FEDER-000024), and Programa Operacional Competitividade e Internacionalização (COMPETE) (PTDC/QEQ-QAN/1742/2014 - POCI-01-0145-FEDER-016530).

References

1. Firdhouse, M.J., et al. Biosynthesis of silver nanoparticles and its applications. *J Nanotechnol*, 2015, **2015**:18.
2. Garcia-Barrasa, J., et al. Silver nanoparticles: synthesis through chemical methods in solution and biomedical applications. *Cent Eur J Chem*, 2011, **9**(1):7–19.
3. You, C., et al. The progress of silver nanoparticles in the antibacterial mechanism, clinical application and cytotoxicity. *Mol Biol Rep*, 2012, **39**(9):9193–9201.
4. dos Santos, C.A., et al. Silver nanoparticles: therapeutical uses, toxicity, and safety issues. *J Pharm Sci*, 2014, **103**(7):1931–1944.
5. Wei, L., et al. Silver nanoparticles: synthesis, properties, and therapeutic applications. *Drug Discovery Today*, 2015, **20**(5):595–601.
6. Quang Huy, T., et al. Silver nanoparticles: synthesis, properties, toxicology, applications and perspectives. *Adv Nat Sci: Nanosci Nanotechnol*, 2013, **4**(3):033001.
7. Beer, C., et al. Toxicity of silver nanoparticles: nanoparticle or silver ion? *Toxicol Lett*, 2012, **208**(3):286–292.

8. Likus, W., et al. Nanosilver - does it have only one face? *Acta Biochim Pol*, 2013, **60**(4):495–501.
9. McShan, D., et al. Molecular toxicity mechanism of nanosilver. *J Food Drug Anal*, 2014, **22**(1):116–127.
10. Liu, W., et al. Impact of silver nanoparticles on human cells: effect of particle size. *Nanotoxicology*, 2010, **4**(3):319–330.
11. Bartłomiejczyk, T., et al. Silver nanoparticles - allies or adversaries? *Ann Agric Environ Med*, 2013, **20**(1):48–54.
12. Buzea, C., et al. Nanomaterials and nanoparticles: sources and toxicity. *Biointerphases*, 2007, **2**(4):MR17–MR71.
13. Greulich, C., et al. Cell type-specific responses of peripheral blood mononuclear cells to silver nanoparticles. *Acta Biomater*, 2011, **7**(9):3505–3514.
14. Soares, T., et al. Size-dependent cytotoxicity of silver nanoparticles in human neutrophils assessed by multiple analytical approaches. *Life Sci*, 2016, **145**:247–254.
15. Grosse, S., et al. Silver nanoparticle-induced cytotoxicity in rat brain endothelial cell culture. *Toxicol in Vitro*, 2013, **27**(1):305–313.
16. Park, J., et al. Size dependent macrophage responses and toxicological effects of Ag nanoparticles. *Chem Commun*, 2011, **47**(15):4382–4384.
17. Gliga, A.R., et al. Size-dependent cytotoxicity of silver nanoparticles in human lung cells: the role of cellular uptake, agglomeration and Ag release. *Part Fibre Toxicol*, 2014, **11**:11.
18. Bastos, V., et al. The influence of citrate or PEG coating on silver nanoparticle toxicity to a human keratinocyte cell line. *Toxicol Lett*, 2016, **249**:29–41.
19. Nguyen, K.C., et al. Comparison of toxicity of uncoated and coated silver nanoparticles. *J Phys Conf Ser*, 2013, **429**:1–15.
20. Suresh, A.K., et al. Cytotoxicity induced by engineered silver nanocrystallites is dependent on surface coatings and cell types. *Langmuir*, 2012, **28**(5):2727–2735.
21. Cui, W., et al. Effects of aggregation and the surface properties of gold nanoparticles on cytotoxicity and cell growth. *Nanomed Nanotechnol Biol Med*, 2012, **8**(1):46–53.
22. Lubick, N., Nanosilver toxicity: ions, nanoparticles-or both? *Environ Sci Technol*, 2008, **42**(23):8617.
23. Sotiriou, G.A., et al. Antibacterial activity of nanosilver ions and particles. *Environ Sci Technol*, 2010, **44**(14):5649–5654.

24. Park, E.-J., et al. Silver nanoparticles induce cytotoxicity by a Trojan-horse type mechanism. *Toxicol in Vitro*, 2010, **24**(3):872–878.
25. Hsiao, I.L., et al. Trojan-horse mechanism in the cellular uptake of silver nanoparticles verified by direct intra- and extracellular silver speciation analysis. *Environ Sci Technol*, 2015, **49**(6):3813–3821.
26. Ribeiro, D., et al. Proinflammatory pathways: the modulation by flavonoids. *Med Res Rev*, 2015, **35**(5):877–936.
27. Paino, I.M., et al. Poly(vinyl alcohol)-coated silver nanoparticles: activation of neutrophils and nanotoxicology effects in human hepatocarcinoma and mononuclear cells. *Environ Toxicol Pharmacol*, 2015, **39**(2):614–621.
28. Singh, R.P., et al. Cellular uptake, intracellular trafficking and cytotoxicity of silver nanoparticles. *Toxicol Lett*, 2012, **213**(2):249–259.
29. Freitas, M., et al. Optical probes for detection and quantification of neutrophils' oxidative burst: a review. *Anal Chim Acta*, 2009, **649**(1):8–23.
30. Yang, E.-J., et al. Inflammasome formation and IL-1β release by human blood monocytes in response to silver nanoparticles. *Biomaterials*, 2012, **33**(28):6858–6867.
31. Liu, H., et al. Comparative study of respiratory tract immune toxicity induced by three sterilisation nanoparticles: Silver, zinc oxide and titanium dioxide. *J Hazard Mater*, 2013, **248–249**:478–486.
32. Barbasz, A., et al. Cytotoxic activity of highly purified silver nanoparticles sol against cells of human immune system. *Appl Biochem Biotechnol*, 2015, **176**(3):817–834.
33. Romoser, A.A., et al. Distinct immunomodulatory effects of a panel of nanomaterials in human dermal fibroblasts. *Toxicol Lett*, 2012, **210**(3):293–301.
34. Nishanth, R.P., et al. Inflammatory responses of RAW 264.7 macrophages upon exposure to nanoparticles: role of ROS-NFkB signaling pathway. *Nanotoxicology*, 2011, **5**(4):502–516.
35. Stepkowski, T.M., et al. Silver nanoparticles induced changes in the expression of NF-kB related genes are cell type specific and related to the basal activity of NF-kB. *Toxicol in Vitro*, 2014, **28**(4):473–478.
36. Prasad, R.Y., et al. Investigating oxidative stress and inflammatory responses elicited by silver nanoparticles using high-throughput reporter genes in HepG2 cells: effect of size, surface coating, and intracellular uptake. *Toxicol in Vitro*, 2013, **27**(6):2013–2021.

37. Nguyen, T., et al. The Nrf2-antioxidant response element signaling pathway and its activation by oxidative stress. *J Biol Chem*, 2009, **284**(20):13291–13295.

38. Aueviriyavit, S., et al. Mechanistic study on the biological effects of silver and gold nanoparticles in Caco-2 cells--induction of the Nrf2/HO-1 pathway by high concentrations of silver nanoparticles. *Toxicol Lett*, 2014, **224**(1):73–83.

39. Eom, H.J., et al. p38 MAPK activation, DNA damage, cell cycle arrest and apoptosis as mechanisms of toxicity of silver nanoparticles in Jurkat T cells. *Environ Sci Technol*, 2010, **44**(21):8337–8342.

40. Hsin, Y.H., et al. The apoptotic effect of nanosilver is mediated by a ROS- and JNK-dependent mechanism involving the mitochondrial pathway in NIH3T3 cells. *Toxicol Lett*, 2008, **179**(3):130–139.

41. Rinna, A., et al. Effect of silver nanoparticles on mitogen-activated protein kinases activation: role of reactive oxygen species and implication in DNA damage. *Mutagenesis*, 2015, **30**(1):59–66.

42. Park, H.S., et al. Attenuation of allergic airway inflammation and hyperresponsiveness in a murine model of asthma by silver nanoparticles. *Int J Nanomed*, 2010, **5**:505–515.

43. Kononenko, V., et al. Nanoparticle interaction with the immune system. *Arh Hig Rada Toksikol*, 2015, **66**(2):97–108.

44. Murphy, A., et al. Silver nanoparticles induce pro-inflammatory gene expression and inflammasome activation in human monocytes. *J Appl Toxicol*, 2016, **36**(10):1311–1320.

45. Kim, S., et al. Phagocytosis and endocytosis of silver nanoparticles induce interleukin-8 production in human macrophages. *Yonsei Med J*, 2012, **53**(3):654–657.

46. Martinez-Gutierrez, F., et al. Antibacterial activity, inflammatory response, coagulation and cytotoxicity effects of silver nanoparticles. *Nanomedicine*, 2012, **8**(3):328–336.

47. Carlson, C., et al. Unique cellular interaction of silver nanoparticles: size-dependent generation of reactive oxygen species. *J Phys Chem B*, 2008, **112**(43):13608–13619.

48. Nguyen, K.C., et al. Toxicological evaluation of representative silver nanoparticles in macrophages and epithelial cells. *Toxicol in Vitro*, 2016, **33**:163–173.

49. Kaur, J., et al. Evaluating cell specific cytotoxicity of differentially charged silver nanoparticles. *Food Chem Toxicol*, 2013, **51**:1–14.

50. Samberg, M.E., et al. Evaluation of silver nanoparticle toxicity in skin in vivo and keratinocytes in vitro. *Environ Health Perspect*, 2010, **118**(3):407–413.

51. Sun, C., et al. Silver nanoparticles induced neurotoxicity through oxidative stress in rat cerebral astrocytes is distinct from the effects of silver ions. *Neurotoxicology*, 2016, **52**:210–221.

52. Hackenberg, S., et al. Silver nanoparticles: evaluation of DNA damage, toxicity and functional impairment in human mesenchymal stem cells. *Toxicol Lett*, 2011, **201**(1):27–33.

53. Greulich, C., et al. Studies on the biocompatibility and the interaction of silver nanoparticles with human mesenchymal stem cells (hMSCs). *Langenbeck's Arch Surg*, 2009, **394**(3):495–502.

54. Miethling-Graff, R., et al. Exposure to silver nanoparticles induces size- and dose-dependent oxidative stress and cytotoxicity in human colon carcinoma cells. *Toxicol in Vitro*, 2014, **28**(7):1280–1289.

55. Abbott Chalew, T.E., et al. Toxicity of commercially available engineered nanoparticles to Caco-2 and SW480 human intestinal epithelial cells. *Cell Biol Toxicol*, 2013, **29**(2):101–116.

56. Suliman, Y.A., et al. Evaluation of cytotoxic, oxidative stress, proinflammatory and genotoxic effect of silver nanoparticles in human lung epithelial cells. *Environ Toxicol*, 2015, **30**(2):149–160.

57. Arai, Y., et al. Difference in the toxicity mechanism between ion and nanoparticle forms of silver in the mouse lung and in macrophages. *Toxicology*, 2015, **328**:84–92.

58. Haberl, N., et al. Cytotoxic and proinflammatory effects of PVP-coated silver nanoparticles after intratracheal instillation in rats. *Beilstein J Nanotechnol*, 2013, **4**:933–940.

59. Ebabe Elle, R., et al. Dietary exposure to silver nanoparticles in Sprague-Dawley rats: effects on oxidative stress and inflammation. *Food Chem Toxicol*, 2013, **60**:297–301.

60. Park, E.J., et al. Repeated-dose toxicity and inflammatory responses in mice by oral administration of silver nanoparticles. *Environ Toxicol Pharmacol*, 2010, **30**(2):162–168.

61. Frankova, J., et al. Effects of silver nanoparticles on primary cell cultures of fibroblasts and keratinocytes in a wound-healing model. *J Appl Biomater Funct Mater*, 2016, **14**(2):e137–e142.

62. Wang, Z., et al. Mechanisms of nanosilver-induced toxicological effects: more attention should be paid to its sublethal effects. *Nanoscale*, 2015, **7**(17):7470–7481.

63. Vianello, A., et al. The mitochondrial permeability transition pore (PTP): an example of multiple molecular exaptation? *Biochim Biophys Acta*, 2012, **1817**(11):2072–2086.

64. Ma, W., et al. Silver nanoparticle exposure induced mitochondrial stress, caspase-3 activation and cell death: amelioration by sodium selenite. *Int J Biol Sci*, 2015, **11**(8):860–867.

65. Piao, M.J., et al. Silver nanoparticles induce oxidative cell damage in human liver cells through inhibition of reduced glutathione and induction of mitochondria-involved apoptosis. *Toxicol Lett*, 2011, **201**(1):92–100.

66. Tran, Q.H., et al. Silver nanoparticles: synthesis, properties, toxicology, applications and perspectives. *Adv Nat Sci: Nanosci Nanotechnol*, 2013, **4**(3):033001.

67. Zhang, R., et al. Endoplasmic reticulum stress signaling is involved in silver nanoparticles-induced apoptosis. *Int J Biochem Cell Biol*, 2012, **44**(1):224–232.

68. Kim, S., et al. Silver nanoparticle-induced oxidative stress, genotoxicity and apoptosis in cultured cells and animal tissues. *J Appl Toxicol*, 2013, **33**(2):78–89.

69. Dubey, P., et al. Perturbation of cellular mechanistic system by silver nanoparticle toxicity: cytotoxic, genotoxic and epigenetic potentials. *Adv Colloid Interface Sci*, 2015, **221**:4–21.

Index